Photovoltaic Systems

second edition

James P. Dunlop

AMERICAN TECHNICAL PUBLISHERS, INC.
ORLAND PARK, ILLINOIS 60467-5756

IN PARTNERSHIP WITH NJATC

American Technical Publishers, Inc. Editorial Staff

Editor in Chief:
    Jonathan F. Gosse
Vice President–Production:
    Peter A. Zurlis
Art Manager:
    James M. Clarke
Technical Editor:
    Julie M. Welch
Copy Editor:
    Valerie A. Deisinger
    Talia J. Turner
Cover Design:
    Jennifer M. Hines

Illustration/Layout:
    Thomas E. Zabinski
    Jennifer M. Hines
Multimedia Coordinator:
    Carl R. Hansen
CD-ROM Development:
    Gretje Dahl
    Daniel Kundrat
    Nicole S. Polak

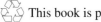 

Photovoltaic Systems

Acknowledgments

National Joint Apprenticeship and Training Committee for the Electrical Industry

Author

Jim Dunlop, PE

Technical Editor

Todd W. Stafford, NJATC Staff

Technical Reviewers

Matt Cotter, Electrical Training Institute (City of Commerce, CA)

Greg Creal, IBEW Local 280 (Tangent, OR)

Brian Crise, NECA-IBEW Electrical Training Center (Portland, OR)

Joe Estrada, Fresno Electrical JATC

John Hardwick, Sharp Electronics Corp.

Jerry Ventre, formerly of the Florida Solar Energy Center

Case Studies

Baltimore Electrical JATC

Cleveland JATC

Detroit Electrical JATC

IN-TECH Electrical Training Center (Chicago, IL)

IBEW Local 103 (Boston, MA)

IBEW Local 363 (Harriman, NY)

San Diego Electrical Training Center

Tri-City JATC (Latham, NY)

Companies and Organizations that Provided Technical Information and Photographs

AMtec Solar

Daystar, Inc.

Department of Energy National Renewable Energy Laboratory (DOE/NREL)

Direct Power and Water Corporation

East Penn Manufacturing Co., Inc.

Enphase Energy

Eppley Laboratory, Inc.

Fluke Corporation

Fronius USA LLC

Kipp & Zonen, Inc.

Library of Congress

LI-COR Environmental

Lucent Technologies Inc./Bell Labs

Meteocontrol GmbH

Morningstar Corp.

NASA/JPL-Caltech

National Fire Protection Association

Panduit Corp.

PV Measurements, Inc.

Schott Solar

Sharp Electronics Corp.

SMA America, Inc.

SMA Technologie AG

SOHO (ESA & NASA)

Solar Pathfinder

SolarWorld Industries America

Solmetric Corporation

Southwest Technology Development Institute/ New Mexico State University

SPG Solar, Inc.

SunWize Technologies

Surrette Battery Company

UniRac, Inc.

United Solar Ovonic LLC

Xantrex Technology Inc.

Photovoltaic Systems

Contents

Photovoltaic Systems

Contents

CD-ROM Contents

* Using the CD-ROM
* Quick Quizzes®
* Illustrated Glossary
* Media Clips
* Flash Cards
* Solar Radiation Data Sets
* Sun Path Charts
* Forms and Worksheets
* Solar Time Calculator
* ATPeResources.com

Introduction

Photovoltaics, along with other renewable-energy technologies, is a rapidly growing sector of the energy market. *Photovoltaic Systems*, Second Edition, is a comprehensive guide to the design, installation, and evaluation of residential and commercial photovoltaic (PV) systems. The textbook covers the principles of photovoltaics and how to effectively incorporate PV systems into stand-alone or interconnected electrical systems. The content includes system advantages and disadvantages, site evaluation, component operation, system design and sizing, and installation requirements and recommended practices. Common scenarios and procedures are discussed throughout.

Each chapter begins with an overview of the covered material in the form of an introduction and chapter objectives. At the end of each chapter is a summary of key concepts, a listing of all definitions from that chapter, and review questions. Answers to odd-numbered review questions are included at the end of the book.

This new edition adds more detail to many illustrations, updates photographs, includes more review questions, and features the following significant content changes:

• Updated electrical requirements in accordance with the 2008 edition of the National Electrical Code®
• Expanded solar radiation and sun position content
• New and expanded coverage of various shading analysis methods
• Expanded safety and personal protective equipment (PPE) content

The *Photovoltaic Systems*, Second Edition, CD-ROM included at the back of the book features interactive resources for independent study to enhance learning, including Quick Quizzes®, an Illustrated Glossary, Media Clips, Flash Cards, several solar resources, and a link to ATPeResources.com. The Quick Quizzes® provide an interactive review of key topics covered in each chapter. The Illustrated Glossary is a helpful reference to textbook definitions, with select terms linked to illustrations and media clips that augment the definition provided.

The solar resources are electronic resources for evaluating potential installation sites and for sizing PV systems, and include the following:

• Solar Radiation Data Sets, which include the National Renewable Energy Laboratory's complete solar radiation data for 239 sites around the United States in easily printable PDF format.
• Sun Path Charts, which are printable PDF charts of solar positions for selected latitudes. The related sun path chart spreadsheet can calculate additional charts as needed.
• Forms and Worksheets, which include a sample site survey form, inspection checklist, maintenance plan, and a set of sizing and financial analysis worksheets, all in PDF format. The sizing and analysis worksheets are also included in spreadsheet files that automatically perform calculations.
• Solar Time Calculator, which facilitates the conversion between standard time and solar time that is required for some solar analyses.

ATPeResources.com provides access to additional online technical content and a list of Internet links to manufacturer, organization, government, and American Tech resources. Clicking on the American Tech web site button (www.go2atp.com) or the American Tech logo accesses information on related training products.

The Publisher

Photovoltaic Systems

CD-ROM references identify additional resources on the included CD-ROM.

Chapter introductions provide an overview of chapter content.

Detailed, full-color illustrations explain the principles and operation of PV systems.

Chapter objectives list learning goals for the chapter.

Factoids provide interesting technical tips or background information.

Installation photographs depict best practices.

Features

Vignettes supplement the text with additional technical, historical, or safety information.

Case studies spotlight unique or chapter-relevant aspects of installed PV systems.

Summaries highlight the key concepts of the chapter.

Key terms are listed at the end of the chapter.

Review questions test for chapter comprehension.

Mr. Jim Dunlop was encouraged early to pursue a career in the electrical field by his high school electrical shop teacher George Patrick Shultz (author of several early NJATC textbooks), and when he attended the University of Florida, he planned to study electrical engineering. Jim's early experiences, however, led him into renewable energy systems. He was also influenced by courses on solar energy systems and direct energy conversion taught by renewable energy pioneer Dr. Erich Farber, founder of the University of Florida's Solar Energy Research Park.

Beginning his career at the Florida Solar Energy Center (FSEC), Jim became involved in the development and evaluation of photovoltaic (PV) systems, which had recently been included in the 1984 National Electrical Code®. Over his twenty-year career at FSEC under Director Jerry Ventre, Jim was involved in developing curricula and delivering numerous training programs on PV systems. In 2001, he led FSEC's efforts in attaining national accreditation for its PV training programs.

From 2001 to 2002, Jim served as the charter technical committee chairperson for the North American Board of Certified Energy Practitioners (NABCEP). As of 2005, Jim is again leading this committee. This position involves developing candidate entry requirements, task analyses, examinations, and study guides for a national PV-installer certification program.

In 2001, Jim became the first Master Trainer in PV systems certified by the Institute for Sustainable Power worldwide. In addition to creating training manuals and instructor materials, he has published over fifty papers and technical reports on the design, testing, and evaluation of PV systems and equipment. Jim recently served as a Curriculum Specialist for the NJATC and continues to be involved in the PV industry through providing training and consulting services for the specification, design, installation, and evaluation of PV systems.

1

Introduction to Photovoltaic Systems

With growing concerns about the future and security of the world's energy supply, renewable resources such as solar power are becoming increasingly important. Various solar energy technologies have been used through millennia of human history. However, practical photovoltaics—the direct conversion of solar energy into electricity—has a history of only about 50 years. This field of study and the resulting industry have been rapidly growing and improving and are expected to become a significant part of the world's energy future.

Chapter Objectives

* *Compare the advantages and disadvantages of installing a PV system.*
* *Understand some of the factors that have motivated the growth of PV technology worldwide.*
* *Evaluate the design priorities for PV systems in different types of applications.*
* *Describe the primary levels of the PV industry and how they interact.*
* *Understand why it is important for installers to be well trained.*
* *Differentiate between flat-plate collectors and concentrating collectors.*
* *Understand how the different types of solar energy technologies utilize solar radiation.*

PHOTOVOLTAICS

Photovoltaics is a solar energy technology that uses the unique properties of certain semiconductors to directly convert solar radiation into electricity. Photovoltaic (PV) systems use wafers, typically made of crystalline silicon, that are sensitive to sunlight and produce a small direct current when exposed to light. When these PV cells, also known as solar cells, are combined into larger arrangements called modules, they produce an appreciable amount of electrical power with no moving parts, noise, or emissions.

A *photovoltaic (PV) system* is an electrical system consisting of a PV module array and other electrical components needed to convert solar energy into electricity usable by loads. These components can be arranged in many ways to design PV systems for different situations, but the most common configuration is a utility-connected system, which is found on commercial and residential buildings. **See Figure 1-1.** These PV systems may or may not include battery storage. The array is usually mounted on a rooftop or nearby on the ground.

Electrical components, such as inverters, charge controllers, and disconnects, control and condition the DC power from the array and either direct it to DC loads or convert it to AC power for use by AC loads. Some of these component functions may instead be combined together into one power conditioning unit (PCU). A *load* is a piece of equipment that consumes electricity. Examples of loads include lights, pumps, heaters, motors, and electronics.

DOE/NREL, Altair Energy

PV systems are also commonly known as solar electric systems.

Advantages

Electricity supplied by a PV system displaces electricity from some other power-generating technology. If the alternative is very expensive, such as a utility connection to a remote location, then the PV system may save the consumer a great deal of money. However, other advantages and benefits of PV systems add value beyond the potential financial savings. Many PV system owners place a high importance on producing clean "green" energy. Photovoltaics is an environmentally friendly technology that produces energy with no noise or pollution. For some owners, operating a PV system makes a statement about protecting the environment and conserving nonrenewable energy sources.

Also, PV systems are very flexible and can be adapted to many different applications. The modular nature of PV arrays and other components make systems easy to expand for increased capacity. Since there are no moving parts, PV systems are extremely reliable and last a long time with minimal maintenance.

PV systems also offer energy independence. A supplemental PV system reduces the consumer's vulnerability to utility power outages, and a stand-alone system eliminates it. Furthermore, sunlight is a renewable energy source that is free and readily available. As conventionally produced electricity is expected to become more expensive and PV system costs are generally decreasing, PV systems can also be used to hedge against future energy rate increases.

Disadvantages

There are, however, some disadvantages to PV systems that have somewhat limited their use. Currently, the most significant issue is the high initial cost of a PV system compared to prices for competing power-generating technologies (when available). PV systems also require a relatively large array area to produce a significant amount of power. The available solar radiation resource at a particular location determines the feasibility of producing appreciable amounts of power.

There is also a lack of knowledge among some groups and in some areas about the potential of solar energy systems, particularly photovoltaics. Consumers may not know what types of systems are available, or even that a solar energy system could be successful for their application or location. Consumers who wish to install a PV system may discover that the industry infrastructure in their area is not yet fully developed to support their installation. It may be difficult to find a qualified installer in their area, or their installation may be hindered by limited knowledge of PV systems and their requirements among the local utilities and code officials.

Many leaders in the PV industry are addressing these issues in an effort to promote PV systems. Public and private organizations, particularly state and federal governments, are subsidizing PV installations in an effort to offset the additional costs and promote the use of "green" electrical power. Research institutions and manufacturers are working on new PV technologies to increase the efficiency of cells and modules so that more energy can be produced by smaller arrays or in locations with a less-favorable solar radiation resource.

Most importantly, the industry as a whole is involved in educating the public and related organizations about solar energy, through publicity, training programs, cooperative projects, legislation, incentives, and other activities.

Electricity Distribution

Most electricity is distributed through an electrical utility grid to millions of customers from a relatively small number of large power plants. A *utility* is a company that produces and/or distributes electricity to consumers in a certain region or state. The *grid* is the utility's network of conductors, substations, and equipment that distributes electricity from its central generation point to the consumer. **See Figure 1-2.** The grid fans out from the power plants to thousands of homes and businesses within a region. Electricity may travel hundreds of miles before it reaches the end user. The grid regions may be connected together so that consumers still have power if part of the distribution system breaks down. Outages, though rare, do still occur, often due to an overloaded system or severe weather events.

Typical Utility-Connected PV System

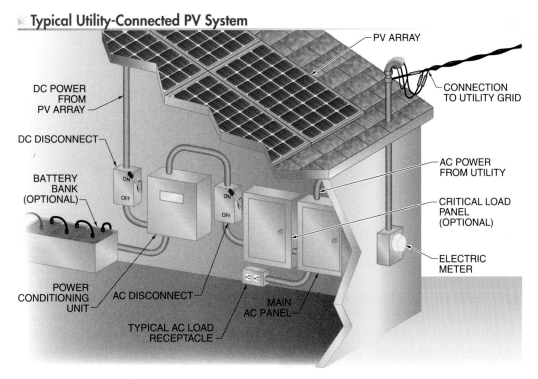

Figure 1-1. *A utility-connected PV system is the most common system configuration. Various electrical components control, condition, and distribute the power to on-site loads.*

Centralized Electricity Distribution

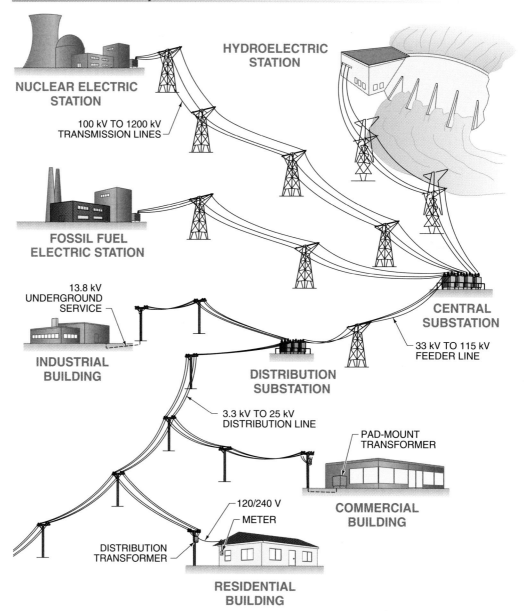

Figure 1-2. *An electric utility produces electricity at a power plant and distributes it to consumers through power lines, substations, and transformers.*

Distributed generation is a system in which many smaller power-generating systems create electrical power near the point of consumption. The electricity may travel only a few feet to the loads, which avoids the losses from long transmission lines. Distributed generation systems can include PV systems, wind turbines, engine generators, or other relatively small-scale power systems. **See Figure 1-3.** A distributed generation system may serve as the only source of power for the consumer (a stand-alone system), or as backup or supplemental power for a utility grid connection. If consumers are connected to the utility grid, excess power can be distributed to the grid if it is not needed by the on-site loads.

◈ Distributed Generation

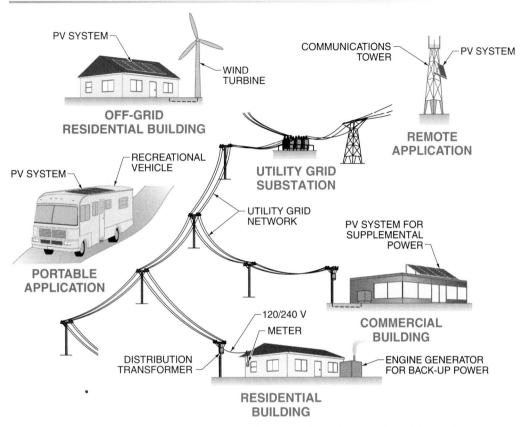

Figure 1-3. *Distributed generation systems produce electricity close to where it is used.*

Development

PV technology has been developing for more than 160 years, but has progressed exponentially in the last few decades. Photovoltaics has only recently become a practical technology for power generation.

Invention. Edmund Becquerel, a French physicist, is credited with discovering the photovoltaic effect in 1839. The nineteen-year-old scientist observed an increase in electron emissions between a pair of electrodes submerged in a conductive solution when the experiment was exposed to light. However, there was little practical application for electricity at the time, and the discovery went unutilized.

In 1873, British engineer Willoughby Smith observed the light sensitivity of selenium while testing materials for underwater telegraph cables. In the absence of light, selenium exhibited high resistance, but when exposed to light, selenium became highly conductive.

The resistance was inversely proportional to the intensity of the light. **See Figure 1-4.** This observation of photoconductivity led to experimentation on how to use selenium to utilize solar energy.

DOE/NREL, Canyonlands Needles Outpost
Distributed generation systems provide electrical power in remote locations.

Direct energy conversion systems produce electrical power in one process. For example, fuel cells use electrochemical processes to convert hydrogen and oxygen into electrical energy, and photovoltaics use semiconductor properties to produce power from sunlight.

However, most power is generated by converting energy from one form to another in multiple processes until it ultimately becomes useful energy in the form of electricity. For example, coal-fired power plants change chemical energy (energy in the chemical bonds of coal) into heat energy by combustion. The heat energy is applied to water, which becomes steam, another form of heat energy. The steam is routed to turbines where it moves the blades, which rotate a shaft, producing mechanical energy. The shaft drives a generator that uses the mechanical energy together with magnetic energy to produce electricity.

Steam is the most common method of converting heat energy into mechanical energy and ultimately electricity and is used in many types of power plants. Nuclear power plants convert atomic energy into heat energy, which creates electricity by way of steam. Many other fossil-fuel-powered processes also use steam for energy conversion.

Energy sources that begin with mechanical energy, however, do not include steam or heat energy in the conversion cycle. In hydroelectric plants, water pressure and flow drives the turbine-generators. Turbine-generators can also be driven by wind or tidal power, or even the movement of fluids in solar thermal systems.

No energy conversion is 100% efficient. That is, there is always some energy that does not convert to a new form or is lost through byproducts or waste heat. The energy is not destroyed, but it escapes the system and does no useful work.

Selenium Photoconductive Cells

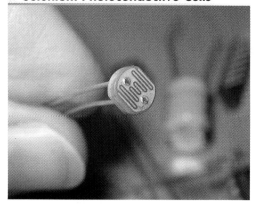

Figure 1-4. *Selenium photoconductive cells are commonly used in light-sensing electronics, such as exposure timing circuits in cameras.*

The first true PV cells were developed by American inventor Charles Fritts in 1883. He covered a selenium wafer with transparent gold film, which produced a tiny current. However, its maximum efficiency never exceeded 1%, so it was considered impractical for power generation. Also, selenium is a relatively rare element, making the production of cells prohibitively expensive.

In 1954, Darryl Chapin and other researchers at Bell Laboratories were investigating the use of PV cells as a power source for remote telephone service stations, but their efforts to improve selenium cells were unsuccessful. At the same time and independently, fellow Bell researchers Calvin Fuller and Gordon Pearson were investigating silicon for use in transistors and rectifiers. They discovered that not only could adding certain impurities improve the desired electrical qualities for the transistors and rectifiers, but that this modified silicon rectifier produced an appreciable electric current when exposed to light. Fuller, Chapin, and Pearson subsequently collaborated and improved the silicon cell into the first useful PV cell. **See Figure 1-5.** The first cells had efficiencies of up to 6%—unimpressive by today's standards, but a significant improvement over selenium. Bell Laboratories called the invention a "solar battery" and conservatively envisioned its application as powering small or remote electrical systems.

Growth. The space race in the 1950s and 1960s spurred the development of space technologies, including photovoltaics. PV cells were ideal power generators for satellites and spacecraft because of the complexity of supplying power by other means and the abundant solar resource available outside Earth's atmosphere. Cells were first developed for the Vanguard I satellite in 1958 and have been used on nearly every spacecraft and satellite since.

Invention of the PV Cell

Lucent Technologies Inc./Bell Labs

Figure 1-5. *The first practical photovoltaic cell was invented at Bell Laboratories in 1954.*

The first common Earth-based applications were in rural telephone systems and radio transmitters. **See Figure 1-6.** These systems brought communications to remote communities, especially in developing countries where the lack of electrical infrastructure made solar energy an ideal solution. Many applications of photovoltaics significantly improved life in developing countries, such as by providing power to pump water for sanitation and irrigation, filter water for drinking, light schools, and refrigerate medicines.

Further improvements in cell efficiency and refinements in cell manufacturing made PV technology a viable option for larger systems. Off-grid homeowners living in remote areas began incorporating PV systems into their homes as a way to enjoy the convenience of modern appliances while maintaining their secluded lifestyle. There were also consumers who saw merit beyond financial factors in "green" PV electricity. These markets were relatively small, but fostered steady growth in PV technology for many years.

Early PV Applications

Lucent Technologies Inc./Bell Labs

Figure 1-6. *Rural communications systems in the 1950s were the first terrestrial applications of PV technology.*

Energy Crises. Oil shortages in the 1970s increased interest in Earth-based PV applications. The U.S. government initiated PV research and development projects and established what would become the National Renewable Energy Laboratory (NREL). Significant developments were made in PV technology, materials, inverters, and interactive systems, and federal legislation introduced tax credits to promote renewable energy production. Significant financial incentive programs are available to consumers in the United States, Europe, and Japan to encourage use of PV technology.

In 1960, solar cells were handmade and cost about $1000 per watt. Manufacturing improvements have reduced end-user costs to less than $4 per watt, while production has grown exponentially. **See Figure 1-7.** Meanwhile, fossil fuel costs have increased as finite supplies have dwindled or become more difficult to reach. The increasing costs for conventional electricity have helped the PV industry grow. **See Figure 1-8.**

Today, PV technology is used in diverse applications including communications networks, grid-connected homes and commercial buildings, and rural and remote lighting and water pumping systems.

Tens of thousands of American homes are now located completely outside the reach of the utility grid and produce their own power with PV systems.

PV APPLICATIONS

The earliest applications of PV systems were in situations where connections to the utility grid were unavailable or cost prohibitive. As PV efficiency has continued to improve and costs have fallen in recent years, more potential applications for PV technology have emerged. Greater demand and increasing production have accelerated the trend toward cost effectiveness in a wide range of applications.

Retail Module Prices

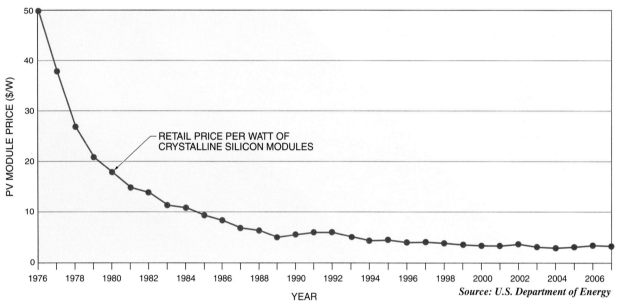

RETAIL PRICE PER WATT OF CRYSTALLINE SILICON MODULES

Source: U.S. Department of Energy

Figure 1-7. *Decades of development and manufacturing improvements have decreased the price per watt for PV systems.*

Module Installation

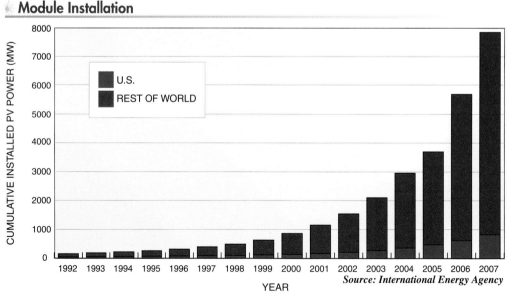

Source: International Energy Agency

Figure 1-8. *Production and installation of PV systems is growing rapidly.*

Today, PV systems can be used in almost any application where electricity is needed and can support DC loads, AC loads, or both. PV systems may be as simple as a PV module directly connected to a load with no other components, or as complex as a utility-interconnected system with multiple power sources.

Space Applications

Space applications are extreme examples of off-grid remoteness, in that there is no possible way to connect satellites and spacecraft to a steady source of terrestrial electricity. The only options are to launch stored electricity, in the form of batteries, with the spacecraft, or to generate electricity in space. Fortunately, the available solar radiation outside Earth's atmosphere is even greater than on the ground.

Satellites were the first practical applications of PV technology and did much to increase the public's knowledge of PV systems. **See Figure 1-9.** However, space-bound PV systems are designed for high efficiency and low weight, with cost being less important. Since the priorities for space-bound PV systems are different from those for Earth-based PV systems, the technologies used for these different applications have diverged somewhat. Even so, the advanced technologies currently used in space applications, though currently cost prohibitive, may someday progress to become practical for terrestrial applications.

Space PV Applications

DOE/NREL, NASA/Smithsonian Institution/ Lockheed Corp.

Figure 1-9. *Nearly every satellite and spacecraft since 1958 has relied on a PV system for power generation.*

SolarWorld Industries America
Camels are used for transportation to remote areas of some countries. Portable PV systems are used to power small refrigeration units for medicines that must be kept cold during a journey.

Portable Applications

Portable PV systems power mobile loads such as vehicles, temporary signs and lighting, and handheld devices. **See Figure 1-10.** The primary advantage of PV systems in these applications is that the load is not permanently connected to a stationary power source and is free to move. Some portable applications require power when the sun is not available, requiring a battery as part of the PV system. Other portable applications, such as battery chargers for small electronics, are simpler systems that operate only under sunlight.

Most portable PV systems are relatively small and can power only modest loads. For ease of transportation, the systems should be relatively light. Portable systems also need to be repositioned each time they are relocated. They can usually be used while in motion, but if not constantly facing the sun, the power output will be reduced.

Vehicles. Recreational vehicles (RVs) and boats can benefit greatly from PV systems. If the vehicle has many loads, such as the appliances in an RV, the PV system is best suited to be a backup to a primary power source, typically an engine generator. A sailboat, however, may use a small, portable PV system as the primary method for charging the batteries for its relatively small number of loads, such as navigation lights and radios.

lications

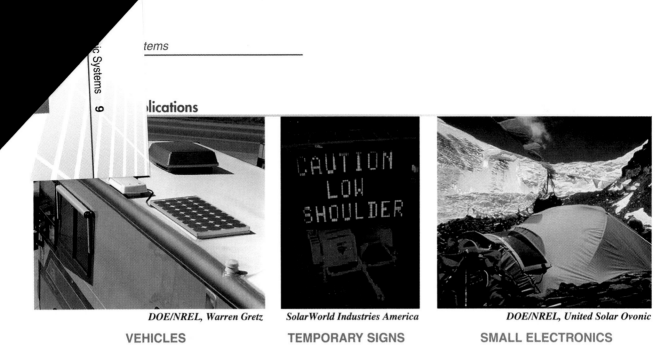

DOE/NREL, Warren Gretz	SolarWorld Industries America	DOE/NREL, United Solar Ovonic
VEHICLES	**TEMPORARY SIGNS**	**SMALL ELECTRONICS**

Figure 1-10. *Portable PV systems tend to be small and intended for specific loads.*

Temporary Signs and Lighting. Electronic construction signs cannot be connected to utility power, either because of remoteness or because of electrical construction hazards, and must be able to move between work areas at a construction site. Small PV systems with batteries can be added to the signs to power them continuously. Temporary lighting for short-term projects or nighttime construction work may be PV-powered because the short duration of the project may not warrant the time or expense of connecting to utility power.

Small Electronics. Small handheld PV systems can be used to directly power or charge small electronic devices such as cell phones, calculators, radios, GPS receivers, and lights, and are common accessories for campers, hikers, and military personnel. Manufacturers have developed these PV systems for ease of use, ruggedness, and portability, but power output is limited by their small size. These arrays are rigid or flexible panels that are unfolded and laid out on a sunny patch of ground. Small arrays have even been sewn into backpacks and jackets. The electronic device to be charged is connected to the PV system in the same way as the usual wall outlet charger.

Emergency Power. During natural disasters when the existing electrical infrastructure has been damaged and is off-line, portable PV power units can be brought in to supply critical services until the utility grid is repaired. Individual emergency PV systems can provide power for street and personal lighting, communications equipment, warning and message signs, water purification, refrigeration of medical supplies and food, or water pumping.

Remote Applications

Remote PV systems power loads that are permanently fixed but too distant to be connected to the utility power grid. A large number of applications require electricity in remote areas, making PV an ideal choice. **See Figure 1-11.**

Off-Grid Residences. The mainstream consumer PV market essentially began with off-grid residences, and this segment still composes a significant portion of the market. As systems have become more affordable, this application has become increasingly common among homeowners who demand value beyond environmental benefits from the systems. PV systems can be sized to power an entire home, but off-grid residences often include other sources of energy, such as wind turbines or engine generators, to supplement or back up the PV system.

Lighting. The availability of low-power DC lamps makes PV energy ideal for remote lighting applications. Lighting needs are clearly greatest at night, making power storage (in the form of batteries) a necessary part of these systems. PV systems can be used to light billboards, highways, information signs, parking lots, and marinas.

Remote Applications

DOE/NREL, Dave Parsons

DOE/NREL, Minnesota Department of Commerce

DOE/NREL, Jerry Anderson, Northwest Rural Public Power District

OFF-GRID RESIDENCES   REMOTE MONITORING   WATER PUMPING

Figure 1-11. *Remote areas where conventional utility-supplied power is out of reach are ideal for the application of PV technology.*

Communications. Radio, television, and telephone signals transmitted over long distances must be amplified by relay towers that are often located in areas inaccessible to utility power. The best sites for communications stations are at higher elevations, making it difficult to provide fuel and maintenance for generators. PV power is ideal for these stations. PV systems can also be used to power communication signals from emergency call boxes and electronic information signs.

Signage and Signals. Small PV systems can provide power for remote signage and signal devices, such as navigational beacons, sirens, highway warning signs, railroad signals, aircraft warning beacons, buoys, and lighthouses.

Remote Monitoring. Data monitoring for scientific research or other purposes is often required at sites far from conventional power sources. PV systems can be used as a power source for remote station monitoring and for transmitting meteorological, seismic, structural, or other data.

Water Pumping. Many water-pumping needs, such as livestock watering, are in remote areas of a farm or ranch and require the most water during the hottest days. These systems can be directly coupled, with the PV system running the pump during the sunniest times of day. PV-powered water pumping is also used to provide water for campgrounds, irrigation, and remote village water supplies.

Utility-Interactive (Supplemental Power) Applications

Systems that are connected to the utility grid and use PV energy as a supplemental source of power offer the greatest flexibility in possible system configurations. **See Figure 1-12.** This is because the size and type of system are not defined by the electricity demand. Being connected to the utility allows the consumer to choose how much supplemental power to derive from alternative sources such as PV systems. The supplemental power offsets a portion of the power needed from the utility, resulting in lower electricity bills.

A PV system may or may not save money in the short-term when competing against relatively inexpensive utility power. However, since installing a PV system for supplemental power is a choice rather than a necessity, owners typically value the advantages of their system beyond purely financial factors. Since utility-interactive systems are the fastest-growing segment of the PV system market, the additional advantages are clearly very important.

PV System Applications

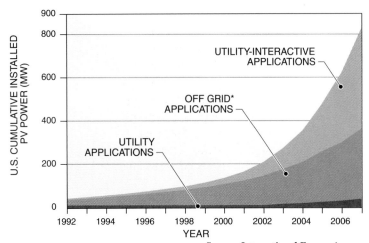

Source: International Energy Agency

* includes portable and remote applications

Figure 1-12. *Utility-interactive systems, most often for single-family homes, are the fastest-growing type of PV system installation.*

PV systems can be used to provide supplemental power to any utility-connected building or structure, including residences, commercial buildings, factories, and institutions. These applications are all similar, except that the systems vary in size. Array mounting systems may also vary because of the differences in construction methods and design of these types of buildings.

Utility-Scale Applications

Utility-scale PV power plants consisting of thousands of modules are not yet common, but are currently being researched as a viable renewable energy option for large-scale electricity generation. **See Figure 1-13.** Initial costs are high and power output varies with the weather, reducing reliability. However, PV systems produce power during daylight hours, when electricity demand is greatest, increasing their value to the utility.

PV technology, as opposed to conventional nuclear, coal-fired, or gas-fired power plants, can provide additional advantages for utility-scale power generation. Utilities can build PV power plants more quickly than conventional power plants because a PV system is simpler and its components are easier to install than those used in other power-generating technologies. The only moving parts of a PV system are

in the tracking system, if one is used. PV power plants can be located closer to populated areas than can conventional power plants because they do not involve hazardous materials or cause air, water, or noise pollution. Also unlike conventional power plants, PV power plants can be easily expanded incrementally as demand increases. However, PV power plants require a large area wherever they are located.

Utility-Scale Applications

DOE/NREL, Warren Gretz

Figure 1-13. *PV technology can be used for large-scale power production, but this application is not yet common.*

Unfortunately, PV electricity still costs considerably more in the United States than electricity generated by conventional plants, and the additional cost must be offset by government subsidies or passed on to the consumer. Yet as the technology continues to improve and fossil fuel costs rise, PV power becomes increasingly cost-effective.

PV INDUSTRY

A diversity of knowledge, skills, and abilities are required to design, install, and commission PV systems. The process involves a number of qualified individuals and organizations, each with an important role in ensuring the safe, reliable, and long-term performance of the PV power system.

Like most industries, the PV industry has several levels of business. **See Figure 1-14.** These levels can include a variety of enterprises: large and small, domestic and foreign, and public and private.

PV Industry

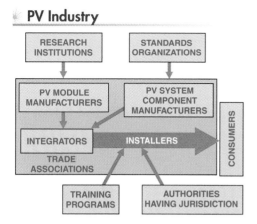

Figure 1-14. *The PV industry is composed of several levels of businesses and organizations.*

World PV Industry

In 1992, the U.S. had 40% of the world's installed PV capacity, the largest share held by any country. By 2007, PV installations in the United States had grown to more than nineteen times the 1992 level, but accounted for only 11% of the world's installed PV capacity. Some other countries have been installing PV systems at a much faster rate than the United States. Currently, the global PV industry is dominated by Japan and Germany. These countries produce and install most of the world's PV modules.

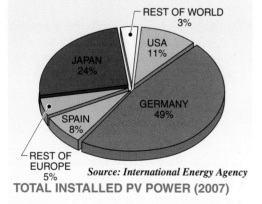

Source: International Energy Agency

TOTAL INSTALLED PV POWER (2007)

The strong growth overseas is driven by progressive incentive policies and high utility rates. Germany currently consumes about half of the worldwide production of solar modules, mostly for utility-connected residences and commercial buildings.

Manufacturers

There are dozens of PV cell and module manufacturers in the world, representing hundreds of individual manufacturing sites. The equipment needed to fabricate and assemble the PV modules is adapted from the semiconductor electronics industry and specialized for PV manufacturing. Additional manufacturers support the PV industry by supplying the other major components and equipment, such as inverters and mounting systems. These products are used in various types of renewable energy systems.

Electrical equipment manufacturers design, develop, and fabricate the balance-of-system (BOS) components. *A balance-of-system (BOS) component* is an electrical or structural component, aside from a major component, that is required to complete a PV system. BOS components include the conductors, connectors, switchgear, fuses, and hardware that connect, support, or interface between the primary devices such as the array, inverter, and battery system. Electrical BOS components are also common outside the PV industry, as they are used in nearly every electrical system.

Integrators

PV manufacturers do not typically sell directly to consumers wishing to install a PV system. PV system components also are not commonly purchased through distributors or retailers that stock other electrical equipment or building hardware. In between PV manufacturers and consumers are integrators. An *integrator* is a business that designs, builds, and installs complete PV systems for particular applications by matching components from various manufacturers. Integrators typically develop relationships with a small group of manufacturers, who may then offer preferred pricing and referrals to the integrator.

Integrators work with homeowners, businesses, organizations, contractors, and utilities to design, install, monitor, and maintain PV systems. They also work with builders and architects to create aesthetically pleasing buildings using PV systems that meet local codes, standards, and regulations.

Silicon

Silicon is the primary raw material for producing PV cells. Silicon makes up one-quarter of Earth's crust and is found in many minerals, including sand, amethyst, granite, quartz, flint, opal, asbestos, and clay. Even so, silicon is difficult to process into a usable form because it is always combined with other elements.

PV cells use many of the same processing and manufacturing techniques as other semiconductor devices, such as computer chips, so the silicon processing and supply chain is well established. However, the two industries compete for the same supply of processed silicon and the combined demand can exceed the supply. After many decades of progressively lower PV module prices per watt, increasing demand has put pressure on the refined silicon supply. The high demand and limited supply has stabilized or even slightly reversed the price trend. For continued growth in both industries, the silicon supply chain will need to respond with significant capacity increases.

Sharp Electronics Corp.

Quartz mining is a significant source of silicon for the PV industry.

Installers

Installers may be directly employed by an integrator, or by an electrical contractor who specializes in PV installations. Safe and quality PV system installations are essential for the success and acceptance of this emerging technology. Installers, both the contractors and the individual electricians, have important roles in ensuring quality PV installations. **See Figure 1-15.** Installers should exhibit quality electrical craftsmanship and, because of the unique aspects of the equipment and interfacing of PV technology, they should be qualified specifically for installing PV systems. All PV system installers should meet the following criteria:

- Complies with applicable building and electrical codes and standards
- Applies for permits and approvals from local building and utility authorities as required
- Knows his or her capabilities and limitations, and seeks outside expertise as required
- Selects and sizes systems and equipment to meet performance expectations
- Recommends well-engineered, quality components

- Ensures equipment is properly labeled and safety hazards are identified
- Locates and orients array to maximize performance and accessibility
- Mounts array with strong, weather-sealed attachments
- Uses accepted utility-interconnection practices and obtains utility approvals as required
- Completes work in a timely manner while practicing safe and orderly work habits
- Employs safe and accepted methods in the installation and use of PV equipment
- Completes inspections, commissioning, and acceptance tests
- Provides owner/operator with appropriate documentation, instructions, and training
- Provides follow-up service for completed work as required

An experienced installer also has considerable design knowledge and familiarity with many types of PV systems and components and can diagnose and troubleshoot even complex systems effectively. Installers are perhaps the most visible members of the PV industry to the consumers, making it vital that installers be professional and qualified individuals.

Skilled Installers

DOE/NREL, Craig Miller Productions

Figure 1-15. *Skilled and well-trained installers are needed to ensure quality PV system installations.*

Consumers

The consumer, or end user, is the owner of the building or structure powered by the PV system. Consumers include homeowners, developers, and businesses. In some cases, integrators and installers work with agents of the consumer, such as architects, engineers, general contractors, or managers. Consumers drive the growth of the PV industry but may or may not be knowledgeable about PV systems, including the types of features that are available, the advantages and disadvantages of PV power, or how to finance a system. Therefore, much of the marketing effort by the PV industry is aimed toward educating potential consumers about various PV options and incentives.

Organizations

Numerous not-for-profit organizations work to promote and further the PV industry, including organizations involved in research, marketing, installer training, and standards development. These organizations also aim to make systems safer for the installer and the owner/operator.

Research Institutions. Around the world, numerous universities and national laboratories are working to further develop the PV industry. Research institutions are the underlying

force to improve existing PV manufacturing techniques and develop the next generation of technologies. These institutions also educate and train the leaders and high-tech workforce of the future PV industry.

Trade Associations. Businesses within an industry, even competitors, often join together to form trade organizations in order to support each other and promote the industry to the consumer, in an effort to increase the market for all involved. Manufacturers, integrators, and other groups in the PV industry have formed national and regional alliances including the Solar Energy Industries Association (SEIA) the Solar Electric Power Association (SEPA), and many others. These organizations host trade shows and conferences to facilitate industry contacts and educate the public. Not all businesses belong to trade associations, but most find them to be mutually beneficial alliances.

Installer Training. Training programs on the installation and operation of PV systems are organized by manufacturers, schools, and trade unions. These programs promote quality installations but can vary widely in length and scope, which can make it difficult for consumers to choose a skilled integrator or installer for a desired system. The North American Board of Certified Energy Practitioners (NABCEP) is an organization seeking to standardize installer qualifications.

DOE/NREL, Ajeet Rohatgi

Universities and research institutions are investigating ways to improve the efficiency of PV technologies.

IBEW-NECA Technical Institute (Chicago)

Location: Alsip, IL (41.7°N, 87.8°W)
Type of System: Utility-interactive
Peak Array Power: 1.1 kW DC
Date of Installation: Winter 2004-2005 (original installation)
Installers: Apprentices and journeymen
Purpose: Training

The IBEW-NECA Technical Institute (IN-TECH) uses a mixture of permanent and temporary PV systems to train apprentices and journeymen, each with its own strengths. The two permanent installations include a rooftop rack array and an awning-type array cantilevered off a wall. These systems feed into the building's electrical system and serve as ongoing monitoring and maintenance learning tools for commercial-type PV systems. Two temporary systems are based on residential-type sloped single roofs and provide hands-on installation experience. One of the temporary roof systems is unique for two reasons.

First, the system is indoors. The training center is located in a former high school, with a gymnasium that provides a large area to accommodate a section of a residential roof. The roof is sloped about 35°. Although it is only a few feet off the ground, the students work as if it were an actual roof and follow all applicable safety rules, including the use of fall protection harnesses.

IBEW-NECA Technical Institute
Students receive training on every part of installing a PV system, including connecting to utility power.

The system provides a complete training exercise, including routing electrical conduit and connecting the system to the utility. At the back of the roof is a framed wall that supports an electrical panel, inverter, and electric meter. The meter is pre-wired into the building's electrical system with a standard electrical service entrance, as if the PV system was on a detached building with a typical utility connection.

The first part of the training exercise includes configuring the mounting system, attaching the PV modules, wiring the modules together into an array, bending conduit and routing wires from the roof to the electrical panels, connecting the inverter and other components, and testing the system. A row of 12 HID lights mounted to the ceiling and aimed toward the roof simulates sunlight so that the system can be tested under typical outdoor conditions. Loads can be powered from the system for demonstration.

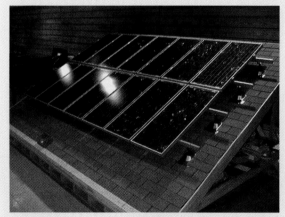

IBEW-NECA Technical Institute
The finished PV system can be tested for performance and used for troubleshooting exercises.

For the second part of the exercise, the instructors test the students' troubleshooting and problem-solving skills. The instructors intentionally create a small problem with the system, such as by disconnecting a conductor or changing the settings on the inverter, and challenge the students to find the problem. The students use test instruments and their knowledge of the system and its operation to determine the cause of the problem and provide the remedy.

The other unique aspect of this system is that the PV installations are infinitely repeatable. After one class has completed training, the system is disassembled for the next class. The modules, mounting system, conductors, conduit, and connectors are completely removed. Only the roof attachment points, empty electrical enclosures, and service entrance remain, and the training exercise begins again for a new class.

NABCEP is composed of representatives from throughout the renewable energy industry and works with the renewable energy and energy efficiency industries and professionals to develop and implement quality credentialing and certification programs for practitioners. NABCEP certification for PV installers requires both experience and education, ensuring that certified professionals meet minimum levels of expertise in PV systems and installation. Certification is awarded upon successful completion of an exam. NABCEP certification is voluntary and does not replace local training and licensing requirements, but it provides a measure of quality for solar energy system installers.

A joint effort by the International Brotherhood of Electrical Workers® (IBEW®) and the National Electrical Contractors Association (NECA) provides nationwide standardized training for union electrical apprentices and journeymen. Training is conducted in local training centers, many of which have PV systems for training and supplemental power purposes.

Standards. Many organizations that develop standards and safety guidelines for the electrical industry are also involved, directly or indirectly, in the PV industry. Several organizations, such as the Institute for Electrical and Electronics Engineers (IEEE®), publish standards and guidelines related to photovoltaics. These documents cover standard terminology, module test procedures and conditions, solar radiation measurement, electrical connections, power conditioning, and other topics.

The National Electrical Code® (NEC®), developed by the National Fire Protection Association® (NFPA®), is a set of rules on safe practices for the installation of electrical equipment. The NEC® applies to jurisdictions (municipalities, counties, or states) that have adopted it as their governing electrical code, which includes most of the United States and some other countries. The code applies to all electrical systems, including both utility-connected and stand-alone PV systems. Article 690 of the NEC® specifically addresses the design and installation of PV systems and equipment. It includes definitions, requirements for sizing conductors and circuit protection, disconnecting means, wiring methods, grounding, marking, and connections to other systems.

Other standards related to general electrical and workplace safety are applicable to PV installations. These include the standard NFPA 70E®, *Electrical Safety in the Workplace*, and Occupational Safety and Health Administration (OSHA) regulations related to electrical work, fall protection, material handling, outdoor work, and general workplace safety.

Product listing organizations test products for safety and conformity to standard requirements. Products that pass are certified, or "listed," as matching the manufacturer's specifications and fulfilling the requirements for safety. These products may then bear the listing organization's mark, which is considered a symbol of quality. Underwriters Laboratories Inc.® (UL) is the most prominent listing organization and publishes safety and quality standards against which products are tested. **See Figure 1-16.**

Product Listing Organizations

Figure 1-16. *The official mark of a listing organization signifies that a product meets the organization's standards for safety and quality.*

Authorities Having Jurisdiction. An *authority having jurisdiction* (AHJ) is an organization, office, or individual designated by local government with legal powers to administer, interpret, and enforce building codes. The local AHJ ensures that installation equipment, materials, and procedures result in a safe and quality system. The primary contact from the AHJ is the building inspector who must be knowledgeable in PV system configurations and the applicable codes and standards.

Utilities

Utilities have a unique place in the PV industry. They are the primary producers of the conventionally generated electricity that PV-generated electricity aims to replace, so they compete with PV installations. Utilities are invested in large, expensive power plants designed for specific generating processes, such as nuclear reactions or burning coal, so it is not easy for them to significantly change these strategies. However, progressive utilities are also potential PV system owners, as they adopt PV and other renewable energy sources for future growth and eventual phase-out of existing power plants. Either way, utilities are a major part of the PV industry.

Government Policies

To establish an energy portfolio that is sustainable for future growth, state and federal governments, and governments of other countries, have enacted various energy polices. This will become increasingly important as more countries develop and industrialize. Energy demand is expected to continue growing exponentially and new technologies must be available to help meet the demand. **See Figure 1-17.**

Future Energy Development

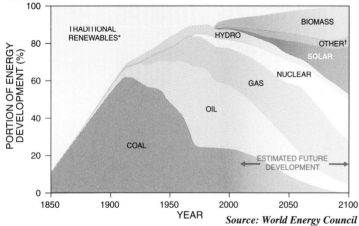

* primarily wood combustion, animal power, and human power
† primarily wind power

Source: World Energy Council

Figure 1-17. *Renewable energy is expected to compose an increasingly larger portion of energy production.*

These policies are primarily aimed at reducing dependency on fossil fuel energy, mandating efficient energy use, and encouraging the development of renewable energy resources. The primary ways in which federal, state, and local governments encourage renewable energy, including photovoltaics, is through incentives and quotas.

Incentives. Incentive programs include grants, rebates, renewable energy credits, low-interest or no-interest loans, sales and property tax exemptions, income tax credits or deductions for individuals and corporations, and cash payments based on energy production. Federal programs are available throughout the United States, while state programs vary widely. Even some utilities offer incentive programs in their areas. Online databases of various incentive programs help consumers research available programs for their area.

SMA Technologie AG
Combinations of energy sources, such as photovoltaics and wind turbines, provide a balanced supply of energy.

Quotas. The quota mechanism is a state policy and is typically called a renewables portfolio standard (RPS). An RPS places an obligation on either the state's utilities or the consumers to source a specified fraction of their electricity from renewable energy sources by a certain deadline. For example, California has committed to producing 20% of its electricity needs with renewable energy by 2010. **See Figure 1-18.** Utilities that fail to meet their obligation are required to pay a penalty fee for each unit of electricity short of the goal. This quota mechanism expands the market for renewable energy systems, which drives competition and is expected to lower costs to the consumer.

Renewables Portfolio Standards (RPS) and Goals

WA:
15% by 2020

OR:
25% by 2025

MT:
15% by 2015

ND:
10% by 2015

SD:
10% by 2015

MN:
25% by 2025

WI

MI

NV:
20% by 2015

UT:
20% by 2025

CO:
20% by 2020

IA:
105 MW

IL:
25% by 2025

OH:
25% by 2025

PA

NY

VT

NH

MA

RI

CT

NJ

DE

MD

ME

CA:
20% by 2010

AZ:
15% by 2025

NM:
20% by 2020

MO:
15% by 2021

VA

NC

DC

TX:
5880 MW by 2015

HI:
20% by 2020

ME: 30% by 2000;
    10% by 2017 new RE
VT: Meets load growth
    by 2012; 20% by 2017
NH: 23.8% by 2025
MA: 15% by 2020 +
    1% annual increase
RI: 16% by 2020
CT: 23% by 2020
NY: 24% by 2013
NJ: 22.5% by 2021
PA: 18% by 2020
DE: 10% by 2019
MD: 20% by 2022
DC: 20% by 2020
VA: 15% by 2025
NC: 12.5% by 2021
MI: 10% + 1100 MW
    by 2015
WI: RPS varies;
    10% by 2015 goal

▨ STATE RPS
▢ STATE GOAL

Source: DSIRE, May 2009

Figure 1-18. *Renewables portfolio standards and goals vary by state.*

SOLAR ENERGY TECHNOLOGIES

Photovoltaics is only one of many technologies for utilizing solar energy. Many other methods can make use of solar energy to produce heat, electricity, light, and even cooling. Some techniques have been used by humans since ancient times. For instance, the sun baked bricks into strong building materials, and dried foods for preservation. Wind, a product of solar energy, moved people across oceans. The modern world has built on these technologies to utilize this abundant resource in more efficient and direct ways.

Collectors

Most solar energy technologies involve special equipment that is added to or incorporated into a structure to collect, convert, and distribute the energy gained from the sun. A *solar energy collector* is a device designed to absorb solar radiation and convert it to another form, usually heat or electricity. Collectors are also sometimes called receivers. Solar collectors are classified into flat-plate and concentrating collectors according to their ability to utilize the different types of solar radiation.

Flat-Plate Collectors. A *flat-plate collector* is a solar energy collector that absorbs solar energy on a flat surface without concentrating it,

and can utilize solar radiation directly from the sun as well as diffuse radiation that is reflected or scattered by clouds and other surfaces. Flat-plate collectors may be installed in a fixed orientation or on a sun-tracking mount. Nearly all commercial and residential solar energy installations use flat-plate collectors. **See Figure 1-19.**

Flat-Plate Collectors

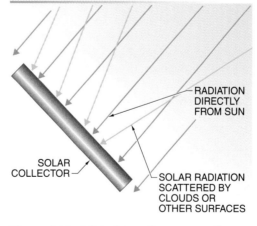

RADIATION
DIRECTLY
FROM SUN

SOLAR
COLLECTOR

SOLAR RADIATION
SCATTERED BY
CLOUDS OR
OTHER SURFACES

Figure 1-19. *A flat-plate collector can utilize any solar radiation, direct or diffuse (reflected), that strikes its surface.*

Renewable energy is a class of energy resources that are replaced rapidly by natural processes. These resources include solar, wind, water, tidal, geothermal, and biomass energy.

Concentrating Collectors. A *concentrating collector* is a solar energy collector that enhances solar energy by focusing it on a smaller area through reflective surfaces or lenses. The high-intensity sunlight is focused onto high-efficiency solar cells or working fluids that transfer thermal energy. Since concentrating collectors utilize only radiation directly from the sun, they must continually track the sun. Concentrating collectors have increased efficiency and reduced size because of the ability to channel more solar radiation onto the desired surface. **See Figure 1-20.**

Solar Architectural Design

Many of today's buildings and homes are built with the conservation of energy and natural resources in mind. Passive solar design techniques incorporated into architecture decrease the amount of energy needed for lighting, heating, cooling, and other loads. The most influential design option is to thoroughly insulate and seal the building envelope. This involves thicker walls, more insulation, joint and transition sealing, double-pane windows, and other features. Walls and roofs made with stone, concrete, or similar materials draw in and hold heat. These measures can significantly reduce the heating and cooling loads necessary to condition the indoor spaces since less energy can escape through the building envelope to the outside environment.

Positioning windows to face the sun allows maximum sunlight and heat into the building. **See Figure 1-21.** If too much sun during the summer is an issue, deciduous trees can be planted to shade windows in warm weather, reducing cooling loads, but allowing sun through in cool weather when the leaves fall. Awnings can also be used to create shade in the summer and allow exposure to sun in the winter.

Solar Architectural Design

DOE/NREL, Warren Gretz

Figure 1-21. *Solar architectural design uses building materials or design techniques to provide light and comfortable temperatures inside a building.*

Concentrating Collectors

RADIATION DIRECTLY FROM SUN

GLASS OR PLASTIC LENS

AREA OF CONCENTRATION

SPECIAL PARABOLA SHAPE

POINT OF CONCENTRATION

PARABOLIC REFLECTOR

LENS CONCENTRATOR

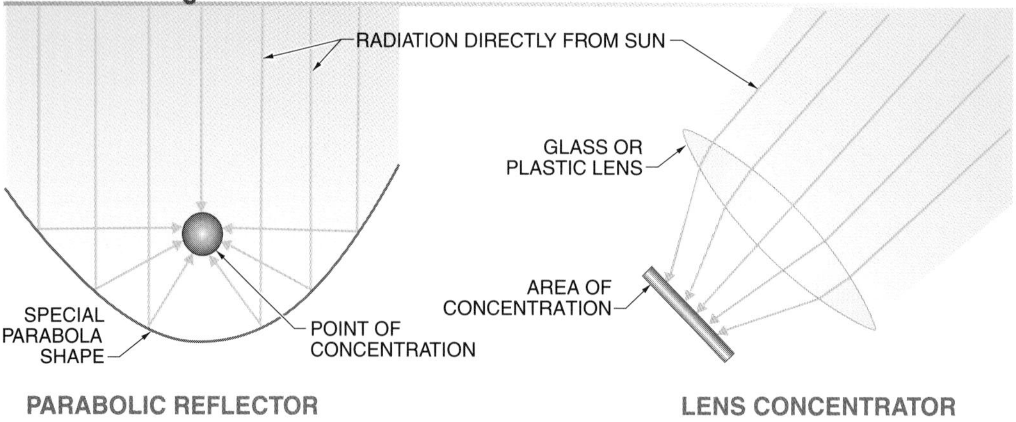

Figure 1-20. *Concentrating collectors focus a large area of direct solar radiation onto a relatively small area.*

Many of these passive solar energy strategies are included in the Leadership in Energy and Environmental Design (LEED®) Green Building Rating System™ to promote reduced energy use. This is an independent rating system that certifies different levels of energy efficiency and environmental protection. The LEED system also evaluates building siting, materials, operations, and occupant comfort.

Solar Thermal Energy

Solar thermal energy systems convert solar radiation into heat energy. Most systems use working fluids that are heated by the sun in solar collectors. Either flat-plate or concentrating collectors can be used. The heat in the fluid is then distributed to a reservoir to store the heat energy, or to other parts of the system to utilize the heat energy.

Solar Thermal Heating. Solar thermal heating collectors are dark-colored containers for the working fluid that convert absorbed solar radiation into heat and transfer the heat to the working fluid. **See Figure 1-22.** The fluid is usually an antifreeze and water solution, but it may be plain water if there is no danger of freezing. If used for space heating, the fluid could even be air. The fluid may be heated in

Solar Thermal Energy Collectors

DOE/NREL, Alan Ford

Figure 1-22. *Solar thermal energy is a relatively simple way to provide domestic hot water or heating.*

batches, in which case the collector also stores the fluid until it is needed. Alternatively, the heated fluid may be pumped through the collectors and stored in a separate reservoir.

Solar thermal heating systems can be scaled for many different applications, but are primarily used in residences to heat water and living spaces. The working fluid may be used to directly warm the home through radiant floor heating, or the heat may be transferred to water or air through heat exchangers. Solar thermal energy can even be utilized in the winter in northern climates, though a supplemental heat source may be needed to fully heat the fluid on cold days.

Solar thermal heating systems can be classified as passive or active. In passive systems, the working fluid circulates through the solar collectors and the rest of the system by convection without the use of mechanical equipment. In active systems, fans and pumps are used to circulate the fluid.

Solar Thermal Cooling. Solar thermal cooling uses the heat energy to power a refrigeration cycle. The heat energy is used to compress a gas, which, when expanded in another part of the system, cools a set of coils. Air blown over the coils cools and is used to air-condition the building. This process requires an extremely hot working fluid, so it is not yet an efficient process except in the sunniest climates, but improvements continue.

Solar Thermal Electricity. Solar thermal energy can also be used to produce electricity, though the processes use working fluids in different ways. The most common types of systems rely on concentrating solar power.

Concentrating solar power (CSP) is a technology that uses mirrors and lenses to reflect and concentrate solar radiation from a large area onto a small area. The energy is then used to heat working fluids or materials that will be used to produce electricity. These systems are complex to build, so they are only feasible for utility-scale power plants. The three main styles of CSP plants include trough collectors, dish collectors, and power towers. **See Figure 1-23.**

Concentrating Solar Power

DOE/NREL, Dave Parsons

TROUGH COLLECTOR

DOE/NREL, Warren Gretz

DISH COLLECTOR

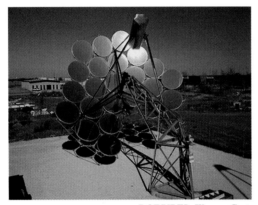

DOE/NREL, Sandia National Laboratories

POWER TOWER

Figure 1-23. *The intense solar radiation needed to produce electricity from thermal energy requires solar concentrating systems.*

Parabolic trough collector systems are the most common and cost efficient. A curved trough focuses sunlight on an absorber tube. The trough is rotated throughout the day to maximize the received solar energy. The absorber tube is filled with fluid, typically oil, which becomes very hot. The oil is then passed through a heat exchanger that transfers heat to water, which boils and produces steam. The steam is used to drive steam turbines that produce electricity.

A dish collector is a large, reflective, parabolic dish that focuses light from a large area onto a single point above the dish. At that point, the intense sunlight heats a working fluid. The resulting heat energy is then converted into electricity in the same way as in trough collector systems.

Power tower systems use a large number of mirrors to concentrate sunlight on a target at the top of a tower. The target, or receiver, is filled with fluid to absorb the resulting intense heat. The fluid is commonly molten salts, which have the ability to hold large amounts of heat. The heat energy can then be transferred to water to produce steam that drives steam turbines. Power tower designs have also been used to directly heat water into steam, eliminating the molten salts step, but this system cannot produce electricity after sunset. The salts, however, hold heat for hours and can continue to produce steam at night.

Solar Cooking. Solar cookers use energy from the sun to heat food or beverages. The container of food or liquids essentially becomes the solar thermal collector. Solar cookers are made with reflectors that direct and concentrate light into the inside of a dish- or box-shaped cooker. The cooking vessel is dark in color to absorb the maximum amount of heat. Food to be cooked is placed inside the solar cooker, just as if it were an ordinary oven.

Solar Chemical Energy

Utilizing solar energy directly through chemical processes is a less-developed technology, but is of great interest as a future energy source. Scientists expect to be able to harness solar energy in chemical reactions in a manner similar to how plants use solar energy through photosynthesis. The energy might then be released through reverse reactions.

However, an indirect use of solar chemical energy is through biofuels. Plants grow and produce chemicals by utilizing solar energy. Many of the oils found in plant seeds are chemically similar to petroleum and can be similarly refined into fuels. For example, biodiesel is a fuel that can be made from sources such as soybeans, waste vegetable oil, or even animal fats, and is compatible with any existing diesel engine. Biodiesel also burns cleaner than petroleum-based diesel. Similarly, ethanol fuel can be produced from corn and burned in modified gasoline engines.

Solar Lighting

Solar lighting includes equipment and techniques to direct natural lighting into a building. Increasing the amount of natural light indoors, or daylighting, is more environmentally sound and easier than using grid electricity. Some of the methods could be considered one aspect of solar architectural design, such as window placement, but solar lighting can also include special materials or components to bring additional light indoors. For example, fiber-optic light pipes can bring sunlight into almost any indoor space and from any direction.

Refer to Quick Quiz® on CD-ROM

Summary

- Photovoltaics (PV) is a solar energy technology that uses the unique properties of semiconductors to directly convert solar radiation into electricity.

- Many advantages and benefits add value to PV systems beyond the potential economic savings.

- Photovoltaics is an environmentally friendly technology that causes no noise or pollution.

- Currently, the most significant disadvantage of PV systems is the high initial cost compared to prices for competing power-generating technologies.

- The photovoltaic effect was discovered long before it was used for the generation of electricity.

- Improvements in cell efficiencies and manufacturing methods are reducing the cost of PV systems.

- Higher costs for conventional electricity-generating technologies help make PV and other renewable energy systems more cost-effective.

- Space applications were the first practical use of PV technology.

- PV systems are particularly well-suited for portable and remote applications.

- Supplemental power systems are the fastest-growing application of PV systems.

- Utility-scale PV power plants are not yet common, but are being studied for future widespread use.

- There are many levels of business and organizations that support the PV industry.

- Skilled installers are critical for quality installations and good public perception of the PV industry.

- Government renewable energy policies support further development and growth of the U.S. PV industry.

- Flat-plate collectors absorb solar energy from many directions on a flat surface without concentrating it.

- Concentrating collectors focus solar radiation from a large area onto a small area that contains special PV cells or heat-absorbing materials.

- Integrating solar energy design techniques into architecture is an example of a passive solar energy technology.

- Solar thermal energy is an efficient use of solar radiation that can be utilized in many ways.

- Solar thermal energy can be converted into electricity, though only on a large scale that is impractical for residential and commercial applications.

- Sunlight can be used for natural lighting in almost any indoor space by using fiber optics.

- *Photovoltaics* is a solar energy technology that uses the unique properties of certain semiconductors to directly convert solar radiation into electricity.

- A *photovoltaic (PV) system* is an electrical system consisting of a PV module array and other electrical components needed to convert solar energy into electricity usable by loads.

- A *load* is a piece of equipment that consumes electricity.

- A *utility* is a company that produces and/or distributes electricity to consumers in a certain region or state.

- The *grid* is the utility's network of conductors, substations, and equipment that distributes electricity from its central generation point to the consumer.

- *Distributed generation* is a system in which many smaller power-generating systems create electrical power near the point of consumption.

- A *balance-of-system (BOS) component* is an electrical or structural component, aside from a major component, that is required to complete a PV system.

- An *integrator* is a business that designs, builds, and installs complete PV systems for particular applications by matching components from various manufacturers.

- An *authority having jurisdiction (AHJ)* is an organization, office, or individual designated by local government with legal powers to administer, interpret, and enforce building codes.

- A *solar energy collector* is a device designed to absorb solar radiation and convert it to another form, usually heat or electricity.

- A *flat-plate collector* is a solar energy collector that absorbs solar energy on a flat surface without concentrating it, and can utilize solar radiation directly from the sun as well as radiation that is reflected or scattered by clouds and other surfaces.

- A *concentrating collector* is a solar energy collector that enhances solar energy by focusing it on a smaller area through reflective surfaces or lenses.

- *Concentrating solar power (CSP)* is a technology that uses mirrors and lenses to reflect and concentrate solar radiation from a large area onto a small area.

1. Compare the advantages and disadvantages of installing a PV system.

2. Compare the design priorities of PV systems for space, portable, and supplemental-power applications.

3. What are some advantages to using PV technology to build utility-scale power plants?

4. What is the role of an integrator?

5. How is installer training vital to the continued growth of the PV industry?

6. How are government policies being used to encourage development of renewable energy technologies?

7. Why are concentrating collectors more efficient than flat-plate collectors?

8. Explain the two main technologies that can convert solar radiation into electricity.

2

Solar Radiation

Understanding solar radiation is important to designing and installing solar energy equipment. Solar radiation resources can vary greatly over time, location, and climate conditions. However, array performance can be maximized for the available solar energy by carefully aligning solar collectors or using sun-tracking systems to follow the sun. Solar radiation resource data is used to determine the proper orientation for solar collectors and to estimate their output.

Chapter Objectives

- Differentiate between solar irradiance (solar power) and solar irradiation (solar energy).
- Identify the factors affecting the quantity and composition of solar energy received on Earth's surface.
- Identify the factors affecting the sun's apparent position and path through the sky.
- Calculate differences between solar time and standard time.
- Evaluate how array orientation affects solar energy received by modules.
- Demonstrate how solar radiation data is used in sizing and estimating performance for PV systems.

THE SUN

The sun is a gaseous body composed mostly of hydrogen, with some helium and traces of heavier elements. The gases swirl and flow under the influence of gravity, magnetic fields, and heat energy. Gravity also causes intense pressure and heat at the core, which initiates nuclear fusion reactions. Fusion is a nuclear reaction that combines the atoms of lighter elements into atoms of heavier elements, and releases enormous quantities of energy. The sun fuses hydrogen into helium at its core and the resulting energy radiates outward. The energy travels to the photosphere, where it escapes into space in the form of visible light and other radiation. *Radiation* is energy that emanates from a source in the form of waves or particles. The photosphere is the visible surface of the sun. **See Figure 2-1.**

An astronomical unit (AU) is the average distance between Earth and the sun (93 million mi) and is used as a measuring unit for other distances in the solar system. Traveling at the speed of light (186,000 mi/s), radiation from the sun takes more than 8 min to reach Earth's surface. **See Figure 2-2.** Earth receives approximately 170 million GW of power from the sun, which is a relatively tiny fraction of the sun's total output, but is millions of times greater than the maximum power demand of Earth's entire population.

> A photon is a fundamental unit of electromagnetic radiation. The high-energy photons released in fusion reactions at the sun's core take a very long time to reach the surface, losing energy along the way. Photon travel time is estimated to be tens of thousands of years. Upon reaching the sun's surface, photons escape as visible light.

SOLAR RADIATION

Fusion reactions in the sun convert mass into energy that radiates outward in all directions. The solar radiation travels through space and a small fraction of it reaches Earth. Various factors affect the quantity and composition of the solar energy that reaches the Earth's surface and is eventually harnessed by PV devices.

Solar Photosphere

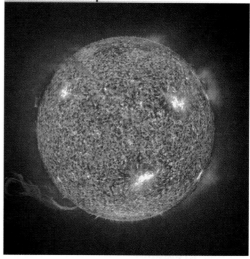

SOHO (ESA & NASA)

Figure 2-1. *A false color image of the sun enhances the turbulent nature of the sun's photosphere, including a roiling surface, sunspots, and giant flares.*

Solar Irradiance (Solar Power)

Solar irradiance is the power of solar radiation per unit area. Solar irradiance is commonly expressed in units of watts per square meter (W/m^2) or kilowatts per square meter (kW/m^2). Irradiance is measured with respect to area as if the solar radiation is striking an imaginary unit surface. **See Figure 2-3.** Since it is a measure of power (rate of energy), solar irradiance is an instantaneous value. For example, the speed (rate of travel) of an object is also an instantaneous value.

DOE/NREL, Powerlight Corporation
PV systems convert radiation from the sun into electricity.

Astronomical Unit

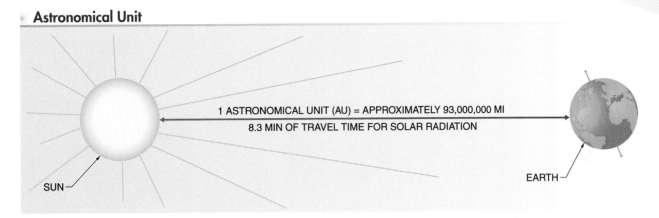

Figure 2-2. *Even over the vast distance, an enormous amount of energy reaches Earth from the sun.*

Solar Irradiance

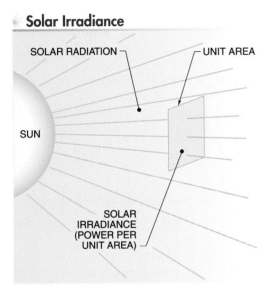

Figure 2-3. *Solar irradiance is solar power per unit area.*

Solar irradiance is used as a reference condition to evaluate the output performance of a solar energy system at a given point in time, or for rating the power output of solar energy utilization equipment such as PV modules.

Solar irradiance varies slightly as the sun goes through normal cycles of maximum and minimum activity. However, distance from the sun has a much greater effect. The *inverse square law* is a physical law that states that the amount of radiation is proportional to the inverse of the square of the distance from the source. This means that at twice the distance of Earth to the sun, solar radiation is only one-fourth the amount on Earth. Likewise, moving three times the distance away from a light source decreases the intensity by a factor of nine. **See Figure 2-4.** This law is used to calculate the irradiance at different locations in the solar system.

Inverse Square Law

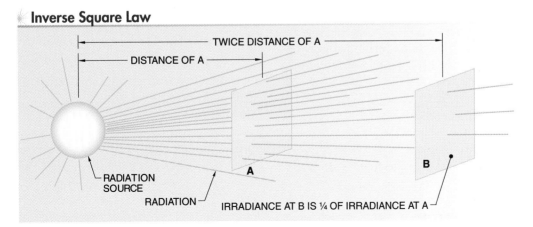

Figure 2-4. *The inverse square law states that irradiance is reduced in proportion to the inverse square of the distance from the source.*

Almost all radiation from the sun is in the form of waves (electromagnetic radiation), but a very small portion consists of subatomic particles (corpuscular radiation), primarily charged protons, neutrons, helium nuclei (alpha particles), and electrons (beta particles). These charged particles form the "solar wind" that streams from the sun. When they reach Earth's magnetic field, they become concentrated near the poles and interact with the atmosphere, resulting in aurora displays known as the Northern (or Southern) Lights. During extreme solar activity, these particles can also interfere with radio and power transmission.

Solar Irradiance and Irradiation

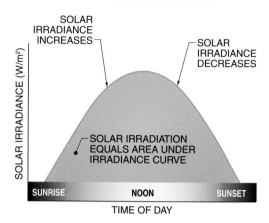

Figure 2-5. *Solar irradiation equals the total solar irradiance over time.*

Solar Irradiation (Solar Energy)

Solar irradiation is the total amount of solar energy accumulated on an area over time. The period of time may be an hour, day, month, or year. Solar irradiation is commonly expressed in units of watt-hours per square meter (Wh/m²) or kilowatt-hours per square meter (kWh/m²). Solar irradiation quantifies the amount of energy received on a surface over time, and is the principal data needed for sizing and estimating the performance of PV systems.

Greater solar irradiance (power) means energy is accumulated faster, which results in greater solar irradiation (total energy). If solar irradiance is comparable to the speed of an object, solar irradiation is comparable to the total distance the object travels over time. Greater speed (irradiance) results in greater total distance (irradiation) over the same amount of time.

At the surface of Earth, the magnitude of solar irradiance changes throughout the day. It begins at zero during nighttime, increases as the sun rises, peaks around noon, and decreases as the sun sets. In a plot of solar irradiance versus time, solar irradiation equals the area under the irradiance curve. **See Figure 2-5.**

Solar irradiation can be calculated by applying the formula:

$$H = E \times t$$

where

H = solar irradiation (in Wh/m²)

E = average solar irradiance (in W/m²)

t = time (in hr)

For example, if the average solar irradiance is 600 W/m² over 8 hr, what is the total solar irradiation over this period?

$$H = E \times t$$

$$H = 600 \times 8$$

$$H = \textbf{4800 Wh/m}^2 \textbf{ or 4.8 kWh/m}^2$$

Extraterrestrial Solar Radiation

Extraterrestrial solar radiation is solar radiation just outside Earth's atmosphere. Extraterrestrial solar radiation is also sometimes known as top-of-atmosphere (TOA) radiation. Knowledge of the quantity and characteristics of extraterrestrial solar radiation is required, for example, in the design of PV arrays for the International Space Station and Earth-orbiting satellites. This benchmark is also important to understanding how Earth's atmosphere affects solar power.

The Solar Constant. In space, solar power remains relatively constant for a given distance from the sun because there is no atmosphere to scatter and absorb the energy. The *solar constant* is the average extraterrestrial solar power (irradiance) at a distance of 1 AU from the sun, which has a value of approximately 1366 W/m². While this is not a true constant because the value varies by a few percent over the year, it is essentially stable for practical purposes.

Estimating Solar Irradiance

The solar constant is an irradiance value calculated for Earth and cannot be used for PV systems outside of Earth's vicinity, such as on satellites orbiting other planets. In fact, interplanetary space probes encounter constantly changing solar irradiance as they travel toward or away from the sun. Using the inverse square law, the solar irradiance for any other location in the solar system can be estimated with the following formula:

$$E = \frac{E_E}{D^2}$$

where

E = solar irradiance at location (in W/m²)

E_E = solar constant for Earth (1366 W/m²)

D = ratio of distance of location from sun to distance of Earth from sun (which also equals the location's distance in AU)

This formula can be used to estimate the solar irradiance for Mars, which orbits the sun at an average distance of 1.52 AU (142 million mi).

$$E = \frac{E_E}{D^2}$$

$$E = \frac{1366}{1.52^2}$$

$$E = \textbf{591 W/m}^2$$

Solar irradiance at Mars is approximately 591 W/m², only 43% of Earth's solar constant.

Since Mars is farther from the sun than Earth, the PV module on the Sojourner rover was designed to maximize electrical power from a reduced solar irradiance.

NASA/JPL–Caltech

Solar Spectrum. Almost all of the energy received from the sun is electromagnetic radiation. *Electromagnetic radiation* is radiation in the form of waves with electric and magnetic properties. The waves vary in length depending on the source and energy level. The wavelength determines the properties of the radiation. Extremely short wavelength (trillionths of a meter) radiation takes the form of gamma rays, which is high-energy radiation produced by sub-atomic reactions. Extremely long wavelength (millions of meters) radiation takes the form of radio waves, which are useful for transmitting data over long distances. In between are X-rays, ultraviolet radiation, visible light, and infrared radiation. The *electromagnetic spectrum* is the range of all types of electromagnetic radiation, based on wavelength. **See Figure 2-6.**

Electromagnetic Spectrum

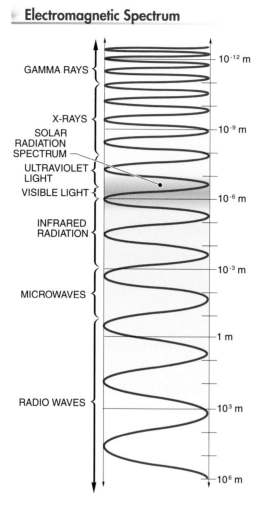

Figure 2-6. *The electromagnetic spectrum is the range of all types of electromagnetic radiation, which vary with wavelength.*

Solar radiation includes more types of energy than just visible light. Most of the energy from the sun covers a continuous portion of the electromagnetic spectrum, from ultraviolet to infrared, though not evenly. The distribution of extraterrestrial solar power at various wavelengths forms a spectral signature unique to the sun. **See Figure 2-7.** A similar spectrum of solar energy reaching Earth's surface is studied to determine how atmospheric effects have reduced the energy levels of certain wavelengths.

Atmospheric Effects

Solar radiation is absorbed, scattered, and reflected by components of the atmosphere, including ozone, carbon dioxide, and water vapor, as well as other gases and particles. Cloud cover and local conditions such as dust storms, air pollution, and volcanic eruptions can also greatly reduce the amount of radiation reaching the surface of Earth. The two major types of radiation reaching the ground are direct radiation and diffuse radiation. *Total global radiation* is all of the solar radiation reaching Earth's surface and is the sum of direct and diffuse radiation. **See Figure 2-8.**

Direct Radiation. Since there is no atmosphere in outer space, solar radiation travels unimpeded through the vacuum of space in a straight line from the sun. The atmosphere allows a portion of this radiation to continue directly through to the surface. *Direct radiation* is solar radiation directly from the sun that reaches Earth's surface without scattering. Direct radiation is sometimes also called beam radiation. The rays of direct radiation are parallel and produce shadows. Extraterrestrial radiation is composed almost entirely of direct radiation.

Diffuse Radiation. *Diffuse radiation* is solar radiation that is scattered by the atmosphere and clouds. Scattering causes radiation to be dispersed in many directions. A point on Earth may receive diffuse radiation from many directions at once, in addition to radiation directly from the sun. The portion of total global radiation that is diffuse radiation varies from about 10% to 20% for clear skies and up to 100% for overcast skies.

Typical flat-plate PV arrays utilize both the direct and diffuse components of the total global radiation reaching the array surface, while concentrating collectors can utilize only the direct radiation component.

A portion of the direct radiation reaching the atmosphere is scattered into diffuse radiation. Another source of diffuse radiation is radiation that reaches the ground and is reflected back up through the atmosphere, where it is scattered or escapes back into space. This reflected radiation is called albedo radiation or reflectance. This type of radiation is not usable by PV devices, except for the fraction that scatters in the atmosphere and adds to the diffuse radiation.

Extraterrestrial Solar Spectrum

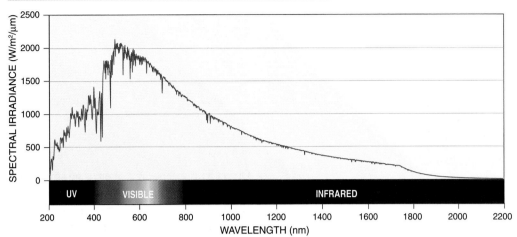

Figure 2-7. *The wavelength distribution of extraterrestrial solar radiation forms a spectral signature unique to the sun.*

Atmospheric Effects

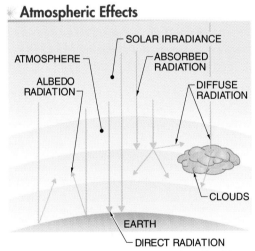

Figure 2-8. *Solar radiation entering Earth's atmosphere consists of direct, diffuse, and albedo radiation.*

Air Mass. The amount of solar radiation that is absorbed or scattered in the atmosphere depends on how much atmosphere it passes through before reaching Earth's surface. When the sun is at zenith, the amount of atmosphere that the sun's rays have to pass through to reach Earth's surface is at a minimum. *Zenith* is the point in the sky directly overhead a particular location. The *zenith angle* is the angle between the sun and the zenith. As the zenith angle increases (the sun approaches the horizon), the sun's rays must pass through a greater amount of atmosphere to reach Earth's surface. This reduces the quantity of solar radiation, and also changes its wavelength composition. **See Figure 2-9.**

Air mass (AM) is a representation of the relative thickness of atmosphere that solar radiation must pass through to reach a point on Earth's surface. By definition, air mass equals 1.0 when the sun is directly overhead at sea level, and is designated as AM1.0. Greater values of air mass indicate greater attenuating effects of the atmosphere. Since there are no attenuating effects outside Earth's atmosphere, air mass for this condition is zero, represented as AM0. All other air mass values are assigned relative to AM1.0.

For example, air mass 2.0 (AM2.0) represents the relative atmospheric path length when the angle between the sun and zenith is 60° (30° above horizon). In this case, the length of the path through the atmosphere is twice the distance of the path for AM1.0.

Air mass depends upon the time of day, the time of year, and the altitude and latitude of the specified location. Time and latitude affect the possible angles of the sun. Altitude affects the amount of atmosphere above the location.

The air mass formula is less accurate for zenith angles greater than about 60°. At these sun positions, very low on the horizon, the atmospheric refraction is also a factor.

☀ Air Mass

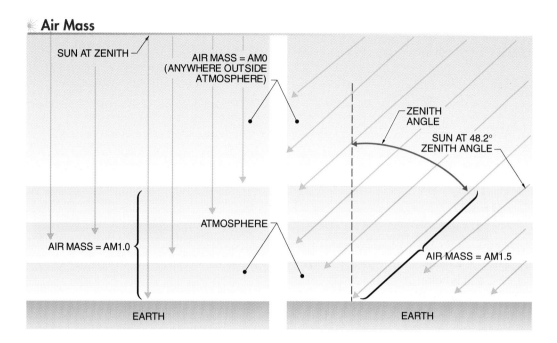

Figure 2-9. *Air mass is a representation of the amount of atmosphere radiation that must pass through to reach Earth's surface.*

Calculating Zenith Angle

The sun's zenith angle for air mass calculations can be determined at any time by using a vertical stake or ruler of known height and measuring the length of shadow cast. The ruler, shadow, and rays of the sun form a triangle. Using trigonometry, the zenith angle can be calculated with the following formula:

$$\theta_z = \arctan\left(\frac{L_S}{L_R}\right)$$

where

θ_z = zenith angle (in deg)

L_S = length of the shadow (in in.)

L_R = length of ruler (in in.)

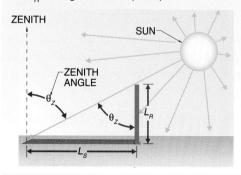

Calculating air mass involves determining the length of a path through the atmosphere at the given angle, and then accounting for the thickness of the atmosphere due to the location's altitude above sea level. The ratio of the local atmospheric pressure to the pressure at sea level is an easy way to make this adjustment. The local pressure varies somewhat with the weather, which correspondingly affects the air mass. However, close approximations of air mass can be calculated with average pressure values, such as those on the solar radiation data sets. Air mass is calculated with the following formula:

$$AM = \frac{p_{local}}{1013 \times \cos\theta_z}$$

where

AM = air mass

p_{local} = local atmospheric pressure (in mbar)

1013 = atmospheric pressure at sea level (in mbar)

θ_z = zenith angle (in deg)

For example, what is the air mass for Denver, CO if the sun is at a 53° zenith angle and the local atmospheric pressure is 836 mbar?

$$AM = \frac{p_{local}}{1013 \times \cos\theta_z}$$

$$AM = \frac{836}{1013 \times \cos 53°}$$

$$AM = \frac{836}{1013 \times 0.602}$$

$$AM = \mathbf{1.37}$$

Air mass 1.5 (AM1.5) is considered representative of average terrestrial conditions in the U.S. and is commonly used as a reference condition in rating PV modules and arrays.

Terrestrial Solar Radiation

Terrestrial solar radiation is solar radiation reaching the surface of Earth. While the solar power outside Earth's atmosphere (extraterrestrial radiation) remains relatively constant, the amount of solar power reaching Earth's surface varies significantly, and because it is reduced by Earth's atmosphere and climate conditions, its value is less than the solar constant.

Peak Sun. Of the total solar energy at Earth's outer atmosphere, about one-third is either reflected from clouds back into space, or scattered and absorbed by the atmosphere before it reaches Earth's surface. Solar irradiance on Earth is approximately two-thirds the value of the solar constant (1366 W/m²). *Peak sun* is an estimate of maximum terrestrial solar irradiance around solar noon at sea level and has a generally accepted value of 1000 W/m².

Greater values of terrestrial solar irradiance can occur at high altitudes, and in very clear, dry climates around solar noon, but 1000 W/m² is a practical value for most locations and climate conditions. This irradiance estimate is often used as the basis for rating terrestrial PV module and system performance.

Peak sun hours is the number of hours required for a day's total solar irradiation to accumulate at peak sun condition. An average day may have only one or two actual hours at

peak sun condition, but the total irradiation for a day may be expressed in units of peak sun hours by dividing by 1000 W/m² (peak sun irradiance). For example, a day with an average irradiance of 600 W/m² over 8 hr may only reach peak sun condition for an hour or less around noon. However, the total irradiation of 4800 Wh/m² (600 W/m² × 8 hr = 4800 Wh/m²) is equivalent to 4.8 peak sun hours (4800 Wh/m² ÷ 1000 W/m² = 4.8 peak sun hr). **See Figure 2-10.**

Peak Sun Hours

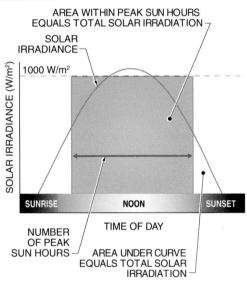

Figure 2-10. *Peak sun hours is an equivalent measure of total solar irradiation in a day.*

Since the electrical power output of PV modules is rated at peak sun condition (1000 W/m²), knowing the number of peak sun hours on a given surface at a given location is used to estimate PV system performance. In fact, most solar irradiation data resources present data in peak sun hour units.

A blackbody is a theoretical body that emits a maximum amount of radiation for a given temperature over all wavelengths. Except for very cold blackbodies, most of the energy is in the visible and near infrared wavelengths. A real body's emissivity indicates how well it radiates energy as compared to a blackbody at a certain temperature. The sun's radiation closely approximates a blackbody profile at 5780°C (approximately 10,000°F).

Refer to solar resources on CD-ROM

Insolation. Another term for solar radiation energy is insolation. *Insolation* is the solar irradiation received over a period of time, typically one day. It is typically expressed as kWh/m²/day or equivalent peak sun hours. Insolation is usually used to rate the solar energy potential of a location by calculating the average energy received on a surface per day. Color-coded maps show how insolation varies for different regions of the country and for different times of the year. **See Figure 2-11.**

Insolation Maps

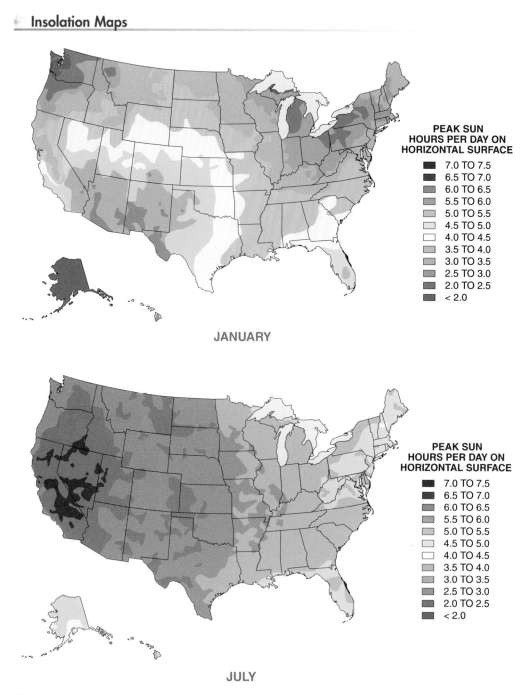

Figure 2-11. *Insolation maps rate locations by their average daily peak sun hours.*

Spectral Distribution. Besides its total power, terrestrial solar radiation also differs from extraterrestrial solar radiation in its spectral distribution, mainly due to the absorption of radiation at certain wavelengths by specific gases in the atmosphere. One of the key atmospheric gases, ozone, plays an important role in blocking harmful ultraviolet radiation from reaching Earth's surface. Other constituents, such as water vapor and carbon dioxide, also absorb solar radiation, principally in the infrared portions of the spectrum. **See Figure 2-12.**

Terrestrial Solar Spectrum

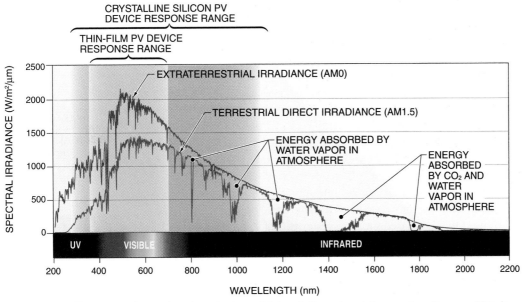

Figure 2-12. *The atmosphere absorbs extraterrestrial radiation at certain wavelengths, resulting in an altered spectral distribution for terrestrial radiation.*

Sunlight Safety

Installers and operators of PV systems often view the sun, either to determine its position or to check the sky conditions when making solar irradiance and performance measurements on PV systems. Looking directly at the sun is not only painful, but can damage the eyes and cause temporary or permanent blindness. While very brief glances at the sun may not cause harm, prolonged viewing with the naked eye heats the retina, potentially damaging it. Even viewing a partially eclipsed sun with the naked eye can cause localized damage to the retina, resulting in permanent blind spots.

With concentrating-type solar energy systems, sunlight intensity on the collector surface may be a hundred or more times the intensity of normal sunlight. Special precautions should be made so as to not accidentally view this concentrated solar radiation when working near these types of collectors. Even brief glances can cause permanent eye damage.

Protecting the eyes and body is important for anyone working on solar energy systems in direct sunlight. Standard practice includes wearing a quality pair of sunglasses that blocks harmful UV rays and glare and using sunscreens with high SPF. Hats and light colored clothing also minimize the personal discomfort of working in direct sunlight. To minimize the risk of heatstroke in extremely warm climates, work in the shade as much as possible and drink plenty of fluids. Viewing the sun with an appropriate welding filter is generally considered safe. Another method that can be used to protect the eyes when making quick glances at the sun is to shield the solar disk with a hand placed at arm's length.

PV systems can be successfully implemented at very high latitudes with relatively low insolation, even Antarctica.

Actual terrestrial spectral distributions can vary depending on atmospheric conditions, location, and air mass. For the purposes of uniform evaluation of modules and calibration of solar simulators, two standard distributions are used that are based on average terrestrial conditions and module orientation. The direct radiation spectral distribution is used for evaluating concentrating-type modules. The total global radiation spectral distribution is used for evaluating flat-plate modules that can utilize both direct and diffuse radiation.

Solar spectral distribution is important to understanding how PV devices operate because each responds to certain parts of the solar spectrum better than others. For example, most silicon-based PV devices, including crystalline silicon (c-Si), respond only to the visible and the near infrared portions of the spectrum, in the range of 300 nm to 1100 nm. Thin-film modules generally have a narrower response range, though some multi-junction designs use different materials in layers to increase spectral response beyond the typical range.

Solar Radiation Measurement

Long-term measurements are the basis for developing historical databases for available solar energy. This information is useful for not only sizing PV systems for a given output, but also for establishing the benchmark for PV system performance expectations over time. Also, solar irradiance (solar power) measurements are essential for installers who must adequately measure and verify that the output of PV systems is consistent with expectations.

Pyranometers. Solar irradiance is typically measured with a pyranometer. A *pyranometer* is a sensor that measures the total global solar irradiance in a hemispherical field of view. Since pyranometers measure both direct and diffuse radiation in a whole-sky view, they are often used to monitor the solar radiation incident on flat-plate type arrays. These sensors are mounted adjacent to arrays, in the same plane (facing the same direction) as the arrays. Solar irradiance values sampled at regular intervals and stored by data acquisition equipment can be used to determine the total solar irradiation for the specific surface orientation and site conditions. **See Figure 2-13.**

Pyranometers

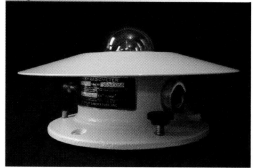

Figure 2-13. *A pyranometer measures total global solar irradiance from the whole sky.*

Diffuse global radiation can be measured by shading a pyranometer, which measures both direct and diffuse radiation, from the direct radiation component. A shadow band pyranometer uses a metal strip in the shape of an arc to shield the pyranometer from direct radiation. To account for changing sun paths, this device must be adjusted daily, either manually or automatically. Other shadowing devices may use a disk that follows the sun to shadow the pyranometer at all times of the day. **See Figure 2-14.** The direct radiation measurement is calculated by subtracting the diffuse measurement from the global measurement.

Pyranometer Shading Devices

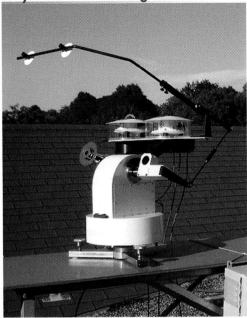

Eppley Laboratory, Inc.

Figure 2-14. *Diffuse solar irradiance can be measured by adding a shadowing device to a pyranometer, which blocks the direct component of total irradiance.*

Precision pyranometers use thermopile sensors (thermocouple arrays that output a voltage proportional to irradiance) and offer the most accurate and consistent response across a range of wavelengths. Pyranometers that use silicon solar cell or photodiode detectors are less precise, but are also less expensive and offer the durability required for most field measurements. Some are even small handheld meters with an easy-to-use digital readout of total solar irradiance. **See Figure 2-15.**

Pyrheliometers. Direct solar radiation is measured with a pyrheliometer. A *pyrheliometer* is a sensor that measures only direct solar radiation in the field of view of the solar disk (5.7°). It does not measure the diffuse radiation component. Because pyrheliometers only measure the direct radiation component, they must be pointed directly at the sun and installed on sun-tracking devices to take measurements of direct radiation over the course of the day. Most precision pyrheliometers use thermopile sensors. **See Figure 2-16.**

Handheld Pyranometers

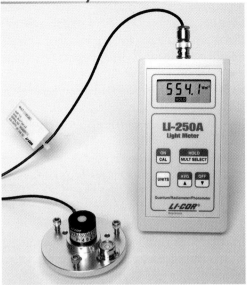

LI-COR Environmental

Figure 2-15. *Handheld pyranometers use less precise sensors than precision pyranometers but are more suitable for field measurements.*

Pyrheliometers

Kipp & Zonen, Inc.

Figure 2-16. *A pyrheliometer measures the direct component of solar irradiance, which is important when installing concentrating collectors.*

The word "pyranometer" is formed from the Greek terms "pyr" (fire), "ano" (sky), and "metron" (measure). The word "pyrheliometer" is formed from the Greek terms "pyr" (fire), "helios" (sun), and "metron" (measure). A pyranometer measures energy from the whole sky while a pyrheliometer measures energy from only the sun.

Reference Cells. A *reference cell* is an encapsulated PV cell that outputs a known amount of electrical current per unit of solar irradiance. Since current output from a PV device varies linearly with the incident solar irradiance, the output current can be used to indirectly measure irradiance. The calibration number is the conversion factor that changes a reference cell's current into solar irradiance with respect to a certain air mass value. Calibration numbers are usually expressed as milliamperes per kilowatt per unit area (mA/kW/m^2) or amperes per kilowatt per unit area (A/kW/m^2), both at AM1.5. The cell is usually encapsulated in an aluminum block with an optical glass cover, and often includes a thermocouple (temperature sensor) attached to the back of the cell to measure and correct the output for temperature variations. **See Figure 2-17.**

Reference Cells

PV Measurements, Inc.

Figure 2-17. *Reference cells output a certain electrical current for each unit of solar irradiance received.*

Reference cells are highly accurate precision instruments, typically used in a laboratory environment to calibrate the intensity of solar simulators, or to measure solar irradiance in flash or pulse type simulators used to measure the output of PV modules.

Calibration. All qualified solar radiation instruments are ultimately traceable to the World Radiometric Reference (WRR), an absolute cavity radiometer maintained by the World Meteorological Organization (WMO). For calibration, instruments are typically compared with a reference standard under a range of conditions, and the output is scaled by an appropriate factor. The Eppley Precision Spectral Pyranometer (PSP) is calibrated against the WRR. Most lower-precision pyranometers are calibrated against a PSP under natural daylight conditions.

Daystar, Inc.

Simple handheld irradiance meters can be useful for field measurements and site assessments.

SUN-EARTH RELATIONSHIPS

Two major motions of Earth affect the apparent path of the sun across the sky: Earth's yearly revolution around the sun and daily rotation about its axis. These motions are the basis for the solar timescale. Also, Earth's axis is tilted at 23.5°. The amount of solar radiation received at a particular location on Earth's surface is a direct result of Earth's orbit and tilt.

Earth's Orbit

In one year, Earth makes a slightly elliptical orbit around the sun. Perihelion is the point in Earth's orbit when it is closest to the sun. Aphelion is the point in Earth's orbit when it is farthest from the sun. Perihelion occurs around January 3 and aphelion occurs around July 4. **See Figure 2-18.**

Ecliptic Plane

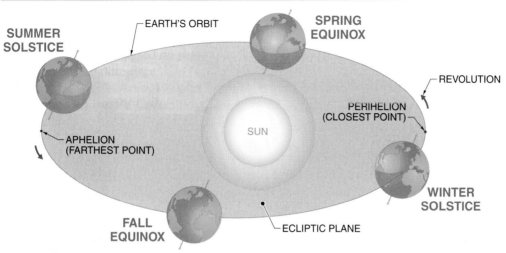

Figure 2-18. *The ecliptic plane is formed by Earth's elliptical orbit around the sun.*

The *ecliptic plane* is the plane of Earth's orbit around the sun. The *equatorial plane* is the plane containing Earth's equator and extending outward into space. Because of Earth's tilt, the angle between these planes is 23.5° and remains constant as Earth makes its annual orbit around the sun. This angle is what causes the seasonal variations in Earth's climate. As Earth revolves around the sun, the Northern Hemisphere tilts away from the sun during winter and toward the sun in the summer. The opposite effect occurs in the Southern Hemisphere, resulting in seasons opposite from those in the Northern Hemisphere.

Solar Declination. *Solar declination* is the angle between the equatorial plane and the rays of the sun. The angle of solar declination changes continuously as Earth orbits the sun, ranging from −23.5° to +23.5° (positive when the Northern Hemisphere is tilted toward the sun). The angle between the ecliptic and equatorial planes does not change, but as viewed from the sun at different times of the year, the equatorial plane appears to change in orientation. It appears to dip below the ecliptic plane (summer in the Northern Hemisphere), become edge-on (fall), tip above the ecliptic (winter), and return to edge-on (spring). **See Figure 2-19.**

Solar Declination

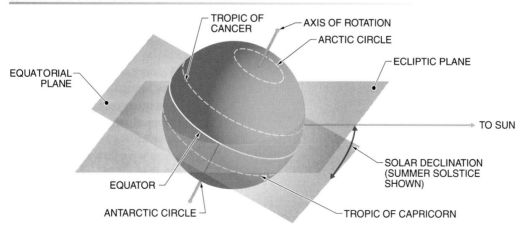

Figure 2-19. *The equatorial plane is tipped 23.5° from the ecliptic plane. As Earth revolves around the sun, this orientation produces a varying solar declination.*

The dates when maximum, minimum, and zero declination occur are used to mark the beginning of seasons. Also, several imaginary lines around the globe represent certain latitudes and the range of sun positions as it appears to an observer on Earth. These include the Equator, the Antarctic Circle, the Arctic Circle, the Tropic of Capricorn, and the Tropic of Cancer.

Solstices. A *solstice* is Earth's orbital position when solar declination is at its minimum or maximum. The summer solstice is at maximum solar declination (+23.5°) and occurs around June 21. The Northern Hemisphere is at its maximum tilt toward the sun. Days are longer than nights in the Northern Hemisphere, with all points south of the Antarctic Circle in total darkness. The sun is at zenith at solar noon at locations at 23.5°N latitude, known as the Tropic of Cancer. **See Figure 2-20.**

The winter solstice is at minimum solar declination (–23.5°) and occurs around December 21. The Northern Hemisphere is at its maximum tilt away from the sun. Days are shorter than nights in the Northern Hemisphere, with all points north of the Arctic Circle in total darkness. The sun is at zenith at solar noon at locations on the Tropic of Capricorn (23.5°S latitude).

At any location in the Northern Hemisphere, the sun is 47° lower in the sky at noon on the winter solstice than at noon on the summer solstice.

Equinoxes. An *equinox* is Earth's orbital position when solar declination is zero. There are two equinoxes in a year. The spring equinox occurs around March 21 and the fall equinox occurs around September 23. At these two points, every location on Earth has equal length days and nights. The sun is at zenith at noon on the equator and rises and sets due east and due west, respectively, everywhere on Earth. **See Figure 2-21.**

Around the equinoxes, the rate of change in declination is large, so the daily change in sun path is at maximum. Conversely, the rate of change in declination and sun path are both at minimum around the solstices. This can be easily seen on an analemma, a type of solar path diagram.

Solstices

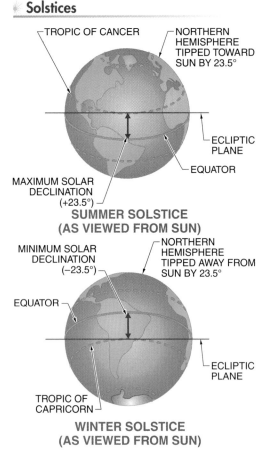

SUMMER SOLSTICE
(AS VIEWED FROM SUN)

WINTER SOLSTICE
(AS VIEWED FROM SUN)

Figure 2-20. *The summer solstice occurs when the Northern Hemisphere is tipped towards the sun. The winter solstice occurs when the Northern Hemisphere is tipped away from the sun.*

Equinoxes

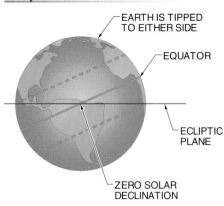

FALL OR SPRING EQUINOX
(AS VIEWED FROM SUN)

Figure 2-21. *The fall and spring equinoxes occur when the sun is directly in line with the equator.*

Earth's Rotation

The rotation of Earth about its axis defines each day. A day is divided into hours, which are standardized for use in the clock time used in everyday life. However, solar energy systems are referenced against solar time, which is based on the apparent motion of the sun across the sky. The two time systems do not always align.

Solar Time. *Solar time* is a timescale based on the apparent motion of the sun crossing a local meridian. A *meridian* is a plane formed by a due north-south longitude line through a location on Earth and projected out into space. The local meridian is the meridian at an observer's exact location. *Solar noon* is the moment when the sun crosses a local meridian and is at its highest position of the day. The sundial is based on the principle of solar time. A *solar day* is the interval of time between sun crossings of the local meridian, which is approximately 24 hr. Most tables or charts of daily sun position are based on solar time.

> Before standard time was introduced, every municipality set its clock by local solar time. This system was adequate until the train made long-distance travel fast enough to require almost constant re-setting of clocks.

Standard Time. *Standard time* is a timescale based on the apparent motion of the sun crossing standard meridians. A *standard meridian* is a meridian located at a multiple of 15° east or west of zero longitude. Zero longitude passes through Greenwich, England and is referred to as the Prime Meridian. Since Earth rotates 360° in approximately 24 hours, each 15° of longitude is equal to one hour of solar time. All standard time zones are at one hour multiples ahead of or behind the time at the Prime Meridian, also referred to as Universal Time Coordinated (abbreviated UTC). **See Figure 2-22.**

Standard time usually differs from solar time, though the difference varies with location and time of year. Thus, actual solar noon generally occurs at a standard time other than 12:00:00 PM. This difference is up to about ±45 min for most locations and is caused by two factors.

Refer to solar resources on CD-ROM

Standard Time

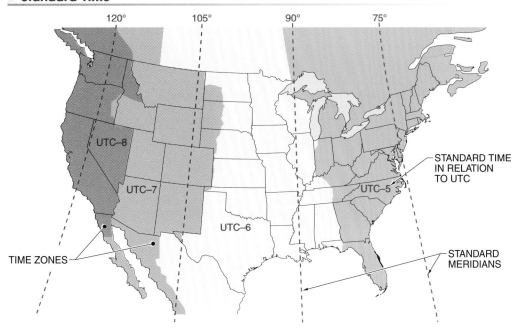

Figure 2-22. *Standard time organizes regions into time zones, where every location in a time zone shares the same clock time.*

Because Earth's orbit is elliptical, the distance between Earth and the sun varies. The closest distance is about 91 million mi and the greatest distance is about 95 million mi.

The primary factor is location. Standard time matches solar time (with regard to the location factor) at locations exactly on the standard meridians. However, locations not on a standard meridian, but within the same time zone, have the same standard time, but different solar time. Solar noon occurs late (after 12:00:00 PM) for locations west of the standard meridian and early (before 12:00:00 PM) for locations east of the standard meridian. This difference increases with distance from the meridian. Most locations are within about ±30 min of their standard meridian. Some locations can be far enough from their designated standard meridian that the difference between standard and solar time can be as much as ±90 min.

A longitude time correction is used to account for the different longitudes of the standard and local meridians. Since 15° of longitude represents 60 min of solar time, each degree of longitude represents 4 min. That is, for each longitudinal degree from the standard meridian, solar time is 4 min before or after standard time. The longitude time correction is calculated with the following equation:

$$t_\lambda = (\lambda_{local} - \lambda_S) \times 4$$

where

t_λ = longitude time correction (in min)

λ_{local} = local longitude (in deg)

λ_S = longitude of standard meridian (in deg)

Another difference between standard and solar time is caused by small eccentricities in Earth's rotation and orbit around the sun. The *Equation of Time* is the difference between solar time and standard time at a standard meridian. This difference varies over the course of a year and can be as much as +16 min or −14 min. The Equation of Time is computed from a formula or determined by looking up the date in a table or on a graph. **See Figure 2-23.**

Using both time corrections, local standard time can be converted to solar time, or vice versa, with the following equations:

$$t_0 = t_S + t_E - t_\lambda$$
$$t_S = t_0 - t_E + t_\lambda$$

where

t_0 = solar time (in hr and min)

t_S = local standard time (in hr and min)

t_E = Equation of Time (in min)

t_λ = longitude time correction (in min)

Equation of Time

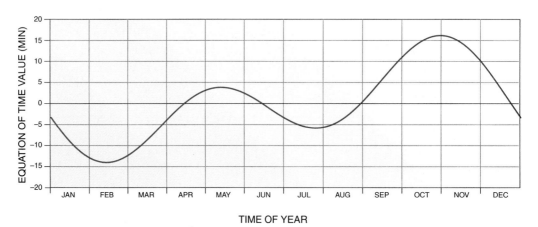

Figure 2-23. *The Equation of Time adjusts for variations in Earth's orbit and rotation that affect solar time.*

For example, at what local standard time does solar noon occur on March 21 in Knoxville, TN? The local longitude is 84°W and Knoxville is in the eastern time zone (standard meridian 75°W).

From the graph, the Equation of Time for March 21 is estimated to be –7.5 min (–00:07:30). The local standard time at which solar noon occurs is calculated by first determining the longitude time correction:

$$t_\lambda = (\lambda_{local} - \lambda_S) \times 4$$
$$t_\lambda = (84 - 75) \times 4$$
$$t_\lambda = \textbf{+36 min (00:36:00)}$$

Then, the local standard time of solar noon can be determined:

$$t_S = t_0 - t_E + t_\lambda$$
$$t_S = 12:00:00 - (-00:07:30) + 00:36:00$$
$$t_S = \textbf{12:43:30 PM}$$

Therefore, solar noon occurs at 12:43:30 PM local time in Knoxville on March 21. The sun will be due south in the sky and at its highest altitude of the day.

Daylight Saving Time. Near the equator, days and nights are nearly the same length throughout the year. However, at higher and lower latitudes than the tropics, there is significantly more daylight in summer than in winter. For this reason, most countries other than those near the equator use Daylight Saving Time (DST), adjusting their clocks during the summer months to make use of additional daylight hours in the evening. Most of the United States observes Daylight Saving Time from the second Sunday of March to the first Sunday of November. When Daylight Saving Time is in effect, one hour is added to local standard time. Arizona and Hawaii do not observe Daylight Saving Time.

Sun Path

At any given location on Earth, the sun's apparent position in the sky depends on the latitude, the time of day, and the time of year. The position of the sun, as observed from the same location and at the same time of day for a year forms a figure eight. An *analemma* is a

diagram of solar declination against the Equation of Time. **See Figure 2-24.**

It is important to understand sun position and sun path when orienting solar collectors and PV arrays, in order to optimize performance and minimize shading from obstructions.

Analemma

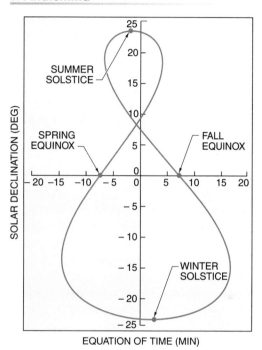

Figure 2-24. *An analemma shows how sun position, at the same time of day, changes throughout the year.*

Sun Position. Two angles are used to define the sun's position, relative to an observer on Earth. The *solar altitude angle* is the vertical angle between the sun and the horizon. During daytime, this angle varies between zero and 90° and complements the zenith angle (the two added together always equal 90°). The *solar azimuth angle* is the horizontal angle between a reference direction (typically due south in the Northern Hemisphere) and the sun. This angle varies between –180° and +180°. Sun position to the east of due south is generally represented as a positive azimuth angle, and to the west as a negative azimuth angle. **See Figure 2-25.**

☀ Sun Position

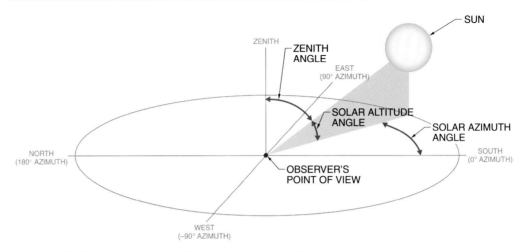

Figure 2-25. *Solar azimuth and altitude angles are used to describe the sun's location in the sky.*

Refer to solar resources on CD-ROM

Sun Path Charts. Tables and charts of solar azimuth angle versus solar altitude indicate the range of sun positions for a particular location. The information is computed from equations used to describe the sun's apparent motion through the sky. The sun position diagrams for the middle latitudes look similar, with the exception that the sun is lower in the sky for latitudes farther north. However, at latitudes below the Tropic of Cancer (23.5°N) and above the Arctic Circle (66.5°N), the diagrams change significantly. For example, the sun path for 15°N latitude (below the Tropic of Cancer) at the summer solstice appears different, since the sun is never in the southern half of the sky (azimuth angles from –90° to +90°). On the other hand, the sun path for 75°N latitude (above the Arctic Circle), the sun never appears above the horizon at the winter solstice and the sun never sets at the summer solstice. **See Figure 2-26. See Appendix.**

☀ Sun Paths

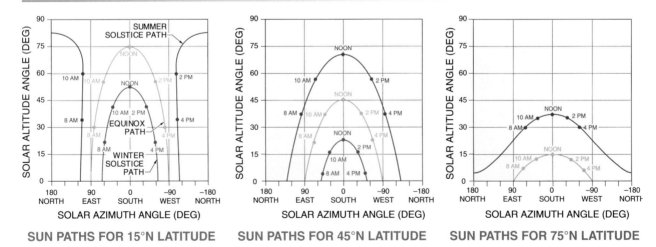

SUN PATHS FOR 15°N LATITUDE   SUN PATHS FOR 45°N LATITUDE   SUN PATHS FOR 75°N LATITUDE

Figure 2-26. *The sun's path across the sky at various times of the year can be illustrated on a diagram. The diagrams change for different latitudes.*

On the equinoxes at any location, the sun rises due east (azimuth +90°) and sets due west (azimuth −90°), which results in days and nights of equal length. In the Northern Hemisphere, daylight lasts longer from March 21 through September 23, with the largest difference at higher latitudes. The opposite effect occurs from September 23 through March 21.

The Solar Window. The *solar window* is the area of sky between sun paths at summer solstice and winter solstice for a particular location. Knowing the solar window at a given site is critical in properly locating and orienting PV arrays to achieve optimal energy performance and to prevent shading from trees and other obstructions. **See Figure 2-27.**

ARRAY ORIENTATION

Geography and the seasonal variation of the sun's path significantly affect the solar radiation received on Earth's surface, and consequently the overall electrical energy-producing capability of PV arrays. The orientation of the array surface also contributes to how much solar radiation can be utilized by a PV system.

Similar to sun position, array orientation is also defined by two angles. The *array tilt angle* is the vertical angle between horizontal and the array surface. The *array azimuth angle* is the horizontal angle between a reference direction and the direction an array surface faces. The reference direction is typically either due north or due south. **See Figure 2-28.**

Solar Window

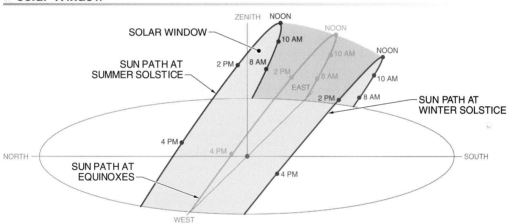

Figure 2-27. *The solar window is the area of sky containing all possible locations of the sun throughout the year for a particular location.*

Array Orientation

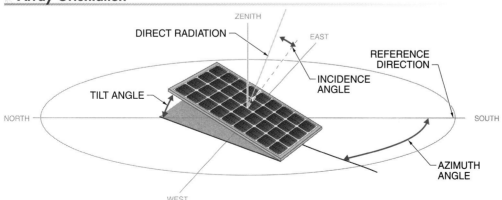

Figure 2-28. *Array orientation can be described using azimuth and tilt angles.*

The sun position and array orientation determine the incidence angle of direct solar radiation. The *incidence angle* is the angle between the sun's rays and a line perpendicular to the array surface. Maximum solar radiation is received when the incidence angle is 0°. For a fixed surface, the incidence angle changes as the sun appears to move through the solar window.

> A surface receives the maximum amount of energy when the direct component of solar radiation is exactly perpendicular to the surface.

Array Tilt Angle

Maximum annual solar energy gain (and therefore, energy performance from arrays) on a fixed surface is achieved by orienting the surface at a tilt angle close to the value of the local latitude. Ideally, arrays in high latitudes should be tilted at steep angles, and arrays in low latitudes should be tilted at very shallow angles.

In some cases, the optimal tilt angle for maximum annual energy gain is exactly equal to the local latitude. However, seasonal sun path changes and local climate variations may affect the optimal tilt angle for an array in a particular location. Longer days during the summer favor a smaller tilt angle for maximizing annual energy gain. This is also reinforced

by the generally overcast weather in the winter in northern climates. Conversely, many southern locations experience very cloudy summer weather, which favors a greater array tilt angle to take advantage of clearer winter weather.

Depending on the system's energy needs, the array can also be optimized for either summer or winter gain by changing the tilt angle. To maximize summer energy production, the array should be mounted at a smaller tilt angle to take advantage of the sun's high arc through the summer sky. This may be desirable for a system with high energy requirements in the summer, such as cooling loads. To maximize winter performance, the array should be mounted at a greater tilt angle to receive more energy when the sun is lower in the winter sky. This may be best for an application with higher energy loads in the winter, such as artificial lighting. **See Figure 2-29.**

For skewing energy performance toward a specific time of year, the tilt angle is typically adjusted from the latitude angle by 15°. During the summer half of the year in the Northern Hemisphere, the declination varies between 0° and +23.5°, for an average summer declination of +15°. **See Figure 2-30**. The declination varies between 0° and –23.5° during the winter half of the year, and the average winter declination is –15°. The averages are closer to the maximum declinations because of the sinusoidal shape of the changing declination curve.

Optimum Array Tilt Angles

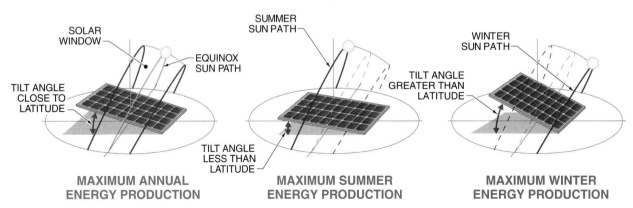

Figure 2-29. *Energy production at certain times of the year can be optimized by adjusting the array tilt angle.*

Seasonal Declination

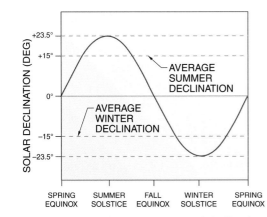

Figure 2-30. *The average seasonal declinations define the optimal tilt angles for those periods.*

The result is an optimal tilt angle for the summer of latitude–15°, so that the array is perpendicular to the average solar path in the sky during the summer. Similarly, an optimal tilt angle for the winter is latitude+15°, so that the array is perpendicular to the average solar path in the sky during the winter. This 15° adjustment neglects the effects of seasonal sun paths and climate, but it is still commonly used, at least in conjunction with other adjustments. Solar radiation resource datasheets for specific sites often includes data for surfaces at latitude, latitude –15°, and latitude +15° title angles.

Array Azimuth Angle

Optimal azimuth angle is due south (or due north in the Southern Hemisphere) for fixed mounts. However, the azimuth angle of array installations on existing structures may be constrained by the orientation or roof pitch of the building. Less solar energy is received when a fixed array is oriented to an azimuth angle other than due south. This reduction increases with larger tilt angles, so orientation is more critical at higher latitudes. At lower latitudes with arrays installed at lower tilt angles, the effect of off-azimuth orientation is much less. In most of the middle and southern U.S. latitudes, tilt surfaces with azimuth orientations of ±45° from due south will receive at least 90% of the annual solar energy received on true south-facing surfaces.

Sun Tracking

Sun tracking is continuously changing the array tilt angle, the array azimuth angle, or both, so the array follows the position of the sun. By following the sun, the incidence angle is always close to 0°, increasing annual solar gain by as much as 40% (depending on the location and type of tracking system).

Sun tracking is classified according to the number and orientation of the axes used to track the sun. *Single-axis tracking* is a sun-tracking system that rotates one axis to approximately follow the position of the sun. The mount may rotate the vertical axis, which changes the azimuth angle, or the north-south axis, which allows the array to follow the sun from east to west. Single-axis tracking brings significant performance gains over fixed surfaces. *Dual-axis tracking* is a sun-tracking system that rotates two axes independently to exactly follow the position of the sun. This further maximizes the amount of solar energy received, though the gain over single-axis tracking is smaller than the gain of single-axis tracking over fixed surfaces.

DOE/NREL, John Thornton
Modules at different orientations produce different amounts of power throughout the day. When used together in the same system, their output is combined on the AC side only.

Optimal Tilt Angle

Logically, the optimal tilt angle for maximum solar radiation received on a south-facing surface throughout a year would point the array toward the center of the solar window. This angle is exactly equal to the latitude of the location. However, the geometry of the solar window is such that the sun is in the sky for longer in the summer than in the winter, lowering the optimal tilt angle. This factor varies with latitude, with the difference being significant at high latitudes and very small at low latitudes.

Furthermore, in some regions, climate and atmospheric factors also result in a slightly lower tilt optimal angle for maximizing the annual energy production. For example, summer skies are usually clearer than winter skies, so that on the days when the sun is visible and solar radiation is receivable, the sun will likely be higher in the sky, promoting a lower tilt angle. There is also less air mass between the sun and a collector when the sun is high in the sky, allowing more solar radiation to reach the surface.

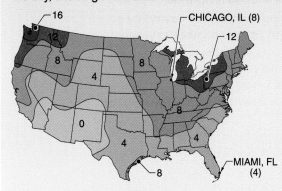

By using computer models and the average climate conditions at weather stations around the country, these factors are condensed down to an optimal tilt angle factor, which is the approximate difference between optimal tilt angle and latitude for a given location. For example, Chicago, Illinois, is located at latitude 42°N. The map indicates that the optimal tilt angle factor for Chicago is 8. Therefore, the optimal tilt angle for maximum annual receivable solar radiation is approximately 34° (42° − 8° = 34°). Using latitude for the tilt angle still provides a good approximation for maximum performance, but installers may wish to use this factor to further optimize tilt angle for critical applications.

Non-Optimal Orientation

The optimal tilt angle factor assumes an azimuth angle of zero (due south). Other azimuth angles produce different results and require complex calculations. Computer models can also be used to calculate the effect of non-optimal array orientations (for tilt and/or azimuth angles) on annual receivable solar radiation.

The annual receivable radiation for a certain orientation is represented as a percentage of the maximum annual receivable radiation possible at the optimal orientation. For example, if an array surface in Chicago is oriented at a 30° tilt and faces 45° west azimuth, it is able to receive only about 95% of the solar radiation that an optimally oriented surface could receive. When mounting surfaces, such as roofs, restrict the possible array orientations, these calculations can predict the effect of a non-optimal orientation on system performance. Fortunately, the reductions in performance are usually within acceptable ranges.

For a visual reference on the effects of non-optimal array orientations, the percentages can be plotted against the possible ranges of azimuth angles and tilt angles. Comparing plots for different locations shows the influence of latitude on the relative importance of tilt and azimuth orientations.

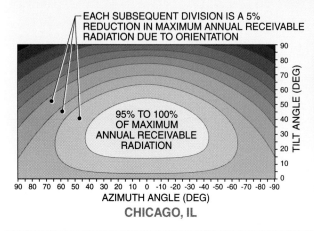

CHICAGO, IL

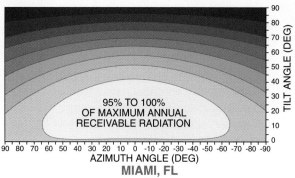

MIAMI, FL

SOLAR RADIATION DATA SETS

Solar radiation data indicates how much solar energy strikes a surface at a particular location on Earth during a particular period of time. PV systems produce electrical energy in direct proportion to the amount of solar energy received on the surface of the array.

Solar radiation is a highly variable resource and significant differences exist among regions in the U.S. For example, the southwestern U.S. receives a greater amount of solar radiation (particularly the direct component) than other regions. In addition, many sites at northern latitudes experience large differences in solar irradiation between winter and summer. These trends can be identified from solar radiation data.

The National Renewable Energy Laboratory (NREL), through its Renewable Resource Data Center (RReDC), provides national assessments of the solar, wind, and other renewable energy resources in the U.S., including maps and average solar irradiation availability for various locations and south-facing surface orientations. The RReDC's *Solar Radiation Data Manual for Flat-Plate and Concentrating Collectors* accounts for variations in site and weather conditions by including monthly average, minimum, and maximum values from data collected over the 30-year period of 1961 through 1990. The database includes data for 239 sites throughout the U.S. **See Figure 2-31. See Appendix.**

Refer to solar resources on CD-ROM

NREL Solar Radiation Data Set Excerpt

Location Knoxville, TN (35.82°N, 83.98°W)   Elevation 299 m   Pressure 983 mbar

SOLAR RADIATION FOR FLAT-PLATE COLLECTORS FACING SOUTH AT A FIXED TILT*

Tilt Angle	Jan	Feb	Mar	Apr	May	Jun	Jul	Aug	Sept	Oct	Nov	Dec	Annual
0	2.3	3.0	4.0	5.2	5.8	6.2	5.9	5.5	4.5	3.7	2.5	2.0	4.2
Latitude −15	3.0	3.7	4.6	5.5	5.8	6.1	5.8	5.7	5.0	4.6	3.3	2.8	4.7
Latitude	3.4	4.0	4.7	5.4	5.5	5.6	5.4	5.5	5.1	4.9	3.7	3.1	4.7
Latitude +15	3.6	4.1	4.6	5.0	4.9	4.9	4.8	5.0	4.8	4.9	3.8	3.3	4.5
90	3.2	3.3	3.2	2.9	2.4	2.2	2.3	2.7	3.2	3.8	3.3	3.0	3.0

SOLAR RADIATION FOR FLAT-PLATE COLLECTORS WITH 1–AXIS (NORTH–SOUTH) TRACKING*

Tilt Angle	Jan	Feb	Mar	Apr	May	Jun	Jul	Aug	Sept	Oct	Nov	Dec	Annual
0	3.1	4.0	5.2	6.7	7.3	7.8	7.3	6.9	5.8	5.0	3.4	2.7	5.5
Latitude −15	3.6	4.6	5.7	6.9	7.4	7.7	7.3	7.1	6.2	5.7	4.0	3.3	5.8
Latitude	3.9	4.8	5.8	6.9	7.1	7.4	7.1	7.0	6.3	5.9	4.3	3.6	5.8
Latitude +15	4.1	4.9	5.7	6.6	6.7	6.9	6.6	6.7	6.1	6.0	4.4	3.7	5.7

SOLAR RADIATION FOR FLAT-PLATE COLLECTORS WITH 2–AXIS TRACKING*

	Jan	Feb	Mar	Apr	May	Jun	Jul	Aug	Sept	Oct	Nov	Dec	Annual
2-Axis Tracking	4.1	4.9	5.8	7.0	7.4	7.9	7.4	7.2	6.3	6.0	4.4	3.8	6.0

DIRECT SOLAR RADIATION FOR CONCENTRATING COLLECTORS*

	Jan	Feb	Mar	Apr	May	Jun	Jul	Aug	Sept	Oct	Nov	Dec	Annual
1-Axis (E–W) Tracking	2.2	2.5	2.7	3.1	3.3	3.6	3.2	3.2	2.8	3.2	2.4	2.1	2.9
1-Axis (N–S)Tracking, Tilt=0	1.8	2.4	3.1	4.1	4.4	4.6	4.2	4.1	3.4	3.3	2.0	1.6	3.2
1-Axis (N–S) Tracking, Tilt=Latitude	2.5	3.0	3.6	4.2	4.2	4.3	4.0	4.1	3.8	4.0	2.8	2.3	3.6
2-Axis Tracking	2.6	3.1	3.6	4.3	4.5	4.7	4.3	4.2	3.8	4.3	2.9	2.5	3.7

CLIMATE CONDITIONS

	Jan	Feb	Mar	Apr	May	Jun	Jul	Aug	Sept	Oct	Nov	Dec	Annual
Temperature†	2.2	4.5	9.4	14.2	18.6	22.9	24.8	24.4	21.2	14.7	9.3	4.5	14.2
Heating Degree Days‡	499	387	276	130	52	0	0	0	10	132	272	429	2187
Cooling Degree Days‡	0	0	0	5	59	137	200	200	95	18	0	0	703

* averages in kWh/m²/day
† in °C
‡ based on 18.3°C

Figure 2-31. *The National Renewable Energy Laboratory (NREL) provides solar radiation data for various locations, times of the year, and south-facing array orientations.*

Each data sheet includes important site information, such as latitude and longitude, elevation, average barometric pressure, and average climate conditions. The site latitude is the basis for array orientation. The barometric pressure may be used in computing the corrected air mass value at the site. Other data can be used in various ways during site surveys and system design.

The solar resource information is presented for four types of systems: fixed flat-plate collectors, single-axis tracking collectors, dual-axis tracking collectors, and concentrating collectors. For each type, average daily receivable solar irradiation is represented in units of kWh/m²/day, which is equivalent to peak sun hours. This data is used to size solar energy utilization systems and estimate their average monthly and annual performance.

Solar Radiation for Flat-Plate Collectors Facing South at Fixed Tilt

The section on flat-plate collectors at fixed tilt is applicable to most installations. It lists data for surfaces facing due south at a variety of tilt angles between horizontal and vertical. Average, minimum, and maximum solar irradiation is given for tilt angles 0° (horizontal), latitude − 15°, latitude, latitude + 15°, and 90° (vertical) for each month and annually. Minimum and maximum values help estimate the potential variation in solar radiation, but the average data is used most often for system sizing and performance estimates.

DOE/NREL, National Center for Appropriate Technology
Sun-tracking arrays maximize solar energy gain by maintaining optimal array orientation at all times.

The data illustrates how array tilt affects the amount of solar radiation received. Typically, as the collector is tilted closer to horizontal, the amount of solar energy received in the winter months decreases, while the amount received in the summer months increases. Increasing the tilt angle toward vertical, the surface receives maximum solar radiation during the winter months and less during the summer.

For example, in the data set for Knoxville, TN, compare the solar radiation data for flat-plate collectors facing south at fixed tilts of 0° and latitude. The table lists the average annual solar radiation on these surfaces as 4.2 kWh/m²/day and 4.7 kWh/m²/day, respectively. A comparison of this data illustrates that tilting the collector to an angle equal to the latitude increases the average daily peak sun hours received by 12% ([4.7 − 4.2] ÷ 4.2 = 0.12). For specific months, the improvement may be even greater. For example, in January the average solar radiation is 2.3 kWh/m²/day and 3.4 kWh/m²/day, respectively. For this month and location, tilting the surface to the latitude angle improves the estimated performance by 48% ([3.4 − 2.3] ÷ 2.3 = 0.48).

Solar Radiation for Flat-Plate Collectors with Single-Axis (North-South) Tracking

The section on single-axis, north-south trackers (the surface moves from east to west during the day) is arranged similar to the data for fixed collectors, though the available solar radiation is generally higher in every category. This data also includes various tilt angles from the horizontal to vertical.

Comparing data between different sections demonstrates how different mounting configurations affect received insolation. For example, the annual average solar irradiation on a south-facing latitude tilt surface is 4.7 kWh/m²/day, while the annual average solar irradiation on a single-axis tracking array is 5.8 kWh/m²/day. Therefore, the performance improvement by mounting an array on a single-axis tracking surface is 23% ([5.8 − 4.7] ÷ 4.7 = 0.23).

Detroit Electrical JATC Training Center

Location: Warren, MI (42.5°N, 83.1°W)
Type of System: Utility-interactive
Peak Array Power: 18 kW DC
Date of Installation: Spring 2006
Installers: Apprentices and journeymen
Purposes: Training, community awareness, supplemental electrical power

In 2006, the Detroit Electrical Joint Apprentice Training Committee (JATC) installed a PV system on the roof of their training center. The system was partly subsidized by the State of Michigan Department of Energy.

The array consists of 90 high-performance modules arranged in tilted rows. The maximum potential power output from the system is 18 kW, but actual output is lower due to the low tilt angle of 5°. While the mounting system does not provide the optimal orientation for annual energy production, the design was chosen because it installs easily, does not penetrate the roof, allows access to every module, and has an aesthetically pleasing profile. In this situation, power output was not the only significant factor in the system design.

Also, the training center is planning to use the opportunity to experiment with system improvements. At each step, the apprentices and journeymen involved in the project will learn how the changes ultimately affect the system output.

The array mounting system currently provides a relatively low tilt angle, though future plans include increasing the tilt angle.

Detroit Electrical JATC
Apprentices and journeymen have been involved in every aspect of PV system installation.

Sensors will measure current, voltage, and other parameters, and this data will be collected and analyzed to determine actual system performance. With the relatively low tilt angle, peak performance is expected during the summer. System performance will be displayed on monitors in the training center lobby and on the training center web site.

After a period of continuous system data monitoring, custom brackets will be added to the mounting system to increase the tilt angle to about 15°. The tilt angle cannot be any greater than that without the rows shading each other, which would significantly reduce the array output. The greater tilt angle is expected to increase annual energy production, which will be quantified with the data monitoring system.

A wind turbine has also been installed in addition to the PV system. Wind turbines complement PV arrays because turbines typically produce power when the solar insolation is low.

In addition to supplementing the facility's electrical supply, the project is being used as a training opportunity for the students. By the time the project is completed, over 100 apprentices and journeymen will have been involved in designing and mounting the array, routing and bending conduit, pulling wires, installing the inverters and disconnects, and completing the final connections. Even more students will be involved in the ongoing data monitoring and system improvements.

The broadest vision is to educate the community about the advantages of solar energy technologies. The training center invites the general public to learn about their system by providing tours and publicly displaying the output data, and as the skilled installers of future PV systems, the electricians and contractors will make quality PV systems more accessible to consumers.

Solar Radiation for Flat-Plate Collectors with Two-Axis Tracking

The section on two-axis trackers represents solar irradiation on a surface that always faces the sun. There is only one category presented since fixed tilt angles are not applicable to two-axis sun tracking mounts.

Two-axis tracking enables greater receiving of solar irradiation than either fixed or single-axis tracking surfaces. The magnitude of improvement depends on the location, but two-axis tracking tends to be a smaller improvement over single-axis tracking than single-axis tracking is over fixed, south-facing surfaces. For Knoxville, the gains are 3% and 28%, respectively.

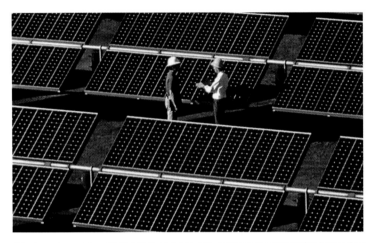

Single-axis, east-west tracking mounts are particularly suited for improving energy production thoughout the year, as the tracking changes the tilt angle.

Direct Solar Radiation for Concentrating Collectors

The section on concentrating collectors provides data for only the direct component of the total global solar radiation with respect to some type of tracking mount. This data is primarily used for concentrating-type collectors that can only utilize the direct radiation component and must be installed on sun-tracking mounts.

ESTIMATING ARRAY PERFORMANCE

Solar radiation data is used to size and estimate the performance for PV systems over time.

Refer to Quick Quiz® on CD-ROM

Simple calculations can be done by hand, or results that are more detailed can be achieved using computer models.

Data Set Calculations

Estimates of the performance of most arrays can be calculated easily from the data provided in solar radiation data sets. For example, assume a PV system (facing due south and at latitude tilt) in Knoxville nominally produces 2 kW AC output at peak sun conditions (1 kW/m²). The annual average irradiation on this surface is 4.7 kWh/m²/day, or 4.7 peak sun hours per day. The array is producing 2 kW per peak sun hour, and since there are an average of 4.7 peak sun hours per day over the entire year, the average daily energy production is 9.4 kWh (2 kW × 4.7 hr/day = 9.4 kWh). Therefore, the annual energy production for this array is estimated at 3431 kWh/yr (2 kW × 4.7 hr/day × 365 days/yr = 3431 kWh/yr).

PVWatts® Modeling

One of the limitations of a solar radiation data set is that it only includes data for south-facing surfaces. Determining the insolation available for off-azimuth orientations require complicated calculations, which are best done with computer models.

A popular software tool for estimating the performance of PV systems is PVWatts®. This on-line tool is provided by the National Renewable Energy Laboratory and is based on solar radiation data that is more detailed than that found on the data set sheets. Along with built-in calculation algorithms, the software can provide detailed performance expectations, even for array orientations other than those in the data sets.

The user selects a location from a map and specifies the PV system size, type of array mount and orientation, and a derating factor (accounting for power conversion efficiency and other system losses). PVWatts calculates the monthly and annual solar irradiation on the array surface, and the resulting electrical energy production. Output data can also be analyzed on an hourly basis.

HOMER Optimization Modeling

The National Renewable Energy Laboratory (NREL) offers free modeling software, known as HOMER, for optimizing the size of distributed-power systems. HOMER is particularly useful for analyzing the design and cost effectiveness of systems with multiple power sources. The software can analyze various combinations of distributed-generation systems, including PV systems, wind turbines, micro-hydroelectric turbines, engine generators, and fuel cells, to determine the best mix for a particular location and application.

The model is based on resource data, system costs, and detailed load analyses for each hour of the year. The software can handle a large amount of input data, which improves the precision of the model. The resulting system optimization data can be displayed in graphical or tabular form. **See Figure 2-32.**

HOMER Optimization Modeling Software

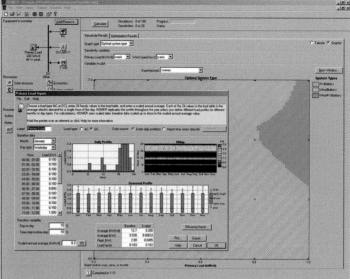

Figure 2-32. *NREL's HOMER software is used to optimize the cost effectiveness of systems with multiple distributed-generation sources.*

Summary

- The sun's radiant energy comes from nuclear reactions at its core.

- Solar irradiance is a measure of solar power per unit area. Solar irradiation is a measure of solar energy received on a surface over time.

- The solar constant is the irradiance reaching Earth's atmosphere.

- Solar radiation is composed of electromagnetic radiation in a range of wavelengths and in various amounts.

- The atmosphere reduces the amount of radiation that reaches Earth's surface and changes the composition of the radiation.

- Peak sun is the estimated maximum solar irradiance at Earth's surface.

- Pyranometers, pyrheliometers, and reference cells can be used to measure solar irradiance.

- Earth's tilt results in changing solar position in the sky throughout the year.

- Solar time is based on the sun's apparent motion (due to Earth's rotation) and can differ significantly from standard time.

- The solar window is the area of sky encompassing the range of possible sun positions throughout the year for a particular location.

- The type, orientation, and mounting configuration of an array affect how much available solar radiation it can utilize.

- Optimal orientation for maximum annual energy production of a flat-plate collector is an azimuth angle of due south and a tilt angle close to the latitude.

- Solar radiation data can be used to optimize the performance of a PV system for a particular location, season, month, array type, array orientation, and mounting configuration.

- *Radiation* is energy that eminates from a source in the form of waves or particles.

- *Solar irradiance* is the power of solar radiation per unit area.

- *Solar irradiation* is the total amount of solar energy accumulated on an area over time.

- *Extraterrestrial solar radiation* is solar radiation just outside Earth's atmosphere.

- The *solar constant* is the average extraterrestrial solar power (irradiance) at a distance of 1 AU from the sun, which has a value of approximately 1366 W/m^2.

- *Electromagnetic radiation* is radiation in the form of waves with electric and magnetic properties.

- The *electromagnetic spectrum* is the range of all types of electromagnetic radiation, based on wavelength.

- *Total global radiation* is all of the solar radiation reaching Earth's surface and is the sum of direct and diffuse radiation.

- *Direct radiation* is solar radiation directly from the sun that reaches Earth's surface without scattering.

- *Diffuse radiation* is solar radiation that is scattered by the atmosphere and clouds.

- *Zenith* is the point in the sky directly overhead a particular location.

- The *zenith angle* is the angle between the sun and the zenith.

- *Air mass (AM)* is a representation of the relative thickness of atmosphere that solar radiation must pass through to reach a point on Earth's surface.

- *Terrestrial solar radiation* is solar radiation reaching the surface of Earth.

- *Peak sun* is an estimate of maximum terrestrial solar irradiance around solar noon at sea level and has a generally accepted value of 1000 W/m^2.

- *Peak sun hours* is the number of hours required for a day's total solar irradiation to accumulate at peak sun condition.

- *Insolation* is the solar energy that reaches Earth's surface over the course of a day.

- A *pyranometer* is a sensor that measures the total global solar irradiance in a hemispherical field of view.

- A *pyrheliometer* is a sensor that measures only direct solar radiation in the field of view of the solar disk (5.7°).

- A *reference cell* is an encapsulated PV cell that outputs a known amount of electrical current per unit of solar irradiance.

- The *ecliptic plane* is the plane of Earth's orbit around the sun.

- The *equatorial plane* is the plane containing Earth's equator and extending outward into space.

- *Solar declination* is the angle between the equatorial plane and rays of the sun.

- A *solstice* is Earth's orbital position when solar declination is at its minimum or maximum.

- An *equinox* is Earth's orbital position when solar declination is zero.

- *Solar time* is a timescale based on the apparent motion of the sun crossing a local meridian.

- A *meridian* is a plane formed by a due north-south longitude line through a location on Earth and projected out into space.

- *Solar noon* is the moment when the sun crosses a local meridian and is at its highest position of the day.

- A *solar day* is the interval of time between sun crossings of the local meridian, which is approximately 24 hr.

- *Standard time* is a timescale based on the apparent motion of the sun crossing standard meridians.

- A *standard meridian* is a meridian located at a multiple of 15° east or west of zero longitude.

- The *Equation of Time* is the difference between solar time and standard time at a standard meridian.

- An *analemma* is a diagram of solar declination against the Equation of Time.

- The *solar altitude angle* is the vertical angle between the sun and the horizon.

- The *solar azimuth angle* is the horizontal angle between a reference direction (typically due south in the Northern Hemisphere) and the sun.

- The *solar window* is the area of sky between sun paths for summer solstice and winter solstice.

- The *array tilt angle* is the vertical angle between horizontal and the array surface.

- The *array azimuth angle* is the horizontal angle between a reference direction and the direction an array surface faces.

- The *incidence angle* is the angle between the sun's rays and a line perpendicular to the array surface.

- *Single-axis tracking* is a sun-tracking system that rotates one axis to approximately follow the position of the sun.

- *Dual-axis tracking* is a sun-tracking system that rotates two axes independently to exactly follow the position of the sun.

Review Questions

1. Explain the difference between irradiance and irradiation.

2. Why is the solar constant specified for Earth's distance from the sun?

3. What is the difference between direct radiation and diffuse radiation?

4. How are the solar constant and peak sun values similar and how do they differ?

5. What is the role of air mass in terrestrial solar radiaton?

6. What is insolation?

7. How can pyranometers be used to measure only direct solar radiation?

8. Explain how the solstices and equinoxes relate to the sun's apparent motion across the sky.

9. Which two factors affect the difference between solar time and standard time?

10. Which factors define the solar window?

11. Why does the optimal array tilt angle change throughout the year?

12. How does the type of collector and mount affect the potential energy gain for an array?

3

Site Surveys and Preplanning

A preliminary assessment establishes the objectives, resources, and requirements of a prospective PV system. A detailed survey of the site conditions evaluates all the site-specific issues related to a potential PV installation. During a site survey, an installer identifies potential array locations, measures distances and angles, evaluates existing structural and electrical infrastructure, documents relevant site information, and may conduct an energy audit. This information is then used to size systems and components, estimate performance, prepare proposals, and plan installations.

Chapter Objectives

- *Identify issues to be discussed to determine customer needs, concerns, and expectations.*
- *Identify factors to consider in a preliminary assessment, including the local solar resource, environmental conditions, and building code and utility interconnection requirements.*
- *Explain the process of determining potential array locations.*
- *Describe methods for determining and diagramming shading patterns.*
- *Discuss considerations in determining the suitability and condition of existing roofing, structural systems, and electrical systems and equipment.*
- *Explain the function of an energy audit and identify opportunities for conservation and energy efficiency.*
- *Identify factors to be considered when preparing a proposal, including estimates for cost, size, performance, and value of a PV system.*

PRELIMINARY ASSESSMENT

A preliminary assessment addresses several fundamental questions needed to decide the the feasibility of a PV system. These include the customer's needs and concerns, the available solar resource, the environmental conditions, and the structural and electrical requirements. Identification of these factors begins the site surveying process, which influences all other subsequent system design and installation tasks, including selecting equipment, preparing bids, conducting value assessments, and planning and completing installations.

Customer Development

The customer may be a home or business owner, or an agent acting on the customer's behalf. If the installation is part of new construction or a large job, then architects, engineers, and builders may be involved in making certain decisions about design or installation requirements. For most smaller commercial and residential jobs, the customer may be directly involved at some or all levels. It is important to establish roles and responsibilities for any individuals and organizations involved in the project.

Salespersons, designers, and installers of PV systems must identify customer needs, concerns, and expectations. Some customers may be well informed and know exactly what they need or want, while others may need explanations of the types of systems and their costs, functions, and performance. Many suppliers and contractors provide literature to help identify and explain options. **See Figure 3-1.**

Costs and funding are usually the customer's primary concern. Fortunately, there are many resources for funding significant portions of a PV system, including financing, state and federal government incentives, and the exchange of renewable energy credits. When discussing various sizes and types of systems with customers, the installer can also present information on applicable funding programs. Later, when the system is sized and configured, the installer may use this information to help the customer apply for these programs.

Customer Consultation

SolarWorld Industries America

Figure 3-1. *The installer should meet with each customer to discuss available PV system options.*

The installer should discuss the installation schedule with the customer. Schedules should have a degree of flexibility to accommodate unavoidable delays, such as weather or shipping problems. However, some installations will have strict completion dates dictated by financing deadlines or other construction. Access to a site may also be restricted to certain times. The installer must coordinate with the overall schedule and minimize potential delays to ensure the work is completed in a timely manner.

Customers should be advised of other potential impacts of the installation, such as restricted access to certain areas, or storage of materials or equipment left on the site. Installers should also be prepared to address maintenance requirements, product and installation warranties, and follow-up service agreements.

Solar Resource

The solar radiation resource should be researched before visiting the site. This information is needed not only for the site survey, but also for designing the system and establishing its performance expectations.

Solar radiation resource information is available as data sets for specific locations, and provides average daily solar radiation for each month on various surfaces. Where no data exists for a certain area, the local resource can be estimated by comparing the data for the closest sites. Solar radiation measurement equipment can be used at the site, though meaningful resource data requires more than a year of collection.

When reviewing the solar resource data, the installer should note the differences for the various array tilt angles, and the seasonal variations and monthly minimum and maximum values for receivable solar energy. This information is important when evaluating potential mounting systems.

Environmental Conditions

Weather and other environmental conditions for the site should also be researched. Temperature, wind speed, and humidity data are included with solar radiation data sets. Data on other factors, such as precipitation, are freely distributed by government or research agencies. The possibility for earthquakes, hurricanes, flooding, or any other natural events at the site must also be considered.

Accurate temperature ranges for a site are particularly important, because this significantly affects the operation and performance of modules, batteries, and inverters. Temperature influences the array's output voltage, so the site temperature extremes are considered when configuring arrays for the appropriate voltage.

Special environmental concerns may dictate the use of different materials, equipment, designs, or installation practices. For example, humid and marine environments require special materials to resist corrosion. Where flooding or storm surge is a possibility, equipment must be located high off the ground or adequately protected from water. Where windstorms or seismic events are possible, special structural support or flexibility may be required. Under any of these circumstances, special building codes and standards may apply.

Since solar radiation data sets generally provide information only on surfaces facing due south and at certain tilt angles, computer models can be used to predict the solar radiation received, and therefore PV system energy production, for other orientations.

The installer may also identify other existing or potential renewable energy supply options, such as wind or solar thermal energy. As with utility or generator power, other renewable energy systems can operate in conjunction with a PV system, but the combination must be carefully designed. Integration of multiple energy sources can affect key decisions early in the process, including the size, cost, design, and integration of the system.

Code Compliance

Installers must understand the building code requirements in their state or region. Building codes determine some of the information that must be gathered during a site survey, and affect subsequent planning, system design, and installation.

Schott Solar

Customers may require that a PV system have another function in addition to producing electricity, such as shading a parking lot.

All system components, interconnections, and attachments must meet applicable local building and electrical codes. Most U.S. states and jurisdictions have adopted the International Building Code® and the International Residential Code® as part of their local building codes. For electrical safety and code compliance, most jurisdictions enforce a recent edition of the National Electrical Code®. Contact the local authority having jurisdiction (AHJ) to determine which specific building and electrical codes apply.

Requirements for connecting small renewable energy systems to a utility's electricity grid may vary. State policies regarding utility interconnection differ, and utilities within the same state may also impose their own connection requirements. Most requirements relate to safety, equipment certification, power quality, contracts, liability insurance, metering, and rates.

Utility Interconnection

For interactive systems, interconnection agreements are typically required by the local electric utility company. Interconnection policies and requirements may vary among states and utilities, and many have special provisions for insurance, disconnects, labeling, metering, and system size.

Understanding the terms and conditions for interconnection is an important first step that dictates many options and decisions concerning system design and installation. In addition, it is highly advisable to establish contacts with the AHJ and utility early in the installation process.

SITE SURVEYS

A *site survey* is a visit to the installation site to assess the site conditions and establish the needs and requirements for a potential PV system. Site surveys identify suitable locations for the array and other equipment. The most appropriate array locations have enough surface area for the size of array needed, permit the best possible orientation, and are not excessively shaded. Other factors include accessibility and proximity to other equipment, structural support, existing electrical infrastructure, architectural appearance, and

Refer to form on CD-ROM

protecting both personnel and equipment from hazards. An audit of the customer's energy use determines how a PV system will contribute to the electrical supply and may reveal opportunities for energy conservation that will improve the value of the system. All this information should be carefully documented for later use in system sizing, preparing estimates, and planning the installation. **See Figure 3-2.**

Survey Safety

All individuals working on PV systems should be familiar with applicable OSHA safety standards contained in Volume 29 of the U.S. Code of Federal Regulations (29 CFR), specifically Part 1926, *Safety and Health Regulations for Construction.* These regulations include several subparts that are applicable to the installation of PV systems. Depending on the site conditions, these regulations may also require certain safety equipment or practices during site surveys.

Personal Protective Equipment (PPE). Personal protective equipment (PPE) for site surveys include appropriate clothing and footwear for working outdoors, such as hats, sunglasses, and sunscreen to protect against sun exposure. Additional equipment, such as safety glasses, hard hats, safety shoes, and fall protection gear, are common types of PPE required for PV system work. **See Figure 3-3.**

Safety glasses, goggles, and face shields provide eye and face protection against hazards such as dust and other flying particles, corrosive gases, vapors, and liquids. The selection of eye protection is based on protection from a specific hazard and its comfort. It must not restrict vision or movement or interfere with other PPE.

A hard hat protects a worker's head against impact from low-hanging structures or falling objects. Hard hats are classified by their level of protection. Class A general service hard hats are suitable for construction, having good impact protection but limited voltage protection. Class B hard hats are intended for electrical and utility work, as they also protect against high-voltage shock and burns. Class C offer limited protection against bumps from fixed objects, but do not protect against falling objects or electrical shock.

Site Survey

PV SYSTEM SITE SURVEY INFORMATION

GENERAL INFORMATION

Customer __Smith Residence__

Site Address __123 Main St., Anytown, IL 60123__

Contact Name __John Smith, homeowner__

Phone __555-0123__   Fax __555-0124__   Email _____

Utility __Regional Power, Inc.__

Contact Name __Beverly Jones__

Phone __555-6789__   Fax __555-6790__   Email __jones@regionalpower.com__

Permitting Authority __Village of Anytown__

Contact Name __Gary Roberts, building inspector__

Phone __555-5215__   Fax __555-5235__   Email __roberts@anytown.vil.us__

Type of System Desired __Utility-interactive__   Output (kW) __5.0__

Critical Loads (W) __N/A__   Autonomy (days) __N/A__

SOLAR RESOURCE & WEATHER

Latitude __41°N__   Longitude __87°W__   Basic Wind Speed (mph) __9.0__

Orientation __Lat__   Insolation (kW/m²/day) AVG __3.9__ MAX __6.3__ MIN __1.5__

Temperatures (°F) AVG __50__   MAX __105__   MIN __-20__

Potential for Extreme Weather __Heavy thunderstorms, heavy snowfall__

PROPOSED ARRAY LOCATION

Area (ft²) __1000__   Slope (°) __33°__   Azimuth Orientation (°) __210° (SSW)__   Height (ft) __10-25__

Accessibility __requires extension ladder or aerial lift, fall protection__

Shading Analysis Results __tree will shade location from 9 AM to 11 AM November through January__

ROOFING

Type of Roofing __Asphalt shingle__   Age (yrs) __10__   Thickness (″) __1.25__

Surface Condition __good, no repairs needed__

Type of Supporting Structure __2 × 6 rafters 12″ O.C.__

Structural Concerns __none__

ELECTRICAL SYSTEM

Primary Service __120/240 V, 1φ, 200 A__   Source __Overhead from utility pole__

Other Sources __none__

Ratings __200 A main breaker__

Annual Electricity Use (kWh) __10,900__   Rate (¢/kWh) __8.6__

Point of Connection __Main service panel__

Proposed Inverter Location __Basement__   Accessibility __good__

Proposed Battery Location __N/A__   Accessibility __N/A__

Proposed Disconnect Locations __Basement, exterior wall__   Accessibility __both good__

Electrical Concerns __none__

Other Comments or Issues __Customer wants to be able to easily upgrade and/or expand system in future.__
__Possibilities include larger array and battery backup system.__

Surveyor __Jane Miller__   Date __7/12__

Signature _____

Figure 3-2. *Information gathered during a site survey should be carefully documented.*

Personal Protective Equipment (PPE)

DOE/NREL, Craig Miller Productions

Figure 3-3. *Common personal protective equipment (PPE) for preparing and installing PV systems includes head, eye, foot, and fall protection.*

Sturdy safety shoes with impact-resistant toes and heat-resistant soles protect the foot from heavy objects that might roll onto or fall on employees' feet; sharp objects such as nails or spikes; hot, wet or slippery surfaces; and other hazards. Nonconductive safety shoes help protect from electrical hazards.

Other PPE may be required for certain tasks or with the use of certain equipment. Each task should be evaluated for potential safety hazards and the required PPE and safety equipment. The proper use of PPE requires the involvement of both the employer and employee.

Electrical Safety. Electrical hazards pose significant potential for personal injury and fire damage. Direct injuries include electrocution (death due to electrical shock), electrical shock, and burns. Indirect injuries include falls due to electrical shock. Other common electrical injuries include concussions resulting from arc explosions as well as eye damage due to arc flash. Working on or near exposed, energized conductors or electrical equipment requires special personal protective

equipment (PPE). The means to assess electrical hazards and the required PPE and other precautions are addressed in NFPA 70E, *Electrical Safety in the Workplace.*

A number of procedures are also used to protect workers from electrical shock. Barriers and guards are used to keep workers away from exposed energized equipment. Workers should pre-plan work, post hazard warnings, use protective measures, and keep working spaces and walkways clear of cords. Ground-fault circuit interrupters (GFCI), switches, and conductor insulation should be checked regularly. Flexible extension cords must be of the 3-wire type (with grounding conductor) and designed for hard use.

Lockout and tagout procedures are used to prevent individuals from unintentionally energizing electrical circuits while they are being serviced or maintained. **See Figure 3-4.** Lockout is the physical locking of a power source disconnect in the OFF position with a padlock. Tagout is the labeling of de-energized controls, equipment, or circuits at all points where they could be energized.

Lockout and Tagout

Panduit Corp.

Figure 3-4. *Lockout and tagout procedures are important parts of an electrical safety program.*

Fall Protection. Arrays installed on buildings or other structures above the ground often involve working at heights where fall protection is required. Many roofs have steeply inclined surfaces, which are particularly hazardous when wet from dew or precipitation. Lifting equipment or special training may be required in some situations. For larger commercial installations, permanent ladders or stairways, access covers, tie-off points, and other features can be installed to improve safety and accessibility.

OSHA requires fall protection for work taking place more than 6′ above the ground, which includes the vast majority of areas where PV modules are installed on buildings. Fall protection can be in the form of equipment worn by a worker to reduce the potential for injury from a fall, or precautions taken in the work area to prevent a fall. Regulation CFR 1926.500 to 1926.503 covers requirements for fall protection methods and equipment.

A full-body harness is a fall protection device that evenly distributes fall-arresting forces throughout the body to prevent injury. **See Figure 3-5.** The harness secures the technician around the legs, torso, and shoulders, and includes a metal D-ring at the back to attach a lanyard. The lanyard secures the harness to a beam, structural member, secured ladder, or lifeline. A shock absorber in the lanyard absorbs most of the force of stopping the fall.

Full-Body Harness

DOE/NREL, Robb Williamson

Figure 3-5. *Full-body harnesses connected to a secure safety line system protect a worker from injuries from falling from sloped roofs or roofs with unprotected edges.*

Guardrail systems can also provide protection against falling. Guardrails can be installed temporarily at a work site, such as at the unprotected edges of rooftops, or may be permanently affixed to elevated platforms such as scaffolds and aerial lifts. This type of fall protection consists of three railings that must be secured to the personnel side of upright structural members.

On large, flat roofs where work will not be required within 6′ of the edge, a warning line system may be used. A warning line is a high-visibility rope marking a controlled-access zone near the edge of the roof. Workers inside the warning line do not need additional fall protection, but workers between the warning line and the edge are required to have some other method of fall protection. A safety monitor watches for fall hazards, allows no workers without additional fall protection within the controlled access zone, and ensures that workers are aware of the warning line if they are working near it.

Ladder Safety. OSHA requires that a stairway or ladder be used at points of access with an elevation change of 19″ or more. Ladders are commonly used in many phases of a PV system installation and must be used in an appropriate and safe manner.

Ladders must be kept in a safe working condition. A competent person must inspect ladders for visible defects, like broken or missing rungs. If a defective ladder is found, it should be immediately marked as defective and withdrawn from service until repaired.

Use ladders only for their designed purpose and load rating. Ladders should be set on stable and level surfaces, and secured to prevent accidental movement. The areas around the top and bottom of a ladder should be clear. Non-self-supporting ladders (those that lean against a wall or other support) must be positioned at an angle where the horizontal distance from the top support to the foot of the ladder is one-quarter the working length of the ladder. The side rails must extend at least 3′ above the upper landing surface.

Double-cleated ladders (with center rail) or multiple ladders, are required for frequent or two-way traffic. If using ladders near exposed and energized electrical equipment, such as transformers or overhead services, the side rails must be made from a nonconductive material such as fiberglass. Some ladders are permanently installed. These fixed ladders may require ladder safety devices, self-retracting lifelines with rest platforms, cages or wells, or multiple ladder sections. **See Figure 3-6.**

Fixed Ladders with Cages

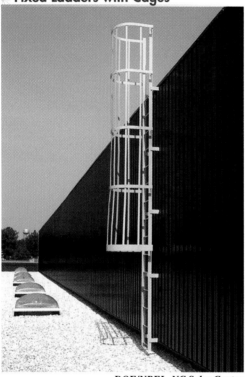

DOE/NREL, NC Solar Center

Figure 3-6. *Fixed ladders often require cages to prevent workers from falling because the vertical orientation can be difficult to climb.*

Employers are responsible for assessing workplace hazards, and providing the appropriate personal protective equipment (PPE) and training in its proper use. Employees are responsible for properly using, inspecting, and maintaining PPE in a clean and reliable condition.

Hand and Power Tool Safety. Like any construction project, the preparation and installation of a PV system involves the use of a number of different hand and power tools. Workers using hand and power tools may be exposed to a number of hazards, including objects or substances that may fall, fly, or splash; harmful dusts, fumes, mists, vapors, and gases; and electrical hazards due to damaged electrical cords, poor connections, or improper grounding.

Basic tool safety includes regular maintenance, appropriate tool choice, inspections for damage and safety devices prior to use, following manufacturers' instructions, and wearing proper PPE. Other specific requirements and hazards depend on the tool type and power source, which includes electric, pneumatic, liquid fuel, hydraulic, and powder-actuated equipment.

Survey Equipment

Site surveys are conducted by observing, photographing, measuring, documenting, and diagramming all applicable site conditions related to a PV installation. Consequently, certain equipment is required to collect this information, including conventional tools, meters, and measuring devices, as well as some special equipment.

Hand Tools. Basic screwdrivers, socket wrenches, pliers, and cutters may be needed to access some areas for inspection, such as electrical panels. Portable flashlights or headlamps are useful to illuminate dark areas, such as attics.

Ladders. Ladders are typically required to access rooftops in order to inspect surfaces, take measurements, and determine suitable areas for array installation. Collapsible combination step/extension ladders can serve a variety of functions, although they may not be tall enough to reach some rooftops. Some sites may require aerial lifts or pole-climbing equipment. Special fall protection equipment or precautions are required at heights over 6′.

Documentation Materials. Notebooks and graph paper are used to record observations, measurements, sketches, and diagrams of the site and layout. Basic calculators with trigonometric functions are useful for computing areas, perimeters, diameters, and angles. A wristwatch with a stopwatch function can be used to mark the time of specific events or measure intervals of time. For example, the power demand of a building can be determined by measuring the revolution rate of a watt-hour meter disk. Compact tape recorders, digital cameras, binoculars, and video equipment can also be used to document site conditions.

Sun Path Calculator. A sun path calculator is used to view the solar window for a particular location, which is needed for assessing shading. **See Figure 3-7.** Other means can be used to evaluate shading, but sun path calculators are usually the quickest and easiest to use.

Sun Path Calculator

Solar Pathfinder

Figure 3-7. *A sun path calculator is used to evaluate shading at potential array locations.*

Electrical Test Equipment. Multimeters are used to measure service voltages, load currents, power, and to check for continuity or energized circuits. Detailed surveys may include load measurements, using watt-hour meters or analyzers to evaluate power quality as necessary. In some cases, temporary load monitoring equipment can be installed at the site for a period of time to document changing load profiles. **See Figure 3-8.**

Measuring Devices. Tape measures, laser distance meters, transits, compasses, levels, protractors, and other measuring devices are often used during detailed site surveys. Required measurements typically include array areas, distances between equipment or components, roof slope, building orientation, and shading obstruction heights. Electronic stud finders identify structural attachment points while conduit or conductor locating devices help avoid hidden electrical conduit or plumbing pipes during installation. An infrared thermometer can measure the temperature of electrical equipment or spaces designated for temperature-sensitive components. A GPS receiver identifies latitude and longitude coordinates, pinpointing the location of site features.

Testing and Measuring Devices

Figure 3-8. *A variety of testing and measuring devices and marking equipment is used during site surveys.*

Marking Equipment. Various means and devices to identify, locate, and label areas related to a PV installation include chalk lines, string, tapes, paint, stakes, and flags. For example, chalk lines can temporarily mark module locations and attachment points on a roof. Temporary marking is also useful to identify proposed equipment locations for customer approval. Construction areas can be cordoned off by barricades or stakes and cautionary tape.

Miscellaneous Hardware. Assortments of fasteners, electrical tape, tie wraps, wire nuts, and other miscellaneous hardware are useful for tasks such as installing meters and monitoring equipment, reconnecting circuits, making small repairs, or neatly arranging wiring.

Consider the safety hazards of working outside during site surveys, such as exposure to direct sunlight and heat, and dehydration. Dangerous insects, snakes, and other animals can hide under eaves or inside electrical boxes and may bite or sting when disturbed. Always have a first aid kit available for minor injuries and work with a partner whenever possible.

Be aware of overhead and underground utilities and document energized power lines during the site survey. Contact the local utility or marking service to have underground conduits, cables, and water lines identified and marked before any construction activity.

Warn bystanders or people walking beneath a work area of hazards from falling tools or materials. If necessary, secure the work area with caution tape, flags, or signage.

Array Location

The primary task of a site survey is determining whether there is a suitable location for a PV array. For many sites, there is only one practical location, typically the most southerly facing portion of a sloped roof. However, some sites may offer multiple array location options, such as a roof with more than one portion facing south, a large flat roof, or a large open field. Each option should be evaluated for suitable area, orientation, shading, accessibility, structural support, and proximity to the electrical infrastructure. The most appropriate location is chosen based on these factors.

Array Area. PV systems are low-density power generators, so large surface areas are required to produce appreciable amounts of power. The required overall area for any given array is based on the desired output, the efficiency of the modules, and how densely the modules are installed in the array. **See Figure 3-9.** For the purposes of the initial site survey, the required array area can be estimated with the following formula:

$$A = \frac{P_{mp}}{\eta_m}$$

where

A = estimated required array area (in m²)

P_{mp} = desired peak array power (in kW DC)

η_m = module efficiency

Allowing for space between the modules for access and maintenance, sloped-roof PV systems with a typical 12% efficiency will require about 9 m² (100 ft²) per kilowatt of peak power.

For example, what is the estimated size of an array using 12% efficient modules that should produce approximately 5 kW at peak sun?

$$A = \frac{P_{mp}}{\eta_m}$$

$$A = \frac{5}{0.12}$$

$$A = \mathbf{42 \ m^2}$$

The required area for mounting an array is typically increased to account for spacing between modules, panels, or rows, which improves access to all parts of the array and prevents modules from shading each other. Protective fencing or unusable space within the potential array location will also increase the total area. The amount of extra space will vary between sites and types of mounting systems. An increase of 10% is common for sloped-roof-mounted arrays, but others may require up to 50%.

Array Orientation. Sun-tracking arrays continuously change orientation to follow the sun. However, fixed arrays depend on one orientation to receive as much solar energy as possible. Optimal fixed array orientations are not always possible, particularly for sloped roofs. The orientation of a roof often limits array orientation options.

The roof slope and orientation must be measured during a site survey for a rooftop installation. The easiest way to measure the slope is with an angle finder, or inclinometer. **See Figure 3-10.** Measuring the rise and run of the roof with a steel square and a level is another method. The slope is then calculated from the following formula:

$$\theta = \arctan\left(\frac{rise}{run}\right)$$

where

θ = roof slope angle (in deg)

rise = rise (vertical) of roof slope (in in.)

run = run (horizontal) of roof slope (in in.)

For example, what is the roof slope angle of a roof surface with a rise of 5″ and a run of 12″?

$$\theta = \arctan\left(\frac{rise}{run}\right)$$

$$\theta = \arctan\left(\frac{5}{12}\right)$$

$$\theta = \arctan(0.42)$$

$$\theta = \mathbf{23°}$$

Array Area Requirements

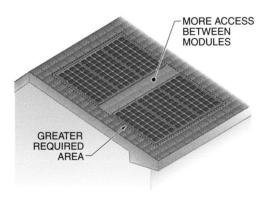

DENSE ARRAY ARRANGEMENT

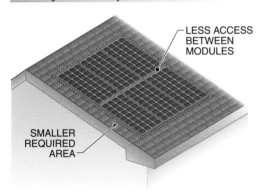

OPEN ARRAY ARRANGEMENT

Figure 3-9. *The density of the module arrangement in an array affects the accessibility and the area required to produce a certain amount of power.*

Measuring Roof Slope

ANGLE FINDER

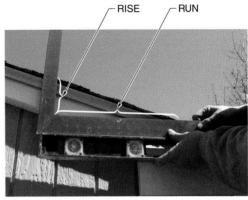

RISE AND RUN

Figure 3-10. *Roof slope is measured with an angle finder or calculated from the rise and run.*

The azimuth orientation of the roof is the direction, such as southwest or due south, that the sloped surface faces, and is determined with a magnetic or electronic compass. The compass is held horizontally and the angle is noted between true south (corrected for magnetic declination if necessary) and the direction the roof is facing. **See Figure 3-11.**

The tools and equipment required to conduct a site survey will vary depending on the type of PV system, location, and existing site conditions. Certain equipment is needed for any site survey. Additional equipment varies depending on the building type or site conditions. Installers learn from previous experience which tools and equipment are required for various situations.

Measuring Roof Orientation

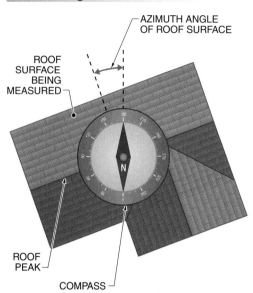

Figure 3-11. *A compass is used to determine the orientation of a sloped roof surface.*

> Based on sun paths alone, the optimal tilt angle for maximum annual solar radiation gain is nearly equal to the latitude. However, since winter skies are often cloudier than summer skies, a slightly smaller tilt angle, which will collect more energy in the summer, may result in greater annual energy production.

When the measured orientation is not optimal, installers and consumers must consider the consequences on receivable solar radiation. Fortunately, some relatively wide ranges of orientations have relatively minor effects on receivable radiation. **See Figure 3-12.**

Array Orientation Effects*

LOCATION	AZIMUTH ANGLE†	TILT ANGLE†				
		Latitude − 20	Latitude − 10	Latitude	Latitude + 10	Latitude + 20
Chicago, IL	0	98%	100%	99%	96%	91%
	±45	94%	95%	93%	90%	85%
	±90	84%	81%	77%	73%	67%
Phoenix, AZ	0	95%	99%	100%	99%	95%
	±45	93%	95%	95%	93%	89%
	±90	86%	84%	81%	77%	72%
Tampa, FL	0	97%	100%	100%	98%	93%
	±45	96%	97%	96%	93%	89%
	±90	92%	90%	86%	82%	77%

* in percentage of solar radiation receivable compared to optimal orientation
† in deg

Figure 3-12. *The potential loss in receivable solar radiation from non-optimal orientations may not be significant.*

In some cases, it may even be desirable to face an array other than due south, such as when more solar energy is available in the morning or afternoon due to the local climate, or to match demand at certain times of the day. Utility-owned PV systems are sometimes installed facing southwest, shifting peak power output to later in the afternoon to coincide with peak demand.

Magnetic Declination. Depending on location, magnetic north may not be the same as true north. Ferromagnetic elements in Earth's core cause variations in the magnetic field that affect compass readings. Electronic navigation equipment and GPS receivers automatically correct for magnetic declination. However, when using a magnetic compass to orient the azimuth angle of solar energy systems, this difference must be applied to compass readings to determine true south.

Magnetic declination is the angle between the direction a compass needle points (toward magnetic north) and true geographic north. If the compass needle points west of true north, this offset is designated as west declination, and the value is negative. If the compass needle points east of true north, this offset is east declination, and the value is positive. **See Figure 3-13.** Maps show values for magnetic declination and rate of change per year for different regions. Magnetic declination values change slowly, but enough that new maps should be consulted every several years. Magnetic declination in 2009 varies in the continental United States by as much as −19° in Maine to +18° in Washington state.

Magnetic Declination Map

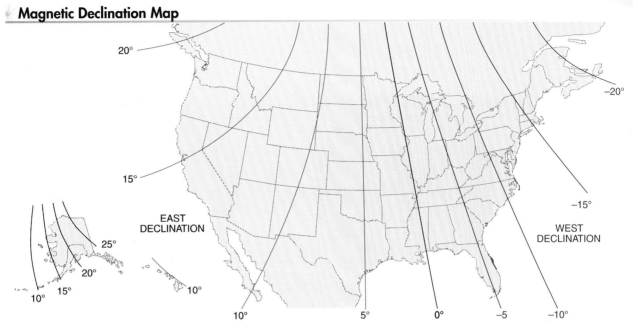

Figure 3-13. *Magnetic declination varies by location and changes slightly over time. Up-to-date maps are used to determine the necessary adjustment to magnetic compass readings.*

Compass Bearings vs. Azimuth Angles

When using a compass, angle designations are called bearings and range from 0° to 360° clockwise around the compass. North is a 0° bearing, east is a 90° bearing, south is a 180° bearing, west is a 270° bearing, and north is a 360° bearing (in addition to 0°).

However, with regard to azimuth angles used with solar energy systems, south may be designated as 0°. East is +90° and west is –90°. The absolute value of the azimuth angle is always the number of degrees from due south. Installers must be aware of the difference between compass bearings and azimuth angles and use each correctly.

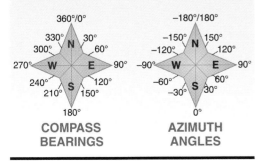

COMPASS BEARINGS

AZIMUTH ANGLES

On a compass, true south is found by subtracting the declination from the direction of magnetic south (usually designated as 180°). For example, near San Francisco, California, the magnetic declination is +15° (or 15° east). If the south end of a compass needle points to 180°, then true south is at 165° (180° – 15° = 165°). Near Providence, Rhode Island, the magnetic declination is –15° (or 15° west). If the south end of a compass needle points to 180°, then true south is at 195° (180° – [–15°] = 195°). **See Figure 3-14.**

Shading Analysis

Shading on solar thermal collectors reduces performance by an amount proportional to the level of shading. However, PV arrays are much more sensitive to shading. Depending on the magnitude and location of the shading, the reduction in output can be disproportionately higher than the percentage of array area shaded. In the worst cases, even a 10% shading can cause the loss of most of the output. **See Figure 3-15.** For this reason, installers must carefully assess the shading potential at an installation site, and be prepared to adjust array position and orientation to minimize shading.

Magnetic Declination

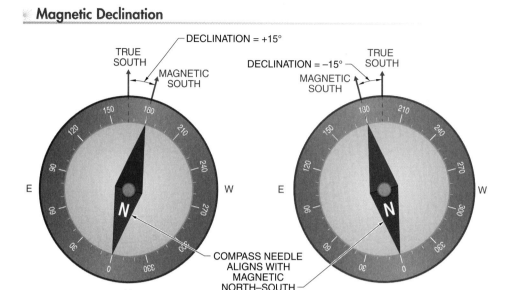

Figure 3-14. *Directional bearings from magnetic compasses must be adjusted for magnetic declination.*

Shading Effects

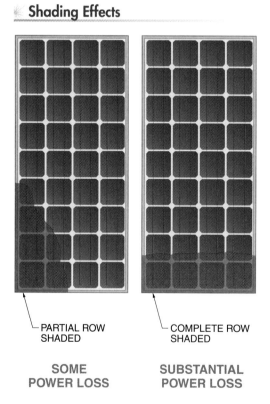

Figure 3-15. *Shading of PV modules and arrays can cause disproportional reductions in power output.*

Preferably, arrays should be installed in a location with no shading at any time. However, site constraints may make this difficult to achieve, especially during winter, early morning, and late afternoon, when the low sun casts long shadows from far away objects. At a minimum, arrays should have an unobstructed solar window from 9 AM to 3 PM (solar time) throughout the summer, when the sun is highest in the sky and the majority of solar radiation is available. **See Figure 3-16.** Also the array circuit design and module choice (for cell circuit configuration) can minimize the effects of unavoidable shading.

Shading can be caused by either natural or man-made obstructions at any time of the day or year. Natural obstructions include trees and other vegetation, as well as geographic features such as hills and mountains. Trees and vegetation are an ongoing shading concern, since they grow larger, causing increasingly more shading. Man-made obstructions include buildings, towers, and poles. Shading can also be caused by parts of the building on which an array is mounted, including plumbing vents, hip roofs, chimneys, cables, or service conductors.

Shading Priority

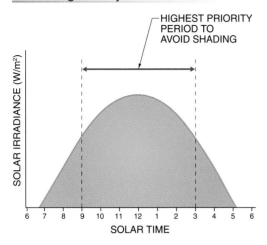

Figure 3-16. *Most of the daily solar energy is received between 9 AM and 3 PM, so avoiding shading during this period is high priority, especially during the summer.*

In the Northern Hemisphere, shading is generally caused by obstructions to the east, south, or west. However, during summer months at low latitudes, the sun can sometimes be in the northern part of the sky, which can cause shading from obstructions immediately north of an array. **See Figure 3-17.**

Options for dealing with shading include trimming or transplanting trees or moving the array. If shading occurs on the lower or southern parts of a proposed array location, moving the array to the north or a few feet higher may reduce or eliminate the shading problem. If shading occurs from the east or west, shading may be reduced by moving the array in the opposite direction.

Sharp Electronics Corp.
Even shading from small obstructions, such as a rooftop vent, may be a significant issue.

Shading from the North

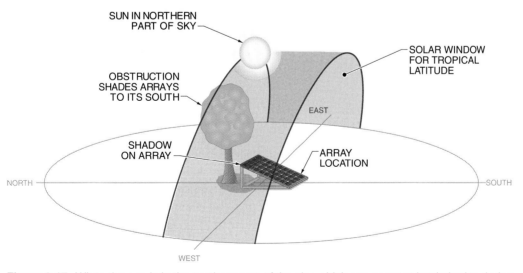

Figure 3-17. *When the sun is in the northern part of the sky, which can occur at low latitudes during the summer, shading can be caused by obstructions immediately north of an array.*

Shading analyses should be conducted at the exact location of the proposed array, including the elevation, because shading can vary greatly within only a few feet. For large arrays, an analysis should be conducted for the entire array area.

The extent of shading at any given location can be determined by comparing the height and location of shading obstructions to the solar window. Any overlap between the obstruction areas and the solar window results in shading. This analysis can be accomplished in various ways. The simplest method is to use a sun path calculator, though other analytical and graphical methods may also be employed.

Sun Path Calculators. A *sun path calculator* is a device that superimposes an image of obstructions on a solar window diagram for a given location. These devices are extremely easy to use and provide a graphical shading analysis for the entire year, which can also be converted into a numerical analysis. Two popular types of sun path calculators are the Solar Pathfinder™ and the Solmetric SunEye™. Both devices work by comparing the local solar window to an image of the surrounding obstructions, which determines the dates and times when shading will occur at that location. When a sun position is overlapped by an obstruction, from that location the sun would appear behind the obstruction, so the location would be shaded.

When using the Solar Pathfinder for shading analysis, the actual position of the sun is irrelevant. However, the device also works like a sundial and can be used to determine solar time based on sun position.

Refer to media clip on CD-ROM

The Solar Pathfinder consists of a latitude-specific sun path diagram covered by a transparent dome. The dome reflects the entire sky and horizon on its surface, indicating the position and extent of shading obstructions. The sun path diagram can be seen through the dome, illustrating the solar window. **See Figure 3-18.** The Solar Pathfinder is placed at a proposed array location and elevation. It is leveled and oriented to true south

with the built-in compass and bubble level. (The compass reading may require adjustment for magnetic declination.) Looking straight down from above, the user observes reflections from the sky and surroundings superimposed on the sun path diagram, which indicates shading obstructions.

Solar Pathfinder

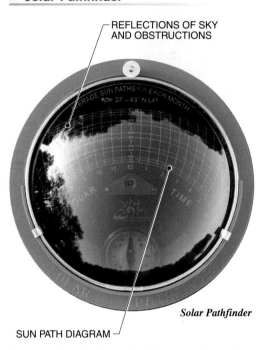
REFLECTIONS OF SKY AND OBSTRUCTIONS

Solar Pathfinder

SUN PATH DIAGRAM

Figure 3-18. *The Solar Pathfinder analyzes shading for potential array locations by comparing the reflections of potential obstructions on the horizon to a sun path diagram of the solar window.*

The Solar Pathfinder allows for two methods of recording the shading obstructions. **See Figure 3-19.** The outline of the obstructions can be drawn onto the sun path diagram, producing a hard copy record of the shading conditions. Alternatively, a digital photo of the device can be loaded into a separate software product where a digital trace of the obstructions is analyzed automatically.

With either method, the resulting trace indicates all the dates and times when a specific location will be shaded. Dates and times indicated outside the trace will be shaded, while those dates and times indicated inside the trace will not be shaded. Numbers within the shaded area can be totaled (manually, or by the software) to quantify the percentage of shading each month. If using the analysis software, users can input additional system, component, and location information to calculate performance expectations and automatically generate custom reports.

Shading Analysis Records

AREAS OF SOLAR
WINDOW WITH
SHADING

Solar Pathfinder

HARD COPY

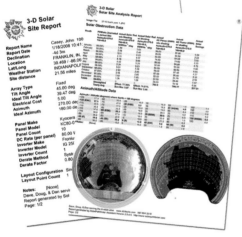

Solar Pathfinder

DIGITAL PHOTO ANALYSIS

Figure 3-19. *A permanent record of the shading for a particular location can be traced on the hard copy diagram, or by photographing the reflections, which can then be analyzed by the Solar Pathfinder Assistant software to generate reports.*

While it uses a similar analysis technique as the Solar Pathfinder, the SunEye directly acquires a digital photo of the sky and includes many analysis features within the handheld hardware device. **See Figure 3-20.** The user first sets the location with the on-board software and levels and orients the device for magnetic south. (Magnetic declination is automatically accounted for by the on-board software, based on the location.) The SunEye includes a digital camera with a special fish-eye lens that photographs the entire hemisphere of the sky and surroundings in one image.

Solmetric SunEye

Solmetric Corporation

Figure 3-20. *The Solmetric SunEye is an electronic device that includes an on-board camera with a fish-eye lens for photographing the entire sky and horizon at once.*

Using the location information, the SunEye generates and displays a solar window diagram specific to the local latitude. The photo is then analyzed by the on-board software to automatically generate a trace of the obstructions. By superimposing the solar window diagram, sky photo, and obstruction trace, the device calculates the monthly solar access (percentage of total insolation that is unshaded). **See Figure 3-21.** The photo and shading data can also be uploaded to a computer for archiving or exporting the data, or generating detailed reports.

Both devices, when using the software features, can make sophisticated calculations and comparisons for different array orientations and some other site options. Also, both the Solar Pathfinder and the SunEye can be used any time of the day or day of the year, in either cloudy or clear weather. In fact, they are often easier to use in the absence of direct sunlight.

SunEye Shading Analysis

Sky02 -- 08/05/06 02:26 -- Southeast corner of roof

Solar Access		Panel
Annual:	63%	Tilt: 38°
May-Oct:	80%	Azim: 180°
Nov-Apr:	41%	

Solmetric Corporation

Figure 3-21. *The SunEye's on-board software can automatically analyze an all-sky photograph for the obstruction that will shade a location at certain times of the day and year.*

Altitude Angle

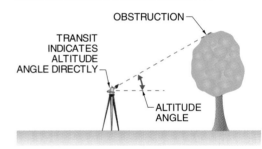

Figure 3-22. *The altitude angle is the vertical angle to the top of an obstruction.*

Altitude Angle Method. A simple method to evaluate shading at a particular location is by measuring or calculating the altitude angle of each obstruction. If the altitude angle of the sun is less than the altitude angle of the obstruction when the sun is at the same azimuth angle as the obstruction, then the location will be shaded. **See Figure 3-22.**

The ranges of azimuth and altitude angles of the sun can be determined from a sun path chart for the local latitude. Obstruction altitude angles can be measured using a transit or a protractor. A properly leveled transit will indicate the profile angle directly on its dials. When using a protractor, the top of the obstruction is sighted along the flat edge of the protractor. The obstruction's altitude angle is equal to the distance between a plumb line and the 90° mark. **See Figure 3-23.** Altitude angles can also be calculated with the following formula:

$$\alpha_{obs} = \arctan\left(\frac{h}{d}\right)$$

where

α_{obs} = altitude angle of obstruction (in deg)

h = height of obstruction (in ft)

d = distance to obstruction (in ft)

Determining Altitude Angle

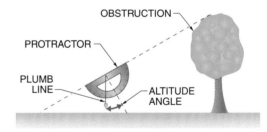

TRANSIT METHOD

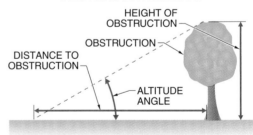

PROTRACTOR METHOD

CALCULATION METHOD

Figure 3-23. *Altitude angles can be determined using a transit, a protractor, or by calculations from measurements.*

For example, what is the altitude angle of a 30′ tree that is 43′ away from a proposed array location at 41°N latitude?

$$\alpha_{obs} = \arctan\left(\frac{h}{d}\right)$$

$$\alpha_{obs} = \arctan\left(\frac{30}{43}\right)$$

$$\alpha_{obs} = \arctan(0.70)$$

$$\alpha_{obs} = \mathbf{35°}$$

Using a compass, it is determined that the tree is 35° east of due south. Comparing these angles to a sun path chart for 41°N latitude shows that for certain times in the morning, during the winter months, the sun will be behind the tree, shading the proposed array location.

This method has limitations. Surroundings with many obstructions of complex shapes are usually analyzed more easily with a sun path calculator. However, this method can be quick and simple for locations with a small number of slender obstructions in the area, such as utility poles or some trees. Spreadsheet programs can also be set up in advance to perform most or all of the calculations and graphically compare the solar window to the locations of obstructions. **See Figure 3-24.**

Magnetic compasses can be used to determine the azimuth angles of obstructions for shading analysis.

Profile Angle Method. Sun path calculators and the altitude angle method are types of shading analyses that determine only if and when a certain location will be shaded by nearby obstructions. Comparisons of the results can be made between potential locations, but these methods cannot directly determine where an array should be located to avoid shading. This type of shading analysis involves calculating the profile angle of the obstruction. The *profile angle* is the projection of the solar altitude angle onto an imaginary plane perpendicular to the surface of an obstruction. This calculation is used to determine the length of a shadow at a certain time, which equals the minimum distance from an obstruction to avoid shading. **See Figure 3-25.** The profile angle is calculated with the following formula:

$$\beta = \arctan\left(\frac{\tan\alpha}{\cos\psi}\right)$$

where

β = profile angle (in deg)

α = solar altitude angle (in deg)

ψ = azimuth angle between solar azimuth and plane of profile angle (in deg)

As the sun moves across the sky, the shadow cast by the obstruction shortens and then lengthens again, changing the profile angle. The profile angle is part of the right triangle formed by the height of the obstruction

Altitude Angle Diagram

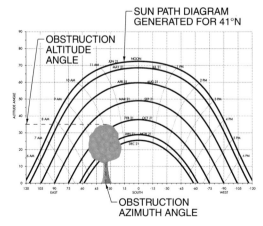

Figure 3-24. *The altitude angle method of shading analysis compares the altitude angle and azimuth angle of potential obstructions to a sun path diagram.*

and the length of the shadow. Therefore, the profile angle is also calculated by the following formula:

$$\beta = \arctan\left(\frac{h}{d}\right)$$

where

β = profile angle (in deg)

h = height of obstruction (in ft)

d = length of shadow (in ft)

The difference between the use of this formula in the altitude angle method and profile angle methods is that the profile angle method accounts for larger obstructions where a range of solar azimuths can shade a particular location. The altitude angle method is actually just a special case of the profile angle method where the plane of the profile angle is the same as the plane (azimuth) of the solar altitude angle, which is true for slender obstructions.

Many resources are available for further study of profile angle geometries and calculations. Profile angles are relevant to several fields other than PV systems, so these resources are often associated with architectural, indoor lighting, and passive solar heating and cooling topics.

The profile angle method is particularly useful when designing arrays consisting of multiple rows of modules mounted at a tilt. If the modules are installed at a tilt and the rows are spaced too close together, the modules in front can shade those behind them. Higher tilt angles increase the potential for this problem, requiring more space between rows.

Treating a row as a shading obstruction to the row behind it, a variation of the profile angle analysis can be used to evaluate the minimum separation distance required between the rows to avoid shading. **See Figure 3-26.** In this case, the minimum distance is the length of the shadow at the time when the shadow just touches the bottom edge of the rear row. The minimum separation distance is calculated by combining the profile angle formulas into the following formula:

$$d = h \times \frac{\cos\psi}{\tan\alpha}$$

where

d = minimum row separation distance (in ft)

h = row height (in ft)

ψ = azimuth angle between solar azimuth and array azimuth (in deg)

α = solar altitude angle (in deg)

Profile Angle

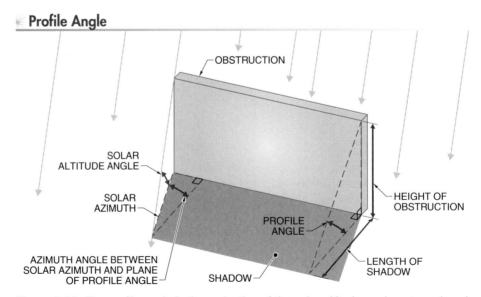

Figure 3-25. *The profile angle is the projection of the solar altitude angle onto an imaginary plane perpendicular to a shading surface. This angle is used to calculate the length of shadows.*

For example, an array consisting of multiple rows of tilted modules is to be installed on a flat rooftop at 40°N latitude. The rows must be spaced to avoid shading each other between 9 AM and 3 PM during the winter solstice. At these times, the solar altitude angle is 14° and the solar azimuth angle is 42°. Also, due to the orientation of the building, the rows are to be installed facing 10° west of due south. Therefore, the azimuth angle between the solar azimuth and array azimuth is 52° at 9 AM (42° + 10°) and 32° at 3 PM (42° − 10°). Both azimuth angle circumstances may need to be calculated to determine which has the greater effect. If the module tilt results in a row height of 1.5′, then what is the minimum row separation distance for these circumstances?

$$d = h \times \frac{\cos \psi}{\tan \alpha}$$

$$d = 1.5 \times \frac{\cos 52°}{\tan 14°}$$

$$d = 1.5 \times \frac{0.62}{0.25}$$

$$d = \mathbf{3.7'}$$

$$d = 1.5 \times \frac{\cos 32°}{\tan 14°}$$

$$d = 1.5 \times \frac{0.85}{0.25}$$

$$d = \mathbf{5.1'}$$

At 9 AM, the minimum distance between rows is 3.7′, but at 3 PM, the minimum distance is 5.1′. The difference is due to the array's azimuth angle. If the array was oriented due south, the two distances, both at 3 hours from solar noon, would be the same. Also, a change in the tilt angle, which would change the row height, would proportionally change the minimum row distance.

Accessibility

A site survey should evaluate the accessibility of each potential array location and all other parts of a PV system for safe and convenient installation, inspection, and maintenance. Convenient access to equipment also helps ensure that it will be operated and maintained properly.

Inter-Row Shading

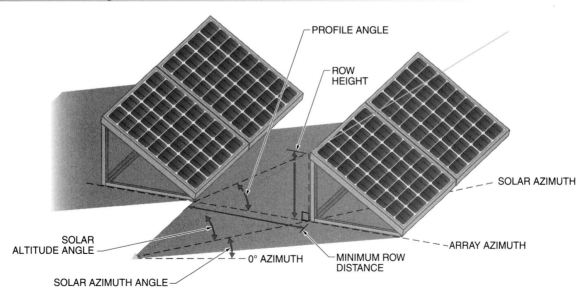

Figure 3-26. *Profile angle calculations are particularly useful for arranging arrays consisting of multiple rows of modules installed at a tilt. The calculation determines the minimum row spacing to avoid modules shading each other at certain times of the day.*

Accessibility also facilitates inspections by building officials and utility personnel. For example, electrical inspectors may require access to junction boxes and other equipment to evaluate its suitability for the application and proper installation, and other building officials may need to look at structural attachments. Utilities may need to evaluate the means and point of interconnection for the system, as well as access to disconnect devices external to a building.

The locations of system components should be arranged conveniently and the connections between them should be neat and professional. Avoid locating components in difficult-to-access areas such as attic spaces or beneath arrays.

Roofing Evaluation

For PV systems to be installed on rooftops, the condition of the roof covering and the underlying structural support must be carefully evaluated during the site survey. For other mounting configurations, any other existing or potential structural supports must be evaluated.

Roof Surfaces. Arrays are installed on many types of roof surfaces, although the type of roof surface and its slope will often dictate the types of mounting system and attachment methods that are feasible. The expected life of the roof covering is also very important. Different types of roof coverings have widely varying life expectancies, depending on the type of material and application environment. Commercial and residential roofing also differs in construction and materials, and some roofing practices vary regionally.

> Regardless of the type of roof, if it is leaking, it probably requires extensive repair or replacement. Signs of leaking roofs include outside light seen through the decking from below, dark spots on rafters or joists, stained or sagging ceilings, buckling or sagging roof surface, and rusty or corroded metal flashings. The installer should make the building owner aware of any potential or existing leaks so that they can be inspected further by a roofing professional.

A primary concern is the condition of the roof covering, particularly its weather-sealing ability. Identifying potential roof degradation early reduces the chances of leak repairs and roof replacement later. When it is determined that a roof is nearing the end of its useful life, or is already deteriorated, the installer should document and make the building owner aware of these findings. It is usually far easier and less expensive in the long term to replace an old or damaged roof before installing an array than to remove and reinstall the array later during re-roofing. Also, documenting any pre-existing leaks ensures that they are not attributed to the array installation.

Signs of deterioration differ for various types of roofing. **See Figure 3-27.** For conventional asphalt shingles, deterioration includes brittleness, cracking, loss of granular coating, warping, or curling up from the shingle edges. Asphalt shingles generally are the least expensive, but have the shortest life of all roof coverings, particularly in hot climates.

For slate, clay, or concrete roof tiles, problems include cracks, misalignment, or flaking material. When only a few tiles are damaged, those can usually be replaced individually. Slate and tile roofs have long life expectancy, but are moderately expensive.

Metal roofing also has a long life, but is expensive. Signs of aging include loss of galvanized coating and rusting on steel roofs, and corrosion and pitting on aluminum roofs. The installer should also note whether metal roofing is architectural (installed over continous decking) or structural (installed across an open rafter system) in nature. This determines how the roof can support workers and affects the design of the array attachment method.

Built-up, membrane, and gravel roofs on flat or low-slope surfaces are common on commercial buildings. Problems can be difficult to identify for these roofs, though membrane brittleness and cracking are obvious signs of deterioration. When it is difficult to determine the age or condition of a roof, a professional roofer should be consulted.

The thickness of the roof decking and covering dictates the appropriate length of fasteners needed to install the array. The thickness can usually be determined by looking under the eave drip edge or flashing along an edge of the roof. Inspecting existing roof penetrations or roof transitions may also help to determine roof thickness. **See Figure 3-28.**

Roof Deterioration

SHINGLE ROOFS

TILE ROOFS

METAL ROOFS

Figure 3-27. *Roofs should be inspected for signs of deterioration during a site survey.*

Measuring Roof Thickness

Figure 3-28. *The thickness of roof decking and covering can be determined by inspecting the edge of the roof under the eaves.*

Structural Support. While detailed structural assessments are conducted by professional engineers, observations can be made during site surveys to identify concerns that may warrant further investigation.

Roofs are prone to loss of structural strength or weather tightness over many years and must be carefully inspected. First, a visual inspection determines the flatness of the roof surface. A string line stretched across the roof in various directions reveals dips wherever there is a gap between the string and the roof surface. **See Figure 3-29.** Sagging roof surfaces generally indicate some underlying problem, such as decking failures, misalignment with structural support, or failing structural members.

Next, the installer should walk carefully across the roof and check for movement of the surface. Any "spongy" areas indicate weakened roof support that must be further inspected. Dry rot, leaks, or insect damage should be noted. The installer should also be alert for any signs of structural modifications and deficiencies due to previous repairs or building additions. In many cases, structural defects can be corrected or additional support added to secure arrays.

Checking Roof Flatness

Figure 3-29. *Noticeable dips on roof surfaces may be a sign of underlying structural defects.*

Arrays must be securely attached to the roof's underlying structural members in order to resist wind and other loads. For this reason, the condition of the structural members, including trusses, rafters, and beams, must be carefully evaluated. Also, the size, layout, and distance between these structural members are required to plan the position and attachment method for the array. Sometimes, this requires access to areas immediately beneath the roof, such as attics.

For ground-mounted arrays, evaluate the terrain, elevations, and soil conditions. Potential problems include loose or lightly compacted soil, standing water, and signs of erosion or past flooding.

Electrical Assessment

The installer must assess the existing electrical systems and equipment, and potential impact of interfacing with a PV system. The condition, size, and ratings of existing electrical services, conductors, overcurrent protection devices, disconnects, and transformers should be examined and documented. Measurements should be taken for service voltages and any special power quality concerns should be investigated. This information determines the limits on system size, appropriate output ratings for inverters, and in some cases, upgrades to existing electrical services may be required to interface with PV systems.

For interactive systems with battery backup, a dedicated subpanel is usually connected to the inverter output. Existing branch circuits with critical loads are relocated from an existing main panelboard to the new subpanel, requiring potentially significant changes to the existing electrical system.

In addition to the array, the installer should identify appropriate locations for other major system equipment and balance-of-system (BOS) components. Locate equipment as close together as possible to avoid long conductor and conduit runs and additional switchgear. **See Figure 3-30.** This also helps to minimize power losses from voltage drops. However, feasible equipment locations may be limited due to accessibility, structural limitations, existing electrical equipment locations, or even architectural appearance concerns.

Equipment Locations

Sharp Electronics Corp.

Figure 3-30. *Inverters and other system components should be located as close together as possible.*

Most inverters can be installed either indoors or outdoors, as long as they are kept cool and dry and have enough space around them for airflow and working space. Inverters should be kept out of direct sunlight or other environments that can raise operating temperatures, limiting power output. Inverters can be very heavy, so plan for heavy-duty brackets or structural attachments. Inverters or other heavy equipment should not be installed to drywall only.

Batteries should be installed within well-ventilated and protective enclosures. These may be in a garage or utility room, an air-conditioned space, outdoors, or buried in in-ground containers. Plans should account for substantial racks and foundations to support the weight of battery banks.

The installer should also evaluate means and locations for conductor and conduit runs between the array and other equipment, as well as the locations for any junction boxes, disconnects, overcurrent protection devices, and other components.

Site Layout Drawings

The layout of the site should be documented thoroughly for later use in the design, planning, and preparing of installations. Site layout diagrams and sketches should identify the shape and dimensions of structures, and the locations and distances between major system components. **See Figure 3-31.** Use photographs, drawings, and notes to document and record all pertinent site details while noting any special conditions affecting the design, installation, or safety of the installation.

> The best locations for PV equipment offer convenient access and proximity to both the array and the existing electrical system infrastructure.

Site Layout Drawing

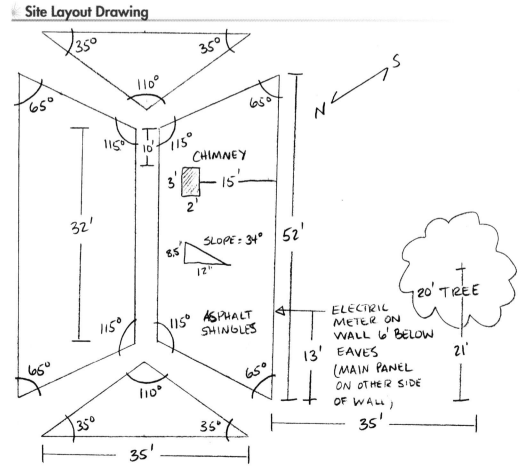

Figure 3-31. *A site layout drawing shows basic building dimensions and locations of major components.*

Refer to worksheet on CD-ROM

Energy Audit

An *energy audit* is a collection of information about a facility's or customer's energy use. An energy audit is a benchmark of energy use used to evaluate energy conservation strategies and size PV systems to meet load requirements. A review of a customer's past utility bills or meter readings can determine the total amount and cost of energy consumed per month. Energy bills from throughout a complete year can be compared to determine seasonal energy requirements for heating or cooling loads.

The installer may need to conduct a detailed electrical load analysis, especially for systems that will include batteries. The analysis should include a list of each load and its estimated or measured peak power demand, average daily time of use, and total energy consumed. **See Figure 3-32.** Sometimes reasonable estimates of power consumption can be determined from equipment nameplate ratings, but measurements give the most accurate information. Special measurements may be required, such as in-rush current draw or power quality factors. Compare the results to utility bills to check estimates or reveal loads not already identified.

Energy Conservation

Energy conservation is a particularly important consideration in the design of stand-alone PV systems because it directly impacts the size and cost of the system required to meet the load requirements. It is sensible to apply conservation measures before installing PV systems because their initial costs and payback periods are usually substantially less than increasing the size of the PV system. However, even consumers without PV systems or other renewable energy sources, benefit from energy conservation measures through reduced energy costs.

Energy conservation measures fall into three general categories: modifying a building or its design to reduce energy consumed or lost, managing existing loads to reduce energy demand and waste, and replacing loads with newer and more efficient versions to reduce energy demand.

The value of any specific conservation measure depends on the type of building, application, climate, and any other implemented conservation measures. However, many measures can realize significant energy savings with low-to-moderate investments in time and money, and some require only a change in energy-use habits.

Load Analysis

AC Load	Quantity		Power Rating (W)		Average Daily Use (hrs/day)		Average Daily Energy Use (Wh/day)
Incandescent Lighting	6	x	60	x	6	=	2160
Refrigerator	1	x	475	x	12	=	5700
Microwave	1	x	1200	x	0.5	=	600
Toaster	1	x	1200	x	0.15	=	180
Dishwasher	1	x	1500	x	0.5	=	750
Furnace Fan	1	x	500	x	2	=	1000
TV	1	x	130	x	3.5	=	455
VCR	1	x	40	x	0.75	=	30
Ceiling Fans	3	x	50		6		900

Figure 3-32. *A load analysis is part of an energy audit, which is used to evaluate a customer's energy use for sizing PV systems utilizing batteries.*

Building Modifications. The efficiency of a building envelope determines how well it isolates the indoor environment from the outdoor environment. This separation affects the heating and cooling loads, which are the most significant energy uses in most buildings. Energy loss or gain through the building envelope can be reduced by adding insulation, improving the weather sealing, and installing radiant barrier systems and efficient windows.

Passive cooling strategies include landscaping and architectural features that shade walls and windows. Light-colored roof surfaces reduce unwanted heat gain in the summer, while passive solar heating techniques can help heat a building in the winter.

Energy Management. Energy management involves changing existing load operations to reduce energy use and waste while maintaining the useful output. Examples of energy management include using automatic controls, adjusting setpoints, and adopting sensible choices in appliance use.

Automatic controls are very effective at using energy efficiently. Daylight sensors, dimmers, photo switches, or occupancy sensors can control lighting equipment. Intelligent controls and programmable thermostats can optimize the operation of HVAC systems for any conditions. Also, each additional degree of heating or cooling uses progressively more energy. Increasing thermostat settings by a few degrees when cooling or reducing settings by a few degrees when heating can significantly reduce energy use. Also, ceiling fans improve air circulation and personal comfort without appreciably increasing cooling loads.

Conventional ovens and ranges can add waste heat into a conditioned space. In the winter, the waste heat is useful and offsets a small portion of the heating load. In the summer under cooling loads, however, it can be more energy efficient to use microwave ovens for cooking to avoid the unwanted heat. Excess heat from clothes dryers can also be avoided in the summer by line drying clothes.

Load Efficiency. It is often cost effective to replace an appliance with a new and more efficient one because lower energy consumption reduces operating costs. Many types of appliances are constantly being improved for greater energy efficiency, including air conditioners, refrigerators, freezers, washing machines, entertainment equipment, and computer monitors. Most new appliances include information on energy consumption that can be used to compare the efficiency of different models. **See Figure 3-33.**

Energy-Use Labels

Figure 3-33. *Energy-use labels on new appliances include information on energy consumption and operating costs.*

Many energy efficient lighting products can be integrated with little or no modification to the lighting system, such as by retrofitting compact fluorescent lamps into incandescent light fixtures. Heat losses from water heaters can be reduced by insulating tanks and distribution piping and lowering the thermostat setting. Conserving hot water also reduces energy use. Solar water heating can also be used to supplement the water heating needs in most locations. Front-loading washing machines consume much less energy and water, and are better at removing moisture from clothes during spin cycles. Newer clothes dryers are better insulated to lose less heat, and have motors that are more efficient. Many other strategies and appliances can be implemented to use energy more efficiently.

Alternative energy sources can be used to offset or substantially reduce electrical loads. Natural gas or propane is more efficient at heating water and air than electricity, and can also be used to fuel compressors for refrigeration and air conditioning.

The ENERGY STAR® label identifies appliances designed to use energy efficiently and reduce energy waste. Suppliers of renewable energy systems are also good sources of information on energy conservation and management.

PREPARING PROPOSALS

After all site survey data has been collected and evaluated, the installer uses the information to prepare a proposal for customer approval. Proposals for any PV system should include detailed information about the scope of work and schedule, a description of the services to be provided, the specific equipment to be installed, and the costs. Other information, including cost breakdowns, subcontracts, performance expectations, and value assessments, may also be provided.

Cost Estimates

Estimates of costs for PV installations depend on a number of factors, including equipment, materials, labor, transportation, and profit margin. Costs for modules, inverters, and other major components are reasonably well established, and in most cases distributed through certain networks from the manufacturer, through suppliers, to the individual contractor or installer. As with many industries, contractors who purchase large amounts of equipment from a single supplier generally receive preferential pricing. The greatest variables in the costs are typically site-specific issues, such as difficult-to-access or hazardous locations, applications requiring special equipment, or custom-engineered or architecturally integrated designs.

In developing cost estimates, be sure to include all applicable external and ancillary costs such as architectural/engineering design, bid preparation and contract administration, rental equipment, permit and interconnection fees, and follow-up service contracts as applicable.

Performance Estimates

Proposals should include performance estimates for monthly and annual energy production. The estimates are a baseline for system performance that can be verified at commissioning and later compared to actual performance. Significant discrepancies between performance estimates and actual performance data may identify system problems requiring maintenance. Depending on the type of PV system, performance estimates may be determined in different ways.

Estimating System Size from Energy Requirements. Some PV systems are specifically designed to produce a certain amount of energy, such as stand-alone systems or interactive systems with battery backup. These systems are sized from a minimum energy requirement that equals the expected electrical load requirements during the lowest average insolation scaled up to account for losses in the system due to inefficiencies. Therefore, performance estimates for these systems are actually the initial specifications from which the rest of the system is designed.

Estimating Performance from System Size. Interactive systems without a battery backup have no minimum output requirement. These systems are often sized based on space and/or cost limitations with a performance estimate that is the peak rating of the array scaled down to account for actual solar radiation, operating conditions, and the efficiencies of the power processing equipment. This is the opposite of the procedure for stand-alone systems or interactive systems with battery backup. The installer should clearly explain these losses to the customer to avoid unrealistic expectations.

Value Assessment

Quotations may provide one or more types of value assessment based on the system performance, though further value analysis is conducted after the system is sized. The methods for assessing the value of a PV system depend on the system type, application, location, cost, performance, and other alternative power supply options.

Rebates, incentives, tax credits, and other PV system subsidies may substantially reduce the customer's actual cost, which increases the value. Value of a PV system can also be based on a comparison with competing power sources, such as a utility company. The value of the PV energy in dollars per kilowatt-hour is estimated from the total system cost (initial and life-cycle costs, minus subsidies) divided by the expected energy use over the lifetime of the system. Payback periods are estimated by dividing the total system costs by the avoided costs per year.

Value for PV systems is also found in its application. For example, PV systems can provide power where no other practical options exist. And for interactive systems with battery backup, the availability of back-up power when the grid is down can have considerable value, especially for critical loads such as refrigeration, communications, or data networks.

INSTALLATION PLANNING

Once a quotation has been accepted and a contract established, formal planning and preparation for the installation begins. Installation planning builds on the information and knowledge gained through the preliminary assessment and site survey.

The installer completes the final design; prepares construction drawings; applies for permits, interconnection approval, and financing; purchases equipment; and makes other preparations to complete the installation in the most efficient and timely manner.

Design Completion

For smaller or simpler applications, experienced installers may design the system and purchase all the equipment individually. An owner's manual must then be assembled by the installer and delivered to the customer at the time of commissioning and final acceptance. This system manual should include safety and hazard warnings, descriptions of major components and system operation, parts and source lists, electrical schematics, mechanical drawings, and information on maintenance, troubleshooting, and warranties.

Alternatively, predesigned systems offered by major suppliers and integrators can simplify planning and preparation. Manuals, mechanical drawings, electrical schematics, and other documents are typically already assembled for these systems. As pre-engineered and well-documented systems, they facilitate approval for code compliance and utility interconnection.

After completing the design, applications and construction drawings must be submitted for permits, utility interconnection, and incentive programs, as applicable. Submit designs to the AHJ and local utility as soon as possible so that any issues or concerns can be resolved early. This minimizes the risks of delays or significant system modifications after failed inspections.

Logistics

Logistics can be a significant issue in some projects, especially for remote, stand-alone systems. Installers should become familiar with local resources, such as the closest electrical supply and hardware businesses, accommodations, restaurants, and emergency facilities. The installer must also consider the time it takes to receive equipment, particularly the backlog of demand for modules, and when and how equipment will be delivered to the site, stored, and secured.

Final Preparations

Final preparations include developing a process-oriented approach for installing the array and other equipment. The installer must identify the appropriate installation practices, potential problems, or any special hardware or equipment needed for assembly. The installer should also anticipate needs for spare or replacement parts, and conduct as much system preassembly as possible prior to installation in the field.

Refer to Quick Quiz® on CD-ROM

Summary

- The installer must determine the customer's requirements and wishes regarding the future PV system.

- The installer should research the available solar resource to determine the optimal array orientation.

- The installer should consider the average weather conditions and potential for severe weather.

- The installer should investigate the local building codes and, when applicable, utility interconnection requirements.

- Site surveys are conducted by observing, photographing, measuring, documenting, and diagramming all applicable site conditions related to a PV installation.

- Certain equipment is required to conduct a site survey and collect the necessary information.

- Potential array locations must have sufficient space for the estimated array area and an adequate orientation.

- Sun path calculators, such as the Solar Pathfinder™ and Solmetric SunEye™, are common method of conducting shading analyses on potential array locations.

- Potential locations for the array and other equipment should be evaluated for accessibility, structural integrity, and existing electrical infrastructure.

- Adjustment for magnetic declination is necessary to determine the direction of due south when using a magnetic compass.

- A site survey should include a sketch of the proposed array location and the surrounding structures and electrical equipment.

- An energy audit reveals magnitudes and patterns of energy use that can be useful when sizing PV systems.

- Energy conservation measures can improve the value of a PV system by reducing overall energy requirements.

- Some systems are sized according to an energy production requirement, while other systems are designed to comply with constraints and have energy production estimated from the size.

Definitions

- A *site survey* is a visit to the installation site to assess the site conditions and establish the needs and requirements for a potential PV system.

- *Magnetic declination* is the angle between the direction a compass needle points (toward magnetic north) and true geographic north.

- A *sun path calculator* is a device that superimposes an image of obstructions on a solar window diagram for a given location.

- The *profile angle* is the projection of the solar altitude angle onto an imaginary plane perpendicular to the surface of an obstruction.

- An *energy audit* is a collection of information about a facility's or customer's energy use.

1. How is the available solar resource for a particular location determined?

2. What environmental concerns can dictate the materials, design, and placement of a PV system?

3. What are some of the acceptable methods of fall protection?

4. How do desired output, module efficiency, and module density affect the required area of an array?

5. When might an optimal array orientation be southwest instead of due south?

6. Describe how magnetic declination is used to determine true south.

7. Why is shading analysis a critical part of a site survey and how does it affect potential array locations?

8. What are some of the methods for conducting a shading analysis?

9. Describe some of the methods used to check for roof deterioration.

10. Describe the three general categories of energy conservation measures.

11. How are performance requirements and estimates related to array size?

4

System Components and Configurations

PV systems are highly versatile modular electrical power generation systems. Every PV system requires components to conduct, control, convert, distribute, and store the energy produced by the array. The specific components required depend on the type of system and functional requirements, but major components such as inverters, batteries, charge controllers, as well as wiring, switchgear and overcurrent protection are typically included. PV systems are broadly categorized by how they are or are not integrated with other electrical systems.

Chapter Objectives

- Describe the purposes and functions of the major components in PV systems.
- Identify the common types of energy storage systems.
- Compare the functions of various power conditioning devices.
- Describe various energy sources that can be interfaced with PV systems.
- Compare the features, requirements, and applications of various system configurations.
- List various electrical and mechanical balance-of-system components.

COMPONENTS

In addition to an array, all PV systems require electrical equipment to conduct, control, convert, distribute, store, and utilize the energy produced by an array. The specific components required depend on the application and operational requirements of the system.

Modules and Arrays

The primary component common to all PV systems is the PV array. An array consists of individual PV modules that are electrically connected to produce a desired voltage, current, and power output. **See Figure 4-1.** Modules and arrays produce DC power, which can be used to charge batteries, directly power DC loads, or be converted to AC power by inverters to power AC loads or interface with the electric utility grid.

The voltage of PV modules varies somewhat with temperature, and the current varies proportionately to solar irradiance, so power output is rarely constant. PV systems usually require means to store or condition power so that it can be used effectively by electrical loads.

Modules and Arrays

Schott Solar

Figure 4-1. *PV modules are connected together to form an array, which is the primary DC power-generating source and principal component in any PV system.*

The term "panel" typically refers to a small group of modules, often three or four, physically and electrically connected together as an installation unit. Panelization is sometimes used for quickly installing many small modules.

Batteries are by far the most common type of energy storage device used in PV systems.

Energy Storage Systems

Any power-generating system, including a PV system, faces the problem of matching energy supply with demand. Demand can fluctuate considerably and changes occur within minutes or seconds. However, the output of most power generation technologies tends to be steadier or limited to certain times of the day. Matching electricity production to demand is either impossible or impractical, which means that an electricity supply does not always coincide with when it is needed. **See Figure 4-2.** Excess energy must be stored for later use, and then recovered when it is needed at high demand. Energy storage systems balance energy production and demand.

There are many types of energy storage systems, though not all are suitable for PV systems. Batteries, particularly lead-acid types, are by far the most common means of energy storage in PV systems. However, other types of energy storage systems can be used, and advanced energy storage technologies are expected to compete with battery systems in the future. Energy storage systems are characterized by the form of energy they store and how it is utilized or converted into other forms.

Batteries. Batteries convert electrical energy into chemical energy when charging and vice versa when discharging. Batteries vary greatly in size, cost, and performance. Although a number of rechargeable battery types can be used in PV systems, lead-acid batteries are most common. Other types, including nickel-cadmium, nickel-metal hydride, and other advanced battery technologies, can also be used in PV systems but are considerably more expensive.

Most PV systems that employ batteries require more capacity than can be supplied by a single battery. A *battery bank* is a group of batteries connected together with series and parallel connections to provide a specific voltage and capacity. **See Figure 4-3.** Batteries in a PV system are charged by the array when sunny and discharged by loads when cloudy or at night.

In addition to energy storage, batteries provide short surge currents for loads with special starting requirements, which PV modules (as current-limited power sources) cannot provide. Most importantly, batteries establish a stable system operating voltage, at which the array and loads both operate. A stable system voltage allows the array to be optimized for maximum power and electrical loads or inverters to operate at their rated voltages, avoiding damage from high voltages or reduced performance from low voltages.

Batteries are characterized by their chemistry, nominal voltage, and energy storage capacity in ampere-hours or kilowatt-hours. They can also be evaluated in terms of their specific energy (kWh/kg), specific power (kW/kg), or cost ($/kWh), when comparing various battery types and other forms of energy storage.

Flywheels. Flywheels are being developed as a practical energy storage solution. Flywheels are large, spinning rotors and are commonly used to transfer power from motor and engines to pumps and other rotational loads. Because flywheels resist changes in rotation speed, they help steady the rotation from uneven power sources like piston engines. Flywheels can also be used to store large amounts of kinetic (motion) energy and release it in a short amount of time.

Energy Supply and Demand

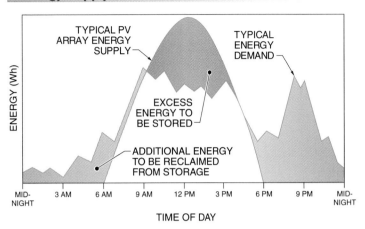

Figure 4-2. *Since the supply of energy from a PV array rarely matches the energy demand at a given time, some type of energy storage is usually required.*

Battery Banks

East Penn Manufacturing Co., Inc.

Figure 4-3. *PV systems with battery storage usually require more than one battery. A battery bank is a group of batteries connected together to provide a specific voltage and capacity.*

Sealed lead-acid batteries, although more costly than open-vent lead-acid batteries, are the most common type of battery in small- to medium-size stand-alone PV systems. This is because they are generally easier to transport, install, and maintain than open-vent batteries.

Excess energy is stored by powering an electric motor that spins the flywheel to high speeds. The flywheel can remain spinning for quite some time with very little loss of speed (kinetic energy). To recover the energy, the motor is switched from a power source to a load. It becomes a generator and produces electricity from the stored kinetic energy. **See Figure 4-4.** The amount of energy stored depends on the mass, shape, and speed of the flywheel. Flywheel energy storage systems are highly reliable and require little or no maintenance. Small-scale systems can store 2 kWh to 25 kWh of power with an efficiency of up to 90%.

Flywheel Energy Storage

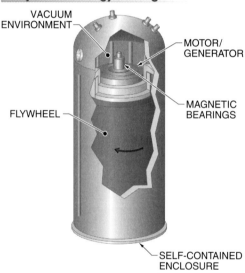

Figure 4-4. *Flywheel energy storage systems convert and store electrical energy as kinetic (motion) energy in the rotation of a heavy rotor.*

Supercapacitors. Capacitors store energy in an electrical field developed by two oppositely charged parallel conductive plates separated by a dielectric (insulating) material. When charging, an accumulation of charge on one plate builds up a voltage between the plates. Capacitance is the ability to store an electrical charge and is measured in farads. Conventional capacitors are common in electronics circuits, although the amount of energy stored is extremely small (corresponding to capacitances of microfarads and millifarads) and is released

quickly. Supercapacitors build on basic capacitor technology by using advanced materials to increase capacitance to hundreds or thousands of farads per cell, resulting in significantly greater energy capacity. **See Figure 4-5.**

Supercapacitors

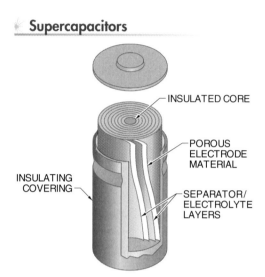

Figure 4-5. *Supercapacitors store energy by accumulating a charge on pairs of thin foil plates. The plates are usually wrapped into a cylinder shape.*

Supercapacitors are a suitable replacement for batteries in many low-power applications. They have fast response times, nearly infinite cycle life without loss of performance, and do not release heat during discharge. They can also smooth out power fluctuations in electrical systems, improving power quality.

Power Conditioning Equipment

Power conditioning equipment converts, controls, or otherwise processes the DC power produced by a PV array to make the power compatible with other equipment or loads. Power conditioning equipment includes inverters, charge controllers, rectifiers, chargers, DC-DC converters, maximum power point trackers, and power quality equipment. These components may be individual devices or may be combined into a single power unit. A *power conditioning unit (PCU)* is a device that includes more than one power conditioning function. However, since an inverter is commonly the major component in a multifunction PCU, the unit is still often called an "inverter."

Large-Scale Energy Storage

There are many ways to store electrical energy, though some are only practical for large-scale applications. However, innovative system designers have been able to adapt some principles used for large-scale energy storage for use in smaller systems. Additionally, new technologies are continuously being developed for possible use in PV applications.

Pumped Water

Pumped water systems use off-peak electricity (when demand is low, usually during the night) to pump water up to elevated reservoirs, in effect storing electricity in the form of potential energy. Potential energy is a source of future energy due to position. Water in an elevated reservoir has potential energy because energy was required to pump it up to an elevation and, if allowed, the water would flow back down. Potential energy is converted back into electrical energy when it is needed (when demand is high, usually during the day) by releasing water and directing its flow through turbine generators.

The energy capacity of water storage systems is determined by the volume of water stored, the maximum flow rate, and the vertical height that the water falls. Efficiencies in these systems are approximately 70% to 85%. That is, about 70% to 85% of the energy used to pump the water can be reclaimed when the water is released. These systems can store many megawatts of power for very long periods, though the energy is generally reclaimed within hours. The disadvantage of this technology is that it requires a specific set of circumstances to be feasible, such as a significant source of water and optimal geography.

Compressed Air

Compressed air systems are another form of large-scale energy storage. Similar to the principle of pumped water energy storage, off-peak power is used to compress air into sealed containers, storing potential energy. In large systems, underground caverns can serve as the container. The compressed air is later released to drive turbine-generators.

Because typical gas turbine generators use as much as two-thirds of the energy produced to drive the compressor, compressed air systems can also be used to improve the performance of gas turbines. The heated compressed air is mixed with oil or gas, burned, and the expanding mixture is used to drive turbine generators and produce electricity.

Inverters. An *inverter* is a device that converts DC power to AC power. In PV systems, inverters convert DC power from battery banks or PV arrays to AC power for AC loads or export to the utility grid. **See Figure 4-6.** Inverters are characterized by power source, power ratings, input and output voltages, waveform, power quality, and power conversion efficiency. Inverters used in PV systems are further classified as either battery-based or utility-interactive types.

Battery-based inverters are used in stand-alone PV systems and operate directly from battery banks as their input source. The DC input corresponds to nominal battery bank voltages, usually 12 V, 24 V, or 48 V. Higher voltages are used for higher output power requirements. The AC output is typically 120 V or 240 V single-phase power, with power ratings from a few hundred watts to over 10 kW. The actual output is determined by the power demand from the loads. Some models may be combined in parallel to produce greater output power levels. Most battery-based inverters also include built-in battery chargers and some may control system loads or other energy sources such as engine generators.

Resistive lighting loads, such as incandescent or halogen lamps, can be powered by either DC or AC, but are less efficient and do not last as long as fluorescent and LED lighting. LEDs can be operated directly by DC power, while fluorescent lighting requires a DC ballast.

Inverters

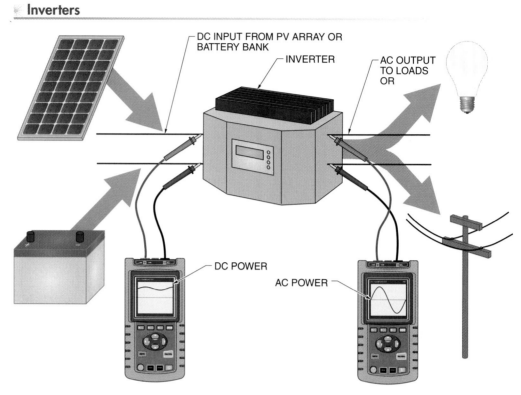

Figure 4-6. *Inverters convert DC power from batteries or arrays into AC power to serve local loads or for export to the utility grid.*

A power conditioning unit is the most complex electronic device in a PV system because it must manage many functions simultaneously, such as inverting, monitoring, and switching.

Utility-interactive inverters draw power directly from PV arrays and operate in parallel with the utility grid. Modern interactive inverters adapt to a wide range of DC input voltages and match AC output to the utility system. Typical AC output voltages are 120 V or 240 V single-phase units with power outputs up to

10 kW, while 208 V and 480 V three-phase units are available from around 10 kW up to a few hundred kW. Utility-interactive inverter output is determined by the DC input from the array, unlike the output from battery-based inverters, which is determined by the load. The output of interactive inverters is supplemental to utility-supplied power, so it must be matched and synchronized to the utility power in order to be combined. Excess power can be fed back to the utility grid if not needed by local loads. In this situation, the utility functions as an infinitely large battery bank for excess electricity storage and additional electricity retrieval, as needed.

Charge Controllers. Nearly every PV system that uses a battery requires a charge controller. **See Figure 4-7.** A *charge controller* is a device that regulates battery charge by controlling the charging voltage and/or current from a DC power source, such as a PV array. Charge controllers protect the battery from overcharge and overdischarge, improving system performance and prolonging battery life.

Charge controllers regulate battery charging by terminating or limiting the charging current when the battery reaches a full state of charge. Charge controllers may also protect batteries from extreme discharges by disconnecting electrical loads when the battery reaches a low state of charge. Charge controllers are characterized by their method of charge regulation, voltage and current ratings, and other features and functions.

Rectifiers and Chargers. A *rectifier* is a device that converts AC power to DC power. A *charger* is a device that combines a rectifier with filters, transformers, and other components to condition DC power for the purpose of battery charging. **See Figure 4-8.** Battery chargers are used in PV systems with an AC power source available, such as the utility grid or an engine generator, to provide supplemental battery charging. A battery charger can be an independent component or an integral part of a battery-based inverter. Battery chargers are characterized by their AC input and DC output ratings, methods of charge regulation, and efficiency.

DC-DC Converters. A *DC-DC converter* is a device that converts DC power from one voltage to another. For example, a power input of 48 V at 10 A can be converted to an output of 24 V at almost 20 A (some losses from conversion). These devices are sometimes used in battery-based PV systems to power loads at voltages higher or lower than the nominal battery voltage. Some are capable of providing more than one output voltage simultaneously. They can also convert high voltage from an array to a lower battery-charging voltage at higher current. DC-DC converters are characterized by power rating, input and output voltages, and power conversion efficiency.

Maximum Power Point Trackers. The voltage and current output from an array can vary with temperature, irradiance, and load. Various combinations of these factors produce power outputs anywhere between the rated (maximum) power level and zero. For any combination of temperature and irradiance, there is a maximum possible power output that corresponds to a certain voltage and current. A *maximum power point tracker (MPPT)* is a device or circuit that uses electronics to continually adjust the load on a PV device under changing temperature and irradiance conditions to keep it operating at its maximum power point.

An MPPT circuit monitors array conditions and dynamically changes its resistance or input voltage to maximize power from an array. The rest of the MPPT works like a DC-DC converter and delivers power at any output voltage required by the load. MPPT functions are included with inverters and some charge controllers.

Charge controllers with MPPT and DC-DC conversion circuits can be beneficial for systems with large arrays or arrays located far from the battery bank. The array is configured at a higher voltage, which lowers the current. Lower current reduces voltage drop and the size and cost of conductors and switchgear needed. The system then converts the voltage to the required charging level.

> Simple interactive inverters are not suitable for providing back-up power because they must be shut down during utility outages for safety reasons. Bimodal inverters are required to provide battery backup during utility outages.

Charge Controllers

Morningstar Corp.

Figure 4-7. *Charge controllers protect batteries in PV systems from overcharge or excessive discharge.*

Power Quality Equipment. Although power quality management is inherent to most inverters, additional equipment may be required in some systems to improve the power quality of the inverter output. Power quality equipment may also be used to address poor power quality of utility power or other sources.

Electrical Loads

An electrical load is any type of device, equipment, or appliance that consumes electricity. They are characterized by purpose, operating voltage, and power demand, and vary widely in performance and other characteristics. Electrical loads typically require either DC or AC power, though some may operate from either, such as resistive loads or universal motors.

Stand-alone PV systems are designed as the sole power source for loads, so the energy consumed by the load determines the size and cost of the required array and battery. In interactive PV systems, the PV power is only supplemental to other power sources, so the size of the loads has little influence on the size, operation, or performance of the PV system. However, in either circumstance, the efficiency of the loads is vital to designing the most practical, efficient, and cost-effective system.

DOE/NREL, Byron Stafford
Small stand-alone PV systems are often used to power medical refrigerators in remote areas. To avoid losing energy in the inverting process, some refrigerators are designed to run on DC power.

Some loads have special characteristics that affect their energy use. For example, most motors require a high starting current. Electrical loads may also be categorized by importance. When power is limited, less critical loads can be disconnected from the system to preserve remaining power for more critical loads.

Rectifiers and Chargers

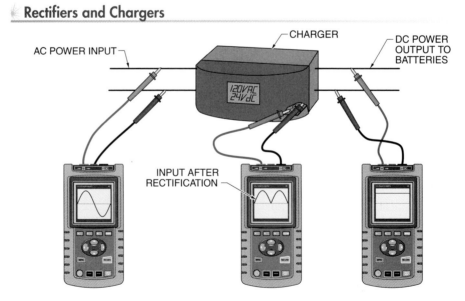

Figure 4-8. *Rectifiers and chargers make AC power from sources such as the utility or engine generators available for charging batteries or other DC loads.*

DC Loads. Since a PV array produces DC power, DC loads are used in many PV applications to avoid having to invert the power to AC. This simplifies the design and installation and eliminates the cost of an inverter. DC loads may be powered directly from arrays or batteries. The most common DC loads used in PV systems are lighting fixtures and motors for fans and pumps. Most DC loads operate at 12 V, 24 V, or 48 V.

Most DC loads and appliances are produced for the marine and RV markets and are readily available from retailers specializing in those areas. However, because they are not intended for use in buildings, they may not meet applicable building codes and regulations. DC appliances should be carefully checked for listing to applicable safety standards.

AC Loads. AC loads are powered from inverters, generators, or the utility grid. AC loads operate at normal service voltages and at a specific frequency. Most residential and commercial loads are AC loads, including refrigerators, air conditioners, televisions, lighting, and motors. AC loads are more widely available and usually less expensive than comparable DC loads. AC loads may require single-phase or three-phase power at voltages from 120 V to 480 V.

Balance-of-System Components

Balance-of-system (BOS) components are all the remaining electrical and mechanical components needed to integrate and assemble the major components in a PV system. **See Figure 4-9.** BOS components are also an important part of safety management.

Mechanical BOS Components. Mechanical BOS components include fasteners, brackets, enclosures, racks, and other structural support for the safe and reliable installation of PV system components. Additional mechanical BOS components, such as fencing, handrails, access ladders, and building or architectural components associated with the PV system, depend upon the particular PV installation. For PV water delivery systems, storage tanks, pumps, and distribution equipment are considered mechanical BOS components.

Electrical BOS Components. Electrical BOS components include conductors, cables, conduits, junction boxes, enclosures, connectors, and terminations needed to make circuit connections between modules, controllers, batteries, inverters, and other electrical systems and equipment. Electrical BOS hardware also includes switchgear, overcurrent protection devices, and grounding equipment needed to safely isolate and protect major components and conductors from overload and fault conditions. Instrumentation and monitoring equipment may also be considered BOS components, as would any special surge protection devices, power quality equipment, related control systems, or dedicated electrical loads.

Balance-of-System (BOS) Components

SolarWorld Industries America
MECHANICAL

ELECTRICAL   *Schott Solar*

Figure 4-9. *Balance-of-system (BOS) components include all the additional mechanical and electrical parts needed to connect and secure the major components.*

DC loads are no more efficient than comparable AC loads. However, using DC power instead of AC power whenever possible avoids the power losses inherent with the inverting process, increasing overall system efficiency.

ELECTRICAL ENERGY SOURCES

Other electrical systems may be integrated with PV systems as additional sources of electrical energy, depending on the type of system and application requirements. Besides the PV array, an electric utility grid is the source of electricity that is by far most commonly connected to PV systems. Other electricity sources include engine generators, gas-turbine generators, uninterruptible power supplies, wind turbines, micro-hydroelectric turbines, and fuel cells.

Engine Generators

An *engine generator* is a combination of an internal combustion engine and a generator mounted together to produce electricity. A *generator* is a device that converts mechanical energy into electricity by means of electromagnetic induction. The arrangement is also sometimes called an engine generator set, or genset.

Generators are typically driven by internal-combustion piston engines. These engines burn a compressed fuel-air mixture, which moves reciprocating pistons that drive a crankshaft. The crankshaft then transfers mechanical power to the generator shaft. The generator uses electromagnetic principles to convert the rotation of the shaft into electricity. **See Figure 4-10.**

Generators can be designed to produce from a few hundred watts of single-phase power to several megawatts of three-phase power. Small portable engine generators are used for emergency residential power or on job sites for temporary power. Standby engine generators are permanent, medium to large systems used for larger back-up and temporary power applications, but are not run continuously as a primary power source. **See Figure 4-11.** Prime-duty engine generators are also medium to large systems, but are designed to operate continually as a primary source of power.

Standby Engine Generators

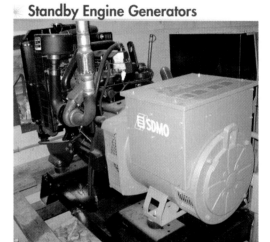

Figure 4-11. *Engine generators are usually installed as a complete, integrated package.*

Engines used with generators are typically fueled with gasoline, diesel, propane, or natural gas. Small systems typically run on gasoline because of its widespread availability, while larger systems can be designed for any of these fuels, though diesel is the most common. The type of fuel affects the load response of the engine generator. Diesel engines are well suited for constant loads, whereas gasoline engines can respond quickly to changing loads. However, diesel engines generally have better partial-load efficiency, run slower, are more robust, last longer, and require less maintenance, so they are predominately used for large stationary applications.

Engine Generators

FUEL-AIR MIXTURE
COMBUSTION
EXHAUST
PISTONS
AC POWER OUTPUT
CRANKSHAFT
MECHANICAL ENERGY (ROTATION)
GENERATOR

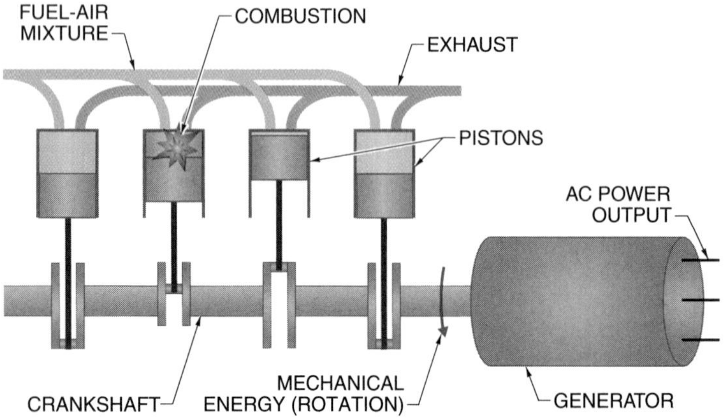

Figure 4-10. *Engines use reciprocating pistons to create mechanical power, which is then converted to electrical power in the generator.*

Engine generators are characterized by their output rating, engine type, fuel source, and fuel-to-power conversion efficiency. Engine generators also require auxiliary equipment to safely and successfully start, operate, and interface with other systems. Some PCUs can control engine generators for battery charging and powering certain AC loads.

Transfer Switches

A transfer switch allows an electrical system to be quickly and safely switched from a primary power source to an alternate power source in order to keep critical loads operating. The most common configuration is an emergency engine generator to back up the utility power. Transfer switches may also be used with PV systems, typically those in off-grid residences or facilities that rely on alternate power from engine generators when insolation is low or the PV system cannot charge the battery system.

Transfer switches are either manual or automatic. With a manual transfer switch, the transfer to alternate power is done by a person physically activating the switch. If the alternate power source is an engine generator, additional start-up procedures are required to bring the engine up to normal operating conditions before making the transfer.

Automatic transfer switches continually monitor the primary power source. When the voltage falls below a certain level, the switch automatically starts a power transfer, such as sending electronic commands to an engine generator to initiate its start-up sequence. When the switch detects the proper voltage and frequency on the alternate source, it completes the transfer. The switch can also manage an automatic and controlled transfer back to primary power when it becomes available again.

In all cases, the primary power source must be completely disconnected from the electrical system before the alternate power source is connected. This prevents the power from back-feeding into the primary system but results is a short power interruption. Depending on the system, the interruption may last from a fraction of a second to several seconds.

Gas Turbine Generators

A *gas turbine* is a device that compresses and burns a fuel-air mixture, which expands and spins a turbine. A *turbine* is a bladed shaft that converts fluid flow into rotating mechanical energy. The shaft power from the turbine can be used to drive a generator to produce electricity. **See Figure 4-12.** Gas turbines are essentially jet aircraft engines developed for stationary applications. Gas turbines are combined with generators to form electricity producing units similar to piston-driven engine generators.

Gas turbine generators produce up to several hundred megawatts of power and are used in large industrial plants, ships, and vehicles such as tanks. Gas turbines have a better power-to-weight ratio than internal combustion engines, which means that they can generate more power from smaller, lighter equipment. The main disadvantages of gas turbine generators are high cost and complexity. They typically run on natural gas, liquefied petroleum gas, diesel, or kerosene fuels. Gas turbine generators are best suited for constant loads.

> Some engine generators can operate on more than one fuel type. For example, a large back-up power system may be connected to a gas line and operate from natural gas under normal outage conditions. However, if the gas supply is interrupted, such as during a natural disaster, the engine may run using an on-site reserve of diesel fuel.

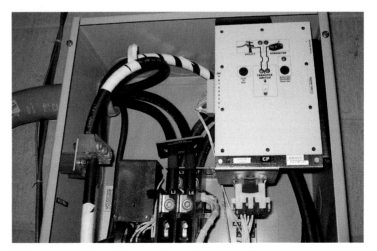

Automatic transfer switches monitor utility power and engage auxiliary power sources, such as an engine generator, if the power falls below preset limits.

Gas Turbine Generators

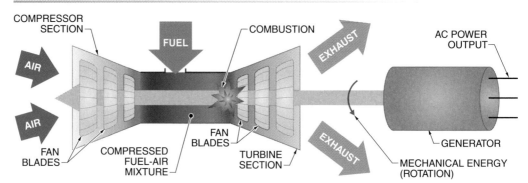

Figure 4-12. *Gas turbines have separate compressor and turbine sections that work together to send mechanical power to a generator.*

Microturbines are small gas turbine generators that produce up to a few hundred kilowatts of power. These have been adapted for portable or distributed power for demand-side applications, as opposed to primary generation. Microturbines are installed at industrial or commercial facilities to supplement the utility during peak loads or provide back-up power. Combined heat and power (CHP) microturbines produce both thermal energy and electricity, improving the overall efficiency of the process.

Uninterruptible Power Supplies

An *uninterruptible power supply (UPS)* is a battery-based system that includes all the additional power conditioning equipment, such as inverters and chargers, to make a complete, self-contained power source. In standby mode, a UPS system keeps its batteries at a full state of charge from utility power. When a utility disturbance occurs, an automatic transfer circuit isolates critical loads from the utility, and supplies AC power inverted from DC power from the battery.

Depending on wind speed, wind turbines turn at variable speeds, which normally produces AC power with variable frequency. Some wind turbines are designed to match utility frequency by using utility power in the generator's exciter circuit. Others convert the AC power to DC and then back to AC, matching the frequency of the utility grid.

Small UPS systems are commonly used by businesses and homeowners to back up computers, security systems, office equipment, and other critical loads for short periods (minutes) in the event of utility outages. Large systems can be used to power large equipment or entire facilities for extended periods (hours). The size of the battery bank determines the magnitude and duration of critical load operation.

UPS systems are often used to supply power during the transition from utility power to a back-up power source, such as engine generators. UPS systems rely on an external power source to charge the batteries. A PV array can charge the batteries when the primary source of power is lost, extending the time that loads can be supplied with power.

Wind Turbines

A *wind turbine* is a device that harnesses wind power to produce electricity. The wind rotates aerodynamic surfaces and the resulting mechanical energy drives a generator to produce electricity. **See Figure 4-13.** Wind power increases significantly with wind speed and wind turbines are viable in areas with an average wind speed greater than about 12 mph. Small wind turbines from 1 kW to 20 kW peak output are often used in conjunction with PV systems as an alternative energy supply. Large commercial wind turbines can produce up to a few megawatts and have rotor diameters up to a few hundred feet.

Wind Turbines

*DOE/NREL, Stuart Van Greuningen,
Idaho Energy Division*

Figure 4-13. *Wind turbines convert the power of the wind into electrical energy.*

Wind turbines are generally classified by size, output, turbine blade type, axis of rotation, and electrical generation, control, and regulation characteristics. Most wind turbines use horizontal-axis rotors with variable yaw and pitch, and electronic clutch mechanisms to load and unload the machine and for stowing during excessive winds.

Wind turbines can be integrated with other electrical systems and equipment in very similar ways to PV systems. They can operate in parallel with the utility grid, and supply supplementary power to site loads. In addition, they may be used to charge batteries that are used to power system loads as required.

Micro-Hydroelectric Turbines

A *micro-hydroelectric turbine* is a device that produces electricity from the flow and pressure of water. Moving water acts against turbine blades, rotating a shaft that operates an electrical generator. The principles are the same as for large-scale hydroelectric plants, such as the Hoover Dam, but on a much smaller scale. Micro-hydroelectric systems are those with electrical outputs less than 100 kW. Many rural and off-grid consumers have successfully implemented micro-hydroelectric turbines in nearby streams and rivers as an auxiliary power source. A significant advantage to these systems is a very steady supply of power.

Fuel Cells

A *fuel cell* is an electrochemical device that uses hydrogen and oxygen to produce DC electricity, with water and heat as byproducts. Although similar to batteries, fuel cells are different in that they require a continual replenishment of the fuel and oxidizer reactants (such as hydrogen and oxygen).

A typical fuel cell element consists of a cathode and anode separated by an electrolytic membrane material. **See Figure 4-14.** As hydrogen gas flows across the anode, electrons are stripped from the hydrogen and flow through an external circuit, re-entering the fuel cell at the cathode. At the same time, positively charged hydrogen ions migrate across the membrane to the cathode, where they combine with oxygen and the returning electrons to form water and heat. Typical fuel cell efficiency is about 50% to 70%, though it may be higher if the heat byproduct is also utilized, such as for water heating.

Fuel Cells

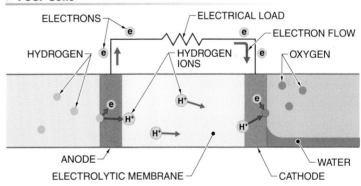

Figure 4-14. *Fuel cells use reactants such as hydrogen and oxygen in a process that transfers electrons, with only water and heat as byproducts.*

An *electrolyzer* is an electrochemical device that uses electricity to split water into hydrogen and oxygen. This is the opposite process as is used in a fuel cell and is analogous to charging a battery. An electrolyzer may be coupled to a PV system to produce hydrogen and oxygen, which can be stored and later used in fuel cells to generate electricity. Hydrogen can also be generated from traditional fuels, such as oil and natural gas products, using a device called a reformer.

PV SYSTEM CONFIGURATIONS

The simplest PV system configuration is a PV module or array connected directly to a DC load. Other components can be added to produce increasingly complex and sophisticated systems. A large number of combinations are possible for PV systems. The optimal configuration for an application depends on load usage, load type, solar resource, auxiliary power sources, and many other factors.

Stand-Alone Systems

A *stand-alone PV system* is a type of PV system that operates autonomously and supplies power to electrical loads independently of the electric utility. Stand-alone PV systems are most popular for meeting small- to intermediate-size electrical loads, and are extensively used when other energy sources are cost-prohibitive or impossible. Stand-alone systems may be designed to power DC and/or AC electrical loads, and typically store energy in batteries.

A stand-alone system may use a PV array as the only power source or may include one or more additional power sources, such as engine generators or wind turbines. However, most stand-alone systems do not interact with the utility grid.

Stand-alone PV systems are sized and designed to power a specific electrical load using the solar radiation resource at a given location. Since the size and cost of any stand-alone system is related to the magnitude and duration of the electrical load and the solar resource, energy efficiency is critical. For these reasons, a thorough load analysis is the first step required in the design and installation of any stand-alone system.

Stand-alone PV systems are classified by their major components and the manner in which they operate in combination with other energy sources.

Direct-Coupled Systems. A *direct-coupled PV system* is a type of stand-alone system where the output of a PV module or array is directly connected to a DC load. These systems do not include any power conditioning equipment or electrical energy storage. The

only other components are BOS components such as mounting systems and disconnects. **See Figure 4-15.** Without batteries, direct-coupled systems are appropriate only when energy demand coincides with energy generation (when the sun is shining). Because PV output varies with irradiance, the load must also be capable of operating over a range of voltages.

Direct-Coupled Systems

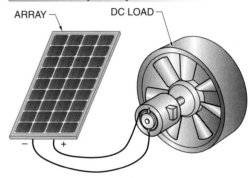

Figure 4-15. *The simplest type of PV system is the direct-coupled system, consisting of only an array and a DC load.*

DC motors are the most common loads for direct-coupled systems, including water pumps and ventilation fans. Direct-coupled PV systems are common for pumping potable and agricultural water supplies. In most of these applications, the water is either used as it is pumped or is stored in reservoirs or tanks. Direct-coupled ventilation fans provide airflow in proportion to irradiance, which is useful for applications such as attic ventilation. Circulation pumps for solar water-heating systems are also ideal applications of direct-coupled PV systems because the need to circulate hot water from a solar collector to a storage tank coincides with PV output. In this sense, direct-coupled PV circulation pumps provide inherent system control, eliminating the need for temperature controls to start and stop the circulation pump.

While direct-coupled systems are the simplest form of any PV system in terms of equipment, they are perhaps the most complex to design properly due to the lack of energy storage or system control. Matching the electrical load to the array is a critical part of designing a well-performing direct-coupled system. Because the output of the array changes with temperature and irradiance, the load does not always keep the array at its maximum power point. DC-DC converters or MPPT devices can be added to help better utilize the array power output.

Self-Regulated Systems. Most stand-alone PV systems require some type of energy storage, and batteries are the most common choice. Batteries are sized based on the expected magnitude and duration of electrical loads, and system operation characteristics. The PV array (and any auxiliary energy sources) must then be properly sized to fully charge the battery under the conditions of highest load and lowest solar insolation for the given application and site.

A *self-regulating PV system* is a type of stand-alone PV system that uses no active control systems to protect the battery, except through careful design and component sizing. It consists of only an array, battery, and load. Fewer components means a less expensive and more reliable system, but only if the system is properly designed. Otherwise, the system is vulnerable to overcharging or undercharging the battery. **See Figure 4-16.**

In order to prevent damaging the battery or charging it incompletely, the size of each component must be matched to the others. The magnitude and duration of the load must be known to a high degree of accuracy. The load must have a consistent load current and day-to-day energy requirement or it must be controlled automatically. Consequently, these designs are not suitable for applications where the load is highly variable or externally controlled by a user or operator.

To protect the battery from overcharge, the battery system must be oversized in relation to the size of the array, which keeps charging current low. Specially designed modules also prevent overcharge by having a voltage range that closely matches the voltage of a fully charged battery.

Charge-Controlled Systems. If loads are variable or uncontrolled, charge control is required to prevent damage to the battery from overcharge or overdischarge. Charge control typically involves interrupting or limiting the charging current to a fully charged battery to prevent overcharge. Alternatively, the charging current can be diverted to a useful, but noncritical, auxiliary load to avoid wasting the excess power. **See Figure 4-17.** Charge control is required by the NEC® if the array current is equal to 3% or more of the rated battery capacity in ampere-hours. For example, a 100 Ah battery would require charge control if the maximum array charging current exceeds 3 A.

Self-Regulating Systems

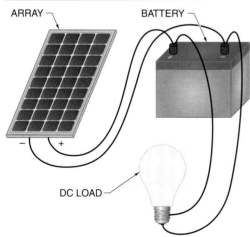

Figure 4-16. *Self-regulating systems avoid the complexity of adding charge control components by precisely sizing the battery and array.*

Self-Regulating Navigational Aids

Perhaps the most successful application of a self-regulating PV system is the design of solar-powered marine navigational beacons. The electrical load is an incandescent lamp that has well-defined current draw and flash duration. The battery is large in relation to the size of the array and the load in order to meet energy requirements of the lamp under conditions of highest load (long winter nights) and lowest insolation (short winter days). Most importantly, low-voltage modules limit the charging current as the battery reaches full charge, preventing overcharge. The U.S. Coast Guard and other maritime authorities have deployed tens of thousands of these navigational aids on buoys and channel markers around U.S. coasts and inland waterways.

Charge-Controlled Stand-Alone Systems

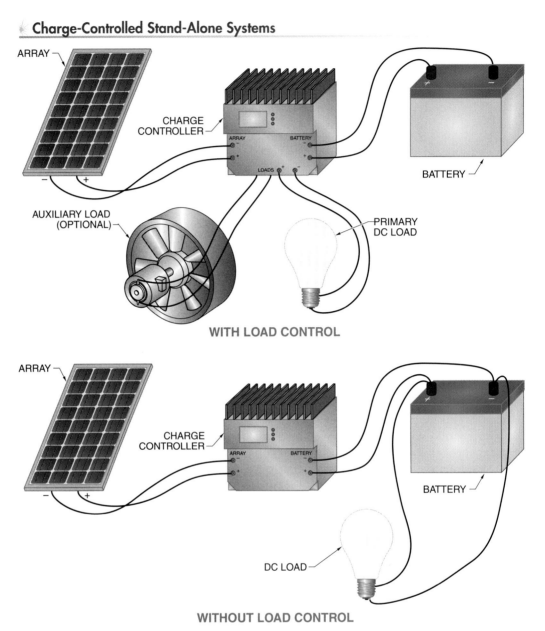

WITH LOAD CONTROL

WITHOUT LOAD CONTROL

Figure 4-17. *Systems with charge control regulate the charging current into the battery. Regulation may involve disconnecting or limiting the array current or diverting the excess current into an auxiliary load.*

Stand-alone systems with battery storage also need to protect the batteries from overdischarge. Overdischarge protection typically involves disconnecting loads at a predetermined low voltage, corresponding to a low state of charge. Charge controllers may include this type of load control, or a separate load controller may be required.

When typical AC loads and appliances are used in stand-alone PV systems, an inverter converts DC power from the batteries to AC power at suitable voltage and power levels. However, the inverter adds system complexity and cost, and some power is lost in the conversion process. Typically, the DC input for stand-alone inverters is connected directly to the battery bank, and the AC output is connected to a distribution panel to power the AC loads. DC loads may also be powered directly from the battery bank. **See Figure 4-18.**

Stand-Alone Systems for AC Loads

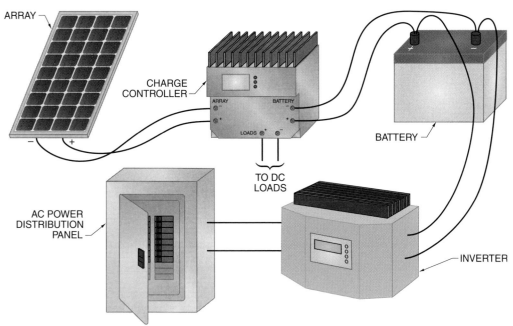

ARRAY

CHARGE CONTROLLER

ARRAY   BATTERY

LOADS

BATTERY

TO DC LOADS

AC POWER DISTRIBUTION PANEL

INVERTER

Figure 4-18. *Stand-alone systems for AC loads must include an inverter, which draws DC power from the battery bank and changes it to AC power for distribution.*

Utility-Interactive Systems

A *utility-interactive system* is a PV system that operates in parallel with and is connected to the electric utility grid. These systems are sometimes called "grid-connected" or simply "interactive" systems. These systems are the simplest and least-expensive PV systems that produce AC power because they require the fewest components and do not use batteries. The primary component in a utility-interactive system is the inverter, which directly interfaces to the array and the electric utility network and converts DC output from an array to AC power that is synchronous with the utility grid. **See Figure 4-19.** Utility-interactive systems are modular, so large systems can be designed as multiple smaller systems with individual inverters, which are then connected together at a common utility connection.

Utility-interactive systems have a bidirectional interface with the utility at the distribution panel. When the PV system does not produce enough power to meet system loads, additional power is imported from the utility. If there is an excess of PV power, the excess power is fed back (exported) to the grid.

Net Metering. If the PV system uses net metering, the utility's electricity meter runs backwards when power is exported from the PV system to the utility. *Net metering* is a metering arrangement where any excess energy exported to the utility is subtracted from the amount of energy imported from it. **See Figure 4-20.** Using this system, energy supplied to the utility from a PV system is effectively credited to the customer at full retail value.

Customers are usually not paid for net excess generation, but credit may be carried over from month to month. This works out very well for systems that produce excess power during the high-insolation summer months but require supplemental power from the utility during the low-insolation winter months. If the PV system and loads are perfectly balanced for annual energy production and demand, the result is a net energy bill of zero. However, some net metering energy credits expire after a certain period, typically one year. Net metering arrangements are not available for all PV systems or with all utilities, but they are becoming increasingly common.

Utility-Interactive Systems

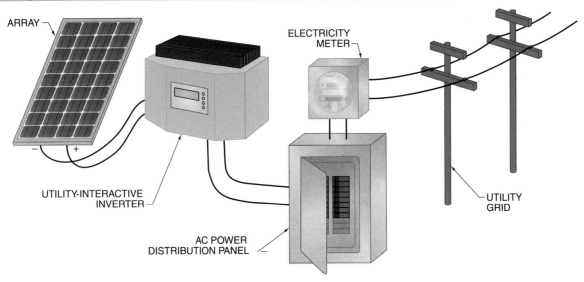

Figure 4-19. *A utility-interactive system is controlled by the inverter, which adds AC power converted from DC power to the utility grid power at the main AC power distribution panel.*

Electricity Exporting

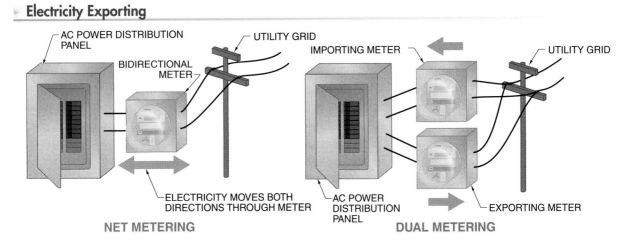

Figure 4-20. *Utility-interactive systems have either net-metering or dual-metering arrangements for exporting electricity to the utility grid.*

Dual Metering. Some large interactive distributed generation systems are not net metered, but dual metered. *Dual metering* is the arrangement that measures energy exported to and imported from the utility grid separately. This arrangement allows energy delivered to the consumer to be billed at full retail value, while power delivered to the utility is credited at a lower wholesale value, the cost the utility pays to generate or purchase energy.

Islanding. If the utility experiences an outage, utility-interactive inverters must stop supplying power to the grid. This is because an inverter exporting power to the utility lines is a serious electrocution hazard to any line workers that may be working to restore power. Interactive inverters monitor utility grid conditions and suspend power exporting when the utility power falls out of the normal range in voltage or frequency. *Islanding* is the undesirable condition where a distributed-generation power source, such as a PV system, continues to transfer power to the utility grid during a utility outage. Anti-islanding provisions are required for all listed interactive inverters.

Interconnection Agreements. Permission is required to interconnect PV systems with the utility grid, since the power affects the utility's distribution system. Risks to the utility include poor power quality from the inverter affecting other consumers on the grid and safety concerns for their electricians about islanding. It is the jurisdiction of the local utility to determine specific interconnection requirements, though general provisions are often established by the state. Most utilities have simple interconnection agreements for PV systems up to a certain size, which require little more than that the system be inspected by local officials for code compliance. Other possible provisions include special labeling, external disconnects, insurance, metering, and billing arrangements.

Bimodal Systems

A *bimodal system* is a PV system that can operate in either utility-interactive or stand-alone mode, and uses battery storage. Bimodal systems are sometimes referred to as battery-based interactive systems. The key component in a bimodal system is the inverter, which draws DC power from the battery system instead of the array. In this case, the array simply acts as a charging source for the battery system.

Bimodal systems operate in a manner similar to UPSs and have many similar components. Under normal circumstances, bimodal systems operate in interactive mode, serving the on-site loads or sending excess power to the utility grid while keeping the battery fully charged. However, if the grid becomes de-energized, a transfer switch automatically disconnects from the utility and operates the inverter to supply AC power sourced from the battery bank. **See Figure 4-21.**

If the system must be taken off-line for service or maintenance, a manual transfer switch or bypass circuit can isolate the array, battery, and inverter from the remainder of the system and directly connect the loads to the utility supply.

Bimodal systems are typically used to back up critical loads, but can also be used to manage the energy supply for different times of the day in order to reduce electricity bills.

Critical Load Backup. Bimodal systems are extremely popular for homeowners and small businesses where a back-up power supply is required for critical loads such as computers, refrigerators, water pumps, or lighting when a utility power outage occurs. In this configuration, the AC power from the inverter is routed to an isolated subpanel that serves only the critical loads.

The ability of these systems to provide back-up power is a significant advantage over interactive-only systems, despite the extra costs (primarily for the battery bank). However, when in back-up mode, the array output and battery system capacity are still limiting factors. As with stand-alone systems, careful sizing and load management are essential for the bimodal system to adequately supply the critical loads for the required amount of time.

Utility Rate Management. Some utilities use time-of-use rates to charge more for electricity during peak demand times of the day than during off-peak times. Peak hours are typically during the day when people are using household appliances and businesses are open. Electricity demand is relatively low during the night. If the rates are significantly different, a bimodal system can be used to manage utility energy use for optimal electricity rates.

A bimodal inverter can be programmed to supply the loads with energy from the array and battery bank during peak times, avoiding or minimizing the use of high-priced utility electricity. During off-peak times, the loads can be powered and the battery bank charged with less-expensive utility electricity. Since peak utility rates are typically charged during the middle of the day when insolation is at maximum, any excess PV energy can be sold back to the utility at higher peak retail rates (if the system uses net metering). This type of system can still function as a backup to the electric utility during an outage, though the use of the battery bank must be carefully controlled to leave capacity available for back-up use.

Bimodal Systems

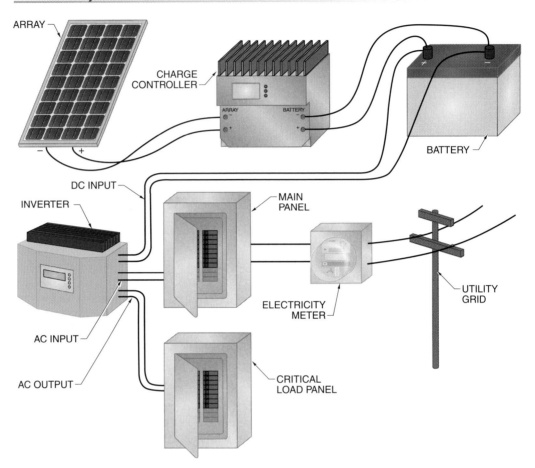

Figure 4-21. *Bimodal systems can operate as either a utility-interactive or a stand-alone system.*

Combined Heat and Power Systems

Most large-scale central power plants burn fuel to drive the generating process, but convert only about one-third of the energy content in the fuel to electrical energy. Most of the remainder is wasted as heat and is dumped, unutilized, into the atmosphere or waterways.

However, distributed generation applications can sometimes use combined heat and power (CHP) technology (also called cogeneration) because they are located near the electrical and thermal loads. Waste heat from power generation is recovered and used for water, space, and process heat applications. CHP systems save fuel that would otherwise be used to produce heat or steam in a separate unit, improving fuel utilization efficiencies to 70% to 90%. Although many types of distributed generators, such as internal combustion engines, rely on nonrenewable fossil fuels, utilizing the fuel more efficiently with CHP applications greatly improves their value.

Small-scale, or micro-CHP, systems for residences and commercial buildings are available but still relatively rare. These systems primarily use internal combustion engines, though new systems based on microturbines or fuel cells are expected to increase the popularity of CHP technology. Solar energy can also be considered a CHP application by using solar thermal systems along with PV systems.

IBEW Local 103 Training School and Union Hall

Location: Dorchester, MA (42.3°N, 71.1°W)
Type of System: Utility-interactive hybrid
Peak Array Power: 5.3 kW DC
Date of Installation: December 2002
Installers: Apprentices and journeymen
Purposes: Training, community awareness, supplemental electrical power

As IBEW Local 103 in the Boston area has learned, PV and wind power complement each other quite well. Being located next to a busy expressway, the Local took the opportunity to showcase its members' technical expertise by installing first a PV system, and later a wind turbine, on its property.

The PV array is mounted on the curtain wall surrounding the air conditioning equipment on the roof of the building. The vertical orientation is not optimal for electricity production, but offers other advantages. The wall is an established surface (requiring little additional mounting structure) that is readily accessible and is easily visible from the ground and the nearby expressway. Since a primary reason for installing the system was to increase public awareness of renewable energy technologies, the array location is ideal. The air conditioning enclosure also protects the junction boxes, disconnects, and inverters. The resulting AC power is routed into the main service panel inside the building.

IBEW Local 103
A representative from Local 103 explains to electrical inspectors how the PV modules generate electricity and contribute to meeting the electrical demand of the building.

In May 2005, Local 103 completed installation of the first wind turbine in the city of Boston. The wind turbine stands over 114′ tall, with three 34′-long blades, and has a rated power output of 100 kW. The generator windings are excited by utility voltage, so the three-phase AC power output is automatically synchronized to utility power. The AC power is then routed to the main electrical service of the building. Since the output from the PV system and the wind turbine combine only at the AC power side at the building's main electrical service, this is an AC bus system.

Data from the main controller of the wind turbine is also routed into the building, where it is monitored and displayed. Statistics from the turbine are available on Local 103's web site.

IBEW Local 103
A Local 103 member assists in aligning the three 34′-long turbine blades on the generator section of the wind turbine.

IBEW Local 103
In addition to meeting a portion of the electrical demand of the building, the PV system and wind turbine showcase the Local's progressive outlook on alternative energy to the community.

Hybrid Systems

A *hybrid system* is a stand-alone system that includes two or more distributed energy sources. (Utility-interactive systems may also use multiple distributed energy sources, although these are not strictly defined as hybrid systems.) Common energy sources used in hybrid systems include PV arrays, engine generators, wind turbines, and microhydroelectric turbines. **See Figure 4-22**. Hybrid systems offer several advantages over PV-only or generator-only systems, including greater system reliability and flexibility in meeting variable loads.

Hybrid Systems

Figure 4-22. *Hybrid systems include power sources other than the PV array and do not interact with the utility grid.*

Refer to Quick Quiz® on CD-ROM

Using a variety of energy sources may also reduce the total system costs. For example, PV-only stand-alone systems are generally cost effective for smaller loads, while engine generator-only systems are usually cost effective for larger loads. In between, at about 20 kWh/day to 100 kWh/day, a hybrid PV system can have a lower initial cost than a PV-only system and a lower life cycle cost than a generator-only system.

Hybrid systems are perhaps the most complex of all PV systems in terms of equipment, system design, and installation. However, they offer the greatest flexibility with regard to configurations and control strategies. The contribution from each source can be optimized for the application while minimizing the initial and operating cost. For example, an array may be designed to power the loads only during the months with the highest insolation, with an engine generator supplying the balance of energy required at other times. The engine generator can be loaded to its maximum output for meeting system loads and charging a battery bank, thereby increasing its fuel efficiency. Also, PV arrays and battery banks can be down-sized in hybrid systems because auxiliary energy sources provide additional power.

Hybrid systems are classified by the way in which the power outputs from the various sources are integrated. A *DC bus hybrid system* is a hybrid system that combines DC power output from all energy sources, including the PV array, for charging the battery bank. The sources are either all DC or rectified to DC before integration. For example, a common DC bus hybrid system integrates a PV array with a small wind turbine.

In some circumstances, AC power sources can instead be integrated on the AC side if they can be configured for synchronized output. An *AC bus hybrid system* is a hybrid system that supplies loads with AC power from multiple energy sources. For example, some inverters can match the frequency of the output from an engine generator so that the two power sources will be compatible.

Many modern inverters are specifically designed to integrate and control multiple energy sources by including built-in battery chargers and automatic transfer switches. For example, when a battery bank is discharged, the inverter control relays may start an engine generator and automatically transfer the loads to the output from the generator. The array power is then used to charge the battery bank.

Summary

- The primary component common to all PV systems is the PV array.

- PV systems usually require means to store or condition PV power so that it can be used efficiently by loads.

- Energy storage systems balance energy production and demand.

- Batteries, particularly lead-acid types, are by far the most common means of energy storage in PV systems.

- Most PV systems require more battery capacity than can be supplied by a single battery, so batteries in PV systems are often connected together to form battery banks.

- Power conditioning components may be individual devices or may be combined into a single power conditioning unit.

- Inverters convert DC power from battery systems or arrays to AC power for AC loads or export to the utility grid.

- Charge controllers manage the charging of batteries from a DC power source, typically a PV array.

- Battery chargers are used when an AC power source, such as the utility grid or an engine generator, is available to provide supplemental battery charging.

- Since a PV array produces DC power, DC loads are used in many PV applications to avoid having to change the power to AC, simplifying the system.

- Balance-of-system (BOS) components are all of the remaining electrical and mechanical components needed to integrate and assemble the major components in a PV system.

- Besides the PV array, the electric utility grid is the most common source of electricity connected to PV systems.

- Other electrical systems may be interfaced with PV systems as additional sources of electricity, depending on the type of system and application requirements.

- Stand-alone PV systems are most popular for meeting small- to medium-sized electrical loads and are extensively used in remote off-grid areas or where extending the utility service is cost-prohibitive or impossible.

- In direct-coupled PV systems, the output of a PV module or array is directly connected to a DC load.

- A self-regulated PV system consists of only an array, battery, and load.

- Whenever loads are variable or uncontrolled, charge control is required to prevent damage to the battery from overcharge or overdischarge.

- Utility-interactive systems operate in parallel with and are connected to the electric utility grid.

- Bimodal systems can operate in either utility-interactive or stand-alone mode and use battery storage.

- Hybrid systems include an energy source other than an array and do not interact with the utility.

- Common energy sources used in hybrid systems include engine generators, wind turbines, and micro-hydroelectric generators.

Definitions

- A *battery bank* is a group of batteries connected together with series and parallel connections to provide a specific voltage and capacity.

- A *power conditioning unit (PCU)* is a device that includes more than one power conditioning function.

- An *inverter* is a device that converts DC power to AC power.

- A *charge controller* is a device that regulates battery charge by controlling the charging voltage and/or current from a DC power source, such as a PV array.

- A *rectifier* is a device that converts AC power to DC power.

- A *charger* is a device that combines a rectifier with filters, transformers, and other components to condition DC power for the purpose of battery charging.

- A *DC-DC converter* is a device that converts DC power from one voltage to another.

- A *maximum power point tracker (MPPT)* is a device or circuit that uses electronics to continually adjust the load on a PV device under changing temperature and irradiance conditions to keep it operating at its maximum power point.

- An *engine generator* is a combination of an internal combustion engine and a generator mounted together to produce electricity.

- A *generator* is a device that converts mechanical energy into electricity by means of electromagnetic induction.

- A *gas turbine* is a device that compresses and burns a fuel-air mixture, which expands and spins a turbine.

- A *turbine* is a bladed shaft that converts fluid flow into rotating mechanical energy.

- An *uninterruptible power supply (UPS)* is a battery-based system that includes all the additional power conditioning equipment, such as inverters and chargers, to make a complete, self-contained power source.

- A *wind turbine* is a device that harnesses wind power to produce electricity.

- A *micro-hydroelectric turbine* is a device that produces electricity from the flow and pressure of water.

- A *fuel cell* is an electrochemical device that uses hydrogen and oxygen to produce DC electricity, with water and heat as byproducts.

- An *electrolyzer* is an electrochemical device that uses electricity to split water into hydrogen and oxygen.

- A *stand-alone PV system* is a type of PV system that operates autonomously and supplies power to electrical loads independently of the electric utility.

- A *direct-coupled PV system* is a type of stand-alone system where the output of a PV module or array is directly connected to a DC load.

- A *self-regulating PV system* is a type of stand-alone system that uses no active control systems to protect the battery, except through careful design and component sizing.

- A *utility-interactive system* is a PV system that operates in parallel with and is connected to the electric utility grid.

- *Net metering* is a metering arrangement where any excess energy exported to the utility is subtracted from the amount of energy imported from it.

- *Dual metering* is the arrangement that measures energy exported to and imported from the utility grid separately.

- *Islanding* is the undesirable condition where a distributed-generation power source, such as a PV system, continues to transfer power to the utility grid during a utility outage.

- A *bimodal system* is a PV system that can operate in either utility-interactive or stand-alone mode and uses battery storage.

- A *hybrid system* is a stand-alone system that includes two or more distributed energy sources.

- A *DC bus hybrid system* is a hybrid system that combines DC power output from all energy sources, including the PV array, for charging the battery bank.

- An *AC bus hybrid system* is a hybrid system that supplies the loads with AC power from multiple energy sources.

Review Questions

1. Why is energy storage needed in most PV systems?

2. Besides energy storage, what advantages do battery systems provide?

3. What is the difference between an inverter and a power conditioning unit?

4. Explain the difference between a charge controller and a charger.

5. How does a maximum power point tracker maximize array output?

6. Which types of electrical energy sources are typically integrated with PV systems?

7. Why is system sizing critically important for stand-alone PV systems?

8. Compare the different types of stand-alone PV systems.

9. Explain the ways in which interactive inverters interface with the utility grid.

10. How can bimodal systems be used to lower electricity costs when utility rates vary by time of day?

11. Compare the advantages and disadvantages of hybrid systems.

5

Cells, Modules, and Arrays

PV systems use cells, modules, and arrays to capture sunlight and convert it into electrical energy. PV systems are modular in nature, meaning that basic building blocks of smaller components and subsystems are integrated to construct larger systems. Individual cells are combined to make modules. Modules are then combined to achieve the desired system voltage and power output, forming an array. The electrical characteristics, responses, and types of connections can all be scaled from cells to modules and arrays.

Chapter Objectives

- *Identify the relationships between PV cells, modules, and arrays.*
- *Describe the photovoltaic effect and the fundamental operation of PV devices.*
- *Understand the current-voltage (I-V) characteristics for PV devices and define the key I-V parameters.*
- *Understand how the electrical load, solar radiation, and operating temperatures affect the electrical output of a PV device.*
- *Translate the voltage, current, and power output of a PV device from a reference condition to another operating condition.*
- *Determine the electrical output of similar and dissimilar PV devices connected in series and in parallel.*
- *Understand the construction and features of PV modules.*
- *Describe the various performance rating conditions for PV modules.*

PHOTOVOLTAIC CELLS

A *photovoltaic cell* is a semiconductor device that converts solar radiation into direct current electricity. Because the source of radiation is usually the sun, they are often referred to as solar cells. Individual PV cells are the basic building blocks for modules, which are in turn the building blocks for arrays and complete PV systems. **See Figure 5-1.**

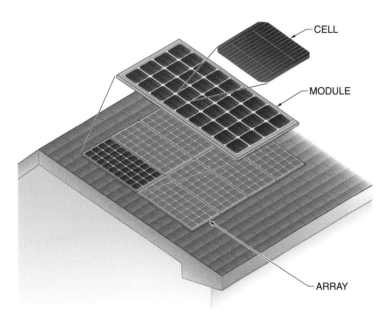

Figure 5-1. *The basic building blocks for PV systems include cells, modules, and arrays.*

The term "photovoltaic" is a combination of the Greek word "phos," meaning "light," and "voltage," which is named after the Italian physicist Alessandro Volta.

Semiconductors

PV cells are made from semiconductor materials. A *semiconductor* is a material that can exhibit properties of both an insulator and a conductor. Semiconductors behave like insulators at very low temperature and their conductivity increases with temperature. At normal temperatures, a semiconductor's electrical conductivity is between that of an insulator and a conductor. Some semiconductors also produce a voltage or exhibit a change in electrical conductivity when exposed to light.

Most PV cells use variations of silicon altered by doping to make them suitable semiconductors. *Doping* is the process of adding small amounts of impurity elements to semiconductors to alter their electrical properties. Pure crystalline silicon has four valence (outer) electrons that each bond with the outer electrons of other silicon atoms to form a crystalline structure. When a small amount of boron, which has three valence electrons, is added to silicon crystals, the boron atoms take the places of a few silicon atoms. The crystalline structure where boron bonds to silicon then has an electron void at the location where a fourth electron is absent. This void is also called a hole, since it can be filled by other electrons. This absence of a negative charge is considered a positive charge carrier. A *p-type semiconductor* is a semiconductor that has electron voids. **See Figure 5-2.**

Semiconductors

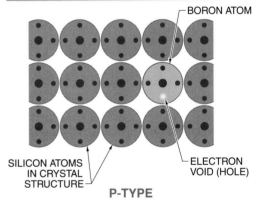

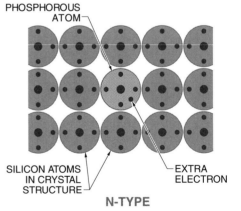

Figure 5-2. *Semiconductor materials with special electrical properties can be made by adding small amounts of other elements to silicon crystals.*

The addition of phosphorous, which has five valence electrons, to silicon results in an electron at the location where the phosphorous bonds to the silicon. This electron is only weakly bound to the phosphorous atom and can be easily induced to move through the material. It is considered a negative charge carrier. An *n-type semiconductor* is a semiconductor that has free electrons.

Photovoltaic Effect

The basic physical process by which a PV cell converts light into electricity is known as the photovoltaic effect. The *photovoltaic effect* is the movement of electrons within a material when it absorbs photons with energy above a certain level. A *photon* is a unit of electromagnetic radiation. Photons contain various amounts of energy depending on their wavelength, with higher energies associated with shorter wavelengths (higher frequency). Photons of light transfer their energy to electrons in the material surface. The extra electrons with enough energy to escape from their atoms are conducted as an electric current. Because of the electric result, the photovoltaic effect is also sometimes called the photoelectric effect. **See Figure 5-3.**

PV cells are wafers made of crystalline semiconductors covered with a grid of electrically-conductive metal traces.

A PV cell is a thin, flat wafer consisting of a p-n junction. A *p-n junction* is the boundary of adjacent layers of p-type and n-type semiconductor materials in contact with one another. When the p-n junction is illuminated, high-energy photons absorbed at the junction impart their energy to extra electrons in the material, moving the electrons to a higher energy state. The electrons gain potential energy and are in a position to do useful work before returning to a lower energy state.

Refer to media clip on CD-ROM

Photovoltaic Effect

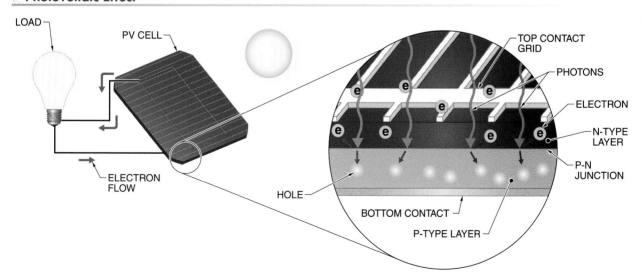

Figure 5-3. *The photovoltaic effect produces free electrons that must travel through conductors in order to recombine with electron voids, or "holes."*

Many of the photons reaching a PV cell have energies greater than the amount needed to excite the electrons into a conductive state. The extra energy imparts heat into the crystalline structure of the cell.

When these electrons are excited, they can move around to other atoms, leaving behind voids, or holes. The holes can act in a similar manner to the electrons, appearing to move when a neighboring electron moves to fill a hole, but they are associated with a positive charge. An electrical field produced by the p-n junction prevents the electrons and holes from immediately recombining, which would accomplish no work. The electrons are repelled from the p-type layer toward the top surface of the cell, and the holes are repelled away from the n-type layer toward the bottom surface. This creates a difference in electrical potential (voltage) between the top and bottom surfaces.

The free electrons are collected by metal contacts on the top surface of the cell and the holes migrate toward the bottom surface. For the electrons and holes to recombine, the electrons must travel from the top surface to the bottom surface. This is accomplished by connecting the surfaces with conductors and loads. The electrons flow through the loads, doing electrical work, then arrive at the back surface of the cell and recombine with the holes.

SolarWorld Industries America
Like other semiconductor devices, PV cells must be manufactured in extremely clean environments. This means that workers must wear special clothing to keep from contaminating the cells or the machines with dust or hair.

This process of electrons and holes being separated by photon energy, and doing work before recombining, occurs continuously while PV cells are exposed to light. There is no way to turn off a PV device, other than completely covering the top surface with an opaque material so that no light reaches the cells.

Cell Materials

PV cells can be produced from a variety of semiconductor materials, though crystalline silicon is by far the most common. The base raw material for silicon cell production is at least 99.99% pure polysilicon, a product refined from quartz and silica sands. Various grades of polysilicon, ranging from semiconductor to metallurgical grades, may be used in PV cell production and affect the quality and efficiency of cells produced.

Crystalline silicon (c-Si) cells currently offer the best ratio of performance to cost compared to competing materials and utilize many of the same raw materials and processes as the semiconductor industry. However, significant research is directed toward developing new PV cell material technologies, as well as improving efficiency and reducing costs of existing technologies. **See Figure 5-4.**

Gallium arsenide (GaAs) cells are more efficient than c-Si cells, but the high cost and toxicity of the GaAs materials have limited their use to space applications. GaAs can be alloyed with indium, phosphorus, and aluminum to create semiconductors that respond to different wavelengths of electromagnetic radiation. This property is utilized to make multijunction cells, producing highly efficient cells attractive for concentrating PV applications. A *multijunction cell* is a cell that maximizes efficiency by using layers of individual cells that each respond to different wavelengths of solar energy. The top layer captures the shortest wavelength radiation, while the longer wavelength components pass through and are absorbed by the lower layers.

Thin-film PV devices are module-based approaches to cell design. A *thin-film module* is a module-like PV device with its entire substrate coated in thin layers of semiconductor material using chemical vapor deposition techniques, and then laser-scribed to delineate individual cells and make electrical connections between cells. Amorphous silicon (a-Si), copper indium gallium selenide (CIGS), and cadmium telluride (CdTe) are among the competing thin-film technologies today. Thin-film modules are less costly to produce and use considerably less raw material than crystalline silicon modules, but most are less efficient than crystalline silicon and may not be as durable in the field.

A *photoelectrochemical cell* is a cell that relies on chemical processes to produce electricity from light, rather than using semiconductors. Photoelectrochemical cells include dye-sensitized cells (Grätzel cells) and polymer (plastic) cells, and are sometimes called organic cells. While the engineering challenges in developing these advanced cells are considerable, some are expected to impact commercial markets within the next decade.

Wafer Manufacturing

The manufacture of commercial silicon modules involves fabricating silicon wafers, transforming the wafers into cells, and assembling cells into modules. A *wafer* is a thin, flat disk or rectangle of base semiconductor material. Wafers are 180 μm to 350 μm thick and are made from p-type silicon. Crystalline silicon cell wafers are produced in three basic types: monocrystalline, polycrystalline, and ribbon silicon. Each type has advantages and disadvantages in terms of efficiency, manufacturing, and costs.

Monocrystalline Silicon. A *monocrystalline wafer* is a silicon wafer made from a single silicon crystal grown in the form of a cylindrical ingot. **See Figure 5-5.** Chunks of highly pure polysilicon are melted in a crucible, along with boron. A small seed crystal is dipped into the molten bath and slowly rotated and withdrawn. Over a period of many hours, the seed crystal grows into a large cylindrical crystal up to 40″ in length and 8″ in diameter. Because the ingot is round, the edges are often cropped to a more rectangular or square shape, which allows cells to be packed more closely in a module. Individual wafers are then cut from the ingot using diamond wire saws. Commercial monocrystalline cells have efficiencies on the order of 14% to 17%, with some laboratory samples having efficiencies as high as about 25%.

Refer to media clip on CD-ROM

SolarWorld Industries America
Robots are often used in the manufacture of PV cells because they perform repetitive tasks easily and do not pose a contamination risk.

PV Material Efficiencies*

MATERIAL	TYPICAL EFFICIENCIES	BEST LABORATORY EFFICIENCY
Multijunction gallium arsenide (GaAs)	33 to 38[†]	40.7[†]
Monocrystalline silicon	14 to 17	24.7
Polycrystalline silicon	11.5 to 14	20.3
Copper indium gallium selenide (CIGS)	9 to 11.5	19.9
Cadmium telluride (CdTe)	8 to 10	16.5
Amorphous silicon (a-Si)	5 to 9.5	12.1
Dye-sensitized (Grätzel)	4 to 5	11.1
Polymer (Organic)	1 to 2.5	5

* in %
[†] in concentrating applications

Source: NREL

Figure 5-4. *Various PV materials and technologies produce different efficiencies.*

Cleveland JATC

Location: Valley View, OH (41.4°N, 81.6°W)
Type of System: Stand-alone portable unit
Peak Array Power: 63 W DC
Date of Installation: April 2004
Installers: Apprentices and journeymen
Purposes: Training

While the installation of large rooftop PV systems provides an excellent opportunity to teach apprentices and journeymen about nearly all aspects of a PV system, typically only a small group of electricians is able to be fully involved in this type of installation process. For subsequent groups, often only demonstrations and walkthroughs of the established and operating PV system are possible, limiting opportunities for hands-on installation training.

Several training centers are experimenting with ways to design and implement repeatable PV systems that can be used to provide PV system installation experience to many classes of students. One method of facilitating training and demonstrations is to scale a complete system down to a portable, self-contained unit, such as the system at the Cleveland JATC training center. This unit is configured as a stand-alone PV system and includes a small battery bank.

There are several benefits to using small, portable PV systems for training. First, the system is self-contained, which allows it to be easily moved indoors or outdoors as needed. Also, with all the components mounted together and very short conductor runs, the system can be easily disassembled or modified for training purposes. Students can practice selecting appropriate conductors and overcurrent protection devices, making electrical connections, configuring electronic controls, and testing and troubleshooting the system. This type of system can also be augmented by the addition of a small wind turbine generator to add DC power to the inverter input circuit and boost the AC output.

The unit's wheels and array mounting allow students to experiment with array tilt and azimuth angles and module shading, observing first-hand how these variables affect power output. The system is configured to operate a variety of loads, such as a battery charge controller, lights, or power tools, or any other AC load connected to the receptacles. To demonstrate the operation of diversion loads, an attached water heater turns excess power into useful hot water.

The only significant drawback to the system is that it cannot provide useful electrical power to the training facility because it is so small and because, in order to maintain its portability, the system is not connected to the on-site electrical system. However, the experience gained from the portable system has encouraged the training center to investigate a future installation of a 5 kW to 6 kW utility-interactive system for the building. In addition to providing an appreciable amount of power to the building, the installation of this system would complement the portable training unit.

Cleveland JATC
The portable training unit includes all the components used in a typical stand-alone PV system, but is more versatile.

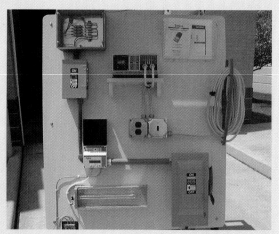

Cleveland JATC
The component configuration allows for easy testing, troubleshooting, and connection of a variety of load types.

Monocrystalline Ingots

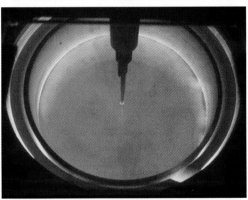

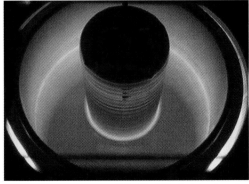

SolarWorld Industries America

Figure 5-5. *Monocrystalline silicon wafers are sawn from grown cylindrical ingots.*

Polycrystalline Silicon. A *polycrystalline wafer* is a silicon wafer made from a cast silicon ingot that is composed of many silicon crystals. **See Figure 5-6.** Molten silicon is poured into a crucible to form an ingot, which is slowly and carefully cooled over several hours. During cooling, many silicon crystals form and grow as the molten material solidifies. The cast ingot is then sectioned with wire saws to form square or rectangular wafers. Polycrystalline wafers can sometimes be distinguished from monocrystalline wafers by their square corners and the grain boundaries appearing on the wafer surface. While polycrystalline cells have slightly lower efficiencies (11.5% to 14%) than monocrystalline cells, their lower manufacturing costs and denser packing in modules makes them competitive with monocrystalline modules.

Polycrystalline Ingots

DOE/NREL, John Wohlgemuth—Solarex

Figure 5-6. *Polycrystalline silicon wafers are sawn from cast rectangular ingots.*

Ribbon Silicon. A *ribbon wafer* is a silicon wafer made by drawing a thin strip from a molten silicon mixture. The melted material is pulled between parallel dies where it cools and solidifies to form a continuous multicrystalline ribbon. The ribbon is then cut at specific intervals to form rectangular-shaped wafers. While cells produced from ribbon silicon wafers have slightly lower efficiencies (11% to 13%) than other silicon cells, this process is less expensive because there is less material waste and it does not require ingot sawing.

Cell Fabrication

Once a crystalline silicon wafer is produced, it must go through additional processing to become a functional PV cell. **See Figure 5-7.** First, the wafers are dipped in a sodium hydroxide solution to etch the surface and remove imperfections introduced during the sawing process. The textured surface increases surface area, allows subsequent coatings to adhere better, and minimizes reflected sunlight.

SolarWorld Industries America

Figure 5-8. *Diffusion of phosphorous gas creates a thin n-type semiconductor layer over the entire surface of a p-type wafer.*

Cell Fabrication

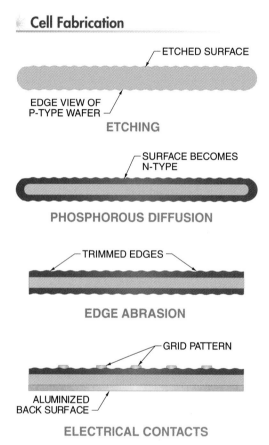

ETCHED SURFACE

EDGE VIEW OF
P-TYPE WAFER

ETCHING

SURFACE BECOMES
N-TYPE

PHOSPHOROUS DIFFUSION

TRIMMED EDGES

EDGE ABRASION

GRID PATTERN

ALUMINIZED
BACK SURFACE

ELECTRICAL CONTACTS

Figure 5-7. *Several steps are involved in turning silicon wafers into PV cells.*

After the wafers are cleaned they are placed on racks and into a diffusion furnace, where phosphorous gas penetrates the outer surfaces of the cell, creating a thin n-type semiconductor layer surrounding the original p-type semiconductor material. **See Figure 5-8.** The edge of the wafer is then abraded to remove the n-type material.

Antireflective coatings are then applied to the top surface of the cell to further reduce reflected sunlight and improve cell efficiency. After the coatings dry, grid patterns are screen printed on the top surface of the cell with silver paste to provide a point for electron collection and the electrical connection to other cells. These grid lines generally include two or more main strips across the cell, with finer lines emanating from the main strips across the cell surface. The configuration of these grid patterns is a critical part of cell design, because they must be of sufficient size and distribution to be able to efficiently collect and conduct current away from the cell, but must be minimized to avoid covering much of the cell surface, which lowers the effective cell surface area exposed to sunlight. Finally, the entire back surface of the cell is coated with a thin layer of metal, typically aluminum, which alloys with the silicon and neutralizes the n-type semiconductor layer on the back surface. This results in the bottom surface of the cell being the positive connection, while the top surface is negative. **See Figure 5-9.**

After cells are produced, each is electrically tested under simulated sunlight and sorted according to its current output. This sorting process largely eliminates problems with current mismatch among series-connected cells and allows manufacturers to produce modules that are of the same physical size but have different power ratings.

CURRENT-VOLTAGE (I-V) CURVES

The *current-voltage (I-V) characteristic* is the basic electrical output profile of a PV device. The I-V characteristic represents all possible current-voltage operating points (and power output) for a given PV device (cell, module, or array) at a specified condition of incident solar radiation and cell temperature.

When voltage is plotted against current for all the operating points, it forms a curve. An *I-V curve* is the graphic representation of all possible voltage and current operating points for a PV device at a specific operating condition. As voltage increases from zero, the current begins at its maximum and decreases gradually until the knee of the curve is reached. After the knee, small increases in voltage are associated with larger reductions in current, until the current reaches zero and the device is at maximum voltage. **See Figure 5-10.**

A PV device can operate anywhere along its I-V curve, depending on the electrical load. At any specific voltage on an I-V curve there is an associated current, and this operating point is known as an I-V pair. Since the product of current and voltage is power, each I-V pair also represents a specific power output, which changes from point to point along the I-V curve.

PV Cells

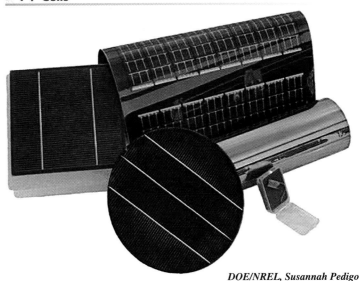

DOE/NREL, Susannah Pedigo

Figure 5-9. *The different materials, processes, and manufacturing steps produce a range of PV cell types.*

I-V Curve

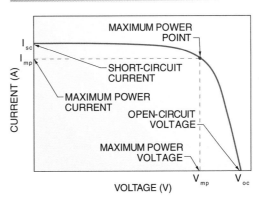

Figure 5-10. *An I-V curve illustrates the electrical output profile of a PV cell, module, or array at a specific operating condition.*

Certain points on an I-V curve are used to rate module performance and are the basis for the electrical design of arrays. The basic I-V curve parameters include the open-circuit voltage (V_{oc}), short-circuit current (I_{sc}), maximum power voltage (V_{mp}), maximum power current (I_{mp}), and maximum power (P_{mp}). It is important to note that the I-V curve changes with cell temperature and irradiance, and that I-V parameters have meaning only when these conditions are specified.

United Solar Ovonic LLC

Some thin-film modules are flexible and portable and are often used to charge batteries in small handheld devices such as cell phones.

Open-Circuit Voltage

The *open-circuit voltage (V_{oc})* is the maximum voltage on an I-V curve and is the operating point for a PV device under infinite load or open-circuit condition, and no current output. Since there is no current at the open-circuit voltage, the power output is also zero. The open-circuit voltage is used to determine maximum circuit voltages for modules and arrays. The open-circuit voltage of a PV device can be measured by exposing the device to sunlight and measuring across the output terminals with a voltmeter or a multimeter set to measure DC voltage. **See Figure 5-11.**

The open-circuit voltage of a PV device is determined by the semiconductor material properties and temperature. Increasing temperature reduces the open-circuit voltage for crystalline silicon. However, the open-circuit voltage is independent of cell area. A large cell can be sectioned into a number of smaller cells, and each will have the same open-circuit voltage. For individual crystalline silicon cells, the open-circuit voltage is typically 0.5 V to 0.6 V at 25°C (77°F). Some thin-film cells have an open-circuit voltage of 1.0 V or higher. Consequently, individual cells are connected in series to build higher, usable voltage levels in modules.

Open-Circuit Voltage Measurement

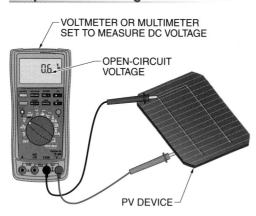

Figure 5-11. *Open-circuit voltage is easily measured with test instruments.*

PV Measurements, Inc.

Special equipment can be used to electronically load PV devices while measuring voltage and current in order to automatically generate I-V curves.

Short-Circuit Current

The *short-circuit current (I_{sc})* is the maximum current on an I-V curve and is the operating point for a PV device under no load or short-circuit condition, and no voltage output. Since voltage is zero at the short-circuit current, the power output is also zero. The short-circuit current of a PV device is used to determine maximum circuit design currents for modules and arrays, and is significantly affected by varying solar irradiance.

Temperature Conversion

As products that are manufactured and traded internationally, specifications for PV modules are often given in metric units, such as degrees Celsius for temperature. Many formulas involving I-V curve parameters also must be calculated with temperature in units of degrees Celsius. Temperature values are easily converted between Celsius and Fahrenheit degrees with the following formulas:

$$F = \left(\frac{9}{5} \times C\right) + 32$$

$$C = \frac{5}{9} \times (F - 32)$$

where

F = temperature (in °F)

C = temperature (in °C)

The short-circuit current can be measured by exposing the device to sunlight and measuring current with an ammeter or multimeter. The measuring procedure depends on the actual current and the type of meter. If the short-circuit current is less than the fused current rating of the meter (typically 1 A or 10 A), the test leads can be connected to the output terminals. The meter short-circuits the PV device with a very small resistance and measures the resulting current. If the current is expected to be close to or higher than the meter rating, this in-line method should not be used. Instead, a conductor with a switch is used to short-circuit the output terminals and a clamp-on ammeter is put around the conductor to measure the resulting current. **See Figure 5-12.**

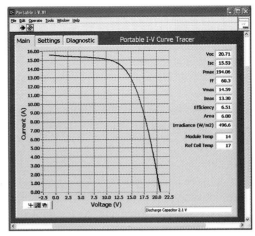

PV Measurements, Inc.

Special test instruments and software can automatically generate I-V curves, calculate key parameters, and produce reports.

Short-Circuit Current Measurement

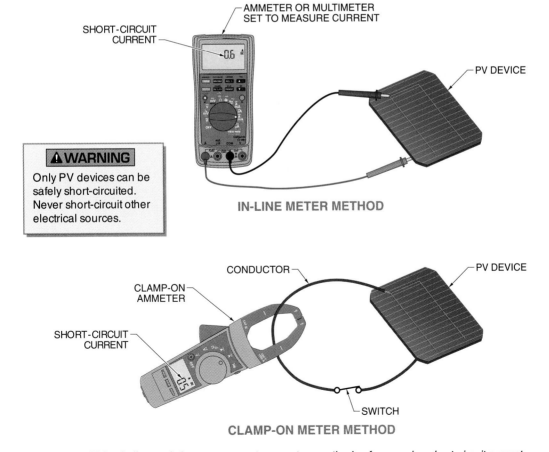

WARNING
Only PV devices can be safely short-circuited. Never short-circuit other electrical sources.

Figure 5-12. *Using in-line and clamp-on ammeters are two methods of measuring short-circuit current.*

Short-circuiting most current sources, such as an AC wall outlet or a battery, is extremely dangerous and can cause immediate damage to components if there is no overcurrent protection. Since there is essentially no resistance, the current flow is very high and can cause a potentially fatal electric shock and extremely high temperatures that may cause electrical burns or fire. Furthermore, short-circuiting a battery can cause it to explode, scattering acid and other toxic materials onto nearby people or components.

PV devices, however, can be safely short-circuited because they are inherently current-limited. In fact, some types of battery charge controllers short-circuit PV arrays as a means of controlling the charging current into battery systems.

A short-circuited PV device will flow current only up to certain point, because the electrons making up the current are not free to flow unless they are released by photons. The number of photons striking the PV device is finite, and only some of those photons transfer enough energy to free an electron. Therefore, the current flow cannot exceed the supply of free electrons. Only greater irradiance can increase the short-circuit current, and that is feasible only up to a point.

The current of a PV device is directly proportional to surface area and solar irradiance. In other words, for a given device, doubling the surface area exposed to solar radiation will double the current output. Likewise, doubling solar irradiance on the device surface will double current.

Maximum Power Point

The operating point at which a PV device produces its maximum power output lies between the open-circuit and short-circuit condition, when the device is electrically loaded at some finite resistance. The *maximum power point* (P_{mp}) is the operating point on an I-V curve where the product of current and voltage is at maximum. A variation of the I-V curve plots power against voltage, which clearly shows the maximum power point. **See Figure 5-13.** Maximum power is often called peak power and the parameter may be designated by W_p for "peak watts."

The I-V pair at the maximum power point is composed of the maximum power voltage and the maximum power current. The *maximum power voltage* (V_{mp}) is the operating voltage on an I-V curve where the power output is at maximum. The *maximum power current* (I_{mp}) is the operating current on an I-V curve where the power output is at maximum. Maximum power is calculated using the following formula:

$$P_{mp} = V_{mp} \times I_{mp}$$

where

P_{mp} = maximum power (in W)

V_{mp} = maximum power voltage (in V)

I_{mp} = maximum power current (in A)

For example, what is the maximum power of a PV module with a maximum power voltage of 23.5 V and a maximum power current of 7.1 A?

$$P_{mp} = V_{mp} \times I_{mp}$$
$$P_{mp} = 23.5 \times 7.1$$
$$P_{mp} = \textbf{166.8 W}$$

I-V Curve with Power

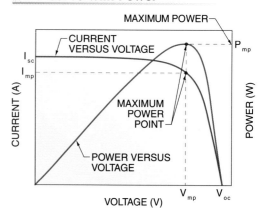

Figure 5-13. *A power versus voltage curve clearly shows the maximum power point.*

The maximum power point is located on the knee of the I-V curve and is the highest efficiency operating point for a PV device for the given conditions of solar irradiance and cell temperature. Due to the shape of the curve, maximum power voltage is typically about 70% to 80% of the value of the open-circuit voltage, while maximum power current is typically about 90% of the value of the short-circuit current. Maximum power voltage and current can be measured only while the PV device is connected to a load that operates the device at maximum power. Alternatively, the maximum power voltage and current can be determined from I-V curve data by multiplying each I-V pair and determining which pair results in the maximum power.

Since horizontal and vertical lines drawn from any current and voltage point to its corresponding axes form a rectangle, and the area of the rectangle equals power, the rectangle with the largest area graphically represents the maximum power point.

The maximum power point is also used to rate the performance of PV devices under specific conditions of solar irradiance and cell temperature.

Fill Factor. *Fill factor (FF)* is the ratio of maximum power to the product of the open-circuit voltage and short-circuit current. Fill factor represents the performance quality of a PV device and the shape of the I-V curve. A higher fill factor indicates that the voltage and current at the maximum power point are closer to the open-circuit voltage and short-circuit current, respectively, producing a more rectangular-shaped I-V curve. **See Figure 5-14.** Fill factor is expressed as a percentage and is calculated with the following formula:

$$FF = \frac{P_{mp}}{V_{oc} \times I_{sc}}$$

where

FF = fill factor

P_{mp} = maximum power (in W)

V_{oc} = open-circuit voltage (in V)

I_{sc} = short-circuit current (in A)

Fill Factor

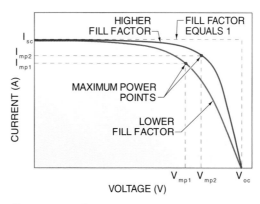

Figure 5-14. *Fill factor represents the shape of an I-V curve.*

For example, what is the fill factor of a PV cell with a maximum power of 3.0 W, an open-circuit voltage of 0.6 V and a short-circuit current of 7.0 A?

$$FF = \frac{P_{mp}}{V_{oc} \times I_{sc}}$$

$$FF = \frac{3.0}{0.6 \times 7.0}$$

$$FF = \frac{3.0}{4.2}$$

$$FF = \textbf{0.714} \text{ or } \textbf{71.4\%}$$

Most commercial crystalline silicon PV cells have fill factors exceeding 70%, while the fill factor for many thin-film materials is somewhat less. For a higher fill factor cell, the current decreases much less with increasing voltage up to the maximum power point, and decreases much more with increasing voltage beyond maximum power. A decrease in fill factor over time indicates problems with PV devices, including degradation of the cells or, more commonly, increased resistance of the wiring or connections in the system.

Efficiency. *Efficiency* is the ratio of power output to power input. **See Figure 5-15.** The efficiency of PV devices compares the solar power input to the electrical power output. Solar irradiance is multiplied by the area of the PV device to determine watts of solar power, which can then be directly compared to watts of electrical power. PV cell efficiencies vary considerably among different PV technologies, and for the same material and technology, efficiencies vary widely between laboratory samples and commercial devices. Efficiency is expressed as a percentage and is calculated with the following formula:

$$\eta = \frac{P_{mp}}{E \times A}$$

where

η = efficiency

P_{mp} = maximum power (in W)

E = solar irradiance (in W/m²)

A = area (in m²)

Efficiency

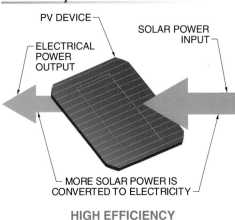

HIGH EFFICIENCY

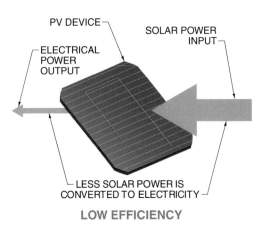

LOW EFFICIENCY

Figure 5-15. *Efficiency is a measure of how effectively a PV device converts solar power to electrical power.*

For example, what is the efficiency of a PV module with a surface area of 1.2 m² and a maximum power output of 160 W when exposed to 1000 W/m² solar irradiance?

$$\eta = \frac{P_{mp}}{E \times A}$$

$$\eta = \frac{160}{1000 \times 1.2}$$

$$\eta = \frac{160}{1200}$$

$$\eta = \textbf{0.133} \text{ or } \textbf{13.3\%}$$

Cells with higher efficiencies require less surface area to produce each watt of power, which saves some costs for raw materials, mounting structures, and other equipment. However, higher efficiency modules are generally no less expensive than less efficient ones, because the price for modules is generally based on the maximum power rating and not on the size.

For modules, efficiencies are often based on the entire module laminate area including the frame, and spacing between individual cells in the module. For individual cells, there is none of this extra area to affect the efficiency. This is one reason why module efficiencies are lower than their associated best cell efficiencies.

Operating Point

The operating point on an I-V curve is determined by the electrical load of the system. For example, if a battery is connected to a PV module, the battery voltage sets the operating voltage of the module. It also establishes the operating current that flows between the device and battery. If an incandescent lamp or DC motor is connected to a PV device, the effective resistance of the lamp filament or motor determines the operating point.

Short-circuit current is associated with zero load resistance and open-circuit voltage is associated with infinite load resistance. Every point in between the two states has a specific load resistance that increases from left to right along the I-V curve.

PV cells operate most efficiently at their maximum power points. However, the maximum power point is constantly changing due to changes in solar irradiance and cell temperature. Consequently, some systems use maximum power point tracking (MPPT) to dynamically match the electrical loads to PV output in order to maximize the performance. This function is included in most interactive inverters and some battery charge controllers.

The electrical load resistance required to operate a PV device at any point can be calculated using Ohm's law. For the maximum power point, the formula is:

$$R_{mp} = \frac{V_{mp}}{I_{mp}}$$

where

R_{mp} = resistance at maximum power point (in Ω)

V_{mp} = maximum power voltage (in V)

I_{mp} = maximum power current (in A)

For example, a module has maximum power voltage of 15 V and maximum power current of 3 A. What resistance is required to operate the module at the maximum power point, and what is its maximum power?

$$R_{mp} = \frac{V_{mp}}{I_{mp}}$$

$$R_{mp} = \frac{15}{3}$$

$$R_{mp} = \mathbf{5\,\Omega}$$

$$P_{mp} = V_{mp} \times I_{mp}$$

$$P_{mp} = 15 \times 3$$

$$P_{mp} = \mathbf{45\,W}$$

System Resistance

A PV device can be modeled by a current source in parallel with a diode, with resistance in series and shunt (parallel). **See Figure 5-16.** Both series and shunt resistances have a strong effect on the shape of the I-V curve.

Series resistance in PV devices includes the resistance of a cell, its electrical contacts, module interconnections, and system wiring. These resistances are in addition to the resistance of the electrical load. Some amount of series resistance in a PV system is unavoidable

because all conductors and connectors have some resistance. However, increasing series resistance over time can indicate problems with electrical connections or cell degradation. Series resistance reduces the voltage over the entire I-V curve. **See Figure 5-17.** Increasing series resistance also decreases maximum power, fill factor, and efficiency. If a PV device is operated at constant voltage (such as for battery charging), increasing series resistance results in decreasing operating current.

Schematic Symbols

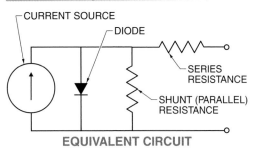

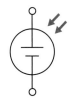

PV CELL SYMBOL

Figure 5-16. *A PV device can be modeled by a current source in parallel with a diode, with resistance in series and parallel.*

Series Resistance

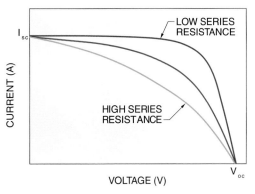

Figure 5-17. *Increasing series resistance in a PV system flattens the knee in the I-V curve, reducing maximum power, fill factor, and efficiency.*

Shunt (parallel) resistance accounts for leakage currents within a cell, module, or array. Shunt resistance has an effect on an I-V curve opposite to the effect of series resistance. Decreasing shunt resistance reduces fill factor and efficiency, and lowers maximum voltage, current, and power, but does not affect short-circuit current. **See Figure 5-18.** Decreasing shunt resistance over time can indicate short-circuits between cell circuits and module frames, or ground faults within an array.

Shunt (Parallel) Resistance

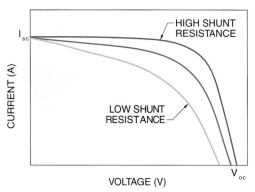

Figure 5-18. *Decreasing shunt resistance reduces fill factor and efficiency and lowers maximum voltage, current, and power, but it does not affect short-circuit current.*

A variable resistive load, such as a rheostat or adjustable resistor, can be used to load a PV device over nearly its entire I-V curve. When combined with meters measuring voltage and current, this method can be used to generate the I-V curves of small PV devices or individual modules, which can then be used to identify the key I-V curve parameters.

An ideal PV device would have no series resistance and infinite shunt resistance, producing a rectangular I-V curve with a fill factor of 100%. In reality, however, both have a finite value and only series resistance can be practically managed. This is why it is important to minimize PV system series resistance as much as possible, especially where long wiring distances are involved.

PV DEVICE RESPONSE

The shape of an I-V curve is determined by the PV device and system properties. However, the magnitude and position of the curve are determined by solar irradiance and the temperature of the PV device. Varying irradiance and temperature produces different voltage-current operating points, which generate different I-V curves for the same PV device. Therefore, a given I-V curve represents only one set of operating conditions for a PV device, at a specified exposure to irradiance and temperature.

Solar Irradiance Response

Changes in solar irradiance have a small effect on voltage but a significant effect on the current output of PV devices. The current of a PV device increases proportionally with increasing solar irradiance. Consequently, since the voltage remains nearly the same, the power also increases proportionally. **See Figure 5-19.** The relationships between irradiance, current, and power can be expressed by the following ratios:

$$\frac{E_2}{E_1} = \frac{I_2}{I_1} = \frac{P_2}{P_1}$$

where

E_2 = solar irradiance 2 (in W/m^2)

E_1 = solar irradiance 1 (in W/m^2)

I_2 = current at irradiance 2 (in A)

I_1 = current at irradiance 1 (in A)

P_2 = power at irradiance 2 (in W)

P_1 = power at irradiance 1 (in W)

This relationship is used to estimate how short-circuit current, maximum power current, or maximum power changes with a change in irradiance:

$$I_{sc2} = I_{sc1} \times \frac{E_2}{E_1}$$

$$I_{mp2} = I_{mp1} \times \frac{E_2}{E_1}$$

$$P_{mp2} = P_{mp1} \times \frac{E_2}{E_1}$$

Solar Irradiance Response

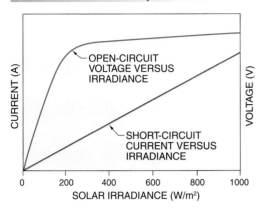

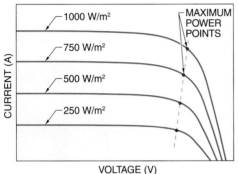

Figure 5-19. *Voltage increases rapidly up to about 200 W/m², and then is almost constant. Current and maximum power increase proportionally with irradiance.*

For example, a PV module has a rated short-circuit current of 6 A and a maximum power of 150 W at 1000 W/m² of solar irradiance. What is the short-circuit current and the maximum power at 600 W/m² of irradiance?

$$I_{sc2} = I_{sc1} \times \frac{E_2}{E_1}$$

$$I_{sc2} = 6 \times \frac{600}{1000}$$

$$I_{sc2} = 6 \times 0.6$$

$$I_{sc2} = \textbf{3.6 A}$$

$$P_{mp2} = P_{mp1} \times \frac{E_2}{E_1}$$

$$P_{mp2} = 150 \times \frac{600}{1000}$$

$$P_{mp2} = 150 \times 0.6$$

$$P_{mp2} = \textbf{90 W}$$

A *family of I-V curves* is a group of I-V curves at various irradiance levels. A family of I-V curves represents the changing output of a PV device with a single changing variable, such as irradiance throughout the day, with others held constant. The family of I-V curves also shows that the maximum power points at various irradiance levels do not follow a line of constant load resistance, which would have to intersect with the origin. Therefore, a resistive load matched to the maximum power point at one irradiance level will not operate the device at maximum power at higher or lower irradiance.

Refer to media clip on CD-ROM

Temperature Response

For most types of PV devices, high operating temperatures significantly reduce voltage output. Current increases with temperature, but only slightly, so the net result is a decrease in power and efficiency. **See Figure 5-20.** Long-term high temperatures can also lead to premature degradation of cells and module encapsulation. For these reasons, it is desirable to install modules and arrays in a manner that allows them to operate as cool as possible.

Temperature Response

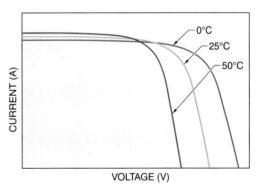

Figure 5-20. *Increasing cell temperature decreases voltage, slightly increases current, and results in a net decrease in power.*

Cell Temperature. The cell temperature of a PV device refers to the internal temperature at the p-n junction. Cell temperature is influenced by ambient temperature, wind speed, solar irradiance, thermal characteristics of the device's packaging, and the way the cell or module is installed or mounted. A cell temperature value is used when calculating how voltage or power are affected by temperature. Cell temperature can be estimated by either directly measuring the cell or module surface temperature or applying the temperature-rise coefficient.

The *temperature-rise coefficient* is a coefficient for estimating the rise in cell temperature above ambient temperature due to solar irradiance. When the temperature-rise coefficient has been established for a particular installation, cell temperature under various conditions can be estimated with the following formula:

$$T_{cell} = T_{amb} + (C_{T\text{-}rise} \times E)$$

where

T_{cell} = cell temperature (in °C)

T_{amb} = ambient temperature (in °C)

$C_{T\text{-}rise}$ = temperature-rise coefficient (in °C/kW/m²)

E = solar irradiance (in kW/m²)

The mounting configuration of arrays has a strong influence on the temperature-rise coefficient. For modules with the back surfaces exposed to wind, the temperature-rise coefficient may be about 15 to 20°C/kW/m² (27 to 36°F/kW/m²). For modules mounted close to or directly on a surface, with the back surface not exposed to wind, the temperature-rise coefficient may be as high as 25 to 30°C/kW/m² (45 to 54°F/kW/m²).A temperature rise of at 20°C to 25°C (36°F to 45°F) over ambient temperature is typical for most array installations at peak sun.

For example, a rooftop-mounted array has a temperature-rise coefficient of 20°C/kW/m². If the ambient temperature is 30°C and the solar irradiance is 1100 W/m² (1.1 kW/m²), what is the cell temperature?

$$T_{cell} = T_{amb} + (C_{T\text{-}rise} \times E)$$
$$T_{cell} = 30 + (20 \times 1.1)$$
$$T_{cell} = 30 + 22$$
$$T_{cell} = \mathbf{52°C}$$

Temperature-Rise Coefficient

With so many factors influencing the magnitude of the effect of irradiance on cell temperature, it is difficult to absolutely determine the temperature-rise coefficient. The coefficient is usually established through field measurements. Actual cell temperature, ambient temperature, and solar irradiance are measured for a particular installation and these values are used in the following formula to calculate the temperature-rise coefficient:

$$C_{T-rise} = \frac{T_{cell} - T_{amb}}{E}$$

The temperature-rise coefficient may be calculated many times under various conditions and averaged to determine an approximate value. This temperature-rise coefficient can then to be used to predict future cell temperature, though this value will be a good estimate only during typical weather conditions. Unusual weather variations, such as high winds, reduce the accuracy of these calculations.

Temperature Coefficients. A *temperature coefficient* is the rate of change in voltage, current, or power output from a PV device due to changing cell temperature. A negative coefficient means the parameter decreases with increasing cell temperature, while a positive coefficient means the parameter increases with increasing cell temperature. Temperature coefficients can be expressed as a unit (absolute) change per degree of temperature change or a percentage (relative) change per degree of temperature change.

Unit change temperature coefficients are generally specific to a PV device manufacturer, and may be based on an individual cell or an entire module. Typical temperature coefficient values for crystalline silicon cells are −2.25 mV/°C/cell (−0.00225 V/°C/cell) and 3.70 μA/°C/cm² (0.0000037 A/°C/cm²).

These temperature coefficients may be specified for certain I-V parameters. For example, a temperature coefficient for open-circuit voltage cannot accurately predict the effects of temperature at other voltages. Module or array coefficients are calculated from cell coefficients with the following formulas:

$$C_V = C_{V\text{-}cell} \times n_s$$
$$C_I = C_{I\text{-}cell} \times n_p \times A$$

where

C_V = module or array absolute temperature coefficient for voltage (in V/°C)

$C_{V\text{-}cell}$ = cell absolute temperature coefficient for voltage (in V/°C/cell)

n_s = number of series-connected cells

C_I = module or array absolute temperature coefficient for current (in A/°C)

$C_{I\text{-}cell}$ = cell absolute temperature coefficient for current (in A/°C/cm²)

n_p = number of parallel-connected cell strings

A = individual cell area (in cm²)

For example, what are the module temperature coefficients for a module with 36 cells (each 144 cm² in area) arranged in two parallel strings of 18 series-connected cells each?

$$C_V = C_{V\text{-}cell} \times n_s$$
$$C_V = -0.00225 \times 18$$
$$C_V = \textbf{-0.0405 V/°C}$$

$$C_I = C_{I\text{-}cell} \times n_p \times A$$
$$C_I = 0.0000037 \times 2 \times 144$$
$$C_I = \textbf{0.00011 A/°C}$$

Percentage change temperature coefficients for crystalline silicon cells are relatively constant among manufacturers or cell fabrication processes and are approximately equal to –0.4%/°C (–0.004/°C) for voltage, 0.1%/°C (0.001/°C) for current, and –0.5%/°C (–0.005/°C) for power. Unlike most unit change coefficients, percentage change coefficients can be used with any point on the I-V curve. The percentages are applied to reference (or rated) values to determine the unit change temperature coefficients using the following formulas:

$$C_V = V_{ref} \times C_{\%V}$$
$$C_I = I_{ref} \times C_{\%I}$$
$$C_P = P_{ref} \times C_{\%P}$$

where

C_V = absolute temperature coefficient for voltage (in V/°C)

V_{ref} = reference (or rated) voltage (in V)

$C_{\%V}$ = relative temperature coefficient for voltage (in /°C)

C_I = absolute temperature coefficient for current (in A/°C)

I_{ref} = reference (or rated) current (in A)

$C_{\%I}$ = relative temperature coefficient for current (in /°C)

C_P = absolute temperature coefficient for power (in W/°C)

P_{ref} = reference (or rated) power (in W)

$C_{\%P}$ = relative temperature coefficient for power (in /°C)

For example, if the open-circuit voltage for a PV module is 21.6 V at 25°C (77°F), what is the unit change temperature coefficient for voltage?

$$C_V = V_{ref} \times C_{\%V}$$
$$C_V = 21.6 \times -0.004$$
$$C_V = \textbf{-0.086 V/°C}$$

DOE/NREL, Warren Gretz
Arrays installed in hot climates, such as deserts, are likely to produce less power than their rating because of high cell temperatures.

Temperature Translations. Temperature translations use temperature coefficients to estimate the change in voltage, current, and power parameters due to cell temperature. Translations determine the potential range of operating voltages and power levels, which must be within certain limits for code compliance and connection to power processing equipment downstream, including batteries, charge controllers, and inverters.

> The small increases in voltage due to greater irradiance are often cancelled by small decreases in voltage due to the associated higher cell temperatures. Therefore, voltage changes due to irradiance are not considered significant for most applications. Changes in cell temperature that are associated with changes in ambient temperatures, however, can have significant effects on voltage.

Temperature translations can be applied across the entire I-V curve, though they are usually applied to open-circuit voltage, maximum power voltage, and maximum power, using reference values at 25°C (77°F). Because the cell temperature effect on current is very small, it is often ignored in field measurements. Voltage and power translations can be calculated using the following formulas:

$$V_{trans} = V_{ref} + (\,[T_{cell} - T_{ref}] \times C_V)$$
$$P_{trans} = P_{ref} + (\,[T_{cell} - T_{ref}] \times C_P)$$

where

V_{trans} = translated voltage at cell temperature (in V)

V_{ref} = reference (or rated) voltage corresponding to T_{ref} (in V)

T_{cell} = cell temperature (in °C)

T_{ref} = reference (or rated) temperature (in °C)

C_V = absolute temperature coefficient of voltage (in V/°C)

P_{trans} = translated power at cell temperature (in W)

P_{ref} = reference (or rated) power corresponding to T_{ref} (in W)

C_P = absolute temperature coefficient of power (in W/°C)

For example, a PV module with 36 series-connected cells has a rated open-circuit voltage of 21.7 V at 25°C. The manufacturer provides a temperature coefficient for open-circuit voltage for individual cells of –2.1 mV/°C/cell (–0.0021 V/°C/cell). The ambient temperature is 30°C, the solar irradiance is 800 W/m² (0.8 kW/m²), and temperature-rise coefficient is 25°C/kW/m². What is the expected open-circuit voltage under these conditions?

First, using ambient temperature and solar irradiance information, estimate the cell temperature using the temperature-rise coefficient formula:

$$T_{cell} = T_{amb} + (C_{T\text{-}rise} \times E)$$
$$T_{cell} = 30 + (25 \times 0.8)$$
$$T_{cell} = \mathbf{50°C}$$

Next, determine the temperature coefficient of open-circuit voltage for the entire module.

$$C_V = C_{V\text{-}cell} \times n_s$$
$$C_V = -0.0021 \times 36$$
$$C_V = \mathbf{-0.076\ V/°C}$$

Finally, the voltage translation formula is used to determine the translated open-circuit voltage:

$$V_{trans} = V_{ref} + (\,[T_{cell} - T_{ref}] \times C_V)$$
$$V_{trans} = 21.7 + (\,[50 - 25] \times -0.076)$$
$$V_{trans} = 21.7 + (25 \times -0.076)$$
$$V_{trans} = 21.7 + (-1.9)$$
$$V_{trans} = \mathbf{19.8\ V}$$

Therefore, at operating temperatures 25°C (45°F) higher than the rated temperature, the voltage is reduced by about 10%. Power is similarly affected. Consequently, higher-than-rated temperatures are one of the primary factors causing arrays to operate at less than their rated power in the field.

MODULES AND ARRAYS

Although a PV cell is the basic device that converts sunlight into electricity, it is not practically useful in this form. Individual cells produce very small amounts of power, so many must be connected together to create appreciable amounts of power. Also, PV cells are delicate and degrade rapidly if exposed to dirt or moisture. Individual cells must be protected from the environment, electrically insulated to protect installers and operators against electrical shock, and packaged in a manner that allows them to be mechanically installed to a support structure and connected to the rest of the electrical system.

A *module* is a PV device consisting of a number of individual cells connected electrically, laminated, encapsulated, and packaged into a frame. **See Figure 5-21.** The PV cells are laminated within a polymer (plastic) substrate to hold them in place and to protect the electrical connections between cells. The cell laminates are then encapsulated (sealed) between a rigid backing material and a glass cover. Some thin-film laminates use flexible materials such as aluminum or stainless steel substrate and polymer encapsulation instead of a glass cover.

Module Construction

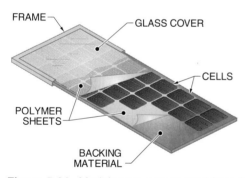

Figure 5-21. *Modules are constructed from PV cells that are encapsulated by several layers of protective materials.*

An *array* is a complete PV power-generating unit consisting of a number of individual electrically and mechanically integrated modules with structural supports, trackers, or other components. **See Figure 5-22.** The term "panel" is also used in relation to modules and arrays. Sometimes panel is used as an alternate term for a module. More commonly, the term

panel refers to an assembly of two or more modules that are mechanically and electrically integrated into a unit for ease of installation in the field. **See Figure 5-23.**

Arrays

DOE/NREL, Craig Miller Productions

Figure 5-22. *An array is a group of PV modules integrated as a single power-generating unit.*

SolarWorld Industries America

Figure 5-23. *Several modules may be connected together to form a panel, which is installed as a preassembled unit.*

Module Assembly

Once crystalline silicon cells are fabricated, tested, and sorted, they are ready for assembly into complete modules. First, the cells are laid out in the specified physical and electrical configuration and soldered together with tin- or silver-coated copper strips. The strips are flexible enough to allow slight movement due to thermal expansion and minor stresses. The conductors for circuit connections are routed to one area on the back of the cell circuit where connections can be made to other modules or system components with external conductors.

Next, the completed cell circuit is placed between thin, clear polymer sheets, with a sheet of tempered glass on top and the back surface material beneath, and these layers are laminated in a press at high temperatures. The intermodule connectors are attached to the back of the module, and the entire laminate is framed with aluminum channel. The frame provides mechanical support, structural mounting features, and electrical grounding for the modules. A final visual inspection and electrical test are performed on all modules before packaging for shipment.

All modules include some means for making intermodule electrical connections, through the use of either pre-wired connectors or a junction box. The junction box may also include bypass diodes and the ability to change the series or parallel configuration of the module cells with certain jumper arrangements. **See Figure 5-24.** For example, a module might be changed from 36 series-connected cells to two parallel strings of 18 series-connected cells. This doubles the current and halves the voltage, but the power output remains the same.

Electrical Connections

PV devices are generally first connected in series to achieve a desired voltage, forming a string. These series strings are then connected in parallel to build current and power. Cells are connected to form modules and modules are connected to form arrays. The same principles are used to build successively larger systems.

Series Connections. Individual cells are connected in series by soldering thin metal strips from the top surface (negative terminal) of one cell to the back surface (positive terminal) of the next. Modules are connected in series with other modules by connecting conductors between the negative terminal of one module to the positive terminal of another module. When individual devices are electrically connected in series, the positive connection of the whole circuit is made at the device on one end of the string and the negative connection is made at the device on the opposite end. **See Figure 5-25.**

Junction Box

Figure 5-24. *A junction box on the back of a module provides a protected location for electrical connections and bypass diodes.*

Only PV devices having the same current output should be connected in series. When similar devices are connected in series, the voltage output of the entire string is the sum of the voltages of the individual devices, while the current output for the entire string remains the same as for a single device. Correspondingly, the I-V curve for a string of similar PV devices is the sum of the I-V curves of the individual devices. **See Figure 5-26.**

When PV devices with dissimilar current outputs are connected in series, the devices with lower current output absorb current from the devices with higher current output. This results in a loss of power, and potential overheating and damage to the lower-current output devices. The current output for a circuit of dissimilar devices in series is limited to the current of the lowest-current output device in the entire string.

However, PV devices with different voltage outputs can be connected in series without loss of power as long as each device has the same current output. As with similar devices, the current output remains the same and the voltage output of the circuit equals the sum of the voltages of the individual devices.

Series Connections

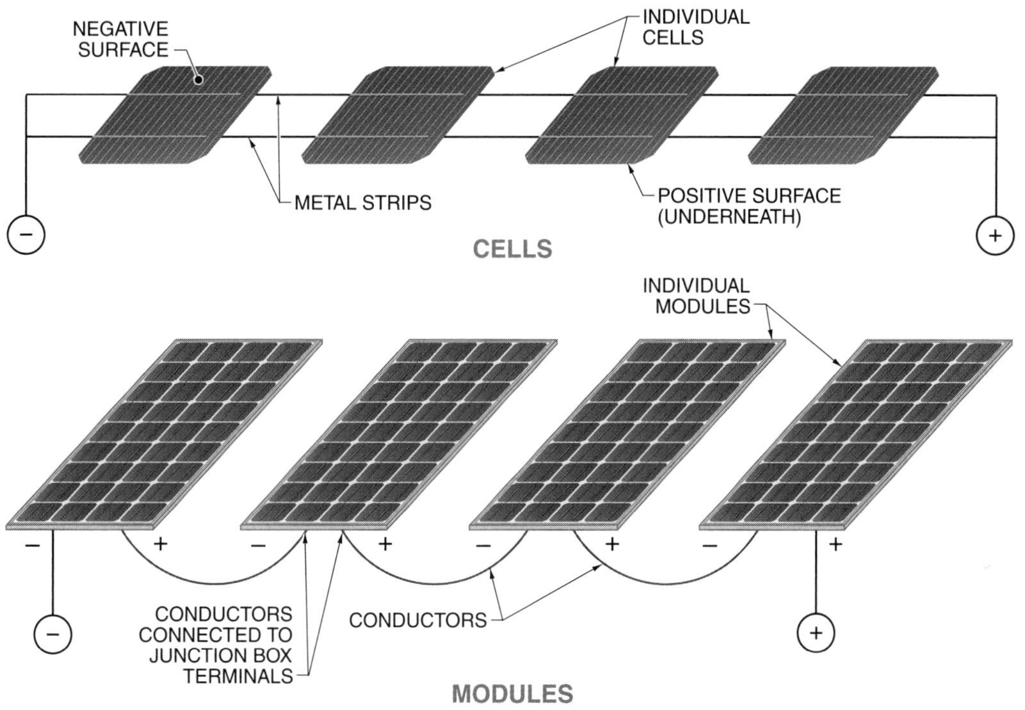

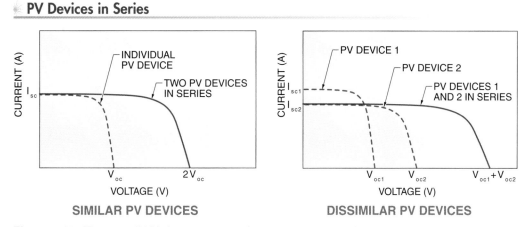

Figure 5-25. *PV cells or modules are connected in series strings to build voltage.*

PV Devices in Series

Figure 5-26. *The overall I-V characteristics of a series string are dependent on the similarity of the current outputs of the individual PV devices.*

Connecting devices with different voltages may be done when the load requires a nonstandard voltage. For example, a circuit with signal lamps designed to operate from a nominal 30 V battery may include two standard 36-cell (12 V) modules connected in series with a third module having only 18 cells (6 V), as long as they have similar current outputs.

The maximum number of modules that may be connected in a series string is limited by the maximum system voltage rating of the modules and other components. Most modules are rated for a maximum system voltage of 600 V.

Parallel Connections. Parallel connections are not generally used for individual PV devices, especially cells, but for series strings of cells and modules. Parallel connections involve connecting the positive terminals of each string together and all the negative terminals together at common terminals or busbars. **See Figure 5-27.**

When similar devices are connected in parallel, the overall circuit current is the sum of the currents of individual devices or strings. The overall voltage is the same as the average voltage of all the devices connected in parallel. **See Figure 5-28.**

As opposed to series connections, PV devices with dissimilar current output may be connected in parallel. This commonly occurs when an existing array is expanded. New module strings are connected in parallel with existing strings having similar voltage but different current output.

PV Devices in Parallel

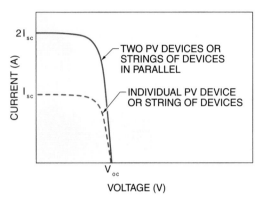

SIMILAR PV DEVICES

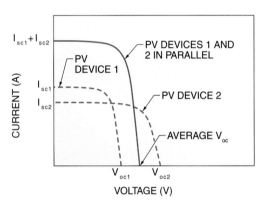

DISSIMILAR PV DEVICES

Figure 5-28. *The overall I-V curve of PV devices in parallel depends on the similarity of the current outputs of the individual devices.*

Parallel Connections

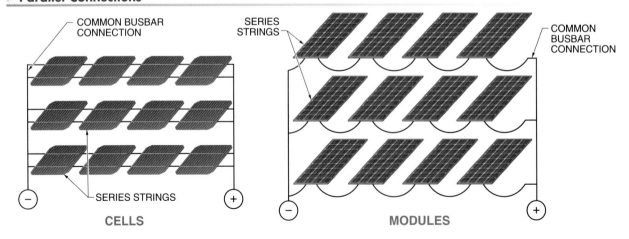

Figure 5-27. *Strings of PV cells or modules are connected in parallel to build current.*

Module Characteristics

PV modules are available in a range of sizes and designed for a variety of applications. **See Figure 5-29.** Smaller modules of less than 50 W are typically used individually for low-power battery charging applications, such as navigational aids, accent lighting, motorist-aid call boxes, and small circulation pumps. Smaller modules are often more expensive per unit watt output than larger ones, and are not typically used to build large arrays due to the large number of intermodule connections and mechanical attachments that would be required.

Larger modules with peak output of 50 W and greater are often referred to as power modules, and are connected in groups to configure arrays with higher voltage and power output. A typical 120 W crystalline silicon module that is 12% efficient has a surface area of about 1 m^2 and weighs approximately 20 lbs to 30 lbs. Production costs and systems integration time are significantly less with larger modules, so the typical power module size is steadily increasing. Modules of about 200 W peak power are now common, with some modules at more than 300 W and 100 lbs each. The factors currently limiting module size include the module's packaging and weight, structural integrity and resistance to high wind loads, and the ability to easily ship and handle the modules in the field.

Modules

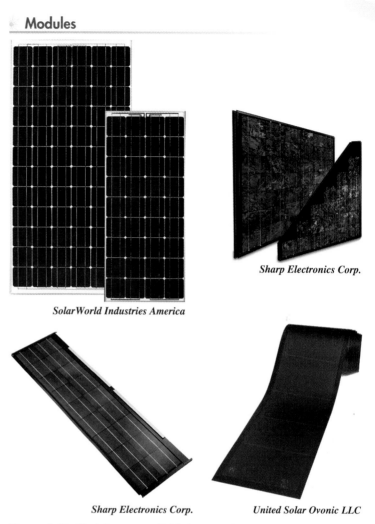

Sharp Electronics Corp.

SolarWorld Industries America

Sharp Electronics Corp.   *United Solar Ovonic LLC*

Figure 5-29. *Modules are available in several sizes and shapes, including squares, rectangles, triangles, flexible units, and shingles.*

SolarWorld Industries America
Robots arrange cells in various configurations and build complete modules with little or no intervention from human workers.

Flat-plate modules are used in the majority of residential and commercial applications. Flat-plate modules are generally rectangular in shape and about 1″ to 2″ thick (including the frame and laminate). Some manufacturers offer other shapes, such as triangles, to better utilize space on hipped or angled roof surfaces and for aesthetic appeal.

Bypass Diodes

Reverse bias is the condition of a PV device operating at negative (reverse) voltage. A reverse-biased device will continue to pass current, but since the voltage is negative, the device will consume power instead of producing power.

Reverse-bias conditions can occur when a PV cell is open-circuited or shaded, or when other conditions result in unequal current output from devices in a series string. Bypass diodes, sometimes called shunt diodes, are used in parallel with groups of cells or modules to prevent a reverse-bias condition.

A *bypass diode* is a diode used to pass current around, rather than through, a group of PV cells. The current is allowed to pass around groups of cells that are shaded or develop an open-circuit or other high resistance condition, preventing an interruption of the continuity of the string. This allows the functional cells or modules in the string to continue delivering power. The consequence, however, is that the string will operate at a lower voltage. **See Figure 5-30.**

If a series string of cells is thought of as a highway for current, bypass diodes can be considered a detour route. Under normal conditions, highway traffic flows smoothly and the detour is not needed. However, if an accident slows or stops highway traffic, traffic can continue flowing by taking the detour route. In a similar way, bypass diodes allow current to flow around obstructed paths.

Without a bypass diode, reverse voltage may decrease until the breakdown voltage is reached. *Breakdown voltage* is the minimum reverse-bias voltage that results in a rapid increase in current through an electronic device. The high currents can result in potentially damaging levels of power dissipation within the module, and under extreme cases, the resulting high temperatures can melt the module laminate and pose a fire hazard. A bypass diode allows a reverse bias of only 0.7 V, which limits the reverse voltage to a level where only a small amount of power may be dissipated. **See Figure 5-31.**

Bypass diodes are either embedded in the module laminate, and therefore are nonserviceable, or are located in the module junction box, where they can be inspected and replaced as required. Bypass diodes are typically installed around groups of 12 to 18 series-connected cells, with most modules of 36 cells or more incorporating two or more bypass diodes. Bypass diodes must be able to handle the maximum operating voltage for the number of cells or modules bypassed, and must be rated in excess of the maximum circuit current.

Refer to media clip on CD-ROM

Bypass Diodes

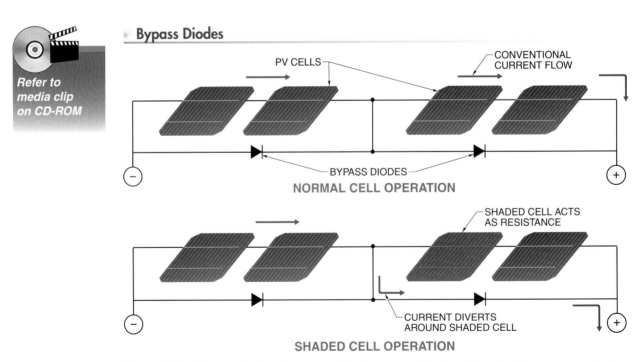

Figure 5-30. *Bypass diodes allow current to flow around devices that develop an open-circuit or high-resistance condition.*

Breakdown Voltage

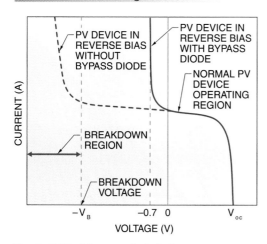

Figure 5-31. *A bypass diode limits reverse current through PV devices, preventing excessive power loss and overheating.*

Module Standards

A number of standards have been developed to address the safety, reliability, and performance of modules, arrays, and other equipment, as well as complete PV systems.

Safety. PV modules are classified as electrical equipment, so they must conform to accepted product safety standards, and must be listed or approved by an OSHA Nationally Recognized Testing Laboratory (NTRL). In the United States, Underwriters Laboratories, Inc. (UL) tests and certifies modules for electrical safety according to UL 1703 *Safety Standard for Flat-Plate Photovoltaic Modules and Panels*. The requirements cover flat-plate modules intended for roof-mounted, ground-mounted, or building-integrated systems with a maximum system voltage of 1000 V or less. The standard also covers components intended to provide electrical system connections or structural mounting for modules.

Installation requirements for modules and arrays is covered primarily by the NEC®. In addition, all PV installers should be familiar with construction standards established by OSHA, which are covered in Chapter 29 of the U.S. Code of Federal Regulations, Part 1926, *Safety and Health Regulations for Construction*.

Reliability. The reliability of modules is a critical factor affecting the performance, lifetime, and costs for PV systems. To promote a high standard for quality, modules produced by leading manufacturers are often tested according to International Electrotechnical Commission (IEC) standards, including the standards IEC 61215 *Crystalline Silicon Terrestrial Photovoltaic (PV) Modules—Design Qualification and Type Approval* and IEC 61646 *Thin-Film Terrestrial Photovoltaic (PV) Modules—Design Qualification and Type Approval*. These standards are commonly used as guidelines for module procurement. Design qualification tests include thermal cycling, humidity and freezing, impact and shock, immersion, cyclic pressure, twisting, vibration and other mechanical tests, and excessive and reverse current electrical tests. These tests are similar to, but more extreme than, the product listing tests.

Design qualification has important implications for product warranties offered by manufacturers. As a result, most major module manufacturers offer warranties of 20 years or more, guaranteeing module peak power output of at least 80% of initial nameplate ratings. This equates to a degradation rate of no more than 1% per year. These exceptionally long warranty periods are not typical among other electrical equipment and appliance warranties, but are offered to assure buyers of the long-term performance of PV systems.

Performance Ratings

Modules are rated for electrical performance as part of product testing. Financial incentives for PV installations are in many cases based on an array's electrical performance rating. Module manufacturers develop performance ratings for their products based on the I-V curves of sample modules under simulated sunlight. Since modules are most often marketed based on their peak power rating, accurate representation of this information is critical to the consumer. ASTM International and the IEC produce standards used by manufacturers and independent laboratories in testing and rating the performance of modules. These standards include test methods, test equipment, calibration, specifications, terminology, and translation of the results.

Standard performance ratings for modules are referred to as nameplate ratings, and they are required by the NEC® to be clearly labeled on every module. At a minimum, each module must be marked with polarity identification, maximum overcurrent device rating, and ratings at specified conditions for key I-V curve parameters, including open-circuit voltage, maximum permissible system voltage, maximum power current, short-circuit current, and maximum power. Additional information may include applicable certifications such as design qualification, fire class rating, ratings at other temperatures, and allowable wire sizes. In addition, all modules must have an installation guide that covers additional requirements for wiring, mounting, and other installation and operation considerations. **See Figure 5-32.**

Manufacturers typically guarantee the peak power ratings for a given product model to within ±10% of the nameplate rating, though actual performance is typically on the lower side of the rating. Module performance on the higher side of the rating is uncommon, but has important implications on module and system safety and for the adequate sizing of conductors and overcurrent protection for array source circuits. Measuring performance in the field can be complicated and expensive and requires special equipment, so performance verification is difficult. However, verification is usually not necessary because most modules perform to within an acceptable level of their nameplate ratings.

Test Conditions

Due to the dynamic nature of module performance and constantly changing operating conditions, the performance specifications of a module or array have meaning only when the rating conditions are given. These reference conditions are the basis for module performance ratings. Output at other conditions can be determined by translating the data using formulas for temperature and irradiance, the two principal factors affecting PV device performance. **See Figure 5-33.**

Module Labels

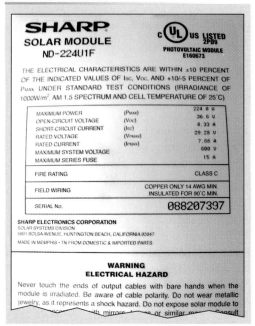

Figure 5-32. *Module labels must include performance ratings for the module and may include other information used to design a PV system.*

Test Conditions

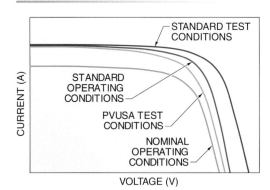

Figure 5-33. *Various test conditions can be used to evaluate module performance and may produce different results.*

Standard Test Conditions (STC). *Standard test conditions (STC)* is the most common and internationally accepted set of reference conditions, and rates module performance at a solar irradiance of 1000 W/m², spectral conditions of AM1.5, and a cell temperature of 25°C (77°F). STC is the basis for module electrical ratings required on module labels by the NEC®.

However, with the exception of clear days in extremely cold climates, modules and arrays seldom operate at STC in the field. Cell temperatures are often 15°C to 30°C (27°F to 54°F) above ambient temperature. Furthermore, peak irradiance levels are only experienced within an hour or so of solar noon on clear days, so modules usually operate under irradiance levels of less than 1000 W/m². For these reasons, other rating conditions are sometimes used.

Standard Operating Conditions (SOC). *Standard operating conditions (SOC)* is a set of reference conditions that rates module performance at a solar irradiance of 1000 W/m², spectral conditions of AM1.5, and at nominal operating cell temperature. *Nominal operating cell temperature (NOCT)* is a reference temperature of an open-circuited module based on an irradiance level of 800 W/m², ambient temperature of 20°C (68°F), and wind speed of 1 m/s. NOCT is product-dependent because it largely depends on a module's construction and heat transfer characteristics.

SOC uses more realistic temperature conditions than STC, and results in lower peak power and voltage ratings than at STC. Manufacturers sometimes list I-V parameters for SOC on module labels or specification sheets.

Nominal Operating Conditions (NOC). *Nominal operating conditions (NOC)* is a set of reference conditions that rates module performance at a solar irradiance of 800 W/m², spectral conditions of AM1.5, and at nominal operating cell temperature. NOC indicates a more typical peak module performance because solar irradiance is closer to 800 W/m² for most of the day. As with SOC, manufacturers may provide NOC ratings in addition to other module specifications.

PVUSA Test Conditions (PTC). *PVUSA test conditions (PTC)* is a set of reference conditions that rates module performance at a solar irradiance of 1000 W/m², ambient temperature of 20°C (68°F), and wind speed of 1 m/s. PTC are nearly identical to SOC, except that the resulting cell temperatures are higher. The SOC rating uses NOCT, which is based on 800 W/m² irradiance, while the cell temperature for PTC is based on 1000 W/m² irradiance. This higher cell temperature makes the PTC peak power ratings a little lower than ratings at SOC for a given module.

Other rating conditions have also been developed in efforts to more realistically estimate actual module performance in the field. Some financial incentive programs may also have means to estimate the AC energy performance for complete PV systems, using specific performance ratings from a given array and interactive inverter, and factoring in the orientation of the array and expected solar energy received on that surface at a given location.

> Due to slight variations in performance between batches of PV cells, manufacturers may offer two or more modules that appear the same in terms of the number and physical size of individual cells used, but that have slightly different performance ratings.

Module Selection

A number of factors may be considered when selecting modules for an array, including price, availability, and warranty, as well as the physical and electrical characteristics. These factors present numerous options and tradeoffs for integrators, installers, and buyers of PV systems.

Electrically, the voltage, current, and power output values are the most important considerations because they define the total number of modules needed to meet the desired energy production requirement. Fewer higher-voltage modules will be required in series, and fewer higher-current modules will be required in parallel, to achieve a desired array configuration. The provisions for electrical connections may also be considered. While the efficiency of modules does not necessarily affect module costs, it does affect the area required for a given power output desired, and consequently may affect balance of system costs, such as space or land area, as well as structural support requirements.

On the physical side, among factors that may be considered for module selection are the overall weight, size, and dimensions of the module, the type of frame and laminate construction, the means for structural attachments, the ability to withstand high mechanical loads, and perhaps even the color and appearance. Larger modules generally require less array field assembly, but may require special handling for lifting and installation. Some types of modules, particularly those intended for building-integrated applications, may use unique means or materials for structural attachments. Building-integrated modules replace conventional building materials such as roofing, glazing, or awnings. Certain sizes and shapes of modules may also be chosen to best utilize the available space for mounting the array.

Arrays

The construction of arrays from modules is similar to the construction of modules from cells. Groups of modules are combined electrically and mechanically, typically first into strings, to achieve the desired array voltage. Strings of series-connected modules may be called panels or subarrays. Connecting modules into series strings before making parallel connections makes it easier to expand the array in the future, replace individual modules, and troubleshoot and diagnose module problems within the array.

The U.S. Coast Guard has special design specifications for modules used in marine navigational aids. These specifications and associated tests require more durable and weatherproof modules for use in harsh marine environments.

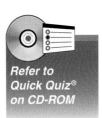

Refer to Quick Quiz® on CD-ROM

The modules or groups of modules are then integrated to form a complete array, using additional series or parallel connections. The result is a complete array that integrates all the modules into a single power-generating unit, with one positive terminal and one negative terminal for connection to other components. **See Figure 5-34.**

In stand-alone systems, the array must be sized to produce enough energy to meet a specific load during the period with the greatest load and lowest insolation, plus some excess energy to account for inefficiencies in battery charging, voltage drop, and other system losses. For interactive PV systems used to supplement normal electrical service, the array is designed to produce a desired amount of peak power or energy over a given period. More typically, the array size is dictated by available financing or the area available for the array, or because interconnection rules or financial incentives otherwise limit the size of the array.

Building an Array

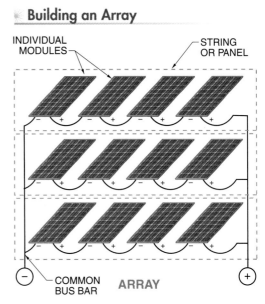

ARRAY

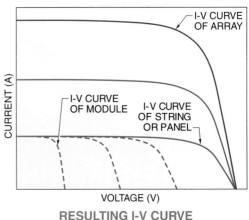

RESULTING I-V CURVE

Figure 5-34. *Modules are added in series to form strings or panels, which are then combined in parallel to form arrays.*

Summary

- Photons striking a PV cell give electrons the energy to move freely, which induces the flow of electrical current.

- Many materials can be made into PV cells, but crystalline silicon is currently the most common and economical material.

- Silicon wafers can be made with monocrystalline, polycrystalline, or ribbon silicon methods.

- The I-V curve illustrates the basic electrical parameters and characteristics of PV devices.

- The maximum power point is the operating point at which the PV device produces the most power and is the most efficient.

- The inherent properties of the PV material and the resistances in the PV system affect the shape of the I-V curve.

- Solar irradiance and cell temperature affect the magnitude and position of the I-V curve.

- Any single I-V curve represents only one set of conditions for solar irradiance and cell temperature.

- Cell temperature is influenced by ambient temperature, wind speed, solar irradiance, thermal characteristics of the device's packaging, and the way the cell or module is installed or mounted.

- I-V curve parameters can be translated from reference irradiance and temperature conditions to other conditions through the use of equations.

- PV cells are integrated electrically and mechanically to build modules. Modules are integrated electrically and mechanically to build arrays.

- The electrical concepts used to build voltage and current through series and parallel connections are scalable from cells to arrays.

- Bypass diodes protect PV devices from damage and excessive loss of power by directing current around shaded or damaged devices.

- Standard test conditions are used with performance figures so that modules can be compared effectively.

- Various standard test conditions have been developed in efforts to more realistically estimate module performance in real-world situations.

Definitions

- A *photovoltaic cell* is a semiconductor device that converts solar radiation into direct current electricity.

- A *semiconductor* is a material that can exhibit properties of both an insulator and a conductor.

- *Doping* is the process of adding small amounts of impurity elements to semiconductors to alter their electrical properties.

- A *p-type semiconductor* is a semiconductor that has electron voids.

- An *n-type semiconductor* is a semiconductor that has free electrons.

- The *photovoltaic effect* is the movement of electrons within a material when it absorbs photons with energy above a certain level.

- A *photon* is a unit of electromagnetic radiation.

- A *p-n junction* is the boundary of adjacent layers of p-type and n-type semiconductor materials in contact with one another.

- A *multijunction cell* is a cell that maximizes efficiency by using layers of individual cells that each respond to different wavelengths of solar energy.

- A *thin-film module* is a module-like PV device with its entire substrate coated in thin layers of semiconductor material using chemical vapor deposition techniques, and then laser-scribed to delineate individual cells and make electrical connections between cells.

- A *photoelectrochemical cell* is a cell that relies on chemical processes to produce electricity from light, rather than using semiconductors.

- A *wafer* is a thin, flat disk or rectangle of base semiconductor material.

- A *monocrystalline wafer* is a silicon wafer made from a single silicon crystal grown in the form of a cylindrical ingot.

- A *polycrystalline wafer* is a silicon wafer made from a cast silicon ingot that is composed of many silicon crystals.

- A *ribbon wafer* is a silicon wafer made by drawing a thin strip from a molten silicon mixture.

- The *current-voltage (I-V) characteristic* is the basic electrical output profile of a PV device.

- An *I-V curve* is the graphic representation of all possible voltage and current operating points for a PV device at a specific operating condition.

- The *open-circuit voltage (V_{oc})* is the maximum voltage on an I-V curve and is the operating point for a PV device under infinite load or open-circuit condition, and no current output.

- The *short-circuit current (I_{sc})* is the maximum current on an I-V curve and is the operating point for a PV device under no load or short-circuit condition, and no voltage output.

- The *maximum power point (P_{mp})* is the operating point on an I-V curve where the product of current and voltage is at maximum.

- The *maximum power voltage(V_{mp})* is the operating voltage on an I-V curve where the power output is at maximum.

- The *maximum power current (I_{mp})* is the operating current on an I-V curve where the power output is at maximum.

- *Fill factor (FF)* is the ratio of maximum power to the product of the open-circuit voltage and short-circuit current.

- *Efficiency* is the ratio of power output to power input.

- A *family of I-V curves* is a group of I-V curves at various irradiance levels.

- The *temperature-rise coefficient* is the coefficient for estimating the rise in cell temperature above ambient temperature due to solar irradiance.

- A *temperature coefficient* is the rate of change in voltage, current, or power output from a PV device due to changing cell temperature.

- A *module* is a PV device consisting of a number of individual cells connected electrically, laminated, encapsulated, and packaged into a frame.

- An *array* is a complete PV power-generating unit consisting of a number of individual electrically and mechanically integrated modules with structural supports, trackers, or other components.

- *Reverse bias* is the condition of a PV device operating at negative (reverse) voltage.

- A *bypass diode* is a diode used to pass current around, rather than through, a group of PV cells.

- *Breakdown voltage* is the minimum reverse-bias voltage that results in a rapid increase in current through an electronic device.

- *Standard test conditions (STC)* is the most common and internationally accepted set of reference conditions, and rates module performance at a solar irradiance of 1000 W/m², spectral conditions of AM1.5, and a cell temperature of 25°C (77°F).

- *Standard operating conditions (SOC)* is a set of reference conditions that rates module performance at a solar irradiance of 1000 W/m², spectral conditions of AM1.5, and at nominal operating cell temperature.

- *Nominal operating cell temperature (NOCT)* is a reference temperature of an open-circuited module based on an irradiance level of 800 W/m², ambient temperature of 20°C (68°F), and wind speed of 1 m/s.

- *Nominal operating conditions (NOC)* is a set of reference conditions that rates module performance at a solar irradiance of 800 W/m², spectral conditions of AM1.5, and at nominal operating cell temperature.

- *PVUSA test conditions (PTC)* is a set of reference conditions that rates module performance at a solar irradiance of 1000 W/m², ambient temperature of 20°C (68°F), and wind speed of 1 m/s.

Review Questions

1. Describe the basic process of manufacturing PV cells.

2. Explain the relationships between PV cells, modules, panels, and arrays.

3. How does the photovoltaic effect limit the short-circuit current in PV devices?

4. Which methods can be used to estimate or calculate the maximum power point on an I-V curve?

5. How does PV device efficiency affect required device area?

6. What effects do series and shunt resistance have in PV systems?

7. Describe how varying solar irradiance affects the I-V characteristics of a PV device.

8. Describe how varying temperature affects the I-V characteristics of a PV device.

9. Describe how key I-V curve parameters are compensated for varying cell temperature.

10. How are cells electrically connected to produce a module with desired voltage and current parameters?

11. What is the voltage and current resulting from series or parallel connections of dissimilar PV devices?

12. How do bypass diodes protect modules from damage and preserve power performance?

13. Why is it important to understand the test conditions used in a module performance evaluation?

6

Batteries

Energy demand does not always coincide with energy production, so many PV systems include electrical storage batteries. Batteries store electrical energy when it is produced by the array and supply energy to electrical loads when it is needed. They also establish and stabilize system voltages and can deliver high surge currents to loads. There are many types of batteries, each with unique design, electrochemistry, maintenance requirements, and performance characteristics.

Chapter Objectives

- Identify major battery components and their functions.
- Differentiate between the basic types and classifications of batteries.
- Understand the operation of batteries and their discharging and charging characteristics.
- Understand how temperature, discharge and charge rates, and electrolyte specific gravity affect battery capacity and life.
- Understand major principles and considerations for designing battery banks.

BATTERY PRINCIPLES

A *battery* is a collection of electrochemical cells that are contained in the same case and connected together electrically to produce a desired voltage. **See Figure 6-1.** A *battery cell* is the basic unit in a battery that stores electrical energy in chemical bonds and delivers this energy through chemical reactions. Battery designs can vary by type and manufacturer, but many share the same basic components and store electricity using similar electrochemical reactions.

Batteries

East Penn Manufacturing Co., Inc.

Figure 6-1. *Batteries are collections of cells that produce electricity through electrochemical reactions. Cells can be configured into batteries of many different shapes and sizes.*

The word "cell" can refer to the smallest electricity-generating unit in both PV modules and batteries. Also, many people use the word "battery" when they mean "cell." For example, AA is a common size of cell, not battery, for consumer electronics.

Battery Design

A battery cell consists of one or more sets of positive and negative plates immersed in an electrolyte solution. A *plate* is an electrode consisting of active material supported by a grid framework. *Active material* is the chemically reactive compound on a battery cell plate. The amount of active material in a battery is proportional to the energy storage capacity of a battery. The *grid* is a metal framework that supports the active material of a battery cell and conducts electricity. **See Figure 6-2.**

In each cell, there can be a single pair or, more commonly, several pairs of positive and negative plates. When there are multiple pairs, all the positive plates are connected together and all the negative plates are connected together. Thicker plates tolerate deeper discharges over long periods while maintaining good adhesion of the active material to the grid, resulting in longer battery life. Thinner plates allow more pairs per cell, maximizing surface area for delivering high currents. However, thinner plates are less durable and are not designed for deep discharges.

Electrolyte is the conducting medium that allows the transfer of ions between battery cell plates. Electrolyte may be in a liquid or gelled form. An insulating separator keeps the plates from making electrical contact and short-circuiting, but is porous to allow the flow of electrolyte ions between the plates.

The chemical reactions during battery charging produce gases. Some batteries include cell vents to allow the gases to escape to the atmosphere. Other batteries are sealed and do not allow gases to escape under normal conditions, but include pressure-relief vents. These vents remain closed under normal conditions, but open when the battery pressure increases beyond a threshold, often the result of overcharging or high-temperature operation.

Surrette Battery Company

Cells are connected to each other inside the battery case with intercell connectors.

Battery Design

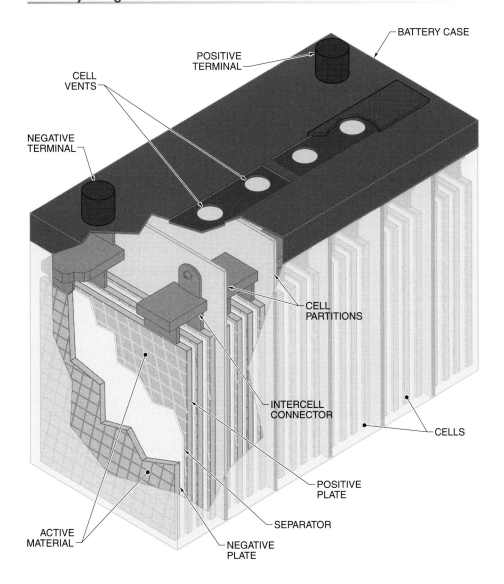

CELL VENTS

POSITIVE TERMINAL

BATTERY CASE

NEGATIVE TERMINAL

CELL PARTITIONS

INTERCELL CONNECTOR

CELLS

POSITIVE PLATE

SEPARATOR

ACTIVE MATERIAL

NEGATIVE PLATE

Figure 6-2. *Many components are common to various battery types.*

Battery cases contain cells in individual chambers and give the battery structure. Cases are typically made from strong and durable plastics, which do not react with the electrolyte. The case includes terminals, vents, cell partitions, and intercell connectors that connect plate assemblies between cells. Terminals are the external electrical connections to a battery. Connections to loads, charging circuits, or other batteries are made at the terminals. Clear battery cases allow easier monitoring of electrolyte levels and battery plate condition.

Steady-State

A cell or battery that is not connected to a load or charging circuit is at steady-state. *Steady-state* is an open-circuit condition where essentially no electrical or chemical changes are occurring. At steady-state, there is electrical potential between the positive and negative terminals, but electrons cannot flow until a load is connected between the terminals. The *open-circuit voltage* is the voltage of a battery or cell when it is at steady-state. The open-circuit voltage of a fully charged lead-acid cell is about 2.1 V.

Capacity

Capacity is the measure of the electrical energy storage potential of a cell or battery. Several physical factors affect the capacity, including the quantity of active material; the number, design, and dimensions of the plates; and the electrolyte concentration. Operational factors affecting capacity include discharge rate, charging method, temperature, age, and condition of the cell or battery. Capacity is commonly expressed in ampere-hours (Ah), but can also be expressed in watt-hours (Wh). For example, a battery that delivers 5 A for 20 hr has delivered 100 Ah. If the battery averages 12 V during discharge, the capacity can also be expressed as 1200 Wh (100 Ah × 12 V = 1200 Wh).

Temperature and discharge rate may affect capacity, especially of lead-acid batteries. Warmer batteries are capable of storing and delivering more charge than colder batteries. **See Figure 6-3.** However, high temperatures decrease the useful life of a battery. Manufacturers generally rate lead-acid battery performance and cycle life at 25°C (77°F). For the best trade-off between capacity and lifetime, the system should be designed for the recommended discharge rate and the battery should be located where the average temperature will be close to the manufacturer's recommendation. Any differences from the rated conditions affect the actual capacity of the battery.

Envelope separators are pockets that encapsulate the entire plate except for the electrical connections at the top. The envelopes prevent pieces of active material from creating internal short circuits if they break away from the grid and fall to the bottom of the cell.

Refer to media clip on CD-ROM

Discharging

Discharging is the process of a cell or battery converting chemical energy to electrical energy and delivering current. Discharging removes energy from a battery. Discharging begins when a load is connected to the battery at the positive and negative terminals. At that moment, a chemical reaction begins that causes electrons to flow from the negative terminal to the positive terminal.

Battery Capacity

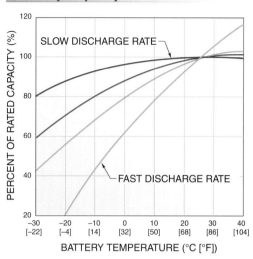

Figure 6-3. *Higher temperatures and slower discharge rates result in increased battery capacity.*

At the negative terminal, the active material in the negative plate reacts with the electrolyte to form a new material that releases excess electrons. At the positive terminal, the active material in the positive plate reacts with the electrolyte to form a new material, which requires extra electrons to complete the reaction. When the battery is discharging, the excess electrons at the negative terminal are conducted outside the battery, through the load, to the positive terminal to complete the reactions at the positive plate. **See Figure 6-4.**

Small, demonstration PV systems may include the batteries to simulate the operation of stand-alone systems.

The voltage between the terminals is highest at the beginning of a discharge cycle and gradually falls. The voltage of a battery system is not constant but ranges from a few volts above its nominal voltage to a few volts below. For example, a nominal 12 V lead-acid battery is between 12.6 V and 13 V at steady-state when fully charged. The battery voltage falls to a voltage between 10.8 V and 11 V during discharge, depending on the discharge current. The *cutoff voltage* is the minimum battery voltage specified by the manufacturer that establishes the battery capacity at a specific discharge rate. Below the cutoff voltage, there is no further usable capacity.

Discharge rate is expressed as a ratio of the nominal battery capacity to the discharge time in hours. For example, a 5 A discharge for a nominal 100 Ah battery would be a C/20 discharge rate. The designation C/20 indicates that $\frac{1}{20}$th of the rated capacity is discharged per hour, or that the battery will be completely discharged after 20 hr. Capacity is directly affected by the rate of discharge. Lower discharge rates are able to remove more energy from a battery before it reaches the cutoff voltage. Higher discharge rates remove less energy before the battery reaches the same voltage. **See Figure 6-5.**

State of Charge (SOC). The *state of charge (SOC)* is the percentage of energy remaining in a battery compared to the fully charged capacity. Discharging a battery decreases the state of charge, while charging increases the state of charge. For example, a battery that has had three-quarters of its capacity removed is at 25% state of charge. **See Figure 6-6.**

Depth of Discharge (DOD). *Depth of discharge (DOD)* is the percentage of withdrawn energy in a battery compared to the fully charged capacity. For example, a battery that has had three-quarters of its capacity removed is at a 75% depth of discharge. By definition, the depth of discharge and state of charge of a battery add up to 100%. Two common qualifiers for depth of discharge in PV systems are the allowable DOD and the average daily DOD.

Discharging Reaction

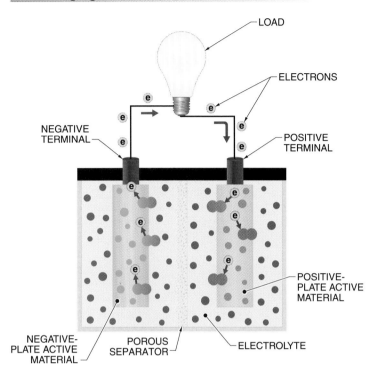

Figure 6-4. *Electrochemical reactions within a cell produce a flow of electrons from the negative terminal to the positive terminal.*

Discharge Rate

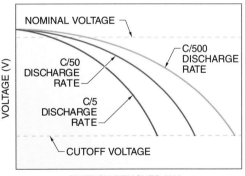

Figure 6-5. *Slower discharge rates remove more energy from a battery than faster discharge rates.*

Battery cells are essentially devices for converting between chemical energy and electrical energy. When a load is connected to the cell or battery, the discharging reaction uses the energy of chemical bonds to free electrons and cause them to flow.

State of Charge vs. Depth of Discharge

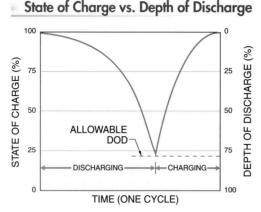

Figure 6-6. *The state of charge and depth of discharge of a battery always add up to equal 100%.*

Allowable DOD. The *allowable depth of discharge* is the maximum percentage of total capacity that is permitted to be withdrawn from a battery. The allowable DOD may be as high as 80% for deep-cycle traction batteries or as low as 15% for automotive batteries. The allowable DOD is determined by the cutoff voltage and discharge rate.

Average Daily DOD. The *average daily depth of discharge* is the average percentage of the total capacity that is withdrawn from a battery each day. **See Figure 6-7.** If the load varies seasonally, such as in a lighting system, the average daily DOD will be greater in the winter months due to greater loads at night. If the loads are constant, the average daily DOD will be greater in the winter due to low temperatures that lower the rated battery capacity. Depending on the rated capacity and the average daily energy load, the average daily DOD may vary from only a few percent in systems designed with high autonomy to as high as 50% for marginally sized battery systems.

Autonomy. In stand-alone PV systems, battery system capacity is generally several times greater than needed for the average daily load requirements in order to account for days with below-normal insolation. *Autonomy* is the amount of time a fully charged battery system can supply power to system loads without further charging. Autonomy is expressed in days. Battery systems are typically sized

for an autonomy period of two to six days. Autonomy may be greater for applications involving a critical load, public safety, or highly variable insolation. Lower average daily DOD or greater battery system capacity results in longer autonomy.

Average Daily Depth of Discharge

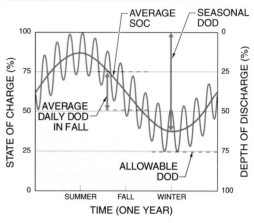

Figure 6-7. *When arrays are used to charge batteries, seasonal insolation variations affect depth of discharge values.*

Self-Discharge. *Self-discharge* is the gradual reduction in the state of charge of a battery while at steady-state condition. Self-discharge is also referred to as standby or shelf loss. Self-discharge is a result of internal electrochemical mechanisms and losses. The rate of self-discharge differs among battery types and increases with battery age. Self-discharge rates are typically specified in percentage of rated capacity per month. Higher temperatures result in higher self-discharge rates, particularly for lead-antimony designs. **See Figure 6-8.**

In operation, self-discharge is usually insignificant compared to normal system loads. However, if the PV system is small or self-discharge is otherwise a concern, continuous application of a very small charge current counters the effects of self-discharge and keeps a battery at full charge.

Charging

A *cycle* is a battery discharge followed by a charge. *Charging* is the process of a cell or battery receiving current and converting the electrical energy into chemical energy. Charging is done by applying an electrical current to the cell or battery in a direction opposite to the discharge. In order for the battery to accept current, the voltage of the charging source must be higher than the battery voltage. For example, a nominal 12 V lead-acid battery is charged with about 14.4 V and reads about 12.6 V when fully charged.

Self-Discharge Rates

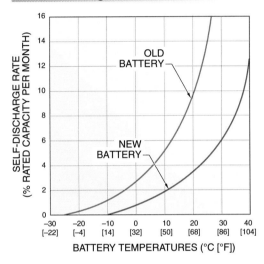

Figure 6-8. *Batteries exhibit high self-discharge rates at higher temperatures.*

The electrons passing through in the opposite direction reverse the chemical reactions and restore the active materials and electrolyte to their original compositions. **See Figure 6-9.** Charge rate is quantified in the same way as discharge rate. For example, a charging rate of C/50 to a 100 Ah battery applies 2 A of current until the battery reaches a specific fully charged voltage.

During charging, the battery voltage rises sharply then stabilizes. Voltage rises again, first very slowly, then faster, and may pass the gassing voltage. The *gassing voltage* is the voltage level at which battery gassing begins. Voltage stabilizes again at the fully charged voltage level. Batteries can be charged in one stage or multiple stages. Three stages of battery charging are bulk charging, absorption charging, and float charging. Equalizing charging is an additional type of charging.

Charging Reaction

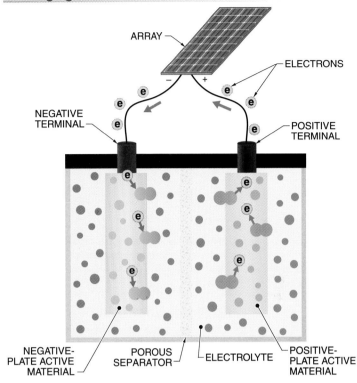

Figure 6-9. *The charging reaction within a cell is the reverse of the discharge reaction.*

Bulk Charging. *Bulk charging* is battery charging at a relatively high charge rate that charges the battery up to a regulation voltage, resulting in a state of charge of about 80% to 90%. **See Figure 6-10.** This is also called normal charging. Single-stage charging is bulk charging only. This mode is simple to implement, but does not fully charge the battery. Multiple-stage charging includes additional charging modes to reach a higher state of charge.

Absorption Charging. Once a battery is nearly fully charged, most of the active material in the battery has been converted to its original form, and current regulation is required to limit the amount of overcharge supplied to the battery. *Absorption charging* is battery charging following bulk charging that reduces the charge current to maintain the battery voltage at a regulation voltage for a certain period. The absorption charging period can be preset or adjustable, and is usually 1 hr to 3 hr. Absorption charging charges another 10% to 15% of battery capacity.

Multiple-Stage Charging

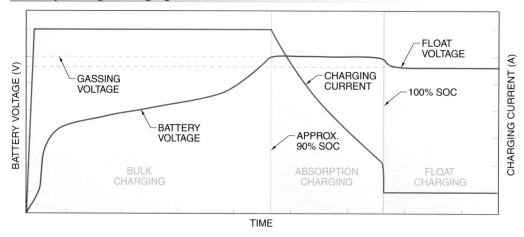

Figure 6-10. *Bulk, absorption, and float charging control battery voltage with the charging current during a multiple-stage charging cycle.*

Float Charging. When the battery reaches nearly 100% state of charge, the charging voltage is lowered slightly to the float voltage and the charging current is set to a very low rate. *Float charging* is battery charging at a low charge rate that maintains full battery charge by counteracting self-discharge. The float charge rate must not exceed the self-discharge rate or the battery will be overcharged. Float charging is also called trickle charging or finish charging. A charge controller remains in float charge mode until the array current is no longer high enough to maintain the battery at float voltage.

> Battery types that cannot be recharged use electrochemical reactions that are not easily reversible. This makes them impractical for PV systems but commonly used in consumer electronics due to their low cost.

Equalizing Charging. An equalizing, or refreshing, charge is used periodically to maintain consistency among individual lead-acid cells. *Equalizing charging* is current-limited battery charging to a voltage higher than the bulk charging voltage, which brings each cell to a full state of charge. Equalizing charging is a controlled overcharge to produce gassing,

ensuring that every cell is fully charged. Equalizing a battery helps to prevent electrolyte stratification, sulfation, and cell voltage inconsistencies that develop during normal battery operation, and maintains battery capacity at the highest possible levels.

For batteries that are deeply discharged on a daily basis, an equalizing charge is recommended every one or two weeks. For batteries less severely discharged, equalizing may only be required every one or two months. An equalizing charge is typically maintained until the cell voltages and specific gravities remain consistent for a few hours. Only flooded open-vent batteries need equalization. Sealed or valve-regulated batteries can be damaged by equalization.

Gassing and Overcharge

When a battery is nearly fully charged, essentially all of the active materials have been converted to their fully charged composition and the cell voltage rises sharply. Further charging at this point results in gassing. *Gassing* is the decomposition of water into hydrogen and oxygen gases as the battery charges. Hydrogen forms at the negative plate and oxygen forms at the positive plate. The gases bubble up through the electrolyte and may escape through cell vents, resulting in water loss from the electrolyte.

Some level of gassing is required to achieve full charge, but gassing must be carefully controlled to prevent excessive water loss. Water may need to be periodically added to the electrolyte to maintain the proper acid concentration, though this cannot be done for all types of batteries. Gassing is useful for gently agitating the electrolyte, which ensures that its concentration is uniform. However, excessive gassing can dislodge active materials from the grids, increase battery temperature, or expose the plates, which can permanently damage the battery.

The gassing voltage is a function of battery chemistry, temperature, charge rate, voltage, and state of charge. As battery temperature decreases, the corresponding gassing voltage increases, and vice versa. At a battery temperature of 0°C (32°F) the gassing voltage of lead-acid cells increases to about 2.5 V. The effect of temperature on the gassing voltage is the reason why the charging process should be temperature compensated. This ensures that batteries are fully charged in cold weather and not overcharged during warm weather.

At a given temperature, the gassing voltage is the same for all charge rates. However, since charge rates affect capacity, gassing begins at a lower state of charge at higher charge rates. For example, at a charge rate of C/20 at 25°C (77°F), the gassing voltage of about 2.35 V per lead-acid cell is reached at about 90% state of charge. At a charge rate of C/5, the gassing voltage is reached at about 75% state of charge.

Overcharge is the ratio of applied charge to the resulting increase in battery charge. For example, a 100 Ah battery may require 110 Ah of charge to account for charge loss due to gassing. The overcharge in this case is 110%. Varying amounts of overcharge are required for different battery types and from different states of charge. Generally, more overcharge is required to charge a battery from a shallow DOD than from a greater DOD, since the final stages of charging are the least efficient. Some batteries may require as much as 120% overcharge while many others require only about 110%.

Electrolyte Concentration

Specific gravity is the ratio of the density of a substance to the density of water. By definition, water has a specific gravity of 1.00. The electrolyte in a fully charged lead-acid battery is a solution of approximately 25% sulfuric acid and 75% water (by volume). Ions from the acid are consumed by the active material during discharging and released during charging, so the concentration of the acid in the water changes with state of charge. Therefore, the specific gravity of the electrolyte is related to the battery state of charge, though it is also affected by temperature.

In a fully charged lead-acid battery, the specific gravity of the electrolyte typically ranges between 1.25 and 1.28 at a temperature of 25°C (77°F). When the battery is fully discharged, the electrolyte is essentially water, with a specific gravity near 1.00. Therefore, electrolyte specific gravity can be used to estimate battery state of charge. However, temperature will also affect specific gravity, so readings must be adjusted to compare to the expected values at 25°C (77°F).

Concentrated sulfuric acid has a very low freezing point (as low as −70°C [−94°F]), while water has a freezing point of 0°C (32°F). Therefore, the freezing point of the electrolyte also varies with the specific gravity of the electrolyte. As the battery becomes discharged, the specific gravity decreases, resulting in a higher freezing point for the electrolyte. **See Figure 6-11.**

Surrette Battery Company

Batteries may be designed so that individual cells can be removed for transport and reasembly if necessary.

Electrolyte Freezing Points

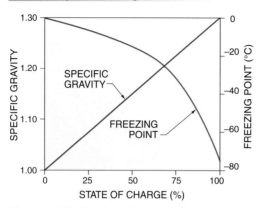

Figure 6-11. *The freezing point of sulfuric acid electrolyte changes at various states of charge because of changes in specific gravity.*

In extreme climates, the specific gravity of the electrolyte is often adjusted outside the typical range to compensate for the effects of temperature extremes.

In warm climates where freezing temperatures do not occur, the specific gravity may be reduced to below 1.25. Lower concentrations also lessen the degradation of the separators and grids, prolonging the useful service life of the battery. However, lower electrolyte specific gravity decreases the storage capacity and discharge rate performance, though these factors are usually offset by the higher operating temperatures.

In very cold climates, the electrolyte may be susceptible to freezing, particularly during winter when the batteries may not be fully charged due to below-average insolation. If the electrolyte freezes, it will expand, causing irreversible damage to the battery. The specific gravity of the electrolyte may be increased slightly, up to 1.30. Increasing the electrolyte concentration lowers the freezing point and accelerates the electrochemical reactions, improving the capacity at low temperatures. However, high specific gravity can reduce the useful service life of a battery.

While the specific gravity can be used to estimate the state of charge of a lead-acid battery, low or inconsistent specific gravity readings between series-connected cells in a battery may also indicate sulfation, stratification, or a lack of charge equalization between cells.

Sulfation. *Sulfation* is the growth of lead sulfate crystals on the positive plate of a lead-acid cell. Sulfation is a side effect of normal battery aging, but can be accelerated by prolonged operation at partial states of charge. Sulfation decreases the electrolyte concentration and the available active material, and therefore the capacity of the cell. **See Figure 6-12.** Sulfation also increases internal resistance within the battery, making it more difficult to charge.

Sulfation

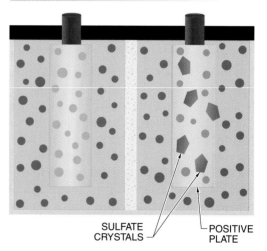

Figure 6-12. *Sulfation reduces the capacity of a lead-acid cell by locking away active material as crystals.*

Battery temperature is often monitored by an electronic sensor placed between adjacent battery cases.

During normal battery discharging, the active materials on both the positive and negative plates are converted to lead sulfate. If the battery is charged soon after being discharged, the lead sulfate converts easily back into the active materials. However, if a lead-acid battery remains at a low state of charge for prolonged periods (weeks), the lead sulfate crystallizes on the positive plate, which inhibits conversion back to active materials during charging. The crystals essentially reduce the amount of active material, decreasing the capacity of the battery. If lead sulfate crystals grow too large, they can also cause physical damage to the plates.

Sulfation is a common problem with lead-acid batteries in some PV systems because the batteries often operate at partial states of charge for extended periods. To minimize sulfation, the array should be sized to charge the battery during the month with the highest load to insolation ratio. Also, the additional energy sources in hybrid systems are effective at keeping batteries fully charged.

Stratification. During charging, electrolyte acid ions form on the plates and gradually descend to the bottom of the cell. Over time, the electrolyte can develop a greater acid concentration at the bottom of the cell than at the top. *Stratification* is a condition of flooded lead-acid cells in which the specific gravity of the electrolyte is greater at the bottom than at the top. **See Figure 6-13.**

If left unmixed, the reaction processes vary between the bottom and the top of the plates, decreasing battery performance and increasing corrosion. If the battery has a transparent case, it may be possible to see more degradation on the bottoms of the plates than the tops. Stratification is generally the result of low charge rates or undercharging, which does not produce the gassing needed to agitate the electrolyte. Tall stationary cells are particularly prone to stratification when charged at low rates. Periodic equalization gassing thoroughly mixes the electrolyte and minimizes stratification problems.

Battery Life

Battery life depends on many design and operational factors, including battery materials,

operating temperatures, cycle frequency, depth of discharges, average state of charge, and charging methods. Battery life is expressed in terms of cycles or years. The end of the useful life of a battery is marked by loss of capacity that increases the daily DOD beyond an acceptable level. End-of-life reductions in battery capacity may occur gradually or abruptly, depending on the type of failure.

Stratification

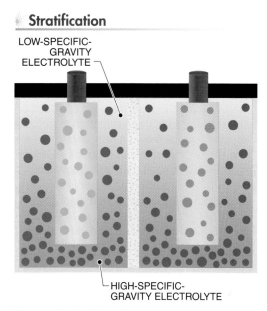

Figure 6-13. *Stratification results when the specific gravity of the electrolyte is higher at the bottom of a cell than at the top.*

For safety and weight reasons, batteries are sometimes drained after initial charging by the manufacturer and shipped in a dry condition. The acid and water are then added during installation.

Exact quantification of battery life in PV systems is difficult due to the number of variables involved, but certain trends can be used to estimate battery life. As long as a battery is not overcharged, overdischarged, or operated at excessive temperatures, the lifetime of a battery is roughly proportional to its average state of charge. For example, a typical flooded lead-acid battery that is maintained above 90% state of charge will provide two to three times more charge/discharge cycles than a battery maintained at an average 50% state of charge.

Battery Disposal and Recycling

Batteries are hazardous items because they contain toxic materials such as plastics, lead, and acids that can harm humans and the environment. For this reason, laws have been established that dictate requirements for battery disposal and recycling. In some cases, battery manufacturers provide guidelines for battery disposal through local distributors. Battery manufacturers may also accept batteries for recycling. In most areas, batteries may be taken to approved recycling centers. Under no circumstances should batteries be disposed of in landfills or burned.

At a recycling center, the battery is broken into pieces and placed in a vat. The lead and heavy materials fall to the bottom, while the plastic rises to the top and the acid is drained off. The materials are separated and each undergoes a different recycling process.

The plastic pieces are washed, dried, and sent to a plastic recycler. There the pieces are melted together and extruded into small plastic pellets of uniform size. The pellets are sold to battery case manufacturers for use in new cases.

The lead plates and other parts are cleaned and then melted in smelting furnaces. The molten lead is poured into ingot molds. Large ingots weighing about 2000 lbs are called hogs. Smaller ingots weighing 65 lbs are called pigs. After a few minutes, the impurities float to the top of the still-molten lead in the ingot molds and are scraped away. When the ingots are cool, they are removed from the molds and sent to battery manufacturers where they are melted again to produce new lead plates and other parts for new batteries. More than 97% of all battery lead is recycled.

Old battery acid can be handled in two ways. The acid can be neutralized with an industrial compound similar to household baking soda. The resulting liquid is mostly water that is treated, cleaned, and tested to ensure that it meets clean water standards. The other method is to process the acid into sodium sulfate, an odorless white powder that is used in laundry detergent, glass, and textile manufacturing.

Operating temperature has a significant affect on battery life. The electrochemical reaction rates double for every 10°C (18°F) increase in temperature. Therefore, battery life decreases by half for every 10°C (18°F) increase. Higher operating temperatures also accelerate corrosion of the positive plate grids, resulting in greater gassing and electrolyte loss. Lower operating temperatures generally increase battery life, but capacity is significantly reduced, particularly for lead-acid batteries. In areas of severe temperature variations, batteries should be located in insulated or otherwise temperature-regulated enclosures.

BATTERY TYPES

Many types and classifications of batteries are manufactured, each with specific design and performance characteristics suited for particular applications. In PV systems, lead-acid batteries are most common due to their wide availability in many sizes, low cost, and well-understood performance characteristics. In a few critical, low-temperature applications, nickel-cadmium batteries are used, but high initial cost limits their use in most PV systems. Other battery designs may be used in small PV applications but are not common.

Electrical storage batteries are either primary or secondary batteries. A *primary battery* is a battery that can store and deliver electrical energy but cannot be recharged. Carbon-zinc and lithium batteries typically used in consumer electronic devices are examples of primary batteries. Primary batteries are not suitable for PV systems. A *secondary battery* is a battery that can store and deliver electrical energy and can be charged by passing a current through it in an opposite direction to the discharge current. Lead-acid batteries commonly used in automobiles and PV systems are examples of secondary batteries.

Battery Classifications

Secondary batteries are often classified as traction; starting, lighting, and ignition (SLI); or stationary batteries. The differences in their design and materials result in different discharge and cycle characteristics. **See Figure 6-14.**

Traction Batteries. *A traction battery* is a class of battery designed for repeated deep-discharge cycle service. Because they are often used in electrically operated vehicles and equipment such as golf carts, forklifts, and hybrid automobiles, traction batteries are also called motive power batteries. These batteries have fewer plates per cell than other types of batteries, and the plates are thick and durable. Traction batteries are very popular in PV systems for their deep-cycle capability, long life, and durability.

Starting, Lighting, and Ignition Batteries. *A starting, lighting, and ignition (SLI) battery* is a class of battery designed primarily for shallow-discharge cycle service. These batteries have many thin plates per cell, allowing the battery to deliver high currents for short periods, and are most often used to power automobile starters. SLI batteries are not recommended for PV applications because they are not designed for deep discharges. However, SLI batteries are sometimes used for PV systems in developing countries where they are the only type of battery available locally. When depth of discharges are limited, SLI batteries may provide up to about two years of useful service in small, stand-alone PV systems.

Stationary Batteries. *A stationary battery* is a class of battery designed for occasional deep-discharge, limited-cycle service. These batteries are designed to last a long time on standby service, so they are commonly used in uninterruptible power supplies (UPSs) to provide back-up power to computers, telephone equipment, and other critical loads or devices. Stationary batteries are not recommended for most PV applications because they are designed for very few charging and discharging cycles.

Battery Classifications

TRACTION BATTERIES

SLI BATTERIES

STATIONARY BATTERIES

Figure 6-14. *Batteries are divided into classes based on discharge and cycle characteristics.*

Manufacturer expectations for battery life may not be applicable in PV systems because most manufacturers do not rate battery performance under the same cycling and depth of discharge conditions as are found in PV systems.

Flooded-Electrolyte Batteries

Flooded electrolyte is electrolyte in the form of a liquid. The electrolyte in flooded lead-acid cells is a sulfuric acid and water solution and the electrolyte in flooded nickel-cadmium cells is an alkaline solution of potassium hydroxide and water. Flooded batteries can be classified as either open-vent or sealed-vent types.

Open-Vent Batteries. Open-vent batteries allow charging gases to freely escape. However, water must be added periodically to replenish water lost through gassing and maintain the correct concentration. Open-vent batteries have removable caps for the addition of water. Distilled or demineralized water must be used to replenish electrolyte because even normal tap water has impurities that can taint a battery and result in premature loss of capacity.

A *catalytic recombination cap (CRC)* is a vent cap that reduces electrolyte loss from an open-vent flooded battery by recombining vented gases into water. CRCs contain particles of a catalyst such as platinum or palladium. A *catalyst* is a substance that causes other substances to chemically react but does not participate in the reaction. The catalyst causes a reaction between the hydrogen and oxygen that are generated by the battery during charging. The gases recombine in the CRC to form water, which drains back into the battery. CRCs on open-vent flooded lead-antimony batteries reduce electrolyte loss by as much as 50% in subtropical climates. **See Figure 6-15.**

CRCs also generate heat during the recombination process. Increasing temperature in CRCs can be used to detect gassing in the battery. If CRCs for different battery cells are at significantly different temperatures during charging (meaning that some cells are gassing and others are not), an equalization charge may be required.

Sealed-Vent Batteries. Sealed-vent flooded batteries have nonremovable caps on the cells that allow only excess charging gases to escape through pressure-relief vents. Under controlled charging, the pressure relief vents remain closed and the process is contained. Excessive overcharge, however, can increase the internal gas pressure to the point of opening the pressure relief vents and releasing gas. This results in a permanent loss of water that cannot be replenished. Manufacturers account for only a small loss of water by including reserve electrolyte.

Catalytic Recombination Caps

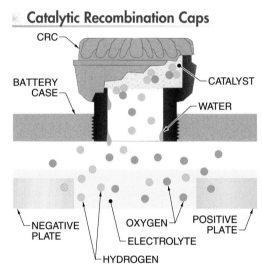

Figure 6-15. *Catalytic recombination caps recombine oxygen and hydrogen into water, which then drains back into the battery.*

Captive-Electrolyte Batteries

Captive electrolyte is electrolyte that is immobilized. **See Figure 6-16.** Batteries with captive electrolyte are sealed and often referred to as valve-regulated lead-acid (VRLA) batteries because they include pressure-relief vents. Captive electrolyte cannot be replenished, so these batteries are intolerant of excessive overcharge. However, captive-electrolyte batteries feature internal gas recombination. This process recombines the oxygen gas from the positive plates and the hydrogen gas from the negative plates back into water, replenishing the electrolyte.

Captive-electrolyte batteries are popular because they are spill-proof and easily transported. Captive electrolyte is also less susceptible to freezing than flooded electrolyte. Since captive-electrolyte batteries do not require water additions, they are ideal for remote applications where maintenance is infrequent or unavailable. The two most common types of captive electrolyte are the gelled electrolyte and absorbed glass mat electrolyte designs.

Captive-Electrolyte Batteries

Figure 6-16. *Captive-electrolyte batteries are sealed and contain electrolyte that is immobilized.*

Gelled Electrolyte. Electrolyte is gelled by adding silicon dioxide to the electrolyte. The mixture is added to the battery as a warm liquid and turns to gel as it cools. Cracks and voids develop within the gelled electrolyte during the first few cycles, providing paths for the gases to move around, facilitating recombination. Excessive gases, however, escape through the pressure-relief vents, drying and shrinking the gel, causing it to lose contact with the plates and reducing capacity. Some gelled electrolyte has a small amount of phosphoric acid added to improve the deep-discharge cycle performance of the battery by minimizing grid oxidation at low states of charge.

Absorbed Glass Mat Electrolyte. The electrolyte in an absorbed glass mat (AGM) battery is a liquid absorbed into fiberglass mats that are sandwiched in layers between the plates. The fiberglass mats also act as separators between the plates. This arrangement also allows the oxygen and hydrogen gases to migrate and recombine as water. AGM batteries are intolerant of high operating temperatures. Recommended charge regulation methods for gelled batteries also apply to AGM batteries.

Lead-Acid Batteries

Lead-acid batteries are the most common type of batteries used in PV systems. Lead-acid batteries are generally inexpensive and widely available in many capacities from 10 Ah to over 1000 Ah. Their deep-cycle character-

istics make them ideal for PV applications, but they do not tolerate extreme temperatures well and may require frequent maintenance. **See Figure 6-17.**

In a lead-acid cell, the active material is lead dioxide (PbO_2) in the positive plates and metallic sponge lead (Pb) in the negative plates. The electrolyte is a solution of sulfuric acid (H_2SO_4) and water. The lead-acid electrochemistry produces a cell with a nominal voltage of 2 V. Six cells are configured in series to produce a battery with a nominal voltage of 12 V.

Lead-acid battery types are differentiated by the elements alloyed with lead in the plate grids, and the physical form of the electrolyte. The most common types of grid alloys for lead-acid batteries are lead-antimony, lead-calcium, and hybrids.

Lead-Antimony Batteries. Lead-antimony battery plate grids include 3% to 6% of antimony (Sb), which provides excellent deep-discharge performance. Lead-antimony grids also limit the shedding of active material and last the longest of the lead-acid batteries at higher temperatures. Disadvantages of lead-antimony batteries include a high self-discharge rate and possibly frequent water additions due to gassing.

Battery Chemistry

Battery chemistry is described with chemical equations, which show how the molecules or ions of the active material and electrolyte combine and/or split into different materials while either releasing or absorbing electrons.

In a lead-acid battery during discharging, the negative plate reaction is the following:

$$PbO_2 + HSO_4^- + 3H^+ + 2e^- \rightarrow PbSO_4 + 2H_2O$$

The positive plate reaction is:

$$Pb + HSO_4^- \rightarrow PbSO_4 + H^+ + 2e^-$$

During charging, the reactions reverse. The materials on the right side of the equation change into the materials on the left. For this reason, the reaction arrows are usually shown as "$\leftrightarrow$."

Therefore, in a nickel-cadmium battery, the negative plate reactions are:

$$Cd + 2OH^- \leftrightarrow Cd(OH)_2 + 2e^-$$

The positive plate reactions are:

$$2NiO(OH) + 2H_2O + 2e^- \leftrightarrow 2Ni(OH)_2 + 2OH^-$$

Lead-Acid Battery Characteristics

	TYPE	COST	AVAILABILITY	DEEP-CYCLE PERFORMANCE	TEMPERATURE TOLERANCE	MAINTENANCE
Flooded electrolyte	Lead-antimony	low	very good	good	good	high
	Lead-calcium open-vent	low	very good	poor	poor	medium
	Lead-calcium sealed-vent	low	very good	poor	poor	low
	Lead-antimony/lead-calcium	low	limited	good	good	medium
Captive electrolyte	Lead-calcium sealed-vent	medium	limited	fair	poor	low
	Lead-antimony/lead-calcium	medium	limited	fair	poor	low

Figure 6-17. *The characteristics of lead-acid batteries vary between different designs.*

Lead-antimony batteries are a robust design with thick plates and are classified as traction batteries. They are well suited to PV applications due to their deep-cycle capability and tolerance of abuse. Most lead-antimony batteries are flooded, open-vent types with removable caps to permit water additions. The frequency of water additions can be minimized by using catalytic recombination caps or having cases with excess electrolyte reservoirs.

Lead-Calcium Batteries. Lead-calcium battery plate grids include 3% to 6% calcium (Ca), which reduces gassing and lowers the self-discharge rate. Less water is lost in lead-calcium batteries than in lead-antimony batteries, so they generally require less maintenance. Disadvantages of lead-calcium batteries include poor charge acceptance after deep discharges, shortened battery life and capacity at higher operating temperatures, and intolerance of overcharging.

Lead-calcium batteries can be either captive-electrolyte or flooded-electrolyte designs. Lead-calcium batteries make up the majority of captive-electrolyte designs and are commonly used in PV applications. Flooded lead-calcium batteries with open vents are stationary batteries, which are not suitable for most PV applications. Flooded lead-calcium batteries with sealed vents are primarily SLI batteries developed as "maintenance-free" automotive starting batteries. They are maintenance-free in the sense that they incorporate sufficient reserve electrolyte to operate over their expected service life without additional water. They have a low cost, but they are designed for shallow

cycles and generally do not last long in most PV applications. These batteries can be used in small stand-alone PV systems such as in rural homes and lighting systems, but must be carefully charged and discharged to achieve useful performance and life.

Hybrid Batteries. A *hybrid battery* is a battery that uses a combination of plate designs to maximize the desirable characteristics of each. The most common type of hybrid battery uses lead-calcium positive plate grids and lead-antimony negative plate grids. This design combines the advantages of the lead-calcium and lead-antimony designs, including good deep-cycle performance, low water loss, and long life. However, stratification and sulfation can be problems with these batteries, and they must be maintained accordingly. These batteries are sometimes used in PV systems with larger capacities and deep cycle requirements. Hybrid batteries are usually flooded designs, although models using gelled electrolyte have been developed.

Nickel-Cadmium Batteries

Nickel-cadmium (Ni-Cd) batteries are secondary batteries with several advantages over lead-acid batteries that make them well suited for PV systems, including long life, low maintenance, excessive discharge tolerance, excellent low-temperature capacity retention, and noncritical voltage regulation requirements. The main disadvantages of nickel-cadmium batteries are high initial cost and limited availability compared to lead-acid designs. **See Figure 6-18.**

Nickel-Cadmium Battery Characteristics

TYPE	COST	AVAILABILITY	DEEP CYCLE PERFORMANCE	TEMPERATURE TOLERANCE	MAINTENANCE
Captive electrolyte sintered plate	high	very good	good	good	none
Flooded electrolyte pocket plate	high	limited	good	excellent	medium

Figure 6-18. *Two types of nickel-cadmium batteries have different performance characteristics.*

A nickel-cadmium cell consists of positive plates of nickel hydroxide (NiO(OH)) and negative plates of cadmium (Cd), immersed in an alkaline potassium hydroxide (KOH) electrolyte solution. The concentration, and therefore the specific gravity, of the electrolyte does not change during discharging or charging reactions.

The nominal voltage for a Ni-Cd cell is 1.2 V, requiring 10 Ni-Cd cells to be configured in series for a nominal 12 V battery. The voltage of a Ni-Cd cell remains relatively stable until the cell is almost completely discharged, then drops off dramatically. Ni-Cd batteries can accept charge rates as high as C/1, and are tolerant of continuous overcharge up to a C/15 rate. Two primary types of Ni-Cd batteries are sintered plate and pocket plate batteries.

Sintered Plate Ni-Cd Batteries. Sintered plate Ni-Cd batteries are made by heat-processing active materials and rolling them into a cylindrical metallic case. The electrolyte is immobilized, preventing leakage and allowing operation in any orientation. These batteries are commonly used in consumer electronic devices. The main disadvantage of sintered plate designs is the memory effect, in which a battery that is repeatedly discharged to only a percentage of its rated capacity will eventually "memorize" this cycle pattern and limit further discharge, resulting in loss of capacity. In some cases, the memory effect can be partially reversed, regaining some capacity, by conducting special charge and discharge cycles. However, sintered plate Ni-Cd batteries are not recommended for PV applications.

Pocket Plate Ni-Cd Batteries. Pocket plate Ni-Cd batteries are large flooded batteries available in capacities up to and over 1000 Ah. These batteries can withstand deep discharges and temperature extremes much better than lead-acid batteries, and they do not experience the memory effect associated with sintered plate batteries, so they are suitable for use in PV systems. Similar to flooded lead-acid designs, these batteries require periodic water additions. The main disadvantage of pocket plate Ni-Cd batteries is high initial cost. However, because of their long lifetimes, the life cycle cost can be the lowest among the battery types for PV applications.

BATTERY SYSTEMS

Battery selection and system design involves many decisions and compromises. Choosing the best battery for a PV application depends on many factors and involves a careful review of the battery specifications with respect to the particular application needs. Some decisions on battery selection may be easy to make, such as required capacity, physical properties, and cost, while other decisions are much more difficult and involve compromises between desirable and undesirable battery features.

Battery Selection

Each battery type has design and performance features suited for particular applications. System designers must consider the advantages and disadvantages of different battery types with respect to the requirements of a particular system. Considerations include lifetime, deep cycle performance, tolerance to high temperatures and overcharge, and maintenance requirements. **See Figure 6-19.** Batteries of different types or ages should not be mixed in the same bank.

Flame arrestor vent caps absorb much of the vented flammable gas in a charcoal filter. These vent caps also help to prevent cell explosions by isolating cells from outside ignition sources.

Batteries are potentially dangerous because they contain hazardous materials and chemicals and store a large amount of electrical energy. Depending on the battery system, certain safety precautions are required.

Personnel Protection
When maintaining batteries, personnel should wear protective clothing such as aprons, ventilation masks, goggles or face shields, and gloves to protect from acid spills or splashes and fumes. Safety showers and eyewashes may be required if batteries are located close to personnel. A fire extinguisher should be located in close proximity to the battery area. In some applications, automated fire sprinkler systems may be required for safety and to protect facilities and expensive load equipment. OSHA regulations, building codes, and other applicable standards must be consulted before installing battery systems.

Electrical Hazards
Batteries can deliver extremely high short-circuit currents, as high as several thousand amperes for moderate-size battery banks. These currents can cause severe electrical burns and can weld tools and other metallic equipment in the circuit. High-voltage battery banks must be isolated into lower-voltage segments for service and maintenance.

To reduce the potential for short circuits, live battery terminals must be protected with plastic or rubber covers. When covers are removed for battery maintenance, service personnel should not wear any watches, rings, bracelets, or other jewelry that could potentially come into contact with the battery terminals. All tools used around batteries should have insulated handles to prevent electrical contact.

Electrolyte Hazards
The electrolyte in lead-acid batteries contains sulfuric acid, which is highly caustic and can destroy clothing and burn skin. For these reasons, protective clothing such as rubber or plastic aprons and face shields should be worn when working with batteries. To neutralize sulfuric acid spills or splashes on clothing, the spill should be rinsed immediately with a solution of baking soda or household ammonia and water. In nickel-cadmium batteries, the electrolyte is a potassium hydroxide solution that can be neutralized with a mixture of vinegar and water.

If electrolyte is accidentally splashed in the eyes, the eyes should be forced open and flooded with cool, clean water for fifteen minutes. If acid electrolyte is ingested, large quantities of water or milk should be drunk, followed by milk of magnesia, beaten eggs, or vegetable oil. A physician should be called immediately.

When preparing an electrolyte solution from concentrated acid and water, the acid should always be poured slowly into the water while mixing. The water should never be poured into the acid. Appropriate nonmetallic funnels and containers should be used when mixing and transferring electrolyte solutions.

Explosion Hazards
During charging, batteries may produce explosive mixtures of hydrogen and oxygen gases that may be present for several hours after charging. Keep sparks, flames, burning cigarettes, or other ignition sources away from batteries at all times. Active or passive ventilation techniques are suggested and often required, depending on the number of batteries located in an enclosure and their gassing characteristics. The use of battery vent caps with a flame arrester feature lowers the possibility of a catastrophic battery explosion. Improper charging and excessive overcharging may increase the possibility of battery explosions.

Batteries should never be connected or disconnected at the terminals while under load. This can cause arcing, which could cause a battery explosion. All loads and charging sources should be disconnected at points away from the batteries before servicing.

Battery Banks
A *battery bank* is a group of batteries connected together with series and parallel connections to provide a specific voltage and capacity. Battery banks can be configured to produce many different voltages. Required battery bank voltage is often determined by load or inverter input voltage requirements. Most PV systems with batteries operate at 12 V, 24 V, or 48 V. For larger loads, it is recommended to use a higher voltage, which lowers the system currents. For example, a 120 W DC load operating from a 12 V battery draws 10 A, but when operating from a 24 V battery, a similar 120 W load draws only 5 A. Lower system current reduces the size and cost of conductors, fuses, disconnects, and other current-handling components in the PV system.

Battery Selection Criteria

SYSTEM REQUIREMENTS

System configuration
Discharge current
Daily depth of discharge
Autonomy
Accessibility
Temperature

BATTERY CHARACTERISTICS

Energy storage density
Sealed or unsealed
Allowable depth of discharge
Charging characteristics
Life cycles
Electrolyte specific gravity
Freezing susceptibility
Sulfation susceptibility
Gassing characteristics
Self-discharge rate
Maintenance requirements
Size and weight
Terminal configuration
Auxiliary hardware availability
Manufacturer reputation
Cost
Warranty

Figure 6-19. *System requirements and characteristics of battery types must be considered when choosing a battery.*

Series Connections. Just like modules in an array, batteries are first connected in series by connecting the negative terminal of one battery to the positive terminal of the next battery, for as many batteries as are in the series string. Because there is only one path for the current to flow, the circuit current remains the same as the individual battery current. For batteries of similar capacity and voltage connected in series, the circuit voltage is the sum of the individual battery voltages, and the circuit capacity is the same as the capacity of the individual batteries. **See Figure 6-20.** If batteries or cells with different capacities are connected in series, the capacity of the string is limited by the lowest-capacity battery.

Parallel Connections. Batteries are connected in parallel by connecting all of the positive terminals together and all of the negative terminals together. Batteries connected in parallel provide more than one path for current to flow, so currents add together at the common connections. The current of the parallel circuit

is the sum of the currents from the individual batteries. The voltage across the circuit is the same as the voltage across the individual batteries, and the overall capacity is the sum of the capacities of each battery. **See Figure 6-21.** Series strings of batteries can also be connected in parallel in the same way.

Batteries in Series

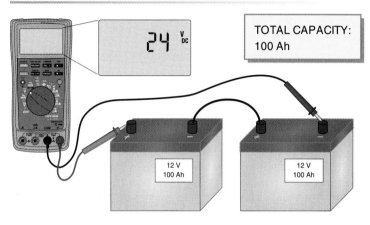

Figure 6-20. *Connecting batteries in series increases system voltage.*

Batteries in Parallel

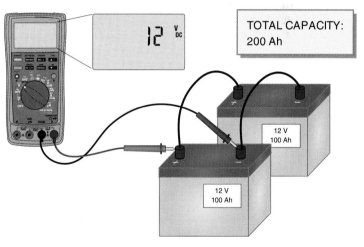

Figure 6-21. *Connecting batteries in parallel increases system capacity.*

Pasted plates are flat grids covered by active material with a peanut butter–like consistency. The active material fills in the spaces between the ribs of the grid. Tubular plates consist of rows of thin vertical rods with active material plated around them, forming tubes.

Baltimore Electrical JATC Training Center

Location: Baltimore, MD (39.3°N, 76.7°W)
Type of System: Utility-interactive with battery backup
Peak Array Power: 960 W DC
Date of Installation: September 2004
Installers: Apprentices and journeymen
Purposes: Training, supplemental electrical power

The Baltimore Electrical JATC facility features a pre-engineered utility-interactive PV system with battery backup. The system includes eight utility-interconnected PV modules and a battery backup system. The system's four 12 V batteries supply power to the building during utility outages.

Training was the primary purpose for installing the system. Almost all of the electrical components are housed inside a rolling cabinet, facilitating demonstrations and instruction on the various parts. The cabinet includes the inverter, charge controller, batteries, breakers, disconnects, and other BOS components. By disconnecting the two main conductors, one for the DC power from the array and the other for AC power out to the building electrical system, the cabinet can be easily moved around for training.

The batteries are housed in a ventilated section of the cabinet so that toxic and explosive gases cannot accumulate. Since the cabinet is located inside the

Baltimore Electrical JATC

Eight modules mounted on the roof supply supplemental power to offset a portion of utility power.

building, the batteries remain close to room temperature. A temperature probe between the batteries helps the charge controller adjust charging operations as needed. Maintenance is minimal.

The impact of the additional PV power on the facility's utility bills is currently small. However, the JATC is considering its options for a larger PV system in the future. The training and experience from the current system might then be applied to the future project to offset a more significant portion of their electricity use and reduce their dependence on the utility.

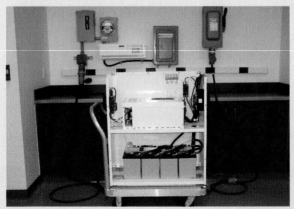

Baltimore Electrical JATC

Most of the electrical components are housed in a cabinet that can be used for teaching purposes.

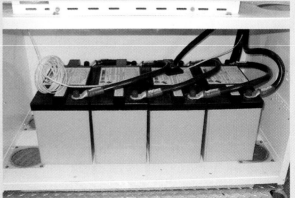

Baltimore Electrical JATC

The PV system includes four 12 V batteries connected in series to supply power to the building in the event of a utility outage.

It is generally recommended that batteries be connected in as few parallel strings as possible. Slight voltage differences between the batteries may occur due to the length, resistance, and integrity of the parallel connections. These voltage differences can lead to inconsistencies in the charge received by each string, eventually causing unequal capacities within the system. The strings with the lowest circuit resistance will be exercised more than the strings of batteries with greater circuit resistance. The batteries in strings that receive less charge may begin to sulfate prematurely.

For PV systems with large capacity requirements, larger batteries allow configurations of one series string rather than several parallel strings. When batteries must be configured in parallel, the connections between the battery bank and the PV power system should be made from opposite sides of the battery bank to improve the distribution of charge and discharge from the bank. **See Figure 6-22.**

Enclosures

Electrical codes and safety standards generally require batteries to be installed in an enclosure separated from controls or other PV system components. Battery enclosures must be of sufficient size and strength to hold the batteries and can be located below ground if necessary to prevent freezing. If the enclosure is located above ground, shading or a reflective coating should be used to limit the direct exposure to sunlight. Avoiding extreme temperature swings improves battery performance, extends battery life, and decreases the need for maintenance.

Large battery banks may be installed in dedicated battery rooms. Batteries may be installed on racks to secure the batteries and provide added structural support.

An enclosure may be insulated or may have cooling or heating mechanisms to protect batteries from extreme temperatures. Passive cooling enclosures reduce battery temperature without the use of active components such as motors, fans, or air conditioners. Active temperature regulation requires electrical power, increasing the complexity, size, and cost of the PV system.

Batteries in Series and Parallel

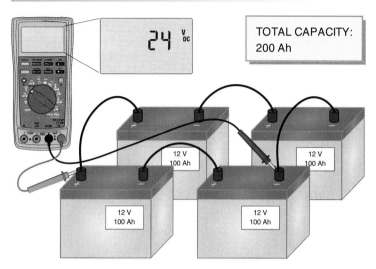

Figure 6-22. *Series and parallel connections can be combined to produce a desired system voltage level and capacity.*

Racks. Battery racks must be made from rigid materials, either metallic or nonmetallic, that are resistant to deterioration from electrolyte. The parts directly contacting the batteries must be electrically insulated or made from nonconductive materials. Batteries must be arranged to provide sufficient working space for inspection and maintenance.

Electrolyte Containment. Battery systems should include some means to contain electrolyte in the event of a spill. Individual batteries may have double-walled cases to contain electrolyte if the inner case is breached. Pans or trays underneath the batteries or liquid-tight battery bank enclosures also provide containment in the event of a spill.

Ventilation. Because batteries can produce toxic, corrosive, and explosive mixtures of gases, ventilation of the battery enclosure is usually required. For small battery systems charged at low rates, accumulation of gases is minimal, and natural ventilation may be adequate. When properly regulated, VRLA batteries may not require special ventilation and may even be used indoors. Otherwise, fans can be used to provide mechanical ventilation. Under no circumstances should batteries be kept in an unventilated area or located in an area accessible to unqualified persons.

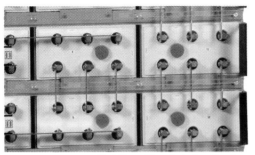

East Penn Manufacturing Co., Inc.

Batteries are connected together in series and parallel to form a battery bank.

Refer to Quick Quiz® on CD-ROM

Overcurrent Protection and Disconnects

Like any other electrical system, battery systems must have proper DC-rated overcurrent protection and disconnects to protect system conductors and to isolate the battery bank from the rest of the system for testing and maintenance. Batteries can deliver extremely high discharge currents, so overcurrent protection devices must have the appropriate interrupting ratings.

Summary

- Batteries are collections of electrochemical cells electrically connected together in series.

- Electrochemical reactions produce a flow of electrons from the negative terminals to the positive terminals of a cell.

- Physical factors affecting capacity include the quantity of active material; the number, design, and dimensions of the plates; and the electrolyte concentration.

- Operational factors affecting capacity include discharge rate, charging method, temperature, age, and condition of the cell or battery.

- Discharging and charging rates are based on the amount of charge applied in one hour.

- Charging is done by applying an electrical current to the cell or battery in a direction opposite to the discharge. In order for the battery to accept current, the voltage of the charging source must be higher than the battery voltage.

- State of charge is the available capacity within a battery, while depth of discharge is the capacity that has been removed.

- Bulk charging quickly charges a battery up to the regulation voltage, and absorption charging slowly completes the charge.

- Equalization is a controlled overcharge that ensures that each cell is fully charged.

- Overcharging causes water in the electrolyte to form gases, which escape through cell vents.

- Water lost from gassing of open-vent batteries must be replaced.

- Water lost from gassing of sealed or captive-electrolyte batteries cannot be replenished.

- Temperature can have significant effects on capacity, electrolyte specific gravity, self-discharge, gassing voltage, voltage setpoints, and battery life.

- The concentration, or specific gravity, of the electrolyte can indicate state of charge and the general health of a cell.

- Sulfation and stratification are common battery problems that can decrease battery performance and life.

- Uncontrolled discharging and charging can result in loss of battery capacity and life.

- The most common batteries used for PV applications are traction batteries.

- Electrolyte can be in a free-flowing liquid form (flooded) or an immobilized form (captive).

- Lead-acid batteries are highly suitable for most PV applications but usually require a lot of maintenance.

- Some nickel-cadmium batteries are suitable for most PV applications and have good temperature tolerance but can be expensive and difficult to obtain.

- Required battery bank voltage is often determined by load or inverter input voltage requirements. Most PV systems with batteries operate at 12 V, 24 V, or 48 V.

- Batteries can be connected in series and parallel combinations to produce a desired system voltage level and capacity.

- It is generally recommended that batteries be connected in as few parallel strings as possible.

- Electrical codes and safety standards generally require batteries to be installed in an enclosure separated from controls or other PV system components.

Definitions

- A *battery* is a collection of electrochemical cells that are contained in the same case and connected together electrically to produce a desired voltage.

- A *battery cell* is the basic unit in a battery that stores electrical energy in chemical bonds and delivers this energy through chemical reactions.

- A *plate* is an electrode consisting of active material supported by a grid framework.

- *Active material* is the chemically reactive compound on a battery cell plate.

- The *grid* is a metal framework that supports the active material of a battery cell and conducts electricity.

- *Electrolyte* is the conducting medium that allows the transfer of ions between battery cell plates.

- *Steady-state* is an open-circuit condition where essentially no electrical or chemical changes are occurring.

- The *open-circuit voltage* is the voltage of a battery or cell when it is at steady-state.

- *Capacity* is the measure of the electrical energy storage potential of a cell or battery.

- *Discharging* is the process of a cell or battery converting chemical energy to electrical energy and delivering current.

- The *cutoff voltage* is the minimum battery voltage specified by the manufacturer that establishes the battery capacity at a specific discharge rate.

- The *state of charge (SOC)* is the percentage of energy remaining in a battery compared to the fully charged capacity.

- *Depth of discharge (DOD)* is the percentage of withdrawn energy in a battery compared to the fully charged capacity.

- The *allowable depth of discharge* is the maximum percentage of total capacity that is permitted to be withdrawn from a battery.

- The *average daily depth of discharge* is the average percentage of the total capacity that is withdrawn from a battery each day.

- *Autonomy* is the amount of time a fully charged battery system can supply power to system loads without further charging.

- *Self-discharge* is the gradual reduction in the state of charge of a battery while at steady-state condition.

- A *cycle* is a battery discharge followed by a charge.

- *Charging* is the process of a cell or battery receiving current and converting the electrical energy into chemical energy.

- The *gassing voltage* is the voltage level at which battery gassing begins.

- *Bulk charging* is battery charging at a relatively high charge rate that charges the battery up to a regulation voltage, resulting in a state of charge of about 80% to 90%.

- *Absorption charging* is battery charging following bulk charging that reduces the charge current to maintain the battery voltage at a regulation voltage for a certain period.

- *Float charging* is battery charging at a low charge rate that maintains full battery charge by counteracting self-discharge.

- *Equalizing charging* is current-limited battery charging to a voltage higher than the bulk charging voltage, which brings each cell to a full state of charge.

- *Gassing* is the decomposition of water into hydrogen and oxygen gases as a battery charges.

- *Overcharge* is the ratio of applied charge to the resulting increase in battery charge.

- *Specific gravity* is the ratio of the density of a substance to the density of water.

- *Sulfation* is the growth of lead sulfate crystals on the positive plate of a lead-acid cell.

- *Stratification* is a condition of flooded lead-acid cells in which the specific gravity of the electrolyte is greater at the bottom than at the top.

- A *primary battery* is a battery that can store and deliver electrical energy but cannot be recharged.

- A *secondary battery* is a battery that can store and deliver electrical energy and can be charged by passing a current through it in an opposite direction to the discharge current.

- A *traction battery* is a class of battery designed for repeated deep-discharge cycle service.

- A *starting, lighting, and ignition (SLI) battery* is a class of battery designed primarily for shallow-discharge cycle service.

- A *stationary battery* is a class of battery designed for occasional deep-discharge limited-cycle service.

- *Flooded electrolyte* is electrolyte in the form of a liquid.

- A *catalytic recombination cap (CRC)* is a vent cap that reduces electrolyte loss from an open-vent flooded battery by recombining vented gases into water.

- A *catalyst* is a substance that causes other substances to chemically react but does not participate in the reaction.

- *Captive electrolyte* is electrolyte that is immobilized.

- A *hybrid battery* is a battery that uses a combination of plate designs to maximize the desirable characteristics of each.

- A *battery bank* is a group of batteries connected together with series and parallel connections to provide a specific voltage and capacity.

1. What is the difference between a battery and a cell?

2. Describe the chemical reactions that take place when a load or conductor is connected to the positive and negative terminals of a battery.

3. What factors affect the capacity of a cell or battery?

4. Explain the relationships between electrolyte specific gravity, freezing point, and state of charge (SOC).

5. Why are traction batteries ideal for most PV applications?

6. What are some relative advantages and disadvantages of captive-electrolyte batteries?

7. How does average daily DOD affect autonomy?

8. How does discharge rate affect battery capacity?

9. How can series and parallel connections be used to design a battery system with a specific voltage and capacity?

7

ARRAY

Charge Controllers

Batteries in PV systems experience a wide range of operating conditions, and must be properly regulated to maximize their performance and lifespan, as well as for safety purposes. Almost every PV system that uses batteries requires a charge controller. Charge controllers manage and monitor battery charging while protecting the batteries from overcharge and overdischarge.

Chapter Objectives

- Identify the principal functions and features of charge controllers.
- Define charge regulation and load control setpoints.
- Explain how temperature compensation affects setpoints.
- Differentiate between various types of charge controller algorithms and switching designs.
- Understand the impacts of control algorithms and setpoints on system and battery performance.
- Identify concerns and requirements for charge controller applications and installation.
- Identify the requirements for PV systems operating without charge control.

CHARGE CONTROLLER FEATURES

A *charge controller* is a device that regulates battery charge by controlling the charging voltage and/or current from a DC power source, such as a PV array. A charge controller in a PV system maintains a battery at its highest possible state of charge while protecting the battery from overcharge by the array and overdischarge by system loads. This maximizes available battery capacity and cycle life. However, charge controllers alone cannot ensure that a battery receives a full charge. The array must be large enough to adequately supply the load and charge the batteries during periods of low insolation.

Charge controllers orchestrate the interaction between major PV system components. They manage energy flow between the array, battery bank, and electrical loads. After system sizing, charge controllers have the greatest impact on overall system performance. **See Figure 7-1.** Some charge controllers also include load control and other energy management features, and may indicate system status through LEDs or LCD displays.

Charge controllers are characterized by their means and methods of charge regulation and the battery conditions (setpoints) used to trigger charge control functions. Charge controllers are rated for specific voltage and current levels. Many include a number of additional features and functions to control system loads, display and record system data, indicate system operational status, or otherwise enhance the ability to manage and optimize the performance of the system.

Battery Charging

Charge controllers manage the array current delivered to a battery, and consequently affect the way a battery accepts charge. *Charge acceptance* is the ratio of the increase in battery charge to the amount of charge supplied to the battery. Charge acceptance is also known as charge efficiency. For example, if a charge cycle delivers 60 Ah in order to gain 54 Ah of additional battery charge, the charge acceptance is 90% (54 Ah ÷ 60 Ah = 0.90 or 90%). Since a given amount of capacity always requires more charge to overcome losses and inefficiencies, charge acceptance is always less than 100%.

Charge Controllers

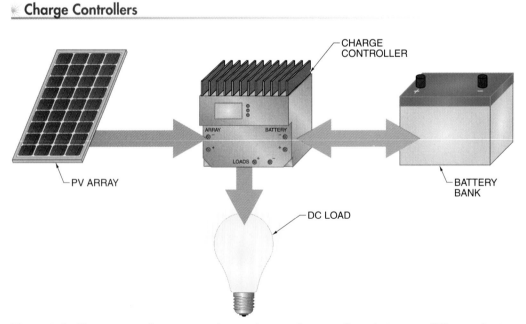

Figure 7-1. *Charge controllers manage interactions and energy flows between a PV array, battery bank, and electrical load.*

Charge controller designs use various methods to regulate current and voltage during battery charging. Single-stage charging processes are simpler, but relatively inefficient. Multistage charging changes the applied voltage and/or current in steps to bring the battery to a higher state of charge. This process is more sophisticated, requiring additional electronics, but it greatly improves charge acceptance. **See Figure 7-2.** Multistage methods can charge batteries faster and limit excessive overcharge and gassing.

Overcharge is the ratio of applied charge to the resulting increase in battery charge. For example, a battery that accepts 60 Ah and regains 54 Ah of charge receives an overcharge of 111% (60 Ah ÷ 54 Ah = 1.11 or 111%). Overcharge is the inverse of charge acceptance.

The amount of overcharge required depends on the type of battery and its condition and state of charge. It also depends on the rate and manner in which charge is delivered to the battery, which is governed by the charge controller. Charge acceptance is generally greater than 90% for most batteries at normal charge rates up to about 90% state of charge. Then it decreases significantly due to internal losses as the battery reaches full charge. Correspondingly, a higher overcharge is required to return a battery to full state of charge from a shallow depth of discharge than from a greater depth of discharge. During the final stages of charging, some of the current supplied to the battery is consumed by gassing and other internal battery losses, lowering charge efficiency.

Overcharge Conditions

It should be noted that the term "overcharge" may be used to describe two different conditions of battery charging. First, normal charging requires that more charge be applied to the battery than the expected gain in state of charge. This type of overcharge is necessary because some of the charge is lost through various battery inefficiencies and gassing. As an analogy, imagine a cup with a small hole at the bottom. As water is poured in (charging), a small amount leaks out (losses). Ultimately, the amount of water in the cup (state of charge) is slightly less than the amount of water that was poured in (applied charge). Therefore, a little extra water (overcharge) is needed to reach a certain level.

The second definition of overcharge applies to batteries that are already fully charged. When the water level reaches the top of the cup (full charge), the pouring (charging current) should be stopped or reduced to a very small amount. (A very small charging current replaces the charge lost through self-discharge.) If the pouring (current) is not stopped or reduced, the cup will overflow (overcharge). This overcharge results in excessive gassing and can damage the battery.

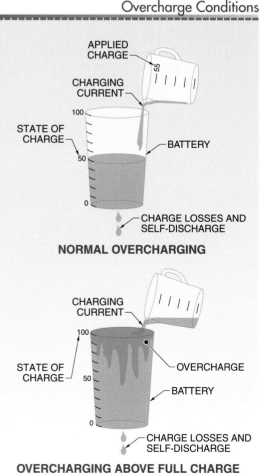

Charging

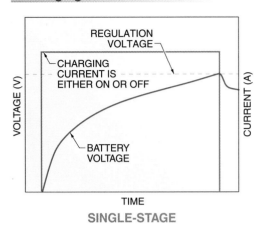

SINGLE-STAGE

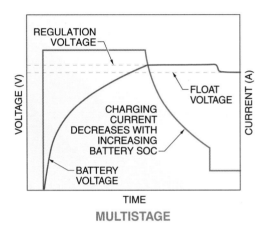

MULTISTAGE

Figure 7-2. *Single-stage battery charging is simpler, but multistage battery charging brings batteries to a higher state of charge.*

Array switching can be done on either the positive or negative leg of the array circuit. However, charge controllers that switch the negative leg may not be grounded to the negative leg, which can affect controller regulation functions.

Undercharged batteries are a major cause of PV system failures and inability to meet system loads. *Deficit charge* is a charge cycle in which less charge is returned to a battery bank than what was withdrawn on discharge. Deficit charge can occur in PV systems during periods of low insolation and high load, particularly for marginally sized systems. It may also be caused by the regulation voltage being set too

low on the charge controller. In properly sized systems, deficit charge should be recoverable when insolation and load conditions return to normal. However, if lead-acid batteries experience repeated deficit charge cycles and are left at partial state of charge for long periods, they will become progressively more difficult to charge due to sulfation and/or stratification. When charged, voltage will rise more quickly in batteries with sulfation or stratification than in normal batteries, falsely indicating a high state of charge. Recovery from this condition can sometimes be accomplished by an equalization charge.

Overcharge Protection

During high insolation periods, energy generated by an array may exceed the electrical load demand. The excess may charge a battery, but if it is already at a high state of charge, then the charging current must be carefully controlled to avoid overcharging. In this case, *overcharge* is the condition of a fully charged battery continuing to receive a significant charging current. A charge controller protects a battery from overcharge through charge regulation. This involves either interrupting or limiting the current flow from the array when the battery approaches a full state of charge. **See Figure 7-3.**

Active means of overcharge protection are required for most PV systems using batteries. According to the NEC®, any PV system employing batteries shall have equipment to control the charging process unless the maximum array charge current multiplied by one hour is less than 3% of the rated battery capacity in ampere-hours. This equates to a charge rate of C/33. For example, a 100 Ah battery requires charge control if the maximum array charging current exceeds 3 A.

Without charge control, the current from the array will flow into a battery proportionally to the irradiance. When a battery is fully charged, unregulated charging will cause the battery voltage to reach exceedingly high levels, causing severe gassing, electrolyte loss, internal heating, and accelerated grid corrosion. If a battery is not protected from overcharge, it will likely lose capacity and fail prematurely.

Overcharge Protection

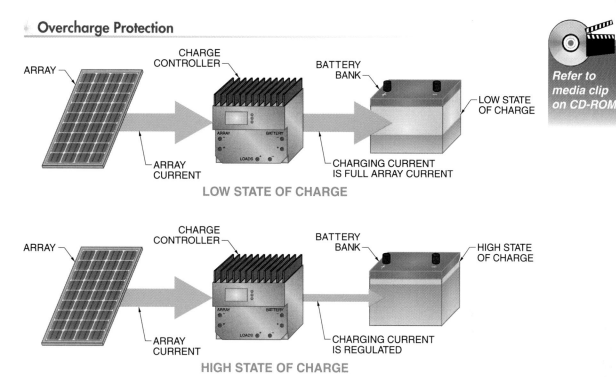

Figure 7-3. *Charge controllers protect batteries from overcharge by terminating or limiting charging current.*

Overdischarge Protection

During periods of low insolation or excessive load demand, a marginally sized array may not produce sufficient energy to charge the battery. *Overdischarge* is the condition of a battery state of charge declining to the point where it can no longer supply discharge current at a sufficient voltage without damaging the battery.

A charge controller protects a battery from overdischarge through load control, disconnecting electrical loads when the battery reaches a low voltage (low state of charge) condition. **See Figure 7-4.** Overdischarge protection limits battery depth of discharge and prevents damage to the battery. Since some loads will operate improperly, or not at all, at lower than normal voltages, this feature also protects them. Under cold conditions, overdischarge protection also prevents a battery from freezing. Once the battery is charged to a certain level, the loads are reconnected to the battery.

Some controllers combine the charge regulation and load control functions in order to protect batteries from both overcharge and overdischarge, while others have only one function. Also, these charge and discharge functions can be combined with other electrical components, such as inverters, to form multifunction power conditioning units (PCUs).

Status Information

Most charge controllers provide information on the operational state of the system. For charge regulation, this may include digital displays or LED lights to indicate when the battery is charging and when it reaches full charge. Other indicators may be used to alert operators of a low battery voltage or low state of charge condition, such as a red light or an audible alarm. **See Figure 7-5.** More advanced controllers may include meters to monitor battery and array currents and voltages or track the amount of energy supplied to and from a battery. This system status information can be extremely useful for improving load management and optimizing system performance.

Overdischarge Protection

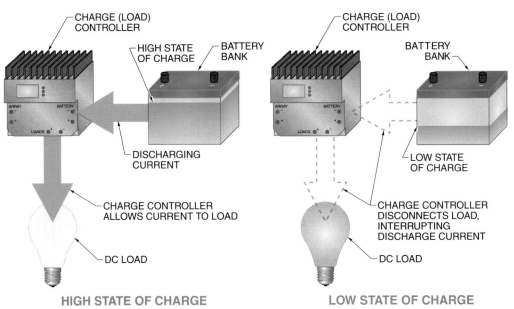

Figure 7-4. *Charge controllers protect batteries from overdischarge by disconnecting loads at low battery voltage.*

Charge Controller Status

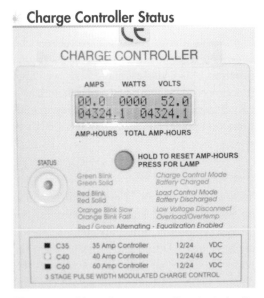

Figure 7-5. *Most charge controllers include displays or LEDs to indicate battery voltage, state of charge, and/or present operating mode.*

In small PV systems, the charge controller may also simplify wiring by providing conductor termination points for the array, battery, and load circuits. Most charge controllers have easy-to use screw terminals for making connections.

Circuit Protection

Some charge controllers incorporate features to protect the unit or loads from damage due to high currents or improper wiring, such as reversed array or battery polarity. Overcurrent devices, such as a fuse or circuit breaker, protect the controller from array or load currents significantly higher than its rating.

Some controllers also protect system loads by sensing when the battery is disconnected and then disconnecting the array, preventing potentially high array open-circuit voltages or unregulated currents from damaging the loads.

Advanced Load Control

Charge controllers also may provide other load control functions besides battery overdischarge protection. For example, charge controllers in PV lighting systems may activate and control the lighting load based on ambient light conditions or time of day. The array can operate as a light sensor, indicating nighttime conditions when the controller measures the array current or voltage drop to a preset low level.

Alternatively, clocks or timers integrated into charge controllers can activate loads for prescribed periods.

CHARGE CONTROLLER TYPES

Charge controllers are characterized by the way they regulate charging current to a battery. A *charge control algorithm* is a programmed series of functions that a charge controller uses to control current and/or voltage in order to maintain battery state of charge. Charge controllers use a variety of algorithms and switching methods to provide charge regulation. In conjunction with the charge control setpoints, these algorithms are selected or modified to optimize battery charging, array energy utilization, load availability, and overall system performance.

Most charge controllers use solid-state switching elements, typically transistors. An *interrupting-type charge controller* is a charge controller that switches the charging current ON and OFF for charge regulation. Interrupting-type charge controllers are simple designs, but not widely used in PV systems. Since the array current is switched abruptly, they have difficulty avoiding excessive overcharge and are best used with the more tolerant flooded, open-vent batteries. A *linear-type charge controller* is a charge controller that limits the charging current in a linear or gradual manner with high-speed switching or linear control. Linear-type charge controllers provide a more consistent charging process, so they can generally charge batteries more efficiently and faster. Linear-type controllers are compatible with more types of batteries.

Shunt Charge Controllers

Unlike batteries, PV devices are current-limited by nature, so PV modules and arrays can be short-circuited without any harm. A *shunt charge controller* is a charge controller that limits charging current to a battery system by short-circuiting the array. The array is short-circuited through a shunt element inside the charge controller, moving the array's op-

erating point on the I-V curve very near the short-circuit condition and limiting the power output. All shunt controllers must also include a blocking diode in series between the battery and the shunt element to prevent the battery from short-circuiting. **See Figure 7-6.**

Because there is some voltage drop through the shunt element and blocking diode, the array is never completely short-circuited, resulting in some power dissipation within the controller. For this reason, most shunt controllers include a heat sink to dissipate power and are generally limited to array source circuits with currents of 20 A or less.

Shunt-Interrupting Charge Controllers. A *shunt-interrupting charge controller* is a charge controller that suspends charging current to a battery system by completely short-circuiting the array. When the battery voltage falls, the controller reconnects the array to resume charging the battery. This ON/OFF cycling of the array current may occur once over a period of several minutes, but at higher charge rates and as the battery approaches full state of charge, the cycling may occur every few seconds.

Shunt-Linear Charge Controllers. A *shunt-linear charge controller* is a charge controller that limits charging current to a battery system by gradually lowering the resistance of a shunt element through which excess current flows. When the battery state of charge and array current are high enough during the day, a shunt-linear controller maintains the battery at the regulation voltage. In some designs, a comparator circuit in the controller monitors the battery voltage, and makes corresponding adjustments to the resistance of the shunt element to regulate array current. A Zener power diode can also be used for simple shunt-linear designs.

Shunt charge controllers are so named because they divert array current to a parallel (or shunt) path when charge regulation is required. This is the same as short-circuiting the array. Although most electrical power sources are extremely dangerous to short circuit, it can be done safely with PV devices because they are inherently current-limited.

Series Charge Controllers

A *series charge controller* is a charge controller that limits charging current to a battery system by open-circuiting the array. As the battery reaches full state of charge, a switching element inside the controller opens, moving the array's operating point on the I-V curve to the open-circuit condition and limiting the power output. This method works in series between the array and battery, rather than in parallel as for the shunt controller. **See Figure 7-7.**

Shunt Charge Controllers

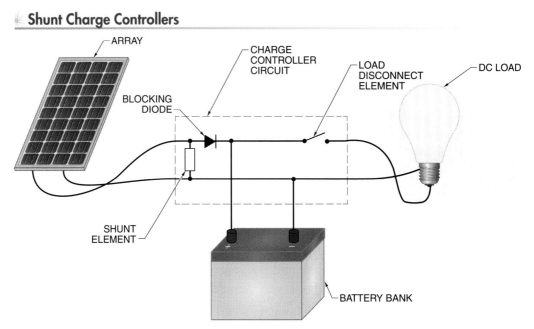

Figure 7-6. *Shunt charge controllers regulate charging current by short-circuiting the array.*

Series Charge Controllers

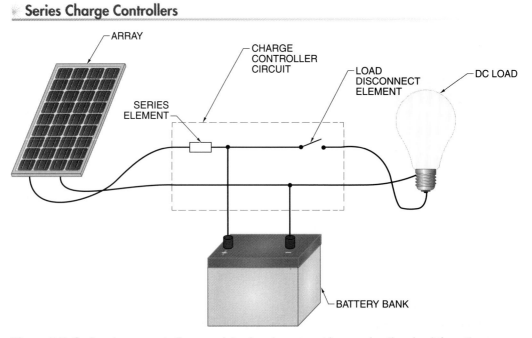

Figure 7-7. *Series charge controllers regulate charging current by opening the circuit from the array.*

Because a series controller open-circuits rather than short-circuits the array, as in a shunt-controller, no blocking diode is needed to prevent the battery from short-circuiting when the controller regulates. Since no blocking diode is used, most series controllers open the circuit between the array and battery at nighttime to prevent reverse current losses from the battery through the array.

Series-Interrupting Charge Controllers. A *series-interrupting charge controller* is a charge controller that completely open-circuits the array, suspending current flow into the battery. When the battery voltage falls, the controller closes the series switching element to reconnect the array and resume charging the battery. Similar to shunt-interrupting charge controllers, the rate of the ON/OFF cycling is dependent on charge rate and battery state of charge.

Series-Linear Charge Controllers. A *series-linear charge controller* is a charge controller that limits charging current to a battery system by gradually increasing the resistance of a series element. This limits current flow into the battery and maintains the battery at the voltage regulation setpoint (as long as array current is sufficiently high). The series regulation element acts like a variable resistor and loads the array at lower current and power levels by operating the array to the right of the maximum power point of its I-V curve. Since series-linear controllers also dissipate power, they generally require a heat sink.

Series-Interrupting, Pulse-Width-Modulated (PWM) Charge Controllers. A *series-interrupting, pulse-width-modulated (PWM) charge controller* is a charge controller that simulates a variable charging current by switching a series element ON and OFF at high frequency and for variable lengths of time. A PWM charge controller pulses the full charging current and varies the width of the pulses to regulate the amount of charge current flowing into the battery. **See Figure 7-8.** The frequency of the pulses is several hundred hertz and the pulses may last only a couple of milliseconds. When the battery is partially charged, the current pulse is essentially ON

all the time. To simulate a lower charging current as the voltage rises, the pulse width is decreased. For example, if the pulses switch the full charging current so that it is ON half the time and OFF half the time, the resulting current effectively simulates a charge current at 50% of the full current.

Pulse-Width Modulation

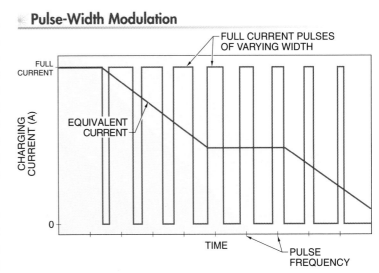

Figure 7-8. *Pulse-width modulation (PWM) simulates a lower current level by pulsing a higher current level ON and OFF for short intervals.*

> Pulse-width modulation can either convey information over a communications channel or control the amount of power sent to a load. Additional applications of PWM include motor controls, telecommunications, audio amplifiers and synthesizers, light dimmers, and voltage regulators.

Similar to the series-linear algorithm in performance, power dissipation within PWM controllers is considerably lower than for other interrupting-type designs. PWM charge controllers can be used with all battery types, and the controlled manner in which power is applied to the battery makes them especially preferred with sealed batteries. PWM charge controllers automatically adjust for varying battery charge acceptance over time, help prevent electrolyte stratification and sulfation, and minimize equalization requirements.

Maximum Power Point Tracking (MPPT) Charge Controllers

Newer charge controllers incorporate the latest in power electronics and microprocessor controls to optimize system performance and allow greater flexibility in system design. A *maximum power point tracking (MPPT) charge controller* is a charge controller that operates the array at its maximum power point under a range of operating conditions, as well as regulates battery charging. An MPPT circuit monitors array output and dynamically changes its resistance or input voltage to move the operating point on the array's I-V curve toward the maximum power point. **See Figure 7-9.** The power is then transformed by a DC-DC converter circuit into another voltage and current required by the load, in this case a battery.

Maximum Power Point Tracking

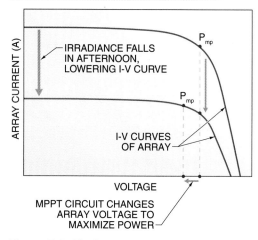

Figure 7-9. *Maximum power point tracking manipulates the load or output voltage of an array in order to maintain operation at or near the maximum power point under changing temperature and irradiance conditions.*

MPPT charge controllers improve the battery-charging performance of PV systems. For example, a nominal 48 V battery bank is typically charged by four 36-cell crystalline silicon modules connected in series to ensure adequate battery-charging voltage under the warmest of operating conditions. Under cool conditions, however, the maximum power voltage of this array may be higher than 65 V, while the battery voltage may be 50 V to 55 V during charging.

Without MPPT control and DC-DC conversion, the array operates at the lower battery voltage, which is a potential power loss up to about 20%. MPPT charge controllers ensure the highest possible array power utilization, and ensure adequate battery-charging voltages under any condition and for a multitude of array and battery voltages.

Diversionary Charge Controllers

Since stand-alone PV systems are designed to supply power to the loads and charge the battery system under worst-case conditions, they often have excess energy available during periods of high insolation or low load. Conventional charge controllers waste this excess power by disconnecting or limiting array current. A *diversionary charge controller* is a charge controller that regulates charging current to a battery system by diverting excess power to an auxiliary load. Instead of wasting excess power, the auxiliary load provides a useful output with the diverted power. **See Figure 7-10.**

The *diversion load* is an auxiliary load that is not a critical system load, but is always available to utilize the full array power in a useful way to protect a battery from overcharge. Examples of diversion loads include loads with inherent energy storage, such as resistance water heaters, and loads that are especially useful under high insolation, such as ventilation fans or irrigation pumps. The inverters in battery-based interactive PV systems are also considered diversion loads.

Since constant availability of the diversion load cannot be guaranteed, the NEC® requires that a second independent means of charge control be used to prevent overcharging the battery wherever diversionary charge control is used. This is usually accomplished by including a conventional series controller. This helps ensure that if any part of a diversionary charge control system fails, the batteries will not be overcharged and create a potentially hazardous condition. NEC® requirements also state that for DC diversion loads, the load voltage rating shall be greater than the maximum battery voltage, and the power ratings of the load shall be at least 150% of the power rating of the array.

Diversionary Charge Controllers

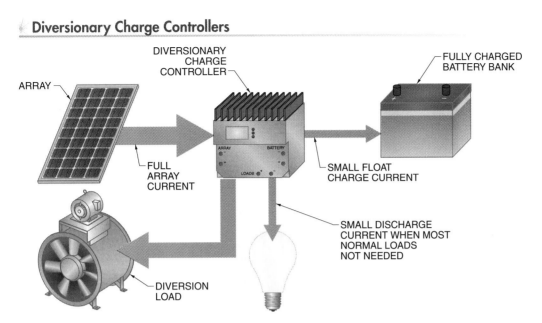

Refer to media clip on CD-ROM

Figure 7-10. *Diversionary charge controllers regulate charging current by diverting excess power to an auxiliary load when batteries are fully charged.*

Hybrid System Controllers

Since the PV array in hybrid systems is not sized to meet all the loads by itself, the auxiliary energy sources are required to provide supplemental battery charging. A *hybrid system controller* is a controller with advanced features for managing multiple energy sources. In many cases, hybrid system control functions are integrated into multifunction PCUs. **See Figure 7-11.**

A principal feature of many hybrid controllers is the ability to start and control an engine generator. Battery voltage is typically used to control engine generator operation, though it can also be controlled based on time of day, load currents, or scheduled cycles of equalization charging or engine exercising. Once a condition requiring engine generator operation occurs, the controller automatically activates the engine starting circuit. For diesel engines, the controller warms the glow plugs for a short period prior to starting. To limit the wear and tear on the starter, the cranking time and the number of starting attempts can also be controlled. If the engine does not start quickly, the controller indicates this condition with an LED or an error message, suggesting that engine service is required.

After the engine starts, the controller may allow a short warm-up period before the full load is connected. The engine generator may operate for a specific amount of time or until the battery voltage reaches a certain level, then shuts down until needed again. If the engine generator is not needed often, the controller runs a regular programmed exercise cycle. The exercise cycle operates the engine for a few minutes at a set interval, such as once a week, to keep it in good working order.

DOE/NREL, Warren Gretz
Hybrid systems can include any combination of alternative energy sources, such as a PV array and a wind turbine.

Hybrid System Controllers

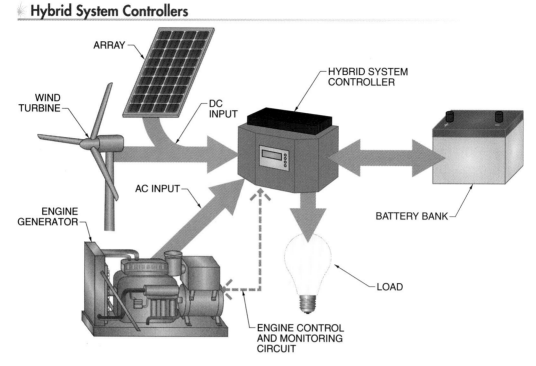

Figure 7-11. *Controllers designed for hybrid PV systems must manage multiple current sources simultaneously.*

The operation of engine generators can be controlled to maximize the use of array energy and set preferred times of operation. For example, the engine generator may not be allowed to operate early in the day or at a high state of charge when it may cause the array controls to regulate. Also, it may not be desirable to have noisy engine generators operating at night. However, excessively high load currents or extremely low battery voltage may temporarily override these controls to permit battery charging or operation of loads directly from the generator.

Ampere-Hour Integrating Charge Controllers

Charge controllers can also regulate charge current based on parameters other than battery voltage. An *ampere-hour integrating charge controller* is a charge controller that counts the total amount of charge (in ampere-hours) into and out of a battery and regulates charging current based on a preset amount of overcharge. **See Figure 7-12.** For example, a lighting

load may discharge 72 Ah from the battery each night. To charge the battery, an ampere-hour charge controller set for 115% overcharge would allow 83 Ah (72 Ah × 115% = 83 Ah) of charge to flow back into the battery before terminating charging.

Ampere-Hour Charge Controllers

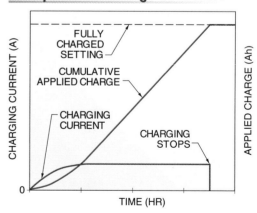

Figure 7-12. *Ampere-hour charge controllers track the cumulative number of ampere-hours applied to a battery bank and discontinue charging at a preset total.*

CHARGE CONTROLLER SETPOINTS

A *charge controller setpoint* is a battery condition, commonly the voltage, at which a charge controller performs regulation or switching functions. Setpoints are associated with either charge regulation or load control. Charge controller setpoints define the range of permissible battery and system voltages. The setpoints greatly affect battery performance, life, and load availability, so their proper specification and adjustment is vital. Simple interrupting-type charge controllers use two basic setpoints for charge regulation and two basic setpoints for load control. Additional setpoints may be available on more advanced charge controllers.

Charge Regulation Setpoints

Optimal charge regulation setpoints ensure that the battery is maintained at the highest possible state of charge without overcharging and over a range of conditions. Two charge regulation setpoints regulate the charging function. A higher voltage setpoint is used to disconnect the array from charging the battery and a lower voltage setpoint is used to reconnect the array and resume battery charging. **See Figure 7-13.**

In small, stand-alone PV systems, such as an electronic sign, the charge controller is especially important.

Voltage Regulation (VR) Setpoint. The *voltage regulation (VR) setpoint* is the voltage that triggers the onset of battery charge regulation because it is the maximum voltage that a battery is allowed to reach under normal operating conditions. The VR setpoint may also be called the regulation voltage. When a battery reaches the VR setpoint, a controller will either discontinue or limit the array charge current to prevent overcharging the battery. For charge controllers that provide multistage charging, the VR setpoint defines the beginning of absorption charge.

The optimal VR setpoint depends on several factors, especially the controller algorithm, temperature, and type of battery. **See Figure 7-14.** VR setpoints for PWM and linear-type charge controllers are generally lower than those for interrupting controllers because they are more efficient at charging.

The VR setpoints used in PV systems are typically much higher than the regulation voltages recommended by battery manufacturers. This is because the battery must be charged within a limited time in a PV system (during sunlit hours), while battery manufacturers allow for the much longer charge times possible in other applications. If the VR setpoint in a typical PV system were set at the manufacturer's recommended regulation voltage, the batteries in most PV applications would never fully charge. A higher setpoint allows a battery to be charged in a shorter time, but must be low enough to avoid excessive overcharge and gassing.

Charge Regulation Setpoints

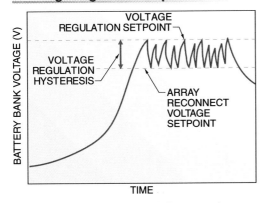

Figure 7-13. *Charge regulation setpoints are the voltage levels at which the charge controller interrupts or reconnects the charging current from the array to the battery bank.*

Typical Voltage Regulation Setpoints*

BATTERY TYPE	INTERRUPTING-TYPE CHARGE CONTROLLER		LINEAR- OR PWM-TYPE CHARGE CONTROLLER	
	Per Cell	Per Nominal 12 V Battery	Per Cell	Per Nominal 12 V Battery
Flooded open-vent lead-acid	2.43 to 2.47	14.6 to 14.8	2.40 to 2.43	14.4 to 14.6
Sealed valve-regulated lead-acid (VRLA)	2.37 to 2.40	14.2 to 14.4	2.33 to 2.37	14.0 to 14.2
Flooded pocket plate nickel-cadmium	1.45 to 1.50	14.5 to 15.0	1.45 to 1.50	14.5 to 15.0

Consult battery and charge controller manufacturer's literature to determine optimal regulation setpoints.
* in V at 25°C

Figure 7-14. *The optimal voltage regulation setpoint depends on the types of battery and charge controller.*

Array Reconnect Voltage (ARV) Setpoint. For interrupting-type charge controllers, once the array current is disconnected at the voltage regulation setpoint, the battery voltage will begin to decrease. When the battery voltage decreases to a predefined voltage, the array is again reconnected to the battery and charging resumes. The *array reconnect voltage (ARV) setpoint* is the voltage at which an interrupting-type charge controller reconnects the array to the battery and resumes charging.

Battery charging in PV systems is considered "opportunity charging" because a battery can be charged only during sunlit hours. Therefore, PV system charge controllers use different battery-charging setpoints than do conventional battery chargers that are connected to a constant power source, such as the utility.

The array current cycles into the battery in an ON/OFF manner because it is disconnected at the VR setpoint and reconnected at the ARV setpoint. As the battery is brought up to a higher state of charge, the duration of each cycle becomes shorter, and the array current remains disconnected from the battery for a longer time.

Linear- and PWM-type charge controllers maintain a battery at the regulation voltage by reducing, rather than completely disconnecting, the charging current. The array current is controlled to hold the battery at the regulation voltage as it completes charging. Therefore, these charge controllers do not have a clearly definable ARV setpoint.

Voltage Regulation Hysteresis (VRH). The *voltage regulation hysteresis (VRH)* is the voltage difference between the voltage regulation (VR) setpoint and the array reconnect voltage (ARV) setpoint. The VRH applies only to interrupting-type charge controllers. For example, a controller with a VR setpoint of 14.5 V and an ARV setpoint of 13.5 V has a voltage regulation hysteresis of 1.0 V.

The VRH is vitally important to the performance of interrupting-type charge controllers. If the VRH is too large, the array current remains disconnected for long periods, wasting energy from the array and making it difficult to fully charge the battery. If the VRH is too small, the array will cycle ON and OFF rapidly, which limits the life of switching elements.

The battery charge and discharge rates also affect the optimal VRH. Battery voltage changes much more quickly at high rates than at lower rates, suggesting a larger VRH value. In general, a smaller VRH is required for PV systems that do not have daytime loads to ensure the array remains charging a sufficient amount of time. Most interrupting-type controllers use VRH values between 0.4 V and 1.4 V based on a nominal 12 V battery system. Since ARV setpoints do not apply to linear or PWM designs, VRH does not either.

Load Control Setpoints

Load control protects batteries from overdischarge by turning the load circuits ON and OFF as needed. **See Figure 7-15.** Load control features are common on charge controllers, or may be added as a separate controller.

Load Control Setpoints

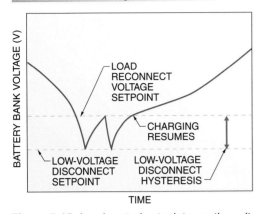

Figure 7-15. *Load control setpoints are the voltage levels at which the charge controller interrupts or reconnects the discharging current from the battery bank to the loads.*

Low-Voltage Disconnect (LVD) Setpoint.
The *low-voltage disconnect (LVD) setpoint* is the voltage that triggers the disconnection of system loads to prevent battery overdischarge because it is the minimum voltage a battery is allowed to reach under normal operating conditions. At given system discharge rates, the LVD setpoint determines the battery's maximum allowable depth of discharge, which affects its life, available capacity, and electrolyte freezing point.

Load disconnect circuits typically use solid-state switches, but since disconnects occur considerably less frequently in PV systems than charge regulation cycles, simple electromechanical relays are sometimes used. If the load current exceeds the rating of the controller's switching element, higher-rated auxiliary relays may be activated by the primary control relay to handle the full load current. Many charge controllers also have a visual or audible signal to alert system operators of the low voltage disconnect condition.

Load disconnections should generally be infrequent in most PV systems. Proper sizing of the battery and array to meet the loads are keys to minimizing load disconnections. If a load disconnect event occurs in a marginally sized system, many days or weeks may be required to fully charge the battery.

The LVD setpoint involves a tradeoff between battery life and load availability. Disconnecting loads from a battery at a voltage that is too low results in excessively deep battery discharges and limits battery life, but improves load availability in the short-term. However, disconnecting loads at too high a voltage protects the battery better, but sacrifices load availability during low insolation periods.

In general, the LVD setpoint in PV systems is selected to discharge deep-cycle batteries to no more than 75% to 80% depth of discharge, based on their rated capacity. Typical LVD setpoint values used for nominal 12 V lead-acid batteries at discharge rates lower than C/30 are between 11.2 V and 11.5 V. High discharge rates require a lower LVD setpoint to achieve the same depth of discharge. For shallow-cycle batteries, such as SLI batteries, or batteries that are subjected to freezing temperatures, a higher LVD setpoint should be used to limit the depth of discharge.

Load Reconnect Voltage (LRV) Setpoint.
The *load reconnect voltage (LRV) setpoint* is the voltage at which a charge controller reconnects loads to the battery system. After the controller disconnects the load from the battery at the LVD setpoint, the battery voltage rises quickly, then more gradually until its open-circuit voltage is reached. As additional charge is provided by the array or another source, the battery voltage rises even more. When a controller senses the battery has reached the LRV, the load is reconnected to the battery.

The LRV setpoint also affects battery health and load availability, and defines the state of charge at which loads are reconnected to the battery. The LRV setpoint should be high enough to ensure the battery has been somewhat charged before reconnecting the loads, but low enough so that loads do not remain disconnected for unreasonably long periods. Some controller designs "lock out" loads until the next day or when the controller senses that the array is again charging the battery. This prevents the unwanted condition of multiple load disconnects and reconnects occurring in one evening, for example, in a PV lighting system.

Some PV systems powering loads of varying importance use multiple LVD setpoints and load controls to disconnect loads at progressively lower battery voltages. This preserves battery charge for the most important loads. A noncritical load is disconnected from the battery system by the charge controller at a higher LVD setpoint to retain some useful battery charge for critical loads. A critical load is connected to power for as long as possible, perhaps even past the normal low-voltage load disconnect point.

A load controller may have separate connections for loads of different priorities, each with different LVD setpoints, or an extremely critical load may be connected directly to the battery system so it cannot be disconnected. However, even critical loads will operate only until the battery system completely discharges. Complete discharge can damage the batteries, so these systems must be carefully designed and operated.

For example, a PV lighting system with several independent lighting fixtures may protect a battery from severe discharges by disconnecting some of the lighting loads when the battery reaches a certain depth of discharge, reducing lighting levels instead of disconnecting the entire system. Multiple load disconnect setpoints are sometimes used for remote medical clinics where vaccine refrigerators are the most critical loads and are set to disconnect at a battery depth of discharge greater than 90%, where noncritical loads are disconnected at a much higher setpoint and state of charge.

LRV setpoints used in small PV systems vary between 12.5 V and 13.0 V (at 25°C) for most nominal 12 V lead-acid batteries. Under typical charge and discharge rates in PV systems, this ensures that the battery reaches 25% to 50% state of charge before the load is reconnected. Systems with dynamic loads or with very high charge or discharge currents may use different setpoints to tailor the settings to the desired battery and load performance.

Low-Voltage Disconnect Hysteresis (LVDH).
The *low-voltage disconnect hysteresis (LVDH)* is the voltage difference between the low-voltage disconnect (LVD) and load reconnect voltage (LRV) setpoints. If the LVDH is too small, the load may cycle ON and OFF rapidly at low battery state of charge, possibly damaging the load or control elements, and extending the time it takes to fully charge the battery. If the LVDH is too large, the load may remain OFF for extended periods while the array more fully charges the battery. The reduced cycling improves battery health, but also limits load availability during low insolation periods. The proper LVDH selection for a given system will depend on load availability requirements, and the battery's charge and discharge rates.

Float Charge Setpoint

Multistage charge controllers with an absorption charge cycle may also include a float charge setpoint, which is a slightly lower voltage than the VR setpoint. When the battery completes the normal charge, the controller lowers the voltage to the float charge setpoint. Float charging holds the battery at a fixed voltage by limiting charge current. Similar to battery manufacturers' specified float voltages, float charge voltages in PV systems can be maintained for an extended

Morningstar Corp.
Dedicated lighting controllers include settings for when a lighting load should be turned ON and OFF.

period without overcharging the battery. Commonly, float voltage for a lead-acid battery is around 13.5 V to 13.8 V (at 25 °C) for a nominal 12 V battery, or 2.25 V to 2.3 V per cell.

Equalization Setpoints

Equalization is a controlled overcharge for a few hours, which is only performed on open-vent batteries. Charge controllers may provide an automated or user-activated equalization charge cycle. During an equalization cycle, the VR setpoint is increased to a higher level, where it remains for a predetermined amount of time before returning to its normal level or the slightly lower float voltage. **See Figure 7-16.** Controllers that rely on only an array for the equalization charge may require several days or weeks to complete the prescribed time at the equalization voltage, especially during periods of high load or low insolation. With oversized arrays or generator-powered chargers, equalization can be accomplished in shorter periods.

Equalization settings include the equalization voltage setpoint, cycle frequency, and duration. Equalization voltages for flooded lead-acid batteries may be as high as 2.5 V to 2.58 V (at 25°C) per cell, which corresponds to 15 V to 15.5 V for a nominal 12 V battery. The frequency and duration of equalization charges depend on the type of battery and its condition, and availability of the array and other DC charging sources. Typically, manufacturers recommend an equalization charge for flooded lead-acid batteries once every few weeks for about 3 hr.

Setpoint Adjustments

Appropriate charge controller setpoints are crucial to maximizing battery life and system performance, and are arguably more important than the type of controller design. A relatively simple design with properly adjusted setpoints will work better than a sophisticated controller with improper settings. A difference of a few tenths of a volt in the VR setpoint can have a significant effect on the battery system.

For some controllers, setpoints are established by the manufacturer for a specific battery type and are not adjustable in the field.

However, many controllers allow adjustment of the setpoints during installation, though the NEC® requires manual means of setpoint adjustment to be inside the charge controller so they are not easily accessible for unqualified persons to make inappropriate changes.

Setpoints may be adjusted with potentiometers that allow a range of settings, or with DIP switches or jumpers for discrete setpoint increments or battery types. **See Figure 7-17.** For microprocessor-based charge controllers, including those integrated with inverters, setpoint adjustments are made through software programming.

Equalization Charge

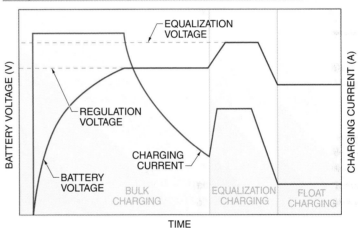

Figure 7-16. *The equalization setpoint brings the battery voltage to a level that is higher than the normal charge regulation voltage.*

Setpoint Adjustments

Figure 7-17. *Setpoints are adjusted with controls inside the charge controller.*

In general, for any type of battery, higher setpoints are required for interrupting-type charge controllers than the more advanced controller algorithms. Since interrupting-type algorithms do not apply charging current as smoothly as more sophisticated algorithms, batteries must be charged to higher voltage to accept charge applied in this way in a similar amount of time. Also, for any type of controller, recommended VR setpoints for flooded open-vent lead-acid batteries are always higher than for sealed VRLA batteries. Flooded batteries can tolerate more overcharge than sealed batteries because water can be added to replenish the electrolyte.

Charge controller and battery manufacturer's data must be reviewed before making any determination on setpoints. However, manufacturer's recommendations are only guidelines, and atypical systems or configurations may require different setpoints. When in doubt, setpoints should be set slightly low to be conservative, and the battery system should be carefully monitored for the first few discharging and charging cycles, preferably under a range of temperature and insolation conditions. Small adjustments can be made until charging and performance is optimized. When the optimal setpoints are established for a particular system, the values should be recorded in the system documentation and on a label near the charge controller.

Temperature Compensation

Temperature affects battery charging characteristics, so some charge controllers automatically adjust setpoints to compensate for this factor. When battery temperature is low, temperature compensation increases the VR setpoint to allow the battery to reach a moderate gassing level and fully charge. When battery temperature is high, the VR setpoint is lowered to minimize excessive overcharge and electrolyte loss. Temperature compensation helps ensure that a battery is fully charged during cold weather and not overcharged during hot weather. Temperature compensation is also used for LVD setpoints.

Charge controllers with temperature compensation use a sensor to measure temperature. For small systems and controls located in the same thermal environment, temperature sensing within the charge controller is generally satisfactory for approximating battery temperature. For larger systems or systems with batteries located in a different thermal environment than the controller, external temperature probes should be used. **See Figure 7-18.** Temperature probes must be securely attached to a battery case to ensure the most accurate battery temperature measurement. If temperature probes become detached from a charge controller, the controller should regulate at the nominal setpoints.

Battery Temperature Sensors

Figure 7-18. *Temperature probes are placed between batteries and connect to a charge controller to compensate the charge regulation setpoint.*

Setpoint voltages are usually based on a nominal battery temperature of 25°C (77°F). Temperature-compensated charge controllers automatically adjust the setpoints based on established coefficients. For lead-acid batteries, the widely accepted temperature coefficient for the regulation voltage is –5 mV/°C/cell (–0.005 V/°C/cell). The temperature compensation coefficient is negative because increases in temperature require a reduction in the setpoint voltages. Temperature coefficients for other battery chemistries differ from that for lead-acid types, but most are negative values.

A setpoint voltage at battery temperatures other than 25°C can be calculated with the following formula:

$$V_{comp} = V_{set} - (C_{Vcell} \times [25 - T_{bat}] \times n_s)$$

where

V_{comp} = temperature-compensated setpoint voltage (in V)

V_{set} = nominal setpoint voltage at 25°C (in V)

C_{Vcell} = temperature compensation coefficient (in V/°C/cell)

T_{bat} = battery temperature (in °C)

n_s = number of battery cells in series

For example, consider a PV system using a 12 V lead-acid battery (6 cells in series) and a charge controller with a nominal VR setpoint of 14.4 V at 25°C. What should the VR setpoint be at 5°C?

$$V_{comp} = V_{set} - (C_{Vcell} \times [25 - T_{bat}] \times n_s)$$
$$V_{comp} = 14.4 - (-0.005 \times [25 - 5] \times 6)$$
$$V_{comp} = 14.4 - (-0.005 \times 20 \times 6)$$
$$V_{comp} = 14.4 + 0.6$$
$$V_{comp} = \textbf{15.0 V}$$

Conversely, if the same battery operates at 40°C (104°F), the VR setpoint is reduced from 14.4 V to 13.95 V. This amounts to a 1.05 V difference in the VR setpoint for a 12 V battery operating over a temperature range of 5°C to 40°C (41°F to 104°F). For higher-voltage battery systems, the compensated VR setpoint will vary even more. For a nominal 48 V system, the VR setpoint will vary by more than 4.2 V over the same 35°C (63°F) temperature range.

CHARGE CONTROLLER APPLICATIONS

Charge controllers used in PV systems must be properly specified, selected, and configured based on the application requirements. They must have appropriate current and voltage ratings, and must be compatible with other equipment in the system. The selection and sizing of charge controllers in PV systems involves consideration of many factors. **See Figure 7-19.**

Charge Controller Selection Criteria

System voltage
PV array and load currents
Battery system type and capacity
Charge controller algorithm
Switching element type
Voltage setpoints
Power requirements
Heat dissipation
Overcurrent protection
Surge suppression
Disconnect devices
Critical load features
Back-up power source control
Environmental conditions
Mechanical design
System indicators, alarms, and meters
Cost
Warranty
Availability

Figure 7-19. *Many criteria should be considered when selecting charge controllers for PV systems.*

Ratings

For NEC® code compliance, charge controllers used in PV systems must be listed to the UL Standard 1741, *Inverters, Converters, and Controllers for Use in Independent Power Systems*. The testing associated with this product listing ensures that charge controllers are safe and suitable for the intended operating conditions.

SolarWorld Industries America

PV systems in developing countries are often small systems consisting of a small array, a few batteries, and a simple charge controller that controls a single load, such as a light.

Charge controllers for PV systems are produced in a variety of system/battery voltage ratings, commonly 12 V, 24 V, and 48 V. Higher voltage charge controllers handle the same power levels at lower currents and are generally more efficient. Many types of charge controllers can be reconfigured for different voltages, making them flexible for use in a variety of system designs. Charge controllers should be sized for greater than the maximum voltage expected from the PV system, usually the open-circuit voltage of the array under the coldest conditions.

Charge controllers are also rated for maximum array and load currents. The charge controller current rating must be specified to 125% of the maximum PV circuit current. For example, a controller rated for 40 A can accommodate an array with short circuit rating of 32 A (40 A ÷ 125% = 32 A). Similarly, a charge controller with load control function must have a load current rating at least 125% of the maximum expected load current. The load current rating may be different from the array current rating.

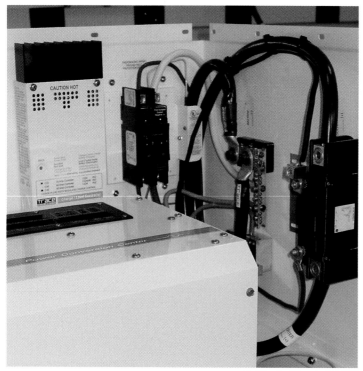

Baltimore Electrical JATC
Installing charge controllers physically near the battery bank and other system components keeps voltage drop low, improving system performance.

Location

When the temperature compensation sensor is inside the charge controller, the charge controller should be as close to the batteries as possible so that they are at the same ambient temperature. If this type of charge controller is located in a cooler space, the setpoints will be too high, and if it is in a warmer space, the setpoints will be too low. However, charge controllers should not be installed in the same enclosure or immediately above flooded lead-acid batteries. Battery gases may deteriorate the controller electronics over time and, in extreme cases, a controller could ignite if an explosive amount of battery gases accumulated without proper ventilation.

To minimize their operating temperature, charge controllers should be installed out of direct sunlight and with unobstructed airflow around the heat sinks. Charge controllers should also be protected from excessive moisture.

Voltage Drop

Installing charge controllers close to batteries also minimizes voltage drop. As charging current increases on the conductors between the charge controller and battery terminals, the voltage drop increases. Since many charge controllers sense battery voltage with the conductors used to deliver charging current, the measured voltage is slightly higher than the actual battery voltage. **See Figure 7-20.** This prematurely activates charge regulation, causing the battery to be undercharged.

Excessive voltage drop also affects the load control setpoints. The discharge current causes the measured voltage from the charge controller to be lower than the actual battery voltage. This results in the loads being disconnected at an actual battery voltage that is slightly higher than the LVD setpoint. The voltage drop during charging and discharging effectively compresses the actual operating voltage range of the battery, reducing the available battery capacity.

Some controllers, particularly ones designed for higher currents, include additional terminals for conductors to sense battery voltage. **See Figure 7-21.** This is called a four-wire or Kelvin measurement. No current flows in these conductors, so there is no voltage drop. Therefore, a precise battery voltage is measured and setpoints are activated at the correct voltages. When a charge controller does not have separate voltage sense leads, voltage drop can be minimized by using oversized conductors and locating the controller close to the batteries.

Multiple Arrays

For large arrays with multiple subarray source circuits, multiple independent charge controllers can be used instead of a single, larger controller. The array charging current is limited in a stepped manner as each controller begins to regulate each subarray. **See Figure 7-22.** This configuration improves overall system reliability in the event of the failure of any individual controller. This design can also facilitate later system expansion by allowing for the addition of a new array source circuit and charge

controller. As long as the charge rate for that circuit is limited to no more than 3% of the battery bank capacity, one source circuit may be left unregulated for float charging.

Multiple Battery Banks

Although rarely, a PV system can be configured to charge multiple separate battery banks with a single array. This situation usually results from a system expansion or from incorporating an isolated emergency standby battery or loads with different operating voltages. The battery banks may be different sizes, types, and ages, or perhaps different voltages. Charge acceptance and charge requirements may be different between the battery banks, leading to one battery bank being undercharged and the other being overcharged if the same charge controller and setpoints are used.

Voltage Drop

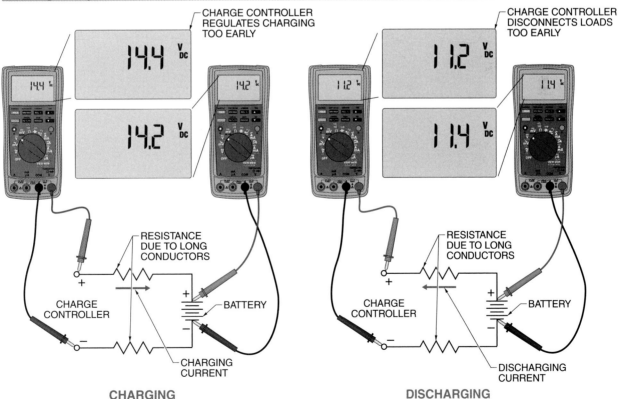

Figure 7-20. *Long conductors between a charge controller and a battery bank have resistance that causes voltage drops. Voltage drops affect the voltage measured at the charge controller, which triggers overcharge and overdischarge protections too early.*

In addition to temperature compensation, some charge controllers also compensate for high charge or discharge currents, which also affect battery voltage.

Careful component selection and system configuration can still effectively optimize the performance and life of multiple battery banks in these situations. Charge controller manufacturer's literature will specify allowable alternative configurations for their equipment.

Separate charge controllers are the best method for charging multiple battery banks from a single array. Each controller must be rated to handle the entire array current, and blocking diodes are required between each charge controller and the array to prevent the battery banks from operating at the same voltage. **See Figure 7-23.** In this configuration, each charge controller can be set for the specific battery types and charge each battery bank appropriately. Since the battery banks are isolated by the blocking diodes and independent charge controllers, system loads may be connected to either battery bank, but not both.

Voltage Sense Terminals

Figure 7-21. *Some charge controllers use additional battery voltage sense conductors to avoid the effect of voltage drop on setpoints.*

Multiple Subarrays and Charge Controllers

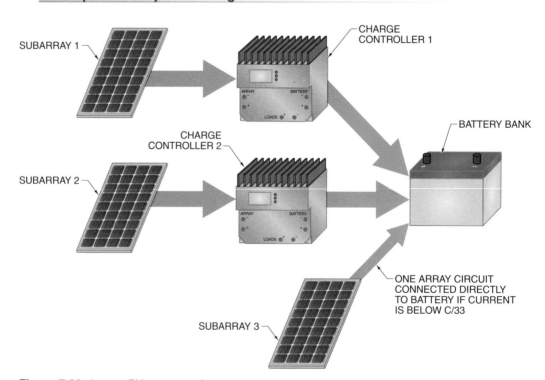

Figure 7-22. *Larger PV systems often use independent charge controllers for each array source circuit.*

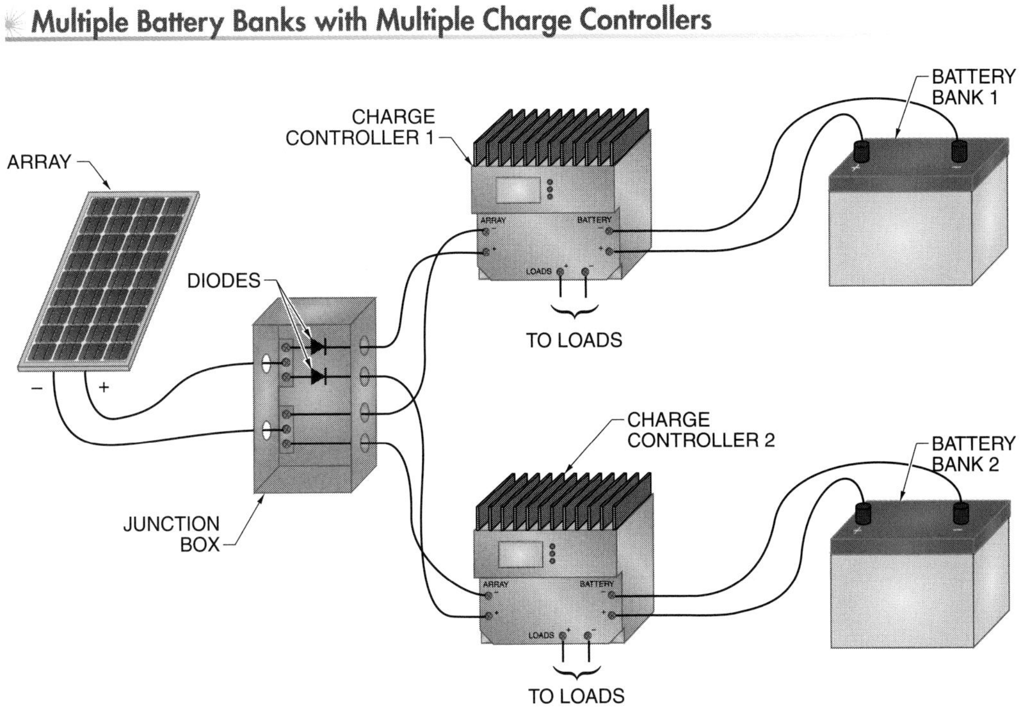

Multiple Battery Banks with Multiple Charge Controllers

Figure 7-23. *Separate charge controllers are usually recommended for charging independent battery banks from a single array.*

Some charge controllers can be used alone to charge multiple battery banks if the voltages are sufficiently close, though this is not recommended by most charge controller manufacturers. In this configuration, a resistor is added between the array and battery positive terminals on the charge controller and blocking diodes are added between the positive battery terminal of the charge controller and the positive terminal of each battery. **See Figure 7-24.** This allows the charge controller to distribute the appropriate charging current to each battery bank. The batteries are also isolated from one another in the controller output circuit, instead of on the array input side when using separate charge controllers.

Multiple battery banks are often used because the system must power loads of different voltages. However, there are alternative strategies for this situation. Lower voltages can be tapped from a higher voltage battery bank, though this can cause equalization problems for anything but very small loads. The best solution is to configure the battery bank for the highest voltage load and use a DC-DC converter to supply lower voltages.

Self-Regulating PV Systems

A charge controller is required in nearly all PV systems using batteries. However, some small systems can be specially designed to operate without a charge controller. A *self-regulating PV system* is a type of stand-alone PV system that uses no active control systems to protect the battery, except through careful design and component sizing. **See Figure 7-25.** When maintenance is infrequent or prohibitively expensive, such as for remote applications, eliminating the need for a sensitive electronic charge controller can simplify the system design, lower costs, and improve reliability.

Self-regulating PV systems must be designed so that the battery is never overcharged or overdischarged under any operating conditions. The expected operating conditions, such as temperature, insolation, and discharge rates, must be well understood for this type of system to function safely and successfully. The energy supply from the array, the energy capacity of the battery, and the energy demand of the load must be carefully balanced.

Multiple Battery Banks with One Charge Controller

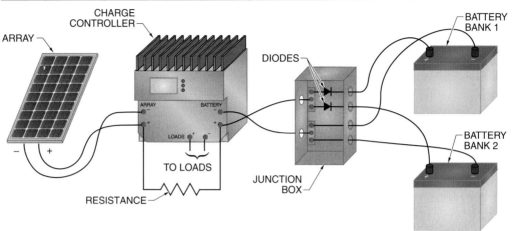

Figure 7-24. *In some cases, a single charge controller may be used to charge independent battery banks.*

Self-Regulating PV Systems

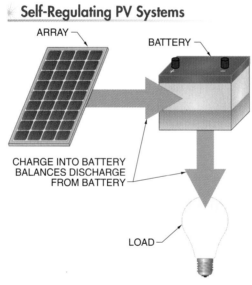

Figure 7-25. *To balance the charge to and from the battery in a self-regulating PV system, the load must be well defined, the battery must be oversized, and the array current must be self-limiting.*

Electrical Load. For self-regulating PV systems to work properly, the electrical load must be well defined and have a consistent day-to-day energy requirement. If the load is not used daily, the array may overcharge the battery. The load must also be consistent and automatically controlled. The load establishes the battery and array sizes needed in any stand-alone PV system, but is particularly important for self-regulating systems.

Environmental Conditions. Applications with widely varying insolation or extreme temperature variations are generally not suitable for self-regulating systems. Seasonal variations in insolation require a larger array to meet the load requirements, which can overcharge the battery. Similarly, temperature changes affect battery charging and array voltage, which cannot be automatically compensated for in self-regulating systems.

Battery and Array Sizing. As with any stand-alone PV system, the array for a self-regulating system must be large enough to supply the loads and charge the battery under the worst-case conditions of high load demand and low insolation. If the array is too small, the battery may discharge completely. However, an array that is too large can overcharge the battery. To prevent overcharge when the battery is fully charged, the maximum array charge rate must be very low. Since the size of the array cannot be reduced, charge rates are reduced by increasing the capacity of the battery bank (adding more batteries). Increasing battery capacity also increases system autonomy, lowers the average daily depth of discharge, and prolongs battery cycle life.

According to the NEC®, any PV system with batteries must employ charge control whenever the charge rate exceeds C/33. More conservatively, charge rates of C/100 or less are considered low enough to be tolerated for extended periods when most types of batteries are already fully charged.

Low-Voltage Modules. Low-voltage modules are an important part of a self-regulating system. Low-voltage modules take advantage of the fact that the module current falls off sharply as the voltage increases above the maximum power point. By matching the maximum power voltage with the battery charge regulation voltage, charge current is naturally limited as the battery approaches full state of charge, helping prevent battery overcharge.

Common standard module designs include 36 cells in series and produce a maximum power voltage of about 17 V at STC. This voltage provides an acceptable charging current for nominal 12 V lead-acid batteries at higher operating temperatures and with moderate voltage drop through conductors and equipment. These modules always operate to the left of the "knee" on their I-V curve, producing maximum power current or greater.

Low-voltage modules include 29 or 30 cells in series and have a maximum power voltage of about 15 V. At typical operating temperatures, the maximum power point falls within the range of typical battery voltages. As a battery approaches full state of charge and the module voltage passes the maximum power point, the module current falls sharply as the voltage increases further. **See Figure 7-26.**

Since module voltage and battery charge voltage requirements both decrease with increasing temperature, the voltages are reasonably matched for moderate temperature variations.

Since PV devices are greatly affected by temperature, the output of a low-voltage module can easily move outside the range of safe battery voltages at various temperatures. In cool weather, the battery may be severely overcharged, and in warm weather, the battery may be undercharged. Self-regulating systems using low-voltage modules should therefore be used only in locations where the temperature variance is small.

Low-Voltage Modules

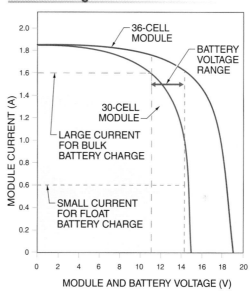

Figure 7-26. *Low-voltage modules control current into a battery bank without a charge controller because their current automatically falls as the battery reaches full charge.*

Refer to Quick Quiz® on CD-ROM

Summary

- Charge controllers manage the interactions between the array, battery bank, and loads; are the heart of stand-alone PV systems; and have a profound affect on battery life, load availability, and overall system performance.

- Charge controllers regulate battery charging and protect batteries from being overcharged or overdischarged.

- Additional features may be incorporated in charge controllers to provide enhanced system control functions.

- Charge controllers use various algorithms and switching designs to manage battery state of charge.

- Interrupting-type charge controllers regulate charging current by switching it ON or OFF.

- Linear-type charge controllers use a variable resistance to regulate charging current in a gradual manner.

- Shunt charge controllers control charging current by short-circuiting the array.

- Series charge controllers control charging current by open-circuiting the array.

- Pulse-width modulation (PWM) simulates a variable current by switching a full current ON and OFF at high speed for varying lengths of time.

- Diversionary charge controllers divert the array current to an auxiliary load when the battery bank is fully charged.

- Charge controller setpoints are the battery conditions that trigger charge control actions.

- Charge regulation setpoints determine when the array's charging current will be applied to the battery bank.

- Load control setpoints determine when the loads will be operating, preventing the battery bank from being overdischarged.

- Optimal charge controller setpoints are determined by the charge control algorithm and type of battery.

- Setpoints should be adjusted for temperature because temperature affects battery charging characteristics.

- Charge controllers must have the appropriate voltage and current rating for the application requirements.

- Long conductors between the charge controller and the battery bank produce a voltage drop that affects the accuracy of voltage measurement by the controller.

- Large arrays with multiple source circuits use individual charge controllers for each circuit.

- Multiple battery banks can be charged from one array, with either multiple charge controllers or one specially configured charge controller.

- Self-regulating PV systems do not require active charge control because the charging and discharging currents are carefully balanced.

Definitions

- A *charge controller* is a device that regulates battery charge by controlling the charging voltage and/or current from a DC power source, such as a PV array.

- *Charge acceptance* is the ratio of the increase in battery charge to the amount of charge applied to the battery.

- *Overcharge* is the ratio of applied charge to the resulting increase in battery charge. *Overcharge* can also refer to the condition of a fully charged battery continuing to receive a significant charging current.

- **Deficit charge** is a charge cycle in which less charge is returned to a battery bank than what was withdrawn on discharge.

- **Overdischarge** is the condition of a battery state of charge declining to the point where it can no longer supply discharge current at a sufficient voltage without damaging the battery.

- A **charge control algorithm** is a programmed series of functions that a charge controller uses to control current and/or voltage in order to maintain battery state of charge.

- An **interrupting-type charge controller** is a charge controller that switches the charging current ON and OFF for charge regulation.

- A **linear-type charge controller** is a charge controller that limits the charging current in a linear or gradual manner with high-speed switching or linear control.

- A **shunt charge controller** is a charge controller that limits charging current to a battery system by short-circuiting the array.

- A **shunt-interrupting charge controller** is a charge controller that suspends charging current to a battery system by completely short-circuiting the array.

- A **shunt-linear charge controller** is a charge controller that limits charging current to a battery system by gradually lowering the resistance of a shunt element through which excess current flows.

- A **series charge controller** is a charge controller that limits charging current to a battery system by open-circuiting the array.

- A **series-interrupting charge controller** is a charge controller that completely open-circuits the array, suspending current flow into the battery.

- A **series-linear charge controller** is a charge controller that limits charging current to a battery system by gradually increasing the resistance of a series element.

- A **series-interrupting, pulse-width-modulated (PWM) charge controller** is a charge controller that simulates a variable charging current by switching a series element ON and OFF at high frequency and for variable lengths of time.

- A **maximum power point tracking (MPPT) charge controller** is a charge controller that operates the array at its maximum power point under a range of operating conditions, as well as regulates battery charging.

- A **diversionary charge controller** is a charge controller that regulates charging current to a battery system by diverting excess power to an auxiliary load.

- The **diversion load** is an auxiliary load that is not a critical system load, but is always available to utilize the full array power in a useful way to protect a battery from overcharge.

- A **hybrid system controller** is a controller with advanced features for managing multiple energy sources.

- An **ampere-hour integrating charge controller** is a charge controller that counts the total amount of charge (in ampere-hours) into and out of a battery and regulates charging current based on a preset amount of overcharge.

- A **charge controller setpoint** is a battery condition, commonly the voltage, at which a charge controller performs regulation or switching functions.

- The **voltage regulation (VR) setpoint** is the voltage that triggers the onset of battery charge regulation because it is the maximum voltage that a battery is allowed to reach under normal operating conditions.

- The **array reconnect voltage (ARV) setpoint** is the voltage at which an interrupting-type charge controller reconnects the array to the battery and resumes charging.

- The *voltage regulation hysteresis (VRH)* is the voltage difference between the voltage regulation (VR) setpoint and the array reconnect voltage (ARV) setpoint.

- The *low-voltage disconnect (LVD) setpoint* is the voltage that triggers the disconnection of system loads to prevent battery overdischarge because it is the minimum voltage a battery is allowed to reach under normal operating conditions.

- The *load reconnect voltage (LRV) setpoint* is the voltage at which a charge controller reconnects loads to the battery system.

- The *low-voltage disconnect hysteresis (LVDH)* is the voltage difference between the low-voltage disconnect (LVD) and load reconnect voltage (LRV) setpoints.

- A *self-regulating PV system* is a type of stand-alone PV system that uses no active control systems to protect the battery, except through careful design and component sizing.

Review Questions

1. What is the difference between charge acceptance and overcharge?

2. How do charge controllers protect battery banks from overcharge and overdischarge?

3. Explain the differences between interrupting-type charge controllers and linear-type charge controllers.

4. Explain the differences between shunt and series charge controllers.

5. How does PWM technology regulate charging current?

6. How does a diversionary charge controller utilize excess energy?

7. Explain four primary charge controller setpoints.

8. What are the possible consequences of a charge controller hysteresis that is too narrow or too large?

9. Why must charge controller setpoints be temperature-compensated?

10. How does the distance between the charge controller and battery bank affect the regulation functions of a charge controller?

11. How do low-voltage modules control charging current?

8

Inverters

The solid-state inverters used in PV systems employ the latest in power electronics to produce AC power from a DC power source that is either a PV array or a battery bank. Inverters can use different circuit designs, switching devices, and control methods to affect the output waveform properties. These properties affect the efficiency and quality of the AC power. As complete power conditioning units, inverters may also include functions for battery charging, monitoring, system control, and maximum power point tracking.

Chapter Objectives

- *Identify basic waveform types and properties.*
- *Compare applications for static inverters.*
- *Explain the basic types of inverters used in PV systems.*
- *Describe the operation of a simple square wave inverter.*
- *Explain how inverters make sine waves from square waves.*
- *Describe the functions and features of power conditioning units.*
- *Understand inverter specifications and ratings.*

AC POWER

Voltage and current, and therefore power, can be either constant or time varying in magnitude. Moreover, time-varying values can either maintain one direction (positive or negative), or alternate between positive and negative directions. **See Figure 8-1.** *Direct current (DC)* is electrical current that flows in one direction, either positive or negative. DC power may be constant or variable, but always maintains one direction.

DC Power vs. AC Power

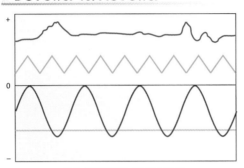

DC POWER SIGNALS

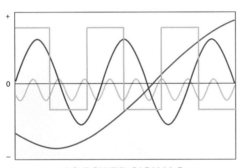

AC POWER SIGNALS

Figure 8-1. *If voltage and current signals are either always positive or always negative, they are DC waveforms. If the signals switch between positive and negative, they are AC waveforms.*

PV modules, and some other power-generating technologies, produce variable DC power. However, most electrical loads operate using AC power, so DC power usually must be transformed into AC power in order to be useful. *Alternating current (AC)* is electrical current that changes between positive and negative directions. AC power is characterized by waveform shape, frequency, and magnitude.

Waveforms

A *waveform* is the shape of an electrical signal that varies over time. Waveforms are used to represent changing electrical current and voltage. A *periodic waveform* is a waveform that repeats the same pattern at regular intervals. A *cycle* is the interval of time between the beginnings of each waveform pattern.

Waveforms cannot be measured or viewed directly. To view the shape of a waveform, the time-varying values (voltage or current) must be plotted against time. Oscilloscopes, or meters with oscilloscope features, are able to display waveforms.

Waves can take a variety of forms, such as smooth curves for gradually changing values, or stepped patterns for abruptly changing values. Sine waves, square waves, and modified square waves are common AC waveforms produced by inverters. **See Figure 8-2.**

AC Waveforms

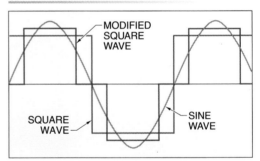

Figure 8-2. *AC waveforms can take a variety of shapes.*

Sine Waves. A *sine wave* is a periodic waveform the value of which varies over time according to the trigonometric sine function. A *sinusoidal waveform* is a waveform that is or closely approximates a sine wave. Sine waves may shift in time or vary in amplitude, but if they retain the shape of the sine wave, they are still sinusoidal.

The rotating generators that provide most of the electrical power on the utility grid naturally produce sine waves, so most loads are designed to operate using sinusoidal AC power. Therefore, interactive inverters produce sine waves for utility synchronization. Other waveforms may damage some loads. However, sine waves are the most complicated type of AC waveform for inverters to produce, so some less-sophisticated inverters approximate a sine wave with square waves or modified square waves.

Square Waves. Square waves and modified square waves are nonsinusoidal waveforms. A *square wave* is an alternating current waveform that switches between maximum positive and negative values every half period. Square waves are inefficient and are not a common inverter output, but are the basis for the improved modified square wave.

Modified Square Waves. A *modified square wave* is a synthesized, stepped waveform that approximates a true sine wave. Also called a modified sine wave or a quasi sine wave, this type of waveform is the typical AC output of many stand-alone inverters. In terms of power quality, a modified square wave is a substantial improvement over a simple square wave. Compared to square wave inverters, modified square wave inverters have lower harmonic distortion, higher peak voltage, higher efficiency, and better surge current capability.

History of Alternating Current

In the late 1880s, electricity was first becoming accessible to homes and businesses. The first electrical power generation and distribution systems were developed by Thomas Edison and used direct current (DC) power. DC power worked well for powering loads, but could not be transmitted more than about a mile without significant power loss from voltage drop. Edison advocated distributed generation as a solution—building many power plants close to where electricity was needed—but this was impractical at the time.

Library of Congress

Thomas Edison's dynamos were the earliest commercial electricity generators.

As an alternative, George Westinghouse and Nikola Tesla devised a system for generating and distributing alternating current (AC) power. AC had many advantages, including safer connections and disconnections without arcing, and the ability to transmit power at high voltage over hundreds of miles with relatively small losses. Edison and Westinghouse became bitter adversaries as each tried to persuade the public that theirs was the better system. Edison tried to convince people that AC power was more dangerous than DC power, even trying to popularize "Westinghoused" as a term for being electrocuted. This controversy became known as the War of Currents.

The Niagra Falls generation project was the first large-scale electrical generation project, and both Edison and Westinghouse competed for the contract. Ultimately, the Niagra Falls Commission chose the AC system. The project was completed in 1896 and was a huge success. Tesla's system of generators, transformers, motors, and conductors set the standards for all future AC power generation, including the 60 Hz frequency.

Edison's DC system did not become the primary electrical distribution system, but some remnants of his attempts lingered for more than a century. Up until 2005, Consolidated Edison, New York City's electric utility company, supplied DC power to 1600 customers in Manhattan, mainly for operating older elevators. Currently, DC power is used in electronics, telephone networks, and in small power networks for off-grid residences and transportation.

Modified square waves are suitable for many AC loads, including linear and switching power supplies used in electronic equipment, transformers, and some motors. However, a load that is sensitive to either peak voltage or zero crossings could experience problems. These inverters may also produce radio frequency (RF) noise that interferes with devices such as radios and TVs. Also, modified square wave inverters should not be used with motor speed controllers or devices that plug directly into a receptacle to charge a battery without a transformer.

Frequency

Frequency is the number of waveform cycles in one second. In the past, frequency was expressed as cycles per second (cps, or simply "cycles"), but it is now commonly expressed in equivalent units of Hertz (Hz). The frequency of the U.S. electric grid is maintained at 60 Hz. Frequency establishes the speed of motors, generators, and some clocks and is one of the most important parameters in synchronizing AC electrical systems.

Period is the time it takes a periodic waveform to complete one full cycle before it repeats. **See Figure 8-3.** Period is the inverse of frequency. For example, a 60 Hz AC waveform repeats 60 times per second. In this case, the period is ⅟₆₀ sec, or 16.7 ms.

Magnitude

There are multiple ways to measure or specify the magnitude of a waveform. *Peak* is the maximum absolute value of a waveform. For example, peak voltage is the maximum value of an AC voltage waveform. *Peak-to-peak* is a measure of the difference between positive and negative maximum values of a waveform.

The *root-mean-square (RMS) value* is a statistical parameter representing the effective value of a waveform. An RMS value is the square root of the mean (average) of the squares of the values. This is not the same as the average value, but it has a special significance. A DC voltage equal to the RMS value of an AC voltage produces the same amount of heat in a resistive circuit as does the AC voltage. Most AC voltage and current measurements are actually RMS values. For example, a typical wall outlet provides about 120 V, which is the RMS value. The peak voltage of the waveform is actually about 170 V.

Three-Phase AC Power

Three-phase AC power includes three separate voltage and current waveforms occurring simultaneously 120° apart. **See Figure 8-4.** Three-phase AC power is commonly used for motors because they can be more efficient and smaller than single-phase motors of the same power output. Many large PV inverters are designed to produce three-phase AC outputs.

Waveform Parameters

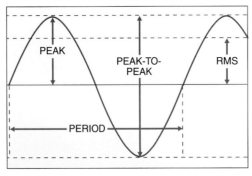

Figure 8-3. *Certain parameters are integral to defining the characteristics of an AC waveform.*

Sine wave patterns occur often in nature, such as in ocean waves, tides, sound waves, and light waves. A plot of average daily temperatures over a year or more also resembles a sine wave.

Three-Phase Power

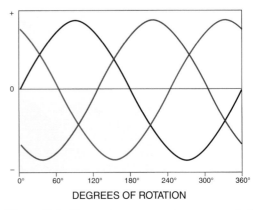

Figure 8-4. *Three-phase power is composed of three separate voltage waveforms that are 120° out of phase.*

True RMS Meters

For a true sine wave, the peak, RMS, and average values of the waveform are always proportional. This means the values can be calculated from one another with simple formulas. These formulas are accurate only for true sine waves.

Calculating the actual RMS value of a waveform is possible with the fast electronics and microprocessors in many meters. A meter that has this capability will be labeled "true RMS." However, a low-end meter shortcuts this process by measuring the peak value and multiplying that value by 0.707, which is a much faster and easier calculation. If the waveform is a true sine wave, the value will be accurate. That is, the calculation will yield the same result as calculating the RMS value directly.

However, if the waveform is not a sine wave or is a distorted sine wave, this shortcut method produces an incorrect RMS value. For example, the RMS and peak voltages are equal for a square wave. This can lead to an improper conclusion that a modified square wave or square wave inverter is not producing its rated voltage or power. A true RMS meter is required to measure these signals accurately. Fortunately, the prices of true RMS meters are falling and the cost advantage of non-true-RMS meters is becoming less significant.

Measuring peak and RMS values with a true RMS meter is an easy way to gain information about the shape of a waveform without using an oscilloscope or graphical multimeter to display the actual waveform. If the calculated RMS value (calculated from the peak value) is less than the actual RMS, then the waveform is wider and flatter, like a square wave. If the calculated RMS value is more than the actual RMS, then the waveform is very narrow.

AC Voltage Conversions

$$V_{RMS} = 0.707 \times V_{PEAK}$$

$$V_{AVG} = 0.638 \times V_{PEAK}$$

$$V_{AVG} = 0.9 \times V_{RMS}$$

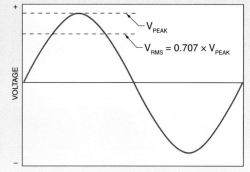

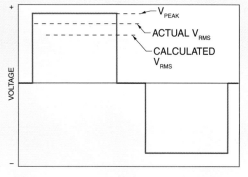

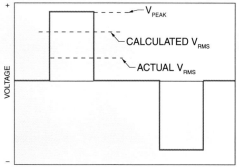

Power Quality

Power quality is the measure of how closely the power in an electrical circuit matches the nominal values for parameters such as voltage, current, harmonics, and power factor. It is common for actual circuit parameters to vary, but allowable ranges are typically very small.

Excessive variations in circuit parameters can cause damage to loads and distribution equipment.

Power quality problems can be caused by the power source, but they can also be caused by loads on the electrical system. It is important to ensure adequate power quality in inverter systems, as in any electrical system. However, most inverters perform much of the power quality monitoring automatically, so problems usually are easy to identify. The inverter may alert the operator to a power quality problem with an alarm or visual display, and may shut down automatically to avoid damage to equipment if the problem is significant. However, an understanding of power quality issues and their common causes is important to avoid problems.

Voltage Variations. Voltage in a power distribution system is typically acceptable within the range of +5% to –10% from the nominal voltage. Small voltage fluctuations typically do not affect equipment performance, but voltage fluctuations outside the normal range can cause circuit and load problems. **See Figure 8-5.**

Voltage Variations

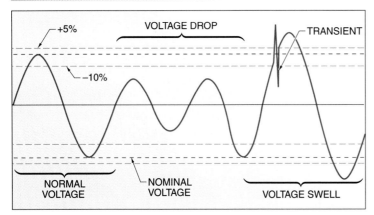

Figure 8-5. *Voltage variations outside allowable ranges include voltage sags, voltage swells, and transients.*

Voltage sags are commonly caused by overloaded transformers, undersized conductors, conductor runs that are too long, too many loads on a circuit, peak power usage periods (brownouts), and high-current loads being turned on. Voltage sags are often followed by voltage swells as voltage regulators overcompensate.

Voltage swells are caused by loads near the beginning of a power distribution system, incorrectly wired transformer taps, and large loads being turned off. Voltage swells are not as common as voltage sags, but are more damaging to electrical equipment.

Transient voltages are temporary undesirable voltages in an electrical circuit, ranging from a few volts to several thousand volts and lasting from a few microseconds up to a few milliseconds. Transient voltages are caused by the sudden release of stored energy due to lightning strikes, unfiltered electrical equipment, contact bounce, arcing, and high current loads being switched ON and OFF. Transient voltages differ from voltage drops and surges by being larger in amplitude, shorter in duration, steeper in rise time, and erratic. High-voltage transients can permanently damage unprotected circuits or electrical equipment.

Voltage Unbalance. *Voltage unbalance* is the unbalance that occurs when the voltages of a three-phase power supply or the terminals of a three-phase load are not equal. Voltage unbalance also results in a current unbalance. **See Figure 8-6.** Voltage unbalance should not be more than 1%. The primary cause of voltage unbalances of less than 2% is too many single-phase loads on one phase of a three-phase system.

Three-Phase Unbalance

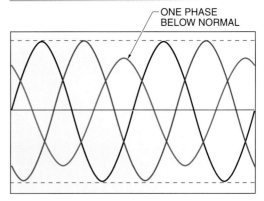

Figure 8-6. *Three-phase voltage and current waveforms are unbalanced if they are not equal in magnitude.*

Single phasing is the complete loss of one phase on a three-phase power supply. Single phasing is the maximum condition of voltage unbalance. Common causes of single phasing include blown fuses, mechanical switching failure, or a lightning strike on the power lines.

Current Unbalance. *Current unbalance* is the unbalance that occurs when current is not equal on the three power lines of a three-phase system. Small voltage unbalances cause large current unbalances. Large current unbalances cause excessive heat, resulting in insulation breakdown. Typically, for every 1% of voltage unbalance, current unbalance is 4% to 8%. Current unbalances should never exceed 10%.

Phase Unbalance. When three-phase power is generated and distributed, the three power lines are 120° out of phase with each other. *Phase unbalance* is the unbalance that occurs when three-phase power lines are more or less than 120° out of phase. Phase unbalance of a three-phase power system occurs when single-phase loads are applied, causing one or two of the lines to carry more or less of the load. Loads must be balanced on three-phase power systems during installation.

Harmonic Distortion. A *harmonic* is a waveform component at an integer multiple of the fundamental waveform frequency. For example, the second harmonic frequency of a 60 Hz sine wave is 120 Hz, the third is 180 Hz, the fourth is 240 Hz, and so on. These higher-frequency harmonic components superimpose on the fundamental frequency, distorting the waveform. **See Figure 8-7.**

Fundamental frequencies and harmonics can add together to form composite waveforms. *Total harmonic distortion (THD)* is the ratio of the sum of all harmonic components in a waveform to the fundamental frequency component. Total harmonic distortion is expressed as a percentage. For example, a current waveform with 5% THD means that 5% of the total current is at frequencies higher than the fundamental.

Harmonics are commonly caused by nonlinear loads, including variable-frequency drives, switching power supplies, and low-quality inverters. Harmonics are also present in square waves and modified square waves, and can be a significant inverter issue. Harmonics cause extra heat in motors and transformers and sometimes create audible noise.

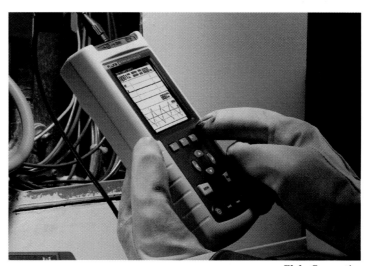

Fluke Corporation
Power quality problems are diagnosed with graphical power quality analyzers.

Power Factor. AC loads are either resistive or reactive loads. A *resistive load* is a load that keeps voltage and current waveforms in phase. *True power* is the product of in-phase voltage and current waveforms and produces useful work. **See Figure 8-8.** True power is also called real power or active power and is represented in units of watts (W).

A *reactive load* is an AC load with inductive and/or capacitive elements that cause the current and voltage waveforms to become out of phase. Inductive loads are the most common reactive

Harmonics

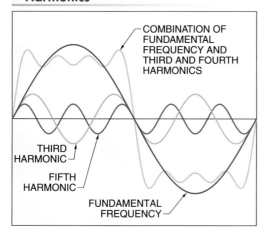

COMBINATION OF FUNDAMENTAL FREQUENCY AND THIRD AND FOURTH HARMONICS

THIRD HARMONIC

FIFTH HARMONIC

FUNDAMENTAL FREQUENCY

Figure 8-7. *Harmonics can add to the fundamental frequency to produce distorted waveforms.*

loads and include motors and transformers. *Reactive power* is the product of out-of-phase voltage and current waveforms and results in no net power flow. Inductive loads momentarily retard current in the process of building magnetic fields and cause the current waveform to lag the voltage waveform in time. Capacitive loads momentarily store voltage and cause the current waveform to lead the voltage waveform in time. Reactive power is represented in units of volt-amperes reactive (VAR).

Power Factor

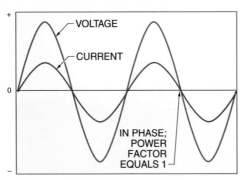

RESISTIVE LOADS

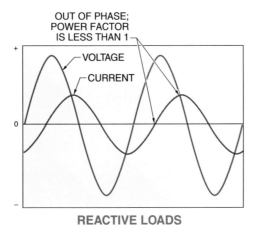

REACTIVE LOADS

Figure 8-8. *Resistive loads keep the voltage and current waveforms in phase, while reactive loads cause the current waveform to lead or lag the voltage waveform.*

Early converters were electromechanical devices that combined a motor with a commutator to produce DC output from an AC input. Inverting the connections resulted in AC output from DC input. The term "inverter" stems from these inverted converters.

Power factor is the ratio of true power to apparent power and describes the displacement of voltage and current waveforms in AC circuits. *Apparent power* is a combination of true and reactive power and is given in units of volt-amperes (VA). For resistive loads, voltage and current waveforms are in phase and apparent power equals the true power, so the power factor equals 1. Reactive load circuits have a power factor of less than 1 because the true power is less than the apparent power.

Low power factor has important consequences for both utility and inverter AC power. Because reactive loads return some power to the source, larger conductors, overcurrent protection, switchgear, and other distribution equipment must be provided for lower-power-factor loads. Consequently, maintaining high power factor minimizes the sizes and costs for this equipment. Furthermore, normal utility kilowatt-hour revenue meters record only true power, not apparent power.

INVERTERS

When AC loads and appliances are to be used in a PV system, an inverter is required to convert DC power to AC power. An *inverter* is a device that converts DC power to AC power. **See Figure 8-9.** The AC output is connected to a distribution panel to power the AC loads. Some inverters allow the flexibility to operate both AC and DC electrical loads. However, inverters add to the complexity and cost of a system, and some power is lost in the conversion process.

Early inverters were electromechanical devices that coupled a DC motor to an AC generator. Electromechanical inverters were noisy and inefficient, and have largely become obsolete. Static (solid-state) inverters change DC power to AC power using electronics and have no moving parts. They are more efficient and much less expensive than electromechanical inverters in most applications. Inverters used in PV systems are exclusively static inverters.

Inverters

SMA America, Inc.

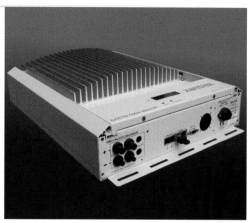

Xantrex Technology Inc.

Sharp Electronics Corp.

Fronius USA LLC

Figure 8-9. *Inverters are available in many different configurations and ratings.*

Electromechanical Rotary Converters

In the late 1900s, rotating electromechanical devices were developed as the first power converters. Depending on their design, they convert DC power to AC power, AC power to DC power, or AC power to AC power at different voltages, phase, or frequency. The input power runs a motor, which produces mechanical power. The mechanical power is then transferred to an electrical generator, which is configured for the desired output power.

Discrete rotary converters are separate motor and generator units that are mechanically coupled by a shaft. This coupling offers power and harmonic isolation, voltage output control, and greater surge protection. Integral rotary converters combine the motor and generator into one self-contained unit. Some of the electrical energy flows directly from input to output, bypassing the mechanical power conversion process, which improves efficiency and allows the unit to be much smaller and lighter than discrete systems.

Electromechanical converters can handle very large loads, but are also heavy and inefficient and require considerable maintenance. These systems are largely obsolete, because solid-state power electronics are smaller, lighter, more efficient, and virtually maintenance free, even when several electronic converters must be connected in parallel to handle large loads.

PV Inverters

Inverters for PV systems are broadly classified as either stand-alone or interactive in operation. The difference involves whether the inverter input is connected to the PV array or the battery bank. PV arrays and batteries have different characteristics, which affects inverter design.

Stand-Alone Inverters. Stand-alone inverters are connected to batteries as the DC power source and operate independently of the PV array and the utility grid. PV arrays charge the batteries but do not directly influence the operation of the inverter. For stand-alone inverters, it is the electrical load connected to the AC output, rather than the DC power source, that affects the performance of the inverter. **See Figure 8-10.** DC loads may also be powered directly from the battery bank.

Xantrex Technology Inc.
Large PV systems may include multiple inverters with their outputs in parallel.

Stand-alone inverters must be sized to meet the total connected AC load for both steady-state and surge-load requirements. Overloading the output of stand-alone inverters raises the temperature of the unit until it automatically shuts down.

Utility-Interactive Inverters. Utility-interactive PV inverters are connected to, and operate in parallel with, the electric utility grid. Sometimes called grid-connected or grid-tie inverters, these inverters interface between the PV array and the utility grid and convert DC output from a PV array to AC power that is consistent and synchronous with the utility grid. Interactive inverters are loaded by the DC source, not the AC output, so AC loads do not directly impact the operation of the inverter. **See Figure 8-11.**

Interactive PV systems are interconnected with the utility at the distribution panel or on the supply side of service entrance equipment. In a sense, the utility acts as an infinitely large energy storage system that accepts excess energy from the interactive system and supplies extra energy when needed. This allows AC power produced by the PV system to either supply on-site electrical loads or to back-feed power to the grid when the PV system output is greater than the site load demand. At night and during other low insolation periods when the electrical loads are greater than the PV system output, the balance of power required by the loads is received from the electric utility.

Stand-Alone Inverters

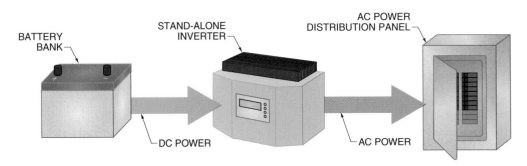

Figure 8-10. *Stand-alone inverters are connected to the battery bank and supply AC power to a distribution panel that is independent of the utility grid.*

Interactive Inverters

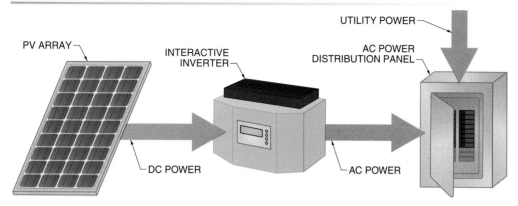

Figure 8-11. *Interactive inverters are connected to the PV array and supply AC power that is synchronized with the utility grid.*

Bimodal Inverters. Bimodal inverters can operate in either interactive or stand-alone mode (though not simultaneously). These inverters are connected to the battery bank like stand-alone inverters, while the PV array charges the batteries. Sometimes referred to as battery-based interactive or multimode inverters, bimodal inverters are popular for small electrical systems where a back-up power supply is required for critical loads such as computers, refrigerators, and water pumps when a utility power outage occurs.

Under normal circumstances and when connected to an energized utility grid, bimodal PV systems operate in interactive mode, serving on-site loads or sending excess power to the grid while keeping the battery fully charged. If a grid outage occurs, the inverter opens the connection with the utility and operates from the battery bank to supply power to loads at a dedicated subpanel. The subpanel loads must not exceed the inverter power rating.

AC Module Inverters. An *AC module* is a PV module that outputs AC power through an interactive inverter attached in place of the normal DC junction box. AC modules are only permitted for interactive operation. The module has no accessible DC wiring and is not subject to normal DC circuit requirements for PV systems, which simplifies system design and installation because far fewer components are needed. The modularity also makes system changes and expansion relatively easy.

Multiple AC module inverters are connected in parallel to form AC branch circuits. Since the inverter electronics must be designed to withstand outdoor weather and temperature extremes, the units can be relatively expensive, but in some cases, the savings from not purchasing and installing DC components may outweigh the additional costs.

Inverters are also available that operate like the inverter portion of fully integrated AC modules, but are separate components that must be added to individual modules. **See Figure 8-12.** However, while the DC portion of the system is minimized, it is still accessible and may be subject to some requirements.

AC Module Inverters

Enphase Energy

Figure 8-12. *AC module inverters are small interactive inverters that are supplied by a single PV module.*

Despite their small size, AC module inverters include many of the sophisticated features of larger inverters, such as anti-islanding protection, maximum power point tracking, and sine wave output. Systems with multiple AC module inverters may even keep track of the operating status and output of individual modules over a data network or power line communications interface, though this data monitoring requires additional equipment.

Switching Devices

Solid-state inverters use electronics to switch DC power and produce AC power. There are many types of electronic components that can perform switching functions. Continuous improvements in semiconductor manufacturing technology and performance are yielding lower-cost, higher-power, and higher-speed electronic power devices. **See Figure 8-13.**

Switching Devices

Figure 8-13. *Solid-state switching devices used in PV inverters include transistors and thyristors.*

Thyristors. Some basic solid-state inverter designs use thyristors, usually of a type called silicon-controlled rectifiers (SCRs). Thyristors have three leads. When a small current is applied to one lead, the thyristor is turned ON and a larger current is allowed to flow between the other two, like closing a switch. Otherwise, the thyristor is like an open switch and allows no current flow. Like a mechanical switch, thyristors can only be completely ON or completely OFF. Large thyristors are used in high-power applications up to several megawatts, such as in HVDC power transmission.

Transistors. Most inverters use transistors, which are similar in switching capability to thyristors, with two main differences. First, a small voltage, rather than a current as in thyristors, activates the transistor. Second, the magnitude of the activating voltage varies the transistor's resistance. This means that in addition to completely ON or completely OFF, transistors can allow every point in between, like a dimmer switch.

There are many kinds of transistors, though inverters commonly use metal-oxide semiconductor field-effect transistors (MOSFETs) or insulated gate bipolar transistors (IGBTs). Power MOSFETs operate at lower voltages with higher efficiency and lower resistance than IGBTs. MOSFETs switch at very high speeds, up to 800 kHz, and are generally used in 1 kW to 10 kW applications. IGBTs handle high current and voltage, but switch at lower speeds, up to 20 kHz, and are more common for large, high-voltage applications that may exceed 100 kW.

Switching Control

Each switching device in an inverter requires a control circuit to activate the switching function. Some circuits are designed to regulate switching automatically from an external signal, while others use microprocessors for timing and control.

Line Commutation. In the simplest inverter designs, switching is controlled automatically by an external source, such as utility power. A *line-commutated inverter* is an inverter whose switching devices are triggered by an external source. Line-commutated inverters alternately turn the switches ON and OFF by the positive and negative half-cycles of the utility voltage, automatically synchronizing the inverter output to the utility. **See Figure 8-14.** Line-commutated inverters use a simple and effective design, but cannot operate independently of the grid.

Line Commutation

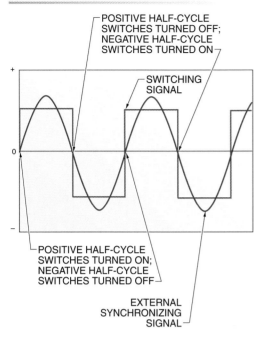

Figure 8-14. *Line-commutated inverters use an external AC signal to activate and deactivate the inverter switching devices.*

Self-Commutation. A *self-commutated inverter* is an inverter that can internally control the activation and duration of its switching. Self-commutated inverters can operate either interconnected with or independent of utility power. This is accomplished by incorporating microprocessors for precise timing and control. Compared to line-commutated inverters, self-commutated inverters can better control AC waveform output, adjust power factor, and suppress harmonics. Most PV inverters manufactured today are self-commutated inverters.

Self-commutated inverters are either voltage-source or current-source types. A voltage-source self-commutated inverter treats the DC input as a voltage source and produces an AC voltage output. A current-source self-commutated inverter treats the DC input as a current source and produces an AC current output. Stand-alone and bimodal inverters are typically of the voltage-source type. Most interactive-only inverters are of the current-source type.

Square Wave Inverters

There can be multiple processing steps to convert DC power into sine wave AC power. The first step in the inverting process is to switch the DC power back and forth to create AC power, producing a square wave. If an inverter performs only this step, it is a square wave inverter. Square wave output is not very efficient and can be detrimental to some loads. More-sophisticated inverters add further processing after this step to produce modified square waves or sine waves.

Switching circuits are typically designed symmetrically. One side of the circuit includes devices to switch and control DC input into the positive half-wave of the AC output. The other side switches and controls a reversed polarity of the DC input into the negative half-wave of the AC output. Two circuit designs for producing square wave output are H-bridge circuits and push-pull circuits.

H-Bridge Inverter Circuit. An H-bridge inverter circuit is very similar to a full-wave rectifying circuit, but the two circuits perform opposite functions. An *H-bridge inverter circuit* is a circuit that switches DC input into square wave AC output by using two pairs of switching devices. One pair is open while the other pair is closed. The two pairs alternate states to change the direction of the DC current flow through the circuit's output. **See Figure 8-15.** This design is known as an H-bridge inverter because the switching array can be drawn in an "H" shape in a circuit diagram.

Inverters control high-power switching devices with low-power electronics to produce AC output.

H-Bridge Inverter Circuits

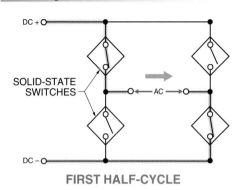

FIRST HALF-CYCLE

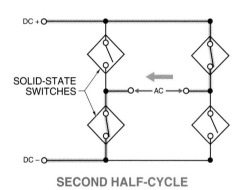

SECOND HALF-CYCLE

Figure 8-15. *H-bridge inverter circuits use two pairs of switching devices to direct a DC input to the output in both directions.*

Push-Pull Inverter Circuits

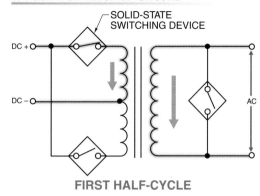

FIRST HALF-CYCLE

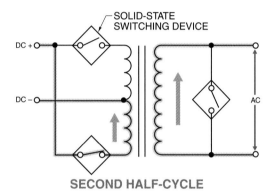

SECOND HALF-CYCLE

Figure 8-16. *Push-pull inverter circuits use one pair of switching devices and a transformer to alternate the direction of direct current.*

Push-Pull Inverter Circuit. A *push-pull inverter circuit* is a circuit that switches DC input into AC output by using one pair of switching devices and a center-tapped transformer. The circuit gets its name from the backward and forward current flow through the circuit. **See Figure 8-16.** First, the top switch closes, allowing current to flow from the DC source through the transformer and back in a clockwise loop. Then, the top switch is opened and the bottom switch is closed. The current flows again from the DC source, this time in a counterclockwise loop. The alternating current in the primary winding of the transformer is in the shape of a square wave, which induces a similar AC output in the secondary winding.

A switching device can be shorted across the output winding to zero the transformer output during part of each half-cycle. This creates an additional step in the output waveform, resulting in a modified square wave.

Low-Frequency Waveform Control

The simple square wave output from inverter circuits can be further refined to improve sine wave approximation. By adjusting the duration of the alternating square pulses, the output becomes a modified square wave. Transformers are used to step the input voltage up to output voltage levels. For shorter pulses, the peak voltage is stepped higher. **See Figure 8-17.** To create a multistepped modified square wave, multiple square wave inverter stages are operated in parallel. The outputs are then combined to produce a stepped waveform that more closely matches a true sine wave. **See Figure 8-18.**

Inverters that use this method are called low-frequency inverters because the frequencies of these waveforms are still only 60 Hz. While they are simple, low-frequency inverters use very heavy transformers and are less efficient than higher-frequency designs.

Low-Frequency Control

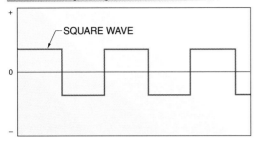

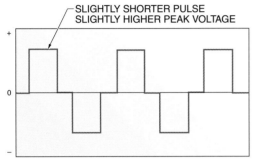

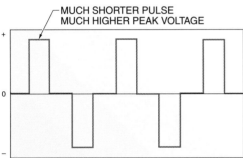

Figure 8-17. *Square waves can be modified by adjusting the duration and magnitude of the pulses.*

Multistepped Modified Square Waves

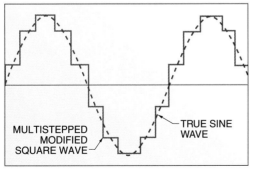

Figure 8-18. *Combining multiple modified square waves with different magnitudes and durations results in a multistepped modified square wave that more closely approximates a sine wave.*

High-Frequency Waveform Control

Pulse-width modulation (PWM) control is used to create a sine wave inverter output. *Pulse-width modulation (PWM)* is a method of simulating waveforms by switching a series device ON and OFF at high frequency and for variable lengths of time. When the pulses are narrow, the current is OFF most of the time, which simulates a low voltage. When pulses are wide, the current is ON most of the time, which simulates a high voltage. **See Figure 8-19.** Some PWM methods also adjust frequency by spacing narrow pulses farther apart and wide pulses closer together.

High-Frequency Pulse-Width Modulation

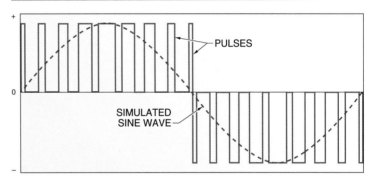

Figure 8-19. *Pulse-width modulation at high frequencies generates the truest approximation of a sine wave.*

By using very high frequencies and gradual changes in pulse width, loads are less affected by abrupt changes in voltage. Most loads operate as if PWM output were a true sine wave. High- frequency switching and power conversion also reduces harmonics in output current, which reduces the size and weight of the transformers. Pulse-width modulation, along with advancements in digital controls and microprocessors, has resulted in very efficient inverter designs.

High-frequency inverters use a DC-DC converter to step DC input voltage up to higher levels or use higher-voltage DC arrays. The DC power is then inverted to AC power at high frequency without the need for a large transformer. Pulse-width modulation is then the final switching stage to produce 60 Hz sine wave AC power.

Transformers are used within inverters and power conditioning units to raise or lower AC voltages and provide electrical isolation.

POWER CONDITIONING UNITS

The physical enclosure that is referred to as an inverter is actually often a power conditioning unit (PCU). Power conditioning units perform one or more power processing and control functions in addition to inverting, such as rectification, transformation, DC-DC conversion, and maximum power point tracking. **See Figure 8-20.** These functions can also be performed by separate components, but this is usually not necessary. Power conditioning units may include system-monitoring capabilities and protective features such as disconnects and overcurrent protection equipment.

Rectifiers

A *rectifier* is a device that converts AC power to DC power. Rectifiers are used in battery chargers and DC power supplies operating from AC power. Many stand-alone and hybrid PV inverters include multifunction, programmable battery chargers with the same charge control and regulation algorithms as separate battery charge controllers.

Transformers

A *transformer* is a device that transfers energy from one circuit to another through magnetic coupling. A transformer consists of two or more coupled windings and a core. Current in one winding creates a magnetic flux in the core, which induces voltage in the other winding. Transformers are used to convert between high and low AC voltages, change impedance, and provide electrical isolation and voltage regulation. **See Figure 8-21.**

Power Conditioning Units

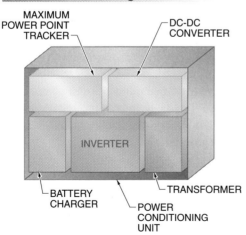

Figure 8-20. *Power conditioning units are inverters that also perform other power control and conversion functions.*

Transformers

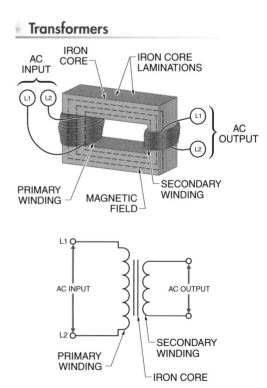

Figure 8-21. *Transformers use induced magnetic fields to transfer AC power from one circuit to another while transforming the power to higher or lower voltages or providing electrical isolation.*

Transformers cannot convert between DC and AC, change the voltage or current of a DC source, or change the frequency of an AC source. However, transformers are integral components of devices that perform these functions, including inverters.

The ratio of the number of turns in the windings of a transformer (turns ratio) determines how voltage and current are stepped up or down. For example, a transformer with a 1:10 turns ratio will step an input voltage up 10 times. Since the power must remain the same (neglecting small losses), the current must be stepped down to one-tenth of the input current. Therefore, a 12 V input at 20 A would be transformed to a 120 V output at 2 A.

An autotransformer is a transformer with only one winding and three or more taps. **See Figure 8-22.** The voltage source is applied to two taps and the load is connected to two taps, one of which is a common connection with the source. Each tap corresponds to a different source or load voltage. In an autotransformer, a portion of the same winding acts as part of both the primary and secondary windings. Autotransformers are an economical and compact way to adjust a voltage up or down slightly. For example, an autotransformer can be used to convert 240 V output from an inverter to 208 V for interconnection to a residential system. Unlike transformers with multiple windings, autotransformers do not provide electrical isolation.

Autotransformers

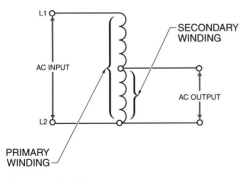

Figure 8-22. *The primary and secondary windings in an autotransformer share some of the same windings.*

DC-DC Converters

A *DC-DC converter* is a device that changes DC power from one voltage to another. Modern switched-mode DC-DC converters use high-frequency switching and transformers to convert DC power to a different voltage, either higher or lower than the source voltage. These devices are lightweight and efficient, and because they use transformers, they provide circuit isolation. A buck converter is a step-down DC-DC converter. A boost converter is a step-up DC-DC converter.

Many PV inverters use DC-DC converters to change the DC input from low voltage to high voltage prior to the power-inverting process. Also, external DC-DC converters may be used on battery-based systems to deliver DC voltage at levels other than the nominal battery voltage. Maximum power point trackers are a form of DC-DC converter. DC-DC converters are characterized by their power rating, input and output voltages, and their power conversion efficiency.

Maximum Power Point Trackers

A *maximum power point tracker (MPPT)* is a device or circuit that uses electronics to continuously adjust the load on a PV device under changing temperature and irradiance conditions to keep it operating at its maximum power point. Since they are connected directly to the array, all interactive inverters include MPPT circuits.

In some system designs with multiple arrays, individual MPPTs are connected to each input source circuit to allow the inverter to extract maximum output. This improves performance if the arrays have different I-V characteristics or are oriented in different directions. As the current output for these arrays varies, the individual MPPTs optimize each array output.

Stand-alone inverters do not directly operate or control the array, so they do not normally include MPPT circuits. However, MPPT battery charge controllers can interface between an array and the battery bank that is powering a stand-alone inverter. This MPPT application can improve array battery-charging potential, but does not affect operation of the inverter.

INVERTER FEATURES AND SPECIFICATIONS

PV system inverters include a number of basic and optional features. Inverter specifications include performance data, operating limits, installation requirements, safety, and maintenance. This information is found on the inverter nameplate and in the product manuals.

Listing and Certifications

Inverters installed in PV systems are required to conform to certain standards for product listing and certifications. These include the safety standard UL 1741 as well as certifications for EMI under FCC Part 15. Inverters must include a listing mark on their nameplate label. **See Figure 8-23.** Inverters not marked as interactive inverters are not permitted to operate in utility-interconnected applications.

Inverter Nameplates

Figure 8-23. *Inverter nameplates include much of the needed information for sizing and operating the inverter.*

UL 1741 *Inverters, Converters, Controllers and Interconnection System Equipment for Use with Distributed Energy Resources* addresses requirements for all types of distributed generation equipment, including inverters and charge controllers for PV systems. IEEE 1547 *Standard for Interconnecting Distributed Resources with Electric Power Systems*, and IEEE 1547.1 *Standard Conformance Test Procedures for Equipment Interconnecting Distributed Resources with Electric Power Systems* are the basis for UL 1741 listing for interactive inverters.

Installation

National codes and standards dictate the safety and installation requirements for inverters used in PV systems. Inverter installation requirements are governed by the NEC® Article 690, "Solar Photovoltaic Systems" and other applicable sections of the NEC®, including overcurrent protection devices, disconnects, grounding, and utility-interconnection. Many of these requirements are based on equipment standards and listing requirements for interactive inverters under UL 1741. Most inverter manufacturers also provide details on requirements for code-compliant installation in their product manuals.

Power Ratings

The principal inverter specification is the output power rating. In the case of stand-alone inverters, the power rating limits the power it can deliver to AC loads. For an interactive inverter, the output power rating limits the power it can handle at its DC input, which limits the size of the PV array.

Interactive inverters are available from many manufacturers in power ratings from 700 W for small residential units to over 500 kW for large commercial and utility-scale installations. Stand-alone inverters are available in sizes down to about 50 W, but are commonly in the 3 kW to 6 kW range. Exceedingly large battery banks and high currents limit stand-alone inverters to a maximum of about 50 kW for large remote-power installations.

Temperature Limitations. Solid-state switching devices are capable of handling only so much current before they overheat and fail. Paralleling switching devices in the design increases the power rating of an inverter. However, thermal management in electronic inverters is still a major concern, and temperature is the primary limiting factor for inverter power ratings.

Manufacturers specify a permissible operating temperature range for their inverters, such as –20°C to +50°C. Inverters may deal with excessive operating temperature in several ways. Many inverters use heat sinks and/or ventilation fans to regulate temperature.

Interactive inverters control high temperatures by limiting the array power delivered to the inverter. The inverter forces the array operating point from maximum power to a higher operating voltage (toward open-circuit voltage), reducing power and current levels. **See Figure 8-24.** Once the inverter temperatures stabilize, the array is again loaded to its full power output. At the higher input voltage, the power output falls below the inverter's rating.

Power Output Limiting

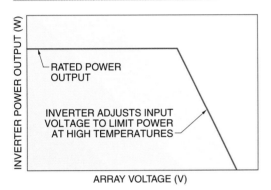

Figure 8-24. *At high temperatures, an interactive inverter may limit current input by raising the input voltage, which also lowers power input and output.*

If the temperature or load limits are exceeded, inverters limit or disconnect their output. Most stand-alone inverters shut down or internally disconnect the AC output if load limits are exceeded. Otherwise overcurrent devices trip if the AC output is excessively loaded. The load must then be decreased (manually) before the inverter can be restarted.

Voltage Ratings

Inverter performance is strongly associated with operating voltages. Voltage ratings are given for the AC output and DC input circuits, and may apply to stand-alone and interactive

inverters in different ways. Voltage specifications are typically given as a range, with minimum and maximum limits for operation.

AC Output. Inverter AC output interfaces with either the utility grid or with electrical loads and appliances, so inverter voltage ratings are consistent with normal utility voltage standards. Smaller inverters (less than 6 kW) typically produce an AC output voltage of 120 V or 240 V nominal single-phase. Larger inverters produce 208 V, 277 V, or 480 V nominal three-phase AC output. Some inverters can be configured for a variety of output voltages at the time of installation.

For interactive inverters, AC voltage output must be maintained at –10% to +5% of the nominal system voltage. In the case of 120 V nominal output, this is a range from 108 V to 126 V. Variations may occur in supply voltage. If the inverter senses that voltage is out of range for more than 30 seconds (1800 cycles), the AC output disconnects within 10 cycles. If the voltage varies more than –30% or +10% for 10 cycles, the inverter must disconnect within 10 cycles. This standard reduces the potential for islanding. Additionally, voltage flicker shall not be allowed to exceed a 3% voltage sag, in accordance with IEEE 519 *Recommended Practice and Requirements for Harmonic Control in Electrical Power Systems.* Inverter specifications may include RMS output voltage regulation.

MOSFET switching devices can become so hot when switching high currents that they must be bonded to large metal heatsinks.

DC Input. DC input voltage ratings are based on the operating characteristics of either a battery bank (for stand-alone inverters) or a PV array (for interactive inverters). Small stand-alone inverters are designed to operate from nominal 12 V lead-acid batteries, or multiples thereof. These inverters operate within a relatively narrow voltage range, between 11 V and 16 V, based on actual battery voltages during normal charging and discharging conditions. For AC output power levels over 1 kW, 12 V inputs are impractical, because increased currents require larger and more expensive conductors and switchgear. Instead, battery banks for larger systems are designed to deliver nominal 24 V (actual range 22 V to 32 V) or 48 V (44 V to 64 V) or higher.

For interactive inverters, the DC input voltage requirements are more complex. Minimum and maximum voltage limits are given for inverter operation and another, narrower voltage range within which the inverter will properly track maximum power from the array. **See Figure 8-25.** Minimum operating voltages are required to perform basic inverter functions and produce an output with sufficient peak and RMS voltage. Below this level, the inverter will not start or operate. Maximum DC voltage avoids exceeding the inverter's voltage-handling capability, and most are limited to less than 600 V for product listing and code compliance reasons.

DC Input Voltage Ranges

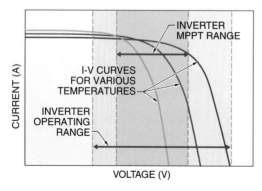

Figure 8-25. *Most inverters operate from a relatively wide range of input voltages, but the range for MPPT operation is smaller.*

There is often a fixed relationship between the utility voltage and the array voltage that the inverter MPPT will track. The required array voltage increases with increasing grid voltage, and the array must have maximum power voltage in this range to permit MPPT operation. **See Figure 8-26.**

Minimum DC Input Voltages

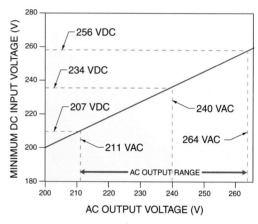

Figure 8-26. *In order to output AC voltage within the specified range, the DC input voltage must meet certain minimum values.*

Some interactive inverters have a large range of DC input voltages, allowing many array sizes and configurations. The DC voltage from an array is also affected by ambient temperature, which complicates sizing. Many inverter manufacturers offer string sizing software to determine the minimum and maximum number of series-connected modules for proper inverter operation. These tools, many available online, include databases of commercial PV module specifications and the manufacturer's inverter ratings. Based on the chosen module information and user-inputted minimum and maximum ambient temperatures for the location, the allowable array configurations are displayed. The highest possible array voltages for the site's record low temperature influence the maximum number of modules. Conversely, a minimum number of modules must be connected in series to achieve sufficient operating voltages under the hottest conditions. A table of results shows how array maximum open-circuit voltage varies by number of modules and ambient temperature.

Frequency Ratings

Inverters for North American markets are designed to produce 60 Hz, while inverters for Europe, Asia, and other parts of the world produce 50 Hz. Loads and systems are more sensitive to variations in frequency than voltage, so only a small variation is allowed from the nominal operating frequency. For nominal 60 Hz operation, the AC output frequency must be maintained between 59.3 Hz and 60.5 Hz.

Current Ratings

Inverters are rated for operating and maximum allowable AC and DC currents, which are determined by the current-handling capabilities of the switching devices. For the DC side, current ratings limit the PV array or battery current that can be applied to the inverter. On the AC side, current ratings limit the AC load for stand-alone inverters and the AC current output for interactive inverters.

Maximum continuous AC output and DC input current ratings are given at a reference temperature, and are the basis for sizing conductors, switchgear, and overcurrent protection for the input and output circuits. DC input current for some interactive inverters decreases with increasing DC input voltage in order to limit inverter output power. **See Figure 8-27.**

DC Input Current

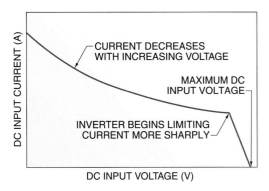

Figure 8-27. *Inverters may limit maximum DC input current with increasing DC input voltage.*

Maximum AC output fault currents may also be specified, and represent the maximum current that an inverter can deliver into a shorted or overloaded circuit. Inverter specifications sometimes include requirements for external AC and DC overcurrent protection devices, usually based on 125% of the maximum continuous rated currents. This equipment is provided for some inverters as part of the power conditioning unit.

Stand-alone inverters draw high DC input currents from low-voltage battery banks, especially at low battery voltages (state of charge). Special circuit design requirements apply to these inverters.

SMA Technologie AG
Inverters for large, building-integrated PV arrays, such as curtain-wall skylights, are fully incorporated into the electrical infrastructure of the building.

Surge Capability. Inverters operating from batteries can deliver high surge and in-rush currents for short periods. Many motors require surge current at startup of as much as 6 to 10 times the normal operating current. Surge capability is given as a maximum AC output current for an amount of time, for example 78 A for 100 ms. This may contrast with a continuous rated output current, such as 33 A. Interactive inverters operating from PV arrays cannot produce surge current because they are loaded by the current-limited PV source.

Harmonics

Utility-interactive inverters must maintain acceptable limits of harmonic distortion at the utility interconnection. In accordance with IEEE 519, total harmonic distortion shall not exceed 5%, and any individual harmonic shall not exceed 3%. Interactive inverters are required to de-energize from the grid if these limits are exceeded.

Power Factor

Interconnection standards require interactive inverters to maintain an output power factor between 0.95 leading and 0.95 lagging. Most interactive inverters produce AC output with a power factor of 1 under all conditions. Output power factor for stand-alone inverters is a function of the load, which may or may not operate with a power factor of 1.

> Harmonic distortion is used to create waveforms of certain shapes. The sum of a series of each integer sine wave harmonic produces a waveform with a sawtooth shape. The sum of a series of odd-numbered harmonics produces a square waveform. The sum of a series of odd-numbered harmonics with every (4n–1)th harmonic inversed produces a waveform with a triangle shape.

Efficiency

Inverter efficiency is the effectiveness of an inverter at converting DC power to AC power. Some input power is used to operate the inverter and some is lost as heat in the switching and power conversion processes. *Stand-by losses* are the power required to operate inverter electronics and keep the inverter in a powered state. Output power depends on the waveform shape. Therefore, the output power is always less than the input power. Inverter efficiency is calculated with the following formula:

$$\eta_{inv} = \frac{P_{AC}}{P_{DC}}$$

where

η_{inv} = inverter efficiency

P_{AC} = AC power output (in W)

P_{DC} = DC power input (in W)

For example, what is the efficiency of an inverter that produces 2000 W of AC output from 2200 W of DC input?

$$\eta_{inv} = \frac{P_{AC}}{P_{DC}}$$

$$\eta_{inv} = \frac{2000}{2200}$$

$$\eta_{inv} = \mathbf{0.91} \text{ or } \mathbf{91\%}$$

Most interactive PV inverters are rated 90% to 95% efficient. Quality stand-alone inverters producing sine wave output have peak efficiencies of about 90%. Lesser-quality modified sine wave inverters may have efficiencies as low as 75% to 85%. In general, high-frequency and high-voltage inverters are more efficient than lower-voltage inverters operating at low frequency.

Inverter efficiency is primarily affected by the inverter load. In stand-alone inverters, the AC load defines the inverter load, and for interactive inverters, the PV array defines the load. Efficiency can also be affected by inverter temperature and DC voltage input. Inverter stand-by losses are nearly constant for all output power levels, so efficiency is lower for low power outputs. **See Figure 8-28.** Some modified square wave stand-alone inverters reach peak efficiency at levels well below their maximum rated power output. Most interactive inverters maintain high efficiency over a wide operating range.

Inverter Efficiencies

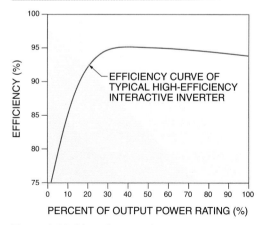

Figure 8-28. *Most sine wave inverters maintain high efficiency over a wide operating-power range.*

Protective Devices

Most inverters include devices to protect the inverter and connected equipment from damage from excessive temperatures, currents, or power levels. **See Figure 8-29.** For example, stand-alone inverters disconnect themselves if DC input voltages become too low, such as from a discharged battery, preserving AC output power quality and preventing the inverter from drawing excessive currents. Nearly all interactive inverters include ground-fault protection. Since most inverters include transformers, they also provide isolation between the DC power source and utility grid or AC output. Many include voltage surge suppression on the DC and/or AC sides. Even when inverters do include these devices, additional equipment may still be required.

Protective Devices

Figure 8-29. *Inverter enclosures may include protective devices such as circuit breakers.*

All interactive inverters must employ protective devices for the utility interface, based on the specified limits of grid operations. These inverters monitor voltage, frequency, power factor, and other parameters, and control output accordingly. If the voltage or frequency of the utility power exceeds preset limits, or a potential islanding condition exists, interactive inverters are required to cease interconnected operations. Once the utility parameters have

stabilized within acceptable limits for at least 5 minutes, the inverter automatically resumes operation.

Physical Characteristics

Physical specifications include the size, weight, and mounting requirements. Low-voltage, low-frequency inverter designs use heavy transformers, so 5 kW inverters may weigh as much as 200 lbs. Special mounting requirements may apply to these inverters. High-frequency inverters use smaller transformers, so they weigh considerably less for equivalent power ratings. Other physical and mechanical characteristics may be provided and applicable to installation.

Data and Control Interfaces

Most modern inverters incorporate microprocessors, and many provide features for data monitoring and communications. Interfaces may include displays and controls on the inverter itself, while others interface with remote units or computers. Status or values can be indicated by LEDs, alphanumeric LCD displays, or graphical LCD displays. Some systems interface with computer software for processing raw data and automatically generating charts, graphs, or graphical displays. **See Figure 8-30.** When connected to a web server, this information can be published on a web site. These systems are particularly flexible for storing and processing system data.

Inverter interfaces typically provide basic system information, including interconnection status, AC output voltage and power, DC input voltage, MPPT status, error codes, fault conditions, and other parameters. Many inverters record energy production on daily and cumulative bases.

Inverter operation parameters are generally not field-adjusted, since this could affect critical safety features and operating limits. However, other power conditioning functions may allow operator control or adjustment, such as battery charger and charge controller settings and the operation and control of generators and other power sources, allowing flexibility for a variety of applications.

Refer to Quick Quiz® on CD-ROM

Inverter Interfaces

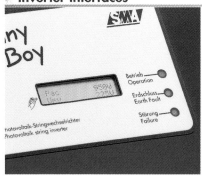

SMA Technologie AG
ON-BOARD DISPLAY

SMA America, Inc.
REMOTE DATA MONITOR

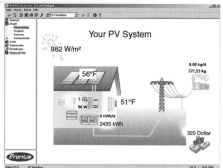

Fronius USA LLC
COMPUTERIZED DATA ACQUISITION

Figure 8-30. *Inverter interfaces include on-board screens, remote data monitors, and computerized data acquisition and processing software.*

Summary

- Sine waves, square waves, and modified square waves are common AC waveforms produced by inverters.

- Sine waves are the most complicated type of AC waveform for inverters to produce, so some less-sophisticated inverters approximate the sine wave with square waves and modified square waves.

- Static (solid-state) inverters change DC power to AC power using electronics, so they have no moving parts and are very efficient.

- AC module inverters are small inverters that take the place of a DC junction box on a PV module.

- Inverters use thyristors and transistors to switch DC power and produce AC power.

- Line-commutated inverters use an external signal to control the switching devices, while self-commutated inverters include microprocessors for precise timing and control of the switching devices.

- By adjusting the duration of the alternating pulses, a square wave becomes a modified square wave.

- Pulse-width modulation (PWM) is used to construct the closest approximation of a sine wave.

- Power conditioning units perform one or more power processing and control functions in addition to inverting, such as rectification, transformation, DC-DC conversion, and maximum power point tracking.

- Transformers are used to convert between high and low AC voltages, change impedance, and provide electrical isolation and voltage regulation.

- The principal inverter specification is the power rating.

- Temperature is the primary limiting factor for inverter power ratings.

- DC input voltage ratings are based on the operating characteristics of either a battery bank (for stand-alone inverters) or a PV array (for interactive inverters).

- Operating and maximum allowable AC and DC current ratings are determined by the current-handling capabilities of switching devices used in the inverter.

- Most inverters include features to protect the inverter and connected equipment from damage from excessive temperatures, currents, or power levels.

- *Direct current (DC)* is electrical current that flows in one direction, either positive or negative.

- *Alternating current (AC)* is electrical current that changes between positive and negative directions.

- A *waveform* is the shape of an electrical signal that varies over time.

- A *periodic waveform* is a waveform that repeats the same pattern at regular intervals.

- A *cycle* is the interval of time between the beginnings of each waveform pattern.

- A *sine wave* is a periodic waveform the value of which varies over time according to the trigonometric sine function.

- A *sinusoidal waveform* is a waveform that is or closely approximates a sine wave.

- A *square wave* is an alternating current waveform that switches between maximum positive and negative values every half period.

- A *modified square wave* is a synthesized, stepped waveform that approximates a true sine wave.

- *Frequency* is the number of waveform cycles in one second.

- *Period* is the time it takes a periodic waveform to complete one full cycle before it repeats.

- *Peak* is the maximum absolute value of a waveform.

- *Peak-to-peak* is a measure of the difference between positive and negative maximum values of a waveform.

- The *root-mean-square (RMS) value* is a statistical parameter representing the effective value of a waveform.

- *Power quality* is the measure of how closely the power in an electrical circuit matches the nominal values for parameters such as voltage, current, harmonics, and power factor.

- *Voltage unbalance* is the unbalance that occurs when the voltages of a three-phase power supply or the terminals of a three-phase load are not equal.

- *Single phasing* is the complete loss of one phase on a three-phase power supply.

- *Current unbalance* is the unbalance that occurs when current is not equal on the three power lines of a three-phase system.

- *Phase unbalance* is the unbalance that occurs when three-phase power lines are more or less than 120° out of phase.

- A *harmonic* is a waveform component at an integer multiple of the fundamental waveform frequency.

- *Total harmonic distortion (THD)* is the ratio of the sum of all harmonic components in a waveform to the fundamental frequency component.

- A *resistive load* is a load that keeps voltage and current waveforms in phase.

- *True power* is the product of in-phase voltage and current waveforms and produces useful work.

- A *reactive load* is an AC load with inductive and/or capacitive elements that cause the current and voltage waveforms to become out of phase.

- *Reactive power* is the product of out-of-phase voltage and current waveforms and results in no net power flow.

- *Power factor* is the ratio of true power to apparent power and describes the displacement of voltage and current waveforms in AC circuits.

- *Apparent power* is a combination of true and reactive power and is given in units of volt-amperes (VA).

- An *inverter* is a device that converts DC power to AC power.

- An *AC module* is a PV module that outputs AC power through an interactive inverter attached in place of the normal DC junction box.

- A *line-commutated inverter* is an inverter whose switching devices are triggered by an external source.

- A *self-commutated inverter* is an inverter that can internally control the activation and duration of its switching.

- An *H-bridge inverter circuit* is a circuit that switches DC input into square wave AC output by using two pairs of switching devices.

- A *push-pull inverter circuit* is a circuit that switches DC input into AC output by using one pair of switching devices and a center-tapped transformer.

- *Pulse-width modulation (PWM)* is a method of simulating waveforms by switching a series device ON and OFF at high frequency and for variable lengths of time.

- A *rectifier* is a device that converts AC power to DC power.

- A *transformer* is a device that transfers energy from one circuit to another through magnetic coupling.

- A *DC-DC converter* is a device that changes DC power from one voltage to another.

- A *maximum power point tracker (MPPT)* is a device or circuit that uses electronics to continually adjust the load on a PV device under changing temperature and irradiance conditions to keep it operating at its maximum power point.

- *Inverter efficiency* is the effectiveness of an inverter at converting DC power to AC power.

- *Stand-by losses* are the power required to operate inverter electronics and keep the inverter in a powered state.

Review Questions

1. Why is sine wave output the preferred inverter output?

2. How do modified square wave inverters compare with square wave inverters in terms of power quality?

3. How is power quality relevant to inverter output?

4. Compare the basic system configurations of stand-alone, interactive, and bimodal inverters.

5. What are the similarities and differences between thyristors and transistors?

6. Why can line-commutated inverters not be used in stand-alone systems?

7. How does higher-frequency control result in a better approximation of a true sine wave?

8. How are inverters different from power conditioning units?

9. Why is it important for inverters to manage their temperature?

10. How do stand-by losses affect inverter efficiency for various power levels?

11. What types of system operating information are typically provided by inverter interfaces?

CRITICAL DESIGN ANALYSIS

Month	Average Daily DC Energy Consumption (Wh/day)	Array Orientation 1		Array Orientation	
		Insolation (PSH/day)	Design Ratio	Insolation (PSH/day)	Desi
January					
February					
March					
April					
May					
June					
July					
August					
September					
October					
November					
December					

9

System Sizing

For electrical systems that use PV arrays as their only source of electricity, system sizing is critical. The size of the array, battery bank, and other major components necessary to adequately meet the load requirements must be carefully calculated. Sizing procedures are an important part of planning any PV system, but are especially stringent for stand-alone systems. Worksheets can be used to organize information and guide system-sizing calculations for most simple systems, though complex or hybrid systems may require computer models or simulation software.

Objectives

- Differentiate between the approaches and methodologies for sizing different types of PV systems.
- Understand the primary factors that affect system sizing.
- Determine the system energy and power requirements from a load analysis.
- Calculate the critical design parameters based on monthly load and insolation information.
- Calculate the size and configuration of the battery bank based on system requirements.
- Calculate the size and configuration of the array based on system requirements.

SIZING METHODOLOGIES

When describing PV systems, it is logical to follow the energy flow from the array side to the loads. However, when sizing a PV system, it is necessary to consider the energy demand before considering the supply. Therefore, PV-system sizing, particularly for stand-alone systems, starts at the load side and proceeds backward to the array. **See Figure 9-1.** The objective is to first determine the requirements of the system loads and then to determine the size of the inverter, battery bank, and array that are needed to meet the requirements.

Since there are many possible PV-system configurations, each with different modes of operation and priorities, the approach and methodology used to size these systems may differ. There may also be a difference in the sizing tolerance; some systems may need to be sized more carefully than others.

Sizing Interactive Systems

Interactive systems require relatively simple calculations and allow the widest variance in component sizing. Since interactive systems operate in parallel with utility service, sizing is not critical because failure of the PV system to produce energy does not affect operation of electrical loads. Additional energy can be imported from the utility at any time.

Sizing interactive systems begins with the specifications of a PV module chosen for the system. Module ratings at Standard Test Conditions (STC) are used to calculate the total expected array DC power output per peak sun hour. This is then derated for various losses and inefficiencies in the system, which includes the following:

• Guaranteed module output that is less than 100%
• Array operating temperature
• Array wiring and mismatch losses
• Inverter power conversion efficiency
• Inverter MPPT efficiency

The result is a final AC power output that is substantially lower but realistically accounts for expected real-world conditions. **See Figure 9-2.** To determine the expected energy production per day, the final AC power output is multiplied by the insolation for the month or year. For example, if the calculated AC power output is 2140 W per peak sun hour and the average annual insolation is 5.1 peak sun hours (kWh/m²/day), then the average energy production is expected to be 10.9 kWh/day.

If the final system power output is not within the desired range, such as above a minimum size requirement for an incentive program, different module and/or inverter choices can be made. Also, various system configurations can be compared with their associated system costs for a value-based analysis.

Refer to worksheet on CD-ROM

Sizing Strategy for Stand-Alone Systems

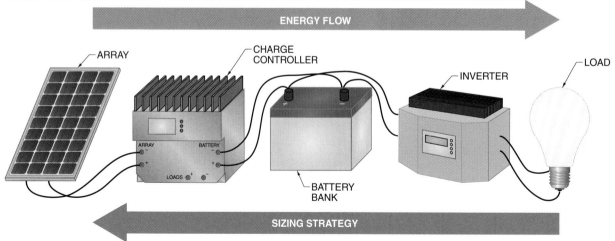

Figure 9-1. *Sizing strategy for stand-alone systems starts at the load side and proceeds backward to the array.*

Interactive System Sizing

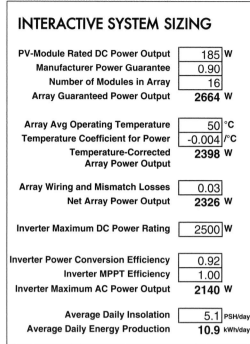

INTERACTIVE SYSTEM SIZING

PV-Module Rated DC Power Output	185	W
Manufacturer Power Guarantee	0.90	
Number of Modules in Array	16	
Array Guaranteed Power Output	**2664**	W
Array Avg Operating Temperature	50	°C
Temperature Coefficient for Power	-0.004	/°C
Temperature-Corrected Array Power Output	**2398**	W
Array Wiring and Mismatch Losses	0.03	
Net Array Power Output	**2326**	W
Inverter Maximum DC Power Rating	2500	W
Inverter Power Conversion Efficiency	0.92	
Inverter MPPT Efficiency	1.00	
Inverter Maximum AC Power Output	**2140**	W
Average Daily Insolation	5.1	PSH/day
Average Daily Energy Production	**10.9**	kWh/day

Figure 9-2. *Sizing interactive systems begins with calculating the peak array DC power output, which is then derated for various losses and inefficiencies in the system to arrive at a final AC power output.*

The size of an interactive system is primarily limited by the space available for an array and the owner's budget. However, financial incentive requirements, net metering limits, and existing electrical infrastructure may also influence system size decisions. Even if short-term periods of high insolation or low demand result in excess electricity, it is not wasted because it can be sold back to the utility for credit against subsequent utility bills. **See Figure 9-3.**

The only exception is a system that is so large that it maintains a net energy export over several months or more. Because many utilities will not carry the credits for more than one year and/or will credit exported electricity at lower wholesale rates, it is not recommended to size an interactive system larger than needed for average annual on-site load requirements. However, since the available space usually cannot accommodate an array this large, this is rarely an issue.

Sizing Interactive Systems

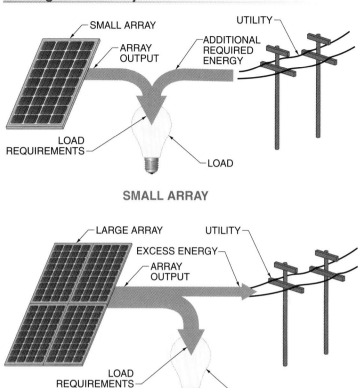

SMALL ARRAY

LARGE ARRAY

Figure 9-3. *Interactive-system sizing is very flexible because the utility can supply extra energy to the system loads and receive excess energy from the PV system.*

Sizing Stand-Alone Systems

Stand-alone PV systems are designed to power specific on-site loads, so the size of these systems is directly proportional to the load requirements. If the system is too small, there will be losses in load availability and system reliability. **See Figure 9-4.** If the system is too large, excess energy will be unutilized and wasted. Therefore, sizing of stand-alone systems requires a fine balance between energy supply and demand.

Because of this necessary balance, sizing stand-alone systems requires more analysis and calculations than are required for interactive systems. Most of these calculations build upon one another as the analyses proceed. Moreover, sizing stand-alone systems is an iterative process. That is, if the final calculations indicate that the components are improperly sized, the starting values must be changed and the calculation process repeated until the system output matches the load requirements.

Sizing Stand-Alone Systems

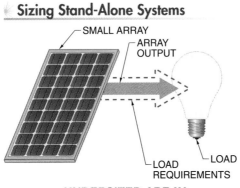

SMALL ARRAY

ARRAY OUTPUT

LOAD REQUIREMENTS

LOAD

UNDERSIZED ARRAY

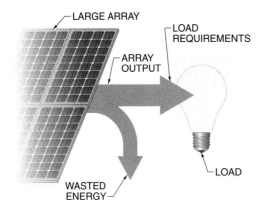

LARGE ARRAY

LOAD REQUIREMENTS

ARRAY OUTPUT

LOAD

WASTED ENERGY

OVERSIZED ARRAY

Figure 9-4. *Stand-alone systems must be carefully matched to load requirements to avoid reducing load availability or producing more energy than is needed.*

Sizing Bimodal Systems

Bimodal systems normally operate as interactive systems, but can operate as stand-alone systems during utility outages. Therefore, the bimodal systems are typically sized according to the stand-alone methodology. However, a significant difference between bimodal systems and true stand-alone systems is that bimodal systems typically supply only a few select critical loads while in stand-alone mode.

The load analysis used for sizing a bimodal system should include calculations for only these critical loads, which are needed during a utility outage. The rest of the subsequent calculations for inverter rating, battery-bank energy storage, and array output are identical to those for stand-alone systems. Similarly, the battery-bank sizing allows for the desired back-up period without grid power from a few hours to several days.

The stand-alone sizing methodology determines the minimum size of a bimodal system. However, since excess energy produced during the normal interactive mode can be exported to the utility grid for credit, there is no penalty for oversizing a bimodal system, at least within the guidelines of sizing conventional utility-interactive systems.

Sizing Hybrid Systems

Hybrid systems are stand-alone, which cannot rely on the utility as a source of electricity. These systems must be able to completely and reliably supply power to their on-site loads. However, there is more than one source of energy, such as a combination of a PV array and an engine generator or wind turbine. The presence of multiple power sources means that the array and battery bank can be smaller while maintaining load availability, especially if one source can provide power on demand, such as an engine-generator.

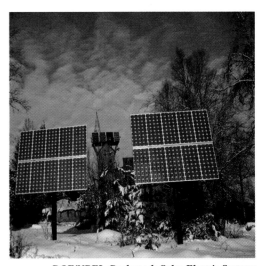

DOE/NREL, Backwoods Solar Electric Systems
Stand-alone PV systems must be carefully designed to meet all the load requirements without excessive oversizing.

The array and battery bank for a PV array and engine generator hybrid system are sized similarly to those for a stand-alone system, with three differences. First, the array is sized to supply only a portion of the total load requirement. For example, the hybrid system may be designed such that 80% of the load demand is supplied by the array and 20% is supplied by the generator on an average day. Second, the sizing calculations do not need to use the worst-case load-to-insolation months for sizing, since the engine generator can be called upon to provide additional power as needed. Average load and insolation values may be used. Finally, battery banks can be sized for a shorter autonomy period (typically only 1 or 2 days) than for PV-only stand-alone systems, also because the generator power is available on demand.

PV array to engine generator output ratios range from 90%:10% to 40%:60%. The optimal ratio is determined by performing sizing calculations using several different ratios and choosing the system that best fits other requirements (such as available array space) and has the lowest expected life-cycle costs. Other factors, such as minimizing the average run time of noisy engines, may also influence sizing.

Combination PV array and engine generator hybrid systems are relatively easy to size because the generator output is completely dependent on demand. Other energy-source combinations, however, such as a PV array and a wind turbine or a PV array and a micro-hydroelectric generator, are much more difficult to design and size adequately because their outputs are less predictable. With so many variables, sizing software is recommended to optimize these systems. Several applications are available free on-line, such as HOMER from NREL.

SIZING CALCULATIONS

Sizing PV systems for stand-alone operation involves four sets of calculations. First, a load analysis determines the electrical load requirements. Then, monthly load requirements are compared to the local insolation data to determine the critical design month. Next, the battery bank is sized to be able to independently supply the loads for a certain length of time, such as if cloudy weather reduces array output. Finally, the PV array is sized to fully charge the battery bank under the critical conditions.

> A hybrid PV and engine generator system utilizes array energy better (wastes less energy) than PV-only systems because more of the available energy is utilized. In some cases, these systems may also cost less overall than PV-only or engine-generator-only stand-alone systems sized for the same load requirements.

Load Analysis

Analyzing the electrical loads is the first and most important step in PV-system sizing. The energy consumption dictates the amount of electricity that must be produced.

All existing and potential future loads must be considered. Underestimating loads will result in a system that is too small and cannot operate the loads with the desired reliability. However, overestimating the load will result in a system that is larger and more expensive than necessary. Comprehensive yet conservative load estimates will ensure that the system is adequately sized.

A detailed load analysis completed during the site survey lists each load, its power demand, and daily energy consumption. **See Figure 9-5.** If load profiles are not nearly identical throughout the year, a load analysis should be conducted for each month. Similar loads can be grouped into categories, such as lighting fixtures with the same power requirements. DC loads, if any, should be listed separately from AC loads. This is because energy for AC loads goes through the inverter, resulting in losses that must be accounted for separately.

Power Demand. Peak-power information is usually found on appliance nameplates or in manufacturer's literature. When this information is not available, peak power demand can be estimated by multiplying the maximum current by the operating voltage, though this is less accurate for reactive loads. Measurements, meter readings, or electric bills may also be used to help establish existing load requirements. **See Figure 9-6.**

Refer to worksheet on CD-ROM

Load Analysis

LOAD ANALYSIS   Month:

AC LOADS

Load Description	Qty	Power Rating (W)	Operating Time (hr/day)	Energy Consumption (Wh/day)

DC LOADS

Total AC Power	W
Total DC Power	W
Total Daily AC Energy Consumption	Wh/day
Total Daily DC Energy Consumption	Wh/day
Weighted Operating Time	hr/day
Inverter Efficiency	
Average Daily DC Energy Consumption	Wh/day

Figure 9-5. *A load analysis tabulates the various kinds of loads and their power and electrical-energy requirements.*

The peak power demands are then summed. The total power demand is considered when determining the required inverter AC-power output rating. While it is not likely that every load would be ON at the same time, it is recommended to size the inverter with extra capacity.

Energy Consumption. Electrical energy consumption is based on the power demand over time. Loads rarely operate continuously, so each load's operating time must be determined. **See Figure 9-7.** This is the total number of hours per day that the load is operating.

Load-Requirement Meters

Figure 9-6. *Load power and energy requirements can be easily measured with inexpensive meters.*

The operating time for loads that cycle on and off automatically is typically determined from the duty cycle. *Duty cycle* is the percentage of time a load is operating. For example, a duty cycle of 40% means that a load is operating 40% of the time, or 9.6 hr/day (40% × 24 hr/day = 9.6 hr/day). Even loads that are plugged in all the time, such as refrigerators and air conditioners, have a variable power requirement based on duty cycle.

User-operated loads are turned on and off manually. Determining the operating time for these loads is simple if they cycle only once per day. However, if loads are switched on and off several times per day, a metering device is probably the easiest method of determining the operating time.

The daily energy consumption for each load is determined by the load's power demand multiplied by the daily operating time. For example, a 60 W light bulb that is on for 4 hr/day consumes 240 Wh of energy (60 W × 4 hr = 240 Wh).

Some loads may not be used every day. In these cases, the average daily operating time is calculated by dividing the total operating time over a longer period by the number of days in the period. For example, a washing machine that operates for 2 hr/wk has an equivalent operating time of 0.29 hr/day (2 hr/wk ÷ 7 days/wk = 0.29 hr/day).

The AC energy consumption and DC energy consumption values are totaled separately. These values are used to determine the total amount of DC energy the system must produce.

Operating Time. Load operating-time data is also used to size the battery bank. For consistent loads that operate for specific periods, calculating the daily operating time is very simple. For example, if the loads are nighttime lighting fixtures that operate for 6 hr each night, the daily operating time is 6 hr.

Load Requirements

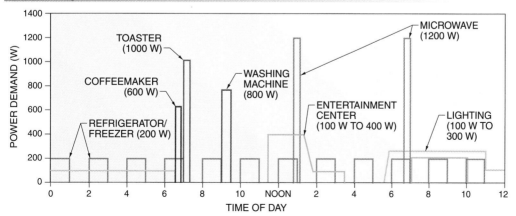

Figure 9-7. *Load requirements include the power demand and electrical-energy consumption for all the expected loads in the system.*

Because the size, and therefore the cost, of a stand-alone PV system is directly proportional to the load energy requirements, energy should be conserved as much as possible in stand-alone systems. This involves reducing wasted energy, reducing load usage, and increasing load efficiency. Even if these measures involve financial costs, it is usually far more practical and cost effective to implement energy conservation measures than to increase the size of the PV system.

In 1980, the Federal Trade Commission began requiring manufacturers of certain home appliances to attach labels that provide an estimate of the product's energy consumption or efficiency. These EnergyGuide labels show the estimated annual operating cost (based on average energy prices) and how this compares with the operating costs for similar models. The EnergyGuide label is required on refrigerators, freezers, clothes washers, water heaters, dishwashers, air conditioners, heating equipment, and other household appliances.

ENERGY STAR is a voluntary federal product labeling and certification program administered by the U.S. Environmental Protection Agency. It recognizes appliances and equipment that have exceeded energy efficiency standards. Products bearing the ENERGY STAR label include many common home appliances and office equipment. Both EnergyGuide and ENERGY STAR labels can be used to identify and select the most efficient electrical loads for PV systems.

Most often, however, there are multiple loads to consider that each operate for various lengths of time. The battery-bank discharge rate will then change as various loads turn ON and OFF during the day. In this case, a weighted average operating time is calculated using the following formula:

$$t_{op} = \frac{(E_1 \times t_1) + (E_2 \times t_2) + \ldots + (E_n \times t_n)}{E_1 + E_2 + \ldots + E_n}$$

where

t_{op} = weighted average operating time (in hr/day)

E_1 = DC energy required for load 1 (in Wh/day)

t_1 = operating time for load 1 (in hr/day)

E_2 = DC energy required for load 2 (in Wh/day)

t_2 = operating time for load 2 (in hr/day)

E_n = DC energy required for nth load (in Wh/day)

t_n = operating time for nth load (in hr/day)

For example, one DC load uses 2400 Wh/day and operates for 4 hr and another DC load uses 1000 Wh/day and operates for 7 hr. What is the weighted average operating time?

$$t_{op} = \frac{(E_1 \times t_1) + (E_2 \times t_2)}{E_1 + E_2}$$

$$t_{op} = \frac{(2400 \times 4) + (1000 \times 7)}{2400 + 1000}$$

$$t_{op} = \frac{9600 + 7000}{3400}$$

$$t_{op} = \textbf{4.9 hr/day}$$

The two loads have a combined effect of a single 694 W load operating for 4.9 hr/day [(2400 Wh + 1000 Wh) ÷ 4.9 hr = 694 W].

If the system includes both AC and DC loads, the AC load energy requirement must be first be converted to equivalent DC energy. This is done by dividing each AC energy consumption amount by the inverter efficiency.

Inverter Selection. If the system includes AC loads, an inverter must be selected. Several factors must be considered when selecting the inverter. First, the inverter must have a maximum continuous power output rating at least as great as the largest single AC load. A slightly oversized inverter is usually recommended to account for potential future load additions. The inverter must also be able to supply surge currents to motor loads, such as pumps or compressors, while powering other system loads.

Inverter voltage output is another consideration. Most stand-alone inverters produce either 120 V single-phase output or 120/240 V split-phase output. Some higher-power inverters for commercial or industrial electrical systems output three-phase power. Alternatively, certain inverters can be stacked and operated in parallel for split-phase output.

The inverter DC-input voltage must also correspond with either the array voltage (for interactive systems) or the battery-bank voltage (for stand-alone systems).

Inverter Efficiency. Inverters are not 100% efficient. Some power is lost in the process of converting DC energy to AC energy. Therefore, more DC energy is required to produce a certain amount of AC energy. Both the AC and DC energy requirements from the load analysis are used to determine how much total DC energy will be required. **See Figure 9-8.** The total amount of DC energy required by the loads is calculated using the following formula:

$$E_{SDC} = \frac{E_{AC}}{\eta_{inv}} + E_{DC}$$

where

E_{SDC} = required daily system DC electrical energy (in Wh/day)

E_{AC} = AC energy consumed by loads (in Wh/day)

η_{inv} = inverter efficiency

E_{DC} = DC energy consumed by loads (in Wh/day)

For example, if a load analysis determines that a system requires 800 Wh/day for the AC loads and 200 Wh/day for the DC loads and the inverter efficiency is 90%, what is the daily DC electrical energy required by the system?

$$E_{SDC} = \frac{E_{AC}}{\eta_{inv}} + E_{DC}$$

$$E_{SDC} = \frac{800}{0.90} + 200$$

$$E_{SDC} = 889 + 200$$

$$E_{SDC} = \textbf{1089 Wh/day}$$

Total DC Energy Requirement

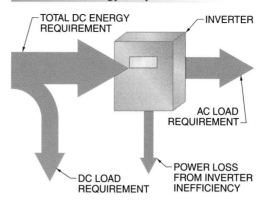

Figure 9-8. *The total DC-energy requirement is determined from the requirements for the DC loads (if any) plus the requirements for the AC loads, taking inverter efficiency into account.*

Inverter efficiency is typically between 80% and 95%. Also, an inverter's efficiency varies with its power output, though usually not more than about 5% over most of its power range. Manufacturer's specifications will typically include efficiency ranges. For sizing calculations, the average efficiency for the expected operating power range should be used.

Critical Design Analysis

A stand-alone system must produce enough electricity to meet load requirements during any month. Therefore, systems are sized for the worst-case scenario of high load and low insolation. A critical design analysis compares these two factors throughout a year, and the data for the worst case is used to size the array. **See Figure 9-9.** The *critical design ratio* is the ratio of electrical energy demand to average insolation during a period. The load data comes from the load analysis, which is usually performed for each month. The insolation data is available from the solar radiation data sets. **See Appendix.** The ratio is calculated for each month.

Refer to worksheet on CD-ROM

Critical Design Month. The *critical design month* is the month with the highest critical design ratio. This is the worst-case scenario, and the associated load and insolation data are used to size the rest of the system.

Critical Design Analysis

CRITICAL DESIGN ANALYSIS

Month	Average Daily DC Energy Consumption (Wh/day)	Array Orientation 1		Array Orientation 2		Array Orientation 3	
		Insolation (PSH/day)	Design Ratio	Insolation (PSH/day)	Design Ratio	Insolation (PSH/day)	Design Ratio
January							
February							
March							
April							
May							
June							
July							
August							
September							
October							
November							
December							

Critical Design Month

Optimal Orientation

Average Daily DC Energy Consumption   Wh/day

Insolation   PSH/day

Figure 9-9. *A critical design analysis compares the load requirements and insolation for each month to determine the critical design month.*

Load Analysis Example: Off-Grid Home in Albuquerque, NM

A secluded home is being constructed in the mountains near Albuquerque, NM. Due to its remoteness, a stand-alone PV system will power the building. Electrical load requirements will be minimized through the use of solar thermal systems, energy-conserving construction materials, and high-efficiency appliances. To size the PV system, a load analysis is conducted for the expected loads. Most of the electrical load requirements are expected to be consistent each month. These include appliances, an entertainment center, a computer, miscellaneous plug loads, and a water pump.

The appliances consist of a refrigerator/freezer, microwave oven, toaster, coffeemaker, and washing machine. The refrigerator consumes 730 kWh/yr, or an average of 2000 Wh/day (730 kWh/yr ÷ 365 days/yr = 2 kWh/day = 2000 Wh/day). The peak power consumption of the refrigerator is measured at 200 W when the compressor operates.

The microwave oven uses 1200 W and operates for about 30 min (0.5 hr) each day. The toaster uses 1000 W and operates for about 3 min (0.05 hr) per day. The coffeemaker uses 600 W and operates for about 15 min (0.25 hr) per day. The washing machine uses an average of 800 W for a complete 30 min (0.5 hr) cycle. If four loads are washed per week, the equivalent is 0.29 hr/day (0.5 hr/cycle × 4 cycles/wk ÷ 7 days/wk = 0.29 hr/day).

The entertainment center consists of a satellite receiver, TV, and DVD player. Together they consume an average of 200 W and are used about 3 hr per day. The computer system operates on 100 W and is used for about 2 hr per day. The miscellaneous plug loads are estimated at 200 W for an average of 1 hr per day. The submersible water pump consumes 800 W and operates about 20 min (0.33 hr) per day.

A few loads, however, will vary depending on the time of the year. Two 50 W ceiling fans will be used primarily during the summer. The lighting loads will be used all year, but will have a longer daily operating time during the winter. Because the total load requirements will change during the year, a load analysis is conducted for each month.

LOAD ANALYSIS

Month: August

AC LOADS

Load Description	Qty	Power Rating (W)	Operating Time (hr/day)	Energy Consumption (Wh/day)
Refrigerator/Freezer	1	200	10	2000
Microwave	1	1200	0.5	600
Toaster	1	1000	0.05	50
Coffeemaker	1	600	0.25	150
Washing Machine	1	800	0.29	232
Entertainment Center	1	200	3	600
Computer System	1	100	2	200
Plug Loads	1	200	1	200
Water Pump	1	800	0.33	264
Ceiling Fans	2	50	24	2400
Fluorescent Lighting	4	15	6	360
Fluorescent Lighting	4	32	4	512

DC LOADS

Total AC Power	5388	W
Total DC Power	0	W
Total Daily AC Energy Consumption	7568	Wh/day
Total Daily DC Energy Consumption	0	Wh/day
Weighted Operating Time	11.2	hr/day
Inverter Efficiency	0.90	
Average Daily DC Energy Consumption	8409	Wh/day

For the month of August, the load analysis yields a total AC-power demand of about 5.4 kW and a daily energy consumption of 7568 Wh/day. If all loads operate at the same time, the inverter must have a continuous power output rating of at least 5.4 kW. Although it is unlikely that all loads will be operating simultaneously, a 5.5 kW inverter is selected to allow for future load additions. Since the efficiency of the inverter is 90%, the total average daily DC energy required is 8409 Wh/day. This number will be used in the critical design analysis, along with energy requirements for every other month, to determine the critical design month. The weighted operating time for the critical design month will be used in the battery-sizing calculations.

DOE/NREL, Altair Energy

Ground mounts usually offer flexibility in orienting an array in the optimal direction and tilt for the critical design month.

If the loads are constant over the entire year, the critical design month is the month with the lowest insolation on the array surface. For most locations in the Northern Hemisphere, this is a winter month, either December or January.

However, when the load requirements vary from month to month, the critical design month must take into account both the loads and the available insolation. Because of these two factors, the critical design month may turn out to be any month of the year.

Sizing for the critical design month typically results in excess energy at other times of the year. If this excess is significant, the system designer may want to consider adding diversion loads or changing to a different system configuration, such as a hybrid system, that better matches the available electrical energy to the loads.

Array Orientation. Since array orientation has a significant effect on receivable solar radiation, array orientation must also be accounted for in a critical design analysis. If the mounting surface restricts the array to only one possible orientation, then the analysis is conducted to determine the critical design factors for that orientation.

However, if multiple orientations are possible, separate analyses are performed for each orientation. A critical design month can be identified for each of the array orientations, since the receivable solar radiation will be different for each. Of the resulting critical design months, the one with the smallest design ratio is the best choice. This orientation minimizes the required array size, while still accounting for the worst-case load-to-insolation situation.

The orientations most commonly used in a critical design analysis are tilt angles equal to the latitude, latitude + 15°, and latitude − 15°, each at an azimuth of due south. The greater array tilt angle maximizes the received solar energy in winter months, and the smaller array tilt angle maximizes the received solar energy in summer months. Insolation data for these orientations is available in the solar radiation data set for the nearest location.

For azimuth angles other than due south, the insolation data must be adjusted to obtain the most accurate results. Computer models are available to predict average monthly insolation for alternate orientations. If tracking systems are to be used, receivable insolation data for the various tracking modes can be used instead of fixed array orientations.

The critical design ratio is calculated for each month for each array orientation or tracking mode. The highest critical design ratio for each orientation corresponds to the critical design month for that orientation. When multiple orientations are considered, the lowest critical design ratio of the resulting critical design months corresponds to the optimal array orientation (of the orientations analyzed). The insolation and load requirements for this month and array orientation are used in subsequent sizing calculations to design the array and battery system.

DC-System Voltage

The DC-system voltage is established by the battery-bank voltage in battery-based systems. This voltage dictates the operating voltage and ratings for all other connected components, including DC loads, charge controllers, inverters, and (for battery-based systems) the array.

Critical Design Analysis Example: Off-Grid Home in Albuquerque, NM

A critical design analysis determines the critical design month and insolation for sizing the PV array. If the insolation values for more than one possible orientation are compared, the critical design analysis indicates the best orientation choice of those analyzed. For the example system at the remote home, the array will be mounted in a ground rack mount. This allows the possible tilt angles of latitude, latitude−15°, and latitude+15°. The insolation values for each month and at each of these orientations are found on the solar radiation data set for Albuquerque. The 8409 Wh/day load requirement for August is included, along with the load requirements for every other month as determined by separate load analyses.

CRITICAL DESIGN ANALYSIS

Month	Average Daily DC Energy Consumption (Wh/day)	Array Orientation 1 Latitude − 15		Array Orientation 2 Latitude		Array Orientation 3 Latitude + 15	
		Insolation (PSH/day)	Design Ratio	Insolation (PSH/day)	Design Ratio	Insolation (PSH/day)	Design Ratio
January	6532	4.6	1420	5.3	1232	5.8	1126
February	6436	5.4	1192	6.0	1073	6.2	1038
March	6254	6.3	993	6.5	962	6.5	962
April	6197	7.3	849	7.2	861	6.6	939
May	6160	7.7	800	7.2	856	6.3	978
June	7568	7.8	970	7.1	1066	6.1	1241
July	8300	7.4	1122	6.9	1203	6.0	**1383**
August	8409	7.2	1168	6.9	1219	6.3	1335
September	7834	6.6	1187	6.8	1152	6.5	1205
October	6160	5.9	1044	6.5	948	6.6	933
November	6327	4.8	1318	5.5	1150	5.9	1072
December	**6578**	4.3	**1530**	**5.0**	**1316**	5.5	1196

Critical Design Month December
Optimal Orientation Latitude
Average Daily DC Energy Consumption 6578 **Wh/day**
Insolation 5.0 **PSH/day**

The critical design ratio is calculated for each month. For each orientation, the highest ratio of load requirement to insolation corresponds to the critical design month. For two of the orientations, the month is December. For the latitude+15° orientation, the month is July. Of the three possible critical design months, the month of December at the latitude orientation produces the lowest ratio. This indicates the optimal orientation. For this designated critical design month, the load requirement value is used for battery-bank sizing and the insolation value is used for array sizing.

DC voltage in battery-based systems is critically important. The DC voltage for battery-based PV systems is usually an integer multiple of 12 V, usually 12 V, 24 V, or 48 V. DC loads, charge controllers, and inverters that operate at these voltages are commonly available.

The selection of the battery-bank voltage affects system currents. **See Figure 9-10.** For example, a 1200 W system operating at 12 V draws 100 A (1200 W ÷ 12 V = 100 A). The same 1200 W system draws only 50 A at 24 V, or 25 A at 48 V. Lower current reduces the required sizes of conductors, overcurrent protection devices, disconnects, charge controllers, and other equipment. Also, since voltage drop and power losses are smaller at lower currents, higher-voltage systems are generally more efficient. Higher-voltage systems also require fewer PV source circuits in the array design.

DC-System Voltage

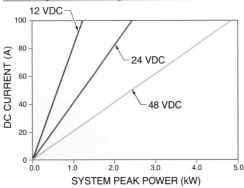

Figure 9-10. *DC-system voltage is chosen in proportion with the array size and to keep the operating current below 100 A.*

As a rule of thumb, stand-alone systems up to 1 kW use a minimum 12 V battery-bank voltage, which limits DC currents to less than 84 A. Similarly, battery voltages of at least 24 V are used for systems up to 2 kW, and at least 48 V for systems up to 5 kW. Very large stand-alone systems may use battery voltages of 120 V, though battery banks over 48 V involve additional code requirements and safety measures.

Stand-alone systems may require large battery banks for the desired system availability.

System Availability

The size of a system in relation to the loads determines its system availability. *System availability* is the percentage of time over an average year that a stand-alone PV system meets the system load requirements. For example, 98% system availability means that a system is able to meet the energy demand about 98% of the time. This means that for 2% of the year, the system cannot meet the load requirements.

No energy-producing system can achieve 100% availability, because of unpredictable events that affect system output. Days or weeks of below-average insolation, such as unusually cloudy weather, will reduce short-term system availability. System availability can also vary between years due to long-term weather patterns. Component failures and lack of maintenance also contribute to system downtime and reduce system availability.

System availability is determined by insolation and autonomy. Accurate estimates of system availability require software to evaluate energy flow in the system on an hour-by-hour basis, but rough estimates are adequate for most PV applications. For a desired system availability, the designer chooses the appropriate length of autonomy. **See Figure 9-11.** *Autonomy* is the amount of time a fully charged battery system can supply power to system loads without further charging. Autonomy is expressed in days. Most stand-alone systems are sized for a system availability of about 95% (about 3 to 5 days of autonomy) for noncritical applications or 99% or greater (about 6 to 10 days or more) for critical applications.

System Availability

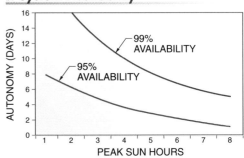

Figure 9-11. *System availability is approximated from the local insolation and the autonomy period.*

However, each percentage-point increase in system availability is increasingly more expensive for larger battery banks and arrays, which is impractical from an economic standpoint for all but the most critical applications. Sizing of stand-alone systems must achieve an acceptable balance between system availability and cost goals for a given application. **See Figure 9-12.** The solar resource for a location also affects the increasing costs of availability. Costs for increasing availability rise more steeply for locations with large seasonal differences in insolation than do costs for locations with more constant insolation.

Availability Costs

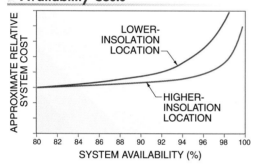

Figure 9-12. *Increasing system availability significantly increases the cost of the system.*

Battery-Bank Sizing

Batteries store excess energy the array produces during periods of high insolation, and supply power to the system loads at nighttime and during periods of low insolation. In stand-alone systems, they also establish the system DC operating voltage and supply surge currents to electrical loads and inverters.

Battery-Bank Required Output. Batteries for stand-alone PV systems are sized to store enough energy to meet system loads for the desired length of autonomy without any further charge or energy contributions from the PV array. **See Figure 9-13.** The amount of battery capacity required for a given application depends on the load requirements and desired autonomy. Greater autonomy requires larger and costlier battery banks, but reduces the average daily depth of discharge, which prolongs battery life.

Battery-Bank Sizing

BATTERY-BANK SIZING

Average Daily DC Energy Consumption for Critical Design Month		Wh/day
DC System Voltage		VDC
Autonomy		days
Required Battery-Bank Output		Ah
Allowable Depth-of-Discharge		
Weighted Operating Time		hrs
Discharge Rate		hrs
Minimum Expected Operating Temperature		°C
Temperature/Discharge Rate Derating Factor		
Battery-Bank Rated Capacity		Ah
Selected Battery Nominal Voltage		VDC
Selected Battery Rated Capacity		Ah
Number of Batteries in Series		
Number of Battery Strings in Parallel		
Total Number of Batteries		
Actual Battery-Bank Rated Capacity		Ah
Load Fraction		
Average Daily Depth-of-Discharge		

Figure 9-13. *The battery-bank sizing worksheet uses information from the load analysis to determine the required size of the battery bank.*

The required battery-bank capacity is determined from the electrical-energy requirements to operate the loads during the critical design month for the length of the autonomy period and at the desired battery-system voltage. The required battery-bank capacity is calculated using the following formula:

Refer to worksheet on CD-ROM

$$B_{out} = \frac{E_{crit} \times t_a}{V_{SDC}}$$

where

B_{out} = required battery-bank output (in Ah)

E_{crit} = daily electrical-energy consumption during critical design month (in Wh/day)

t_a = autonomy (in days)

V_{SDC} = nominal DC-system voltage (in V)

For example, consider a system that requires 450 Wh of energy daily during the critical design month and the nominal DC-system voltage is 24 V. If the system specifies 4 days of autonomy, what is the required capacity to operate those loads from the battery bank?

$$B_{out} = \frac{E_{crit} \times t_a}{V_{SDC}}$$

$$B_{out} = \frac{450 \times 4}{24}$$

$$B_{out} = \frac{1800}{24}$$

$$B_{out} = 75 \, \text{Ah}$$

Therefore, the battery bank will need to supply 75 Ah to the system loads. However, the total of the nameplate ratings of the battery bank must be higher than this, because the usable capacity of a battery is always less than its rated capacity.

> When high availability is required for critical loads, a PV-only system may be prohibitively expensive due to very large array and battery bank requirements. In these cases, a hybrid system may be the best choice.

Battery-Bank Rated Capacity. Three factors affect the amount of usable capacity in a battery. These factors are used to estimate the larger battery-bank rated capacity necessary to supply the required output. **See Figure 9-14.** First, most batteries cannot be discharged to a depth of discharge of 100% without permanent damage. Depending on the battery type, common allowable depths of discharge range from 20% to 80%. Most PV systems use deep-cycle lead-acid batteries, which can be discharged to about 80%. This is the maximum fraction of the total rated capacity that is permitted to be withdrawn from the battery at any time.

Also, low operating temperatures and high discharge rates further reduce battery capacity. Most battery ratings are specified for operation at 25°C (77°F) at a certain discharge rate. At other conditions for these two factors, the usable battery capacity may be lower. For example, a battery operating at –10°C (14°F) and at a discharge rate of C/120 has about only 90% of the capacity it has at 25°C (77°F). These two factors are used together to determine a second capacity-derating factor.

Battery-Bank Capacity

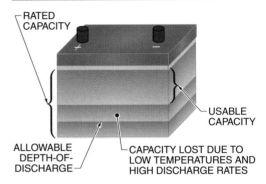

Figure 9-14. *Due to the allowable depth-of-discharge, low temperatures, and high discharge rates, the amount of useful output in a battery bank is less than the rated capacity.*

The operating temperature is the minimum expected operating temperature for the battery bank. This depends on where the batteries will be stored. If they will be stored indoors, the temperature will be relatively steady and equal to the normal indoor temperature. If they will be outside, measures should be taken to minimize large daily and seasonal temperature swings, but the lowest expected temperature in these conditions is used in the analysis.

The average discharge rate is determined from the total operating time over the period of autonomy, taking the allowable depth of discharge into account. Using the daily operating time calculated in the load analysis, the average discharge rate is calculated using the following formula:

$$r_d = \frac{t_{op} \times t_a}{DOD_a}$$

where

r_d = average discharge rate (in hr)

t_{op} = weighted average operating time (in hr/day)

t_a = autonomy (in days)

DOD_a = allowable depth of discharge

For example, if the daily operating time for system loads is 16 hr/day over an autonomy of 3 days, and the allowable depth of discharge is 80%, what is the average discharge rate?

$$r_d = \frac{t_{op} \times t_a}{DOD_a}$$

$$r_d = \frac{16 \times 3}{0.80}$$

$$r_d = \frac{48}{0.80}$$

$$r_d = \textbf{60 hr}$$

The battery bank will discharge at a rate that would completely discharge the batteries in 60 hr. Therefore, the battery-bank average discharge rate is C/60.

With the minimum expected operating temperature and the average discharge rate, the percentage of usable capacity is determined from a graph of discharge rates and operating temperatures. **See Figure 9-15.** Most battery manufacturers report capacity at various discharge rates and temperatures in their specifications.

Battery Capacity Loss vs. Temperature and Discharge Rate

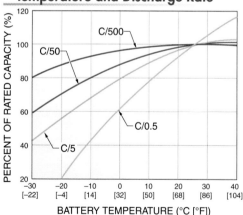

Figure 9-15. *The amount of available capacity from a battery bank depends partly on the operating temperature and discharge rate. These factors may have different effects for different batteries.*

To calculate the total rated capacity of the battery bank, the required battery-bank output is increased proportionally to both the allowable depth of discharge and the temperature and discharge-rate derating factor.

The required capacity is calculated using the following formula:

$$B_{rated} = \frac{B_{out}}{DOD_a \times C_{T,rd}}$$

where

B_{rated} = battery-bank rated capacity (in Ah)

B_{out} = battery-bank required output (in Ah)

DOD_a = allowable depth of discharge

C_{T,r_d} = temperature and discharge-rate derating factor

For example, consider a system that requires a total battery bank output of 500 Ah. The allowable depth of discharge is 75%, the minimum operating temperature is –10°C (–4°F), and the average discharge rate is C/50. From the manufacturer's documentation on battery capacity, this yields a temperature and discharge-rate derating factor of approximately 80%. What is the required battery-bank rated capacity?

$$B_{rated} = \frac{B_{out}}{DOD_a \times C_{T,rd}}$$

$$B_{rated} = \frac{500}{0.75 \times 0.80}$$

$$B_{rated} = \frac{500}{0.6}$$

$$B_{rated} = \textbf{833 Ah}$$

Battery Selection. Individual batteries or cells are selected with enough capacity to avoid or minimize parallel battery connections. Due to wiring resistance and small differences among individual cells, paralleled strings of batteries may not charge and discharge uniformly. A single series-connected string of batteries is preferable, but capacity requirements and the size of batteries available may require more than one string. Generally, the number of parallel battery connections should be limited to no more than 3 to 4 strings. Also, the size and weight of the batteries must be considered with regard to transportation and installation.

The nominal voltage and rated capacity of the selected battery is used to determine the configuration of the battery bank. This information is found on battery nameplates or in manufacturer's literature. **See Figure 9-16.**

Battery Labels

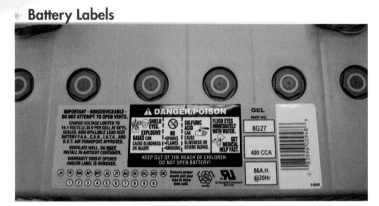

Figure 9-16. *Battery labels list the rated capacity of the battery and important safety information.*

The nominal DC-system voltage divided by the nominal battery voltage determines the number of batteries in a string. This number should calculate evenly. **See Figure 9-17.** The required battery-bank rated capacity divided by the individual-battery rated capacity determines the number of strings to be connected in parallel. This number will likely not be a whole number, but should be rounded up to the nearest whole number. To prevent unnecessarily oversizing the capacity, the battery capacity should be chosen to minimize the amount of rounding.

Battery-Bank Configurations

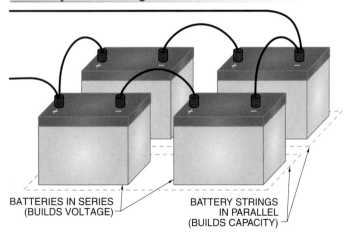

BATTERIES IN SERIES
(BUILDS VOLTAGE)

BATTERY STRINGS
IN PARALLEL
(BUILDS CAPACITY)

Figure 9-17. *Batteries are configured in series and parallel to match the battery-bank rated capacity needed to produce the required output.*

When the battery is chosen and the battery-bank design is configured, the final rated capacity of the battery bank is equal to the rated capacity of an individual battery multiplied by the number of parallel strings.

For example, a battery bank must supply 600 Ah and will operate at 24 V nominal. A nominal 12 V battery is chosen with a rated capacity of 250 Ah. To produce a nominal voltage of 24 V, two 12 V batteries will be connected in series for each string. The number of strings in parallel is calculated to be 2.4 (600 Ah ÷ 250 Ah = 2.4). Rounded up to a whole number of 3 strings, the rated capacity of the battery bank will then be 750 Ah (250 Ah/string × 3 strings = 750 Ah).

This is acceptable, but very conservative, and would result in an unnecessary increase in cost. Choosing a different battery with a rated capacity closer to 200 Ah would be better if the bank were to have 3 strings. A battery with a rated capacity of 300 Ah or slightly higher would be even better, as it would allow a battery bank with only 2 strings. Battery choices may require changes and recalculations to optimize the design of the battery bank.

Battery-Bank Operation. With the final battery-bank configuration and battery choice determined, the predicted average daily depth of discharge is calculated. This will be less than the allowable depth of discharge because the final rated capacity of the battery bank is usually higher than the required rated capacity.

First, the daily load fraction supplied by the battery bank is estimated. At any moment, the power to operate loads in a stand-alone system may be supplied by the array, the battery, or a combination of both. The *load fraction* is the portion of load operating power that comes from the battery bank over the course of a day. For example, a system with only nighttime loads, such as lighting, has a load fraction of 1.0 because all the electrical energy required by the loads is supplied by the battery bank. For daytime-only loads, the average load fraction is zero because the array will normally supply all the required electrical energy to the loads.

For most systems, however, the load fraction is somewhere in between. During periods of high irradiance, the array supplies all of the energy needed by on-site loads (in addition to charging the batteries), but most of the time, energy is supplied in a mix from both the array and the battery bank. With variable loads operating intermittently and for different lengths of time, an accurate calculation of load fraction is complicated. Instead, a load fraction estimate of 0.75 is a common rule of thumb used for most PV systems.

The load fraction estimate does not affect any of the sizing calculations, so the rough estimate provided by the rules of thumb are adequate for the subsequent daily depth-of-discharge estimate calculations.

With the load fraction estimate, the average battery-bank daily depth of discharge is then estimated with the following formula:

$$DOD_{avg} = \frac{LF \times E_{day}}{B_{rated} \times V_{SDC}}$$

where

DOD_{avg} = average battery-bank daily depth of discharge

LF = estimated load fraction

E_{day} = average daily electrical-energy consumption (in Wh)

B_{rated} = total rated battery-bank capacity (in Ah)

V_{SDC} = DC-system voltage (in V)

For example, a 24 V battery bank has a rated capacity of 800 Ah. The estimated load fraction is 0.75, and the average daily electrical-energy consumption is 3900 Wh. What is the predicted average battery-bank daily depth of discharge?

$$DOD_{avg} = \frac{LF \times E_{day}}{B_{rated} \times V_{SDC}}$$

$$DOD_{avg} = \frac{0.75 \times 3900}{800 \times 24}$$

$$DOD_{avg} = \frac{2925}{19,200}$$

$$DOD_{avg} = \textbf{0.15 or 15\%}$$

The array for a stand-alone system must be sized to meet the load requirements during the critical design month.

Array Sizing

Refer to worksheet on CD-ROM

For stand-alone systems, the array must be sized to produce enough electrical energy to meet the load requirements during the critical design month while accounting for normal system losses. This ensures that the battery will always be properly charged and that system availability is high throughout the year. **See Figure 9-18.**

Required Array Output. First, the required array current is calculated from the load requirement and insolation of the critical design month, and the nominal DC system voltage. However, because battery efficiency is less than 100%, more current must be supplied to charge a battery than is withdrawn on discharge. A battery-system charging efficiency factor increases the required array output to a slightly higher value. A value between 0.85 and 0.95 is appropriate for most batteries. The required array current is calculated using the following formula:

$$I_{array} = \frac{E_{crit}}{\eta_{batt} \times V_{SDC} \times t_{PSH}}$$

where

I_{array} = required array maximum-power current (in A)

E_{crit} = daily electrical-energy consumption during critical design month (in Wh/day)

η_{batt} = battery-system charging efficiency

V_{SDC} = nominal DC system voltage (in V)

t_{PSH} = peak sun hours for critical design month (in hr/day)

Array Sizing

ARRAY SIZING

Average Daily DC Energy Consumption for Critical Design Month		Wh/day
DC System Voltage		VDC
Critical Design Month Insolation		PSH/day
Battery Charging Efficiency		
Required Array Maximum-Power Current		A
Soiling Factor		
Rated Array Maximum-Power Current		A
Temperature Coefficient for Voltage		/°C
Maximum Expected Module Temperature		°C
Rating Reference Temperature		°C
Rated Array Maximum-Power Voltage		VDC
Module Rated Maximum-Power Current		A
Module Rated Maximum-Power Voltage		VDC
Module Rated Maximum Power		W
Number of Modules in Series		
Number of Module Strings in Parallel		
Total Number of Modules		
Actual Array Rated Power		W

Figure 9-18. *The array sizing worksheet uses insolation data and load requirements to size the array.*

For example, consider a nominal 24 V system in a location with 4.9 peak sun hours that must supply 1580 Wh per day. The battery-system charging efficiency is estimated at 0.90. What is the required array current?

$$I_{array} = \frac{E_{crit}}{\eta_{batt} \times V_{SDC} \times t_{PSH}}$$

$$I_{array} = \frac{1580}{0.90 \times 24 \times 4.9}$$

$$I_{array} = \frac{1580}{105.8}$$

$$I_{array} = \textbf{14.9 A}$$

SMA Technologie AG
Very large PV systems typically divide the array output among many inverters.

Battery-Bank Sizing Example: Off-Grid Home in Albuquerque, NM

The overall size of the system indicates that the battery bank should be a 48 V system in order to keep DC currents within an acceptable range. Since the loads are not critical and a small engine generator is available for emergencies, the autonomy period is set at 3 days. Using the average daily energy consumption for the critical design month, the required battery-bank output is calculated at 411 Ah.

Based on the allowable depth of discharge, operating temperature, and discharge rate, the rated battery-bank capacity must be higher than 411 Ah. Since deep-cycle batteries will be used, the allowable depth of discharge is about 80% (0.80). A derating factor for temperature and discharge rate is determined from the battery specifications based on the minimum expected operating temperature and the discharge rate. The battery bank will be located in an unconditioned space in the basement, but the operating temperature will never fall below 0°C. The discharge rate is calculated from the daily weighted operating time from the critical design month, the number of days of autonomy, and the allowable depth of discharge. After applying the allowable depth of discharge and the derating factor, the required battery-bank rated capacity is 571 Ah.

A flooded lead-acid 12 V nominal battery is selected with a manufacturer-specified rated capacity of 295 Ah. Four batteries in series are required to provide a system voltage of 48 V. Two series strings of batteries are required to provide the required rated capacity. A total of 8 batteries are needed for the battery bank.

The actual rated capacity will be 590 Ah, just slightly above the required rated capacity.

In this application, the load fraction is estimated at 0.75. The average daily depth of discharge is 17%. From the manufacturer's data, this battery has an expected life of 4000 cycles at 20% average daily depth of discharge. Correspondingly, at least 10 years of service should be expected in this application.

BATTERY-BANK SIZING

Average Daily DC Energy Consumption for Critical Design Month	6578	Wh/day
DC System Voltage	48	VDC
Autonomy	3	days
Required Battery-Bank Output	411	Ah
Allowable Depth-of-Discharge	0.80	
Weighted Operating Time	11.2	hrs
Discharge Rate	42	hrs
Minimum Expected Operating Temperature	0	°C
Temperature/Discharge Rate Derating Factor	0.90	
Battery-Bank Rated Capacity	571	Ah
Selected Battery Nominal Voltage	12	VDC
Selected Battery Rated Capacity	295	Ah
Number of Batteries in Series	4	
Number of Battery Strings in Parallel	2	
Total Number of Batteries	8	
Actual Battery-Bank Rated Capacity	590	Ah
Load Fraction	0.75	
Average Daily Depth-of-Discharge	0.17	

Array Rated Output. Just as with battery banks, certain factors reduce the array output from the factory ratings to actual output values. Therefore, these factors are applied to the required array output to determine the necessary increase in array ratings for sizing and module selection. **See Figure 9-19.**

Soiling is the accumulation of dust and dirt on an array surface that shades the array and reduces electrical output. The magnitude of this effect is difficult to accurately determine, but estimates will account for most of this effect. A derating factor of 0.95 is used for light soiling conditions with frequent rainfall and/or a higher tilt angle, and a derating factor of 0.90 or less is used for heavy soiling conditions with long periods between rainfalls or cleanings. The rated array maximum-power current is calculated using the following formula:

$$I_{rated} = \frac{I_{array}}{C_S}$$

where

I_{rated} = rated array maximum-power current (in A)

I_{array} = required array maximum-power current (in A)

C_S = soiling derating factor

A standard module with 36 series-connected PV cells is particularly suited for battery-based systems. Its maximum power voltage is about 15 V to 16V, which is ideal for charging a 12 V battery system. Correspondingly, multiple modules in series strings, or 72-cell modules, are used for charging higher-order battery banks.

Array Output Loss

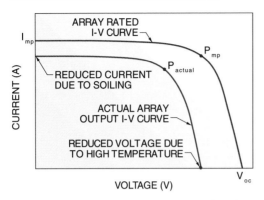

Figure 9-19. *Actual array output is often less than rated output due to soiling and high temperatures.*

High temperature reduces voltage output. A temperature coefficient of $-0.004/°C$ is applied to voltage, indicating that voltage falls by about 0.4% for every degree above the reference or rating temperature, which is usually 25°C (77°F). The maximum module temperature is estimated from the maximum ambient temperature for the location.

Arrays in hot climates produce less than their rated power because of high temperatures.

In addition, the array voltage must be higher than the nominal battery-bank voltage in order to charge the batteries. An array with a 12 V maximum-power voltage will not charge a nominal 12 V battery because the actual voltage of a nearly charged battery is about 14.5 V. The array voltage must be at least 14.5 V to charge a nominal 12 V battery. Therefore, the rated array maximum-power voltage is multiplied by 1.2 to ensure that the array voltage is sufficient to charge the battery bank.

The rated array maximum-power voltage is calculated using the following formula:

$$V_{rated} = 1.2 \times \{V_{SDC} + [V_{SDC} \times C_{\%V} \times (T_{max} - T_{ref})]\}$$

where

V_{rated} = rated array maximum-power voltage (in V)

V_{SDC} = nominal DC-system voltage (in V)

$C_{\%V}$ = temperature coefficient for voltage (in /°C)

T_{max} = maximum expected module temperature (in °C)

T_{ref} = reference (or rating) temperature (in °C)

For example, consider an array for a nominal 24 V DC system that must output 18 A. The soiling conditions are expected to be light and the maximum module temperature is estimated at 50°C. What are the minimum rated maximum-power current and voltage parameters?

$$I_{rated} = \frac{I_{array}}{C_S}$$

$$I_{rated} = \frac{18}{0.95}$$

$$I_{rated} = \mathbf{18.95\ A}$$

$$V_{rated} = 1.2 \times \{V_{SDC} + [V_{SDC} \times C_{\%V} \times (T_{max} - T_{ref})]\}$$

$$V_{rated} = 1.2 \times \{24 + [24 \times -0.004 \times (50 - 25)]\}$$

$$V_{rated} = 1.2 \times [24 + (24 \times -0.004 \times 25)]$$

$$V_{rated} = 1.2 \times (24 - 2.4)$$

$$V_{rated} = \mathbf{25.9\ V}$$

Module Selection. The final step of the sizing process involves selecting a PV module and determining the array configuration based on the current and voltage parameters. For each module, three parameters are needed for sizing: the maximum power, the maximum-power (operating) current, and the maximum-power (operating) voltage. As with batteries, modules should be chosen to result in an array that is as close as possible to the desired array ratings, but slightly higher.

The number of parallel strings of modules required is determined by dividing the rated array current output by the selected module maximum-power current output and rounding up to the next whole number. **See Figure 9-20.**

The number of series-connected modules in each string is determined by dividing the rated array voltage by the selected module maximum-power voltage and rounding up to the next whole number.

The rated array maximum power is calculated by multiplying the rated module maximum power by the total number of modules.

Array Configurations

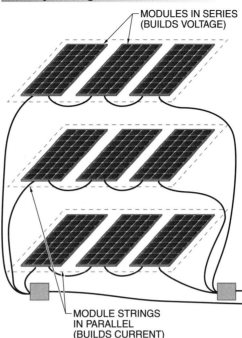

MODULES IN SERIES
(BUILDS VOLTAGE)

MODULE STRINGS
IN PARALLEL
(BUILDS CURRENT)

Figure 9-20. *Modules are configured in series and parallel to match the array rated capacity needed to produce the required output.*

Refer to Quick Quiz® on CD-ROM

Thin Film Module Initial Ratings

Some thin-film PV modules' power and current output are slightly higher for the first few weeks or months of use and then stabilize for the rest of its lifetime. (This does not apply to crystalline silicon modules.) The nameplate or rated parameters for power, current, and voltage may give either the long-term values or the initial values. The nameplate should also specify which is the case. Long-term values can be used without adjustment in sizing calculations.

However, if the module ratings specify only the initial power, current, and voltage parameters, these values may then be derated. Derating results in ratings that are more realistic for sizing calculations. The module label or specifications will likely include information about the guaranteed long-term parameters, which are generally about 90% to 95% of the initial values. If this information is not readily available, derating factors of 0.90 for power and 0.95 for current are generally appropriate.

UNI-SOLAR®
SOLAR ELECTRIC MODULE
Bekaert ECD Solar System LLC
Auburn Hills, MI
Electrical Ratings at 1000 W/m², AM1.5, Cell Temp. 25°C

Max Power:	10.3 W
Current Max Power:	0.62 A
Voltage Max Power:	16.5 V
Short Circuit Current:	0.78 A
Open Circuit Voltage:	23.8 V
Model Type:	USF-11
Max System Voltage:	30 V
Series Fuse:	1.5 A

During initial 8-10 weeks of operation, the module has higher electrical output than rated output. The output power may be higher by 15%, the operating voltage may be higher by 11% and operating current may be higher by 4%.

WARNING - ELECTRICAL HAZARD This solar electric product is a LIVE power source in sunlight. OBSERVE CAUTION and read instructions.

The required array current at maximum power is determined from the daily load requirement and insolation for the critical design month, along with the DC system voltage and an estimated battery-charging efficiency. Battery-charging efficiency is usually approximated at 85% (0.85). The result is 32.2 A. This means the PV array must be sized to produce at least 32.2 A under peak sun conditions.

The required current rating is adjusted upward to 33.9 A to account for reduced current due to a soiling factor of 0.95. The required voltage rating is adjusted upward to 64.1 V to account for reduced voltage due to high temperatures and for adequate battery charging.

The selected PV module has a rated maximum power of 185 W at STC. Rated maximum-power current is 5.11 A, and rated maximum-power voltage is 36.2 V. Two modules in series will provide the necessary array voltage, and 7 parallel strings will provide the necessary current. The array will consist of 14 modules, for a total rated power output of 2590 W, or 2.59 kW.

ARRAY SIZING

Average Daily DC Energy Consumption for Critical Design Month	6578	Wh/day
DC System Voltage	48	VDC
Critical Design Month Insolation	5.0	PSH/day
Battery Charging Efficiency	0.85	
Required Array Maximum-Power Current	**32.2**	A
Soiling Factor	0.95	
Rated Array Maximum-Power Current	**33.9**	A
Temperature Coefficient for Voltage	−0.004	/°C
Maximum Expected Module Temperature	50	°C
Rating Reference Temperature	25	°C
Rated Array Maximum-Power Voltage	**51.8**	VDC
Module Rated Maximum-Power Current	5.11	A
Module Rated Maximum-Power Voltage	36.2	VDC
Module Rated Maximum Power	185	W
Number of Modules in Series	**2**	
Number of Module Strings in Parallel	**7**	
Total Number of Modules	**14**	
Actual Array Rated Power	**2590**	W

Summary

- Sizing analysis for stand-alone systems starts at the load side and proceeds backward to the array.

- Interactive systems are generally sized to be as large as possible within the limits of available space and budget since, in most locations, occasional excess energy can be sold back to the utility.

- Sizing of stand-alone systems involves a fine balance between energy supply and demand. If the system is too small, there will be losses in load availability and system reliability. If the system is too large, excess energy will be unutilized and wasted.

- Bimodal systems are typically sized in the same way as stand-alone systems.

- The PV array and battery bank in a hybrid system can be significantly smaller than in a stand-alone system if the secondary power source is available on demand.

- A detailed load analysis completed during the site survey lists each load, its power demand, and daily energy consumption.

- A weighted average operating time accounts for multiple loads operating for varying lengths of time per day.

- Inverters lose some power in the process of converting DC energy to AC energy, so more DC energy is required to produce a certain amount of AC energy.

- A stand-alone system must produce enough electricity to meet load requirements during any month, so systems are sized for the worst-case scenario of high load and low insolation.

- If the load requirements vary from month to month, the critical design month may turn out to be any month of the year.

- The highest critical design ratio in each orientation corresponds to the critical design month for that orientation. When multiple orientations are considered, the lowest ratio of the resulting critical design months corresponds to the optimal array orientation (of the orientations analyzed).

- The DC voltage for battery-based PV systems is usually an integer multiple of 12 V, usually 12 V, 24 V, or 48 V.

- Noncritical systems are typically designed for 3 to 5 days of autonomy, while critical applications may have 6 to 10 days of autonomy or more.

- Because several factors reduce the useful capacity of a battery, the ratings of battery banks must be higher than the required battery-bank output.

- The nominal voltage and rated capacity of the selected battery are used to determine the configuration of the battery bank.

- Like batteries, several factors reduce the output of a PV module, so the array ratings must be higher than the required array output.

- Each percentage-point increase in system availability costs increasingly more money for larger battery banks and arrays, which is impractical from an economic standpoint for all but the most critical applications.

Definitions

- *Duty cycle* is the percentage of time a load is operating.

- The *critical design ratio* is the ratio of electrical energy demand to average insolation during a period.

- The *critical design month* is the month with the highest critical design ratio.

- *System availability* is the percentage of time over an average year that a stand-alone PV system meets the system load requirements.

- *Autonomy* is the amount of time a fully charged battery system can supply power to system loads without further charging.

- The *load fraction* is the portion of load operating power that comes from the battery bank over the course of a day.

- *Soiling* is the accumulation of dust and dirt on an array surface that shades the array and reduces electrical output.

1. What is involved in sizing interactive systems?

2. Why does the sizing of a stand-alone system require critical calculations and tight tolerances?

3. What methodology is used to size bimodal systems?

4. Explain the three differences between sizing a stand-alone system and sizing the PV portion of a PV array and engine generator hybrid system.

5. Describe the basic analysis procedure for sizing a stand-alone system.

6. Why must AC loads and DC loads be listed separately in a load analysis?

7. What factors are involved in inverter selection?

8. How does sizing for the critical design month improve system availability?

9. How does system availability affect system cost?

10. What three factors affect the required rating of the battery bank in relation to the required battery-bank output?

11. What three factors are used to determine the required voltage and current ratings of the array from the required array voltage and current outputs?

10

Mechanical Integration

The mechanical design and integration of PV systems requires considering the characteristics of the components and the structure and how they will affect each other when integrated. Many factors in the mechanical design process result from information collected during site surveys, including the available structural support and accessibility. Various types of mounting systems and attachment methods are evaluated for their relative advantages and disadvantages and their structural strength. The objective is to produce an economical mechanical installation that is safe, secure, and appropriate for the site and application.

Chapter Objectives

* *Identify the key considerations for integrating arrays on buildings and other structures.*
* *Understand the key factors involved in choosing a mounting system.*
* *Differentiate between the various types of mounting configurations and their features.*
* *Differentiate between the various types of attachment methods.*
* *Compare the various types of structural loads on arrays and the factors that affect each type.*

MECHANICAL CONSIDERATIONS

There are a number of factors to consider in the mounting of PV arrays and other system equipment, and they often depend on the specific application, site conditions, components, and the priorities of the owner. These criteria identify areas of concern, opportunities for enhancement, and possibilities for various products, materials, and installation techniques, helping the designer or installer select the best type of installation for an application.

Physical Characteristics

Some of the first considerations in the mechanical design of an array are the physical characteristics of the modules, such as size, weight, laminate composition, frame type, and mechanical load ratings. The electrical characteristics of modules also influence the array's physical size and mechanical configuration, which affect the choice of the mounting system as well.

Smaller modules with lower output voltage require more connections and hardware to achieve a given nominal power output. Larger modules may require more sophisticated procedures or tools, such as lifting equipment, but they typically involve less installation cost, time, and labor, especially for larger systems. Using larger modules also reduces the number of modules needed for a given power requirement, which reduces the number of connections and, consequently, the number of junction boxes, conduits, conductors, and other balance-of-system (BOS) components. However, there are practical size limitations based on the manufacturing process, effort required for installation, and the mechanical loads permitted on the array structure and attachment points.

Structural Support

Regardless of the type of structure used to support an array, such as a roof, wall, foundation, or even the ground, the strength and rigidity of the structure must be evaluated. This is especially important when installing a PV system as a retrofit, since the structure was not originally designed to support such a system. Most modern buildings are designed and constructed with a significant margin of safety that easily encompasses the relatively small additional loads from a PV system. However, structural evaluation may be required for permit applications, older buildings, or those with nonstandard construction. Analysis of structural loads and fastener strengths also helps determine the best attachment method to use when installing an array.

Accessibility

Accessibility to all parts of a PV system is an important consideration in system planning and design. The selection of an appropriate array mounting system, the location of equipment, and the overall layout of mechanical and electrical components all affect accessibility.

Arrays are most often installed on roofs. These installations frequently involve climbing ladders, operating lifts and buckets, or working at heights where fall protection is required. **See Figure 10-1.** The safest and most practical means to access the array and other equipment should be identified and, for larger commercial installations, provisions such as permanent ladders or stairways, access covers, tie-off points, and other safety features should be considered.

Roof Accessibility

SolarWorld Industries America

Figure 10-1. *Aerial lifts are sometimes required to reach roofs or areas with poor accessibility.*

Arrays mounted on roofs may extend all the way to the roof edges and have little or no space between modules. While this may improve the appearance of arrays on some buildings, it compromises accessibility and may result in higher wind loads. Adequate space to safely install, inspect, clean, remove, replace, or maintain the array should be considered.

Access to attics or other spaces directly beneath roof surfaces must also be considered. Some mounting systems do not require access to these spaces for attachment, but it may be necessary to route electrical conductors through these areas. Attic spaces can be difficult to work in because of high temperatures and lack of working space, so this work should be carefully planned.

Also, if an array mounting system is designed for seasonal tilt adjustment, it is desirable to have convenient access to adjustment points. Owners may be less inclined to make tilt adjustments if access is limited.

Thermal Effects

Temperature affects the electrical performance and lifetime of arrays. Additionally, arrays mounted on a building may impact the building's thermal loads, which should be considered when mechanically integrating PV systems.

Array Performance. Temperature is an important consideration in array mechanical design. High temperatures reduce the power output of crystalline silicon modules. Less is known about thermal effects on module lifespan or on thin-film PV cells, particularly after several years of exposure, but cooler arrays generally are more reliable, last longer, operate with greater efficiency, and produce more power.

For these reasons, array temperatures should be minimized wherever possible. Active cooling means, such as fans and water-circulating pumps, may be used with some concentrating arrays, but are not practical for flat-plate modules. Only passive cooling means are employed for flat-plate modules, such as mounting the array in a way that allows air circulation around the modules. Mounting system design and

installation that allows natural cooling is the principal means of preventing the array from operating at excessively high temperatures.

Keeping modules and arrays clear of obstructions is the easiest way to promote natural cooling. This may include trimming nearby vegetation and avoiding blocking or restricting the open spaces underneath modules with flashings, plates, or other barriers to airflow. Aligning module support channels with the slope of the array or roof surface is a simple way to channel natural airflow, which helps remove heat from the array. Alternately, arranging modules so that the array's longer dimension is lateral (landscape layout) results in lower temperatures because the heat has a shorter distance to escape from under the array. **See Figure 10-2.** These recommendations may not be feasible for some installations, such as building-integrated designs, but tradeoffs between installation design and thermal performance must be considered.

Passive Array Cooling

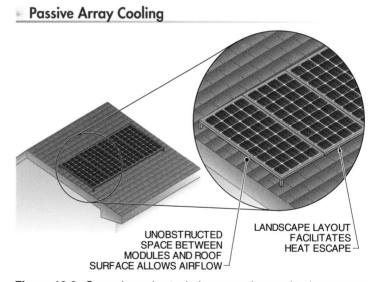

UNOBSTRUCTED SPACE BETWEEN MODULES AND ROOF SURFACE ALLOWS AIRFLOW

LANDSCAPE LAYOUT FACILITATES HEAT ESCAPE

Figure 10-2. *Several passive techniques can be used to keep arrays cool, which improves array performance.*

The temperature-rise coefficient specifies how the temperature of PV cells in an installation increases with ambient temperature and irradiance. The installed nominal operating cell temperature (INOCT) is an estimate of the normal temperature in these situations.

The *installed nominal operating cell temperature (INOCT)* is the estimated temperature of a module operating in a specific mounting system design. This value is based on the module manufacturer's rated nominal operating cell temperature (NOCT) for a particular module, but also takes into account how the mounting method affects module temperature. Some mounting systems affect cell temperature more than other systems. System designers may use a temperature-rise coefficient or INOCT reference values in array calculations. In the field, installers can directly measure cell temperature.

Building Thermal Loads. The impact that an array can have on the heating and cooling loads of a building is an often-overlooked aspect of array mechanical design. Arrays may affect heat transfer into conditioned spaces, depending on the roof type, mounting system, and other factors. **See Figure 10-3.**

Some arrays radiate additional heat into a building. The array absorbs heat energy from direct radiation and conducts it through the roofing materials to the underside of the roof surface. There, the energy heats the interior spaces of the building. This can be a benefit in cold climates, but is a disadvantage in warm climates with heavy air-conditioning loads. Large, well-ventilated attic spaces with adequate insulation can moderate the heat gain into conditioned spaces. Radiant barrier materials applied to the underside of the roof surface can also reduce the heating effect.

Neat and organized installation makes for an aesthetically pleasing array.

Conversely, arrays mounted above building surfaces have little effect on building heat gain, or may even reduce interior building temperatures. These modules shade part of the roof from direct radiation, while the space between the modules and the roof surface keeps the modules from transferring much heat to the roof and allows wind to cool the roof surface. Roofs in these installations are cooler, which reduces heat transfer into conditioned spaces.

Electrical Equipment. Temperature effects must also be considered in the installation and mechanical integration of electrical equipment, including inverters, batteries, and other BOS hardware. Equipment temperatures are usually managed by selecting the best locations for installing the equipment.

Most inverter designs use passive heat sinks or fans to protect critical components. Manufacturers provide specific recommendations for mounting and locating inverters, including instructions for minimizing operating temperatures by avoiding other heat sources, areas where air circulation is obstructed, and direct sunlight.

Batteries are also very sensitive to temperature, both heat and cold. Batteries may be buried in underground containers to mitigate temperature extremes, or installed in enclosures that are well insulated, ventilated, shaded, or possibly air-conditioned.

Conductors in conduits exposed to direct sunlight are a particular concern with PV systems, especially conductors used for array and source circuits, because they may experience temperatures exceeding 60°C (140°F). Locating conduit under roof eaves, beneath the array, or inside the building space are ways of reducing conductor temperatures.

Aesthetics

While the outward appearance of arrays and overall installations has little to do with system functionality or performance, it has a notable influence on consumer acceptance of PV technology. Because the array is the most visible part of any PV system, unattractive array installations may lead to poor public perception of PV systems and unwillingness of others to implement such systems on their own buildings or properties. In many cases, the appearance of an array presents tradeoffs with other system design and installation issues. Design considerations must be appropriately balanced with respect to safety and performance. A number of architectural principles may be applied to PV system design and installation that can improve appearance and public perception without significantly affecting cost or performance. **See Figure 10-4.**

For arrays mounted on sloped roofs, the lines and location of the array should be consistent with building features. Arrays should be mounted parallel to the roof surface, centered and square with the rooflines and edges. Modules should be of the same size and shape and aligned in the same direction. Roof obstructions, such as roof vents, chimneys, or air conditioners, that interrupt contiguous groups of modules should be avoided if possible. Flat rooftops are less visible from the ground and often use racks to tilt the array.

Aesthetic Installations

SolarWorld Industries America

Figure 10-4. *PV systems that match the shape, color, and/or alignment of the mounting surface produce aesthetically pleasing installations.*

Some PV modules are actually windows that produce electricity while letting a portion of sunlight into the building. A further innovation on this idea uses photochromic technology to change the opacity of the window material and regulate interior natural lighting.

Thermal Loads

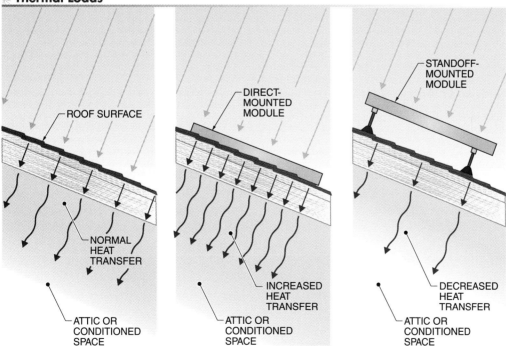

Figure 10-3. *Modules mounted directly on the roof surface increase the heat transfer into a building.*

Color may be a consideration in choosing and integrating modules. The color of PV cells is generally determined by the manufacturing technology, ranging from dark red to gray to bright blue. Sometimes module manufacturers offer color choices for the module frame. If choosing the color of modules is not an option, the colors of building features may be adjusted to blend with the array.

Reflections and glare from arrays can cause annoyances or serious safety issues, affecting nearby buildings, people, traffic, and even aircraft. Reflections can be especially problematic for highly tilted arrays at northern latitudes, when the sun is low in the sky. However, glare is typically limited to specific points of view and times of the day and year. If reflections are a concern, the installer can evaluate solar incidence angles to identify the locations and times with significant glare issues to determine the best mounting option.

Finally, the quality of workmanship on the overall system installation affects the aesthetic appearance of PV systems. Array support structures, hardware, and electrical wiring, conduit, and junction boxes should be as inconspicuous as possible, neatly gathered or concealed beneath the array. The routing of conduits or conductors from the array to other components should be as inconspicuous as possible, especially in locations where it passes through a roof or eave. Installation of a PV system and related equipment in a neat and professional manner improves the overall appearance of the system.

Costs

The mechanical integration and structural installation of an array is often the largest variable in overall system costs, ranging from as little as 10% to more than 40% of initial system costs. Custom, site-specific designs can incur especially high costs, due to additional engineering and architectural requirements and nonstandard installation. However, the use of standard mounting designs and installation techniques is becoming more common, which helps keep mechanical integration costs down. Mechanical integration costs are approximately

proportional to the physical area of the array. Consequently, modules and system designs that are more efficient, and therefore require less area, reduce overall system costs. Other ways to reduce costs for mechanical integration may require a tradeoff with desirable features. For example, alternative materials may cost less than standard materials, but may also be less resistant to corrosion.

Reducing installation time by careful preparation can significantly reduce costs. For example, it is faster and easier to assemble panels or subsystems on the ground before installing them on the roof. **See Figure 10-5.** Assembling components on the ground is also safer.

Preassembly

SolarWorld Industries America

Figure 10-5. *Assembling PV subsystems such as panels before lifting them to the roof is often easier and reduces installation time.*

ARRAY MOUNTING SYSTEMS

An array mounting system must securely hold modules in the most favorable orientation possible. Several types of mounting systems have been developed to meet these requirements for a variety of applications.

Most flat-plate modules have a similar construction and are secured by their frame, either with fasteners through factory-drilled holes or by clamping the module frame to a structural support. Several manufacturers offer universal hardware for installing most types and sizes of modules onto a variety of mounting systems. This standardization makes PV systems versatile, increases installation options, and reduces the costs associated with design, materials, and labor.

The simplest and most common type of array mount for modules is the fixed-tilt type. A *fixed-tilt mounting system* is an array mounting system that permanently secures modules in a nonmovable position at a specific tilt angle. An *adjustable mounting system* is a variation of a fixed-tilt array mounting system that permits manual adjustment of the tilt and/or azimuth angles to increase the array output. **See Figure 10-6.** The tilt angle is decreased for the summer when the sun is higher in the sky, and increased for the winter when the sun is lower. Manual tilt adjustments are usually made monthly or seasonally.

Module Mounting Systems

FIXED-TILT
MOUNT

ADJUSTABLE-TILT
MOUNT

Figure 10-6. *Mounting systems may hold modules at a fixed tilt, or may allow adjustments to be made to the tilt for greater solar energy gain.*

Fixed-tilt and adjustable-tilt mounting systems are further differentiated by whether they mount to a building or the ground, and by the design of the support structure. Each type of mounting system has advantages and disadvantages in terms of cost, performance, ease of installation, and maintenance, and may affect the amount of solar energy received on the array surface.

Building Mounting Systems

Rooftops often offer the best opportunities for the installation of arrays on buildings, because they are usually large unused spaces that are high enough to avoid significant shading from nearby obstructions. Direct mounts, roof rack mounts, and standoff mounts are all designs for mounting arrays on buildings.

Direct Mounts. A *direct mount* is a type of fixed-tilt array mounting system where modules are affixed directly to an existing finished rooftop or other building surface, with little or no space between a module and the surface. **See Figure 10-7.** Direct mounting of conventional flat-plate modules is not recommended because the lack of cooling from natural airflow (breezes) results in high array operating temperatures. INOCT for direct-mounted modules is the highest among building mounting systems. Temperature-rise coefficients for direct mounts can be as high as 40 to 50°C/kW/m² (72 to 90°F/kW/m²). Accessibility to individual modules for installation and maintenance can also be a problem with direct mounts, though their low profile can produce very attractive installations.

Direct Mounts

DOE/NREL, Jim Yost

Figure 10-7. *Direct mounts have little or no space between the modules and the mounting surface.*

Roof Rack Mounts. A *rack mount* is a type of fixed- or adjustable-tilt array mounting system with a triangular-shaped structure to increase the tilt angle of the array. Rack mounts are used on flat or low-pitched roofs, as well as for ground mounts. While rack mounts may be more expensive than other common mounting designs, they are simple to install and are flexible in the module types and tilt angles they can accommodate. **See Figure 10-8.**

Roof Rack Mounts

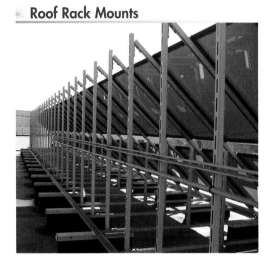

Figure 10-8. *Roof rack mounts secure modules on a triangular trusslike structure that mounts to flat or low-tilt roofs.*

Since rack mounts use an open, trusslike structure, air can circulate freely around the modules and keep them cool. INOCT for rack-mounted arrays is the lowest among building mounting systems, and temperature-rise coefficients are as low as 15 to 20°C/kW/m² (27 to 36°F/kW/m²). This type of structure also allows access to the back surface and electrical terminations of the modules, simplifying installations and maintenance. However, because of their higher profile, rack mounts may experience higher wind loads than do arrays mounted closer to building surfaces.

In some cases, rack mounts may be used on pitched roof surfaces to increase the tilt angle of the array above the roof pitch. If a roof does not face south, rack mounts may be rotated from the east or west to face the sun and increase solar energy received. However, racks mounted obliquely to surfaces should be avoided unless no other options exist.

Standoff Mounts. A *standoff mount* is a type of fixed-tilt array mounting system where modules are supported by a structure parallel to and slightly above the roof surface. **See Figure 10-9.** Most standoff mounts are designed to hold modules to support rails that are attached to the roof by

brackets. Standoff-mounted arrays are by far the most common, preferred, and least-expensive method for installing arrays as a retrofit to existing rooftops, as well as for new construction.

Standoff Mounts

Sharp Electronics Corp.

Figure 10-9. *Standoff mounts allow several inches of space between the modules and the mounting surface.*

Standoff mounts allow air to circulate beneath the array, keeping modules cool and reducing heat gain into buildings. INOCT for standoff arrays is a function of the standoff height. Standoff heights from 1″ to 3″ have a high INOCT, standoff heights from 3″ to 6″ are somewhat cooler, and standoff heights above 6″ are the lowest INOCT for standoff arrays. However, increasing standoff height also increases wind loads and may adversely affect the aesthetic appearance. In general, standoff mounts should be installed between 3″ and 6″ from the top of the module and the roof surface. Temperature-rise coefficients for this height range are around 20 to 30°C/kW/m² (36 to 54°F/kW/m²).

Like most types of mounting systems, standoff arrays are built in a modular fashion by assembling portions of the array into electrical and mechanical panel units, and integrating the assemblies together on the rooftop. These designs also provide reasonable access to the underside of the array for inspection, troubleshooting, and maintenance, by detaching a few module fasteners.

Standoff mounts are attached directly to building structural members, such as roof trusses or rafters, never to decking. In most cases, standard standoff mounting hardware allows for direct attachments from above the roof surface, but some designs may require access to attics or spaces underneath the roof and array. This ensures a strong and reliable structural connection, but can be more time consuming and costly to install.

Structural attachments for smaller standoff panels may use four attachments at corner points, while larger panels might use six or more attachments distributed across the entire length of the panel. One disadvantage of standoff mounts is the relatively large number of structural attachments required, typically around 4 to 6 attachments for every 20 ft^2 to 40 ft^2 of array surface area. Each attachment is a roof penetration that must be properly sealed against weather.

Building-Integrated Systems

A *building-integrated photovoltaic (BIPV) array* is a fixed array that replaces conventional building materials with specially designed modules that perform an architectural function in addition to producing power. BIPV arrays can be integrated into roofing, windows, skylights, curtain wall sections, or complementary architectural features such as awnings, facades, or entranceways. **See Figure 10-10.** Some BIPV systems may replace only the outermost building material, such as roof shingles, while others may replace the entire thickness of a portion of building shell, such as the entire assembly of roof shingles, substrate, and decking. BIPV arrays are most often incorporated into new construction for large commercial buildings, but can be used in residential and small commercial structures as well. Some designs may be available as retrofits.

The cost of the array is partially offset by avoiding the costs of some of the conventional building materials, though high engineering and architectural costs often outweigh the savings. However, BIPV arrays have favorable aesthetic and architectural features because they are so thoroughly integrated into the building's appearance.

Of the many possibilities for BIPV arrays, roofing replacement has been one of the most popular, since roof surfaces are generally well oriented to receive solar energy. Common BIPV roofing products have been designed to replace asphalt shingles, slates, tiles, and metal roofing. In addition, roofing is often replaced at least once during a building's life, making BIPV retrofits an attractive option for reroofing.

Windows, skylights, and other building surfaces are also possible BIPV applications, though they are not as popular because they are often not optimally oriented. Another challenge is that BIPV windows and skylights must have some degree of transparency to retain their dual function. Some use standard cells spaced farther apart in a transparent laminate, utilizing the clear spaces between the cells to allow some light through. However, the low cell density reduces the overall area efficiency. Thin-film products can also be modified to various levels of transparency by changing the backing material. However, since much of the light then passes through, the efficiency of these modules is also reduced. There are many opportunities to improve BIPV module designs, including more efficient modules or new technologies in fully transparent or semitransparent PV materials.

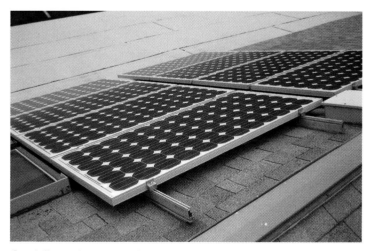

Standoff mounting systems often consist of long rails that hold the array together as a rigid structure and create a gap between the roof and the modules for airflow.

Building-Integrated PV Arrays

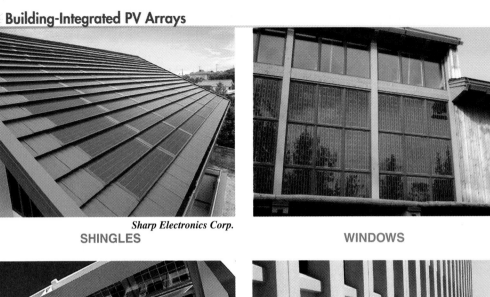

Sharp Electronics Corp.
SHINGLES

WINDOWS

DOE/NREL, Lawrence Berkeley Lab

SKYLIGHTS

DOE/NREL, University of Texas Health Science Center at Houston

AWNINGS

Figure 10-10. *PV modules can be integrated into building exteriors as roof shingles, windows, skylights, awnings, and many other structures.*

Ground Mounting Systems

Arrays are often ground mounted when suitable space on a roof is not available, or for stand-alone remote-power applications where no alternative exists. Arrays can also be installed on structures other than buildings, such as navigational aids, communications towers, signs, bridges, trailers, and vessels. Common ground-mounted array designs include rack mounts and pole mounts. Sun-tracking systems are also usually mounted on the ground.

Ground-mounted arrays generally offer more flexibility than building-mounted arrays in the location and orientation of arrays, allowing for optimal solar energy gain. These systems operate at lower temperatures than do most building mounts, due to greater airflow

around the array. Ground-mounted modules are also more accessible for adjustments and maintenance. As with any array, ground-mounted arrays must be appropriately secured to resist wind loads and other forces acting on the array and its attachment points.

Ground Rack Mounts. Rack mounts are extremely versatile in securing modules in any orientation and tilt, and are used in both building and nonbuilding mounting systems. Rack-mounted arrays can be installed on poles, walls, directly on the ground, or on an intermediate superstructure between the rack and the ground. They can also be installed in multiple rows for larger arrays. For rack-mounted arrays on the ground, the array support structure is generally anchored to the ground with concrete piers, footers, earth anchors, or a foundation of pressure-treated timbers buried in the ground. **See Figure 10-11.** Wood construction is generally avoided in PV system installations, but high-quality pressure-treated timbers or railroad ties may provide adequate long-term performance for mounting foundations. However, local fire codes must be consulted in these situations because this practice may not be allowed in some jurisdictions.

Ground Rack Mounts

SPG Solar, Inc.

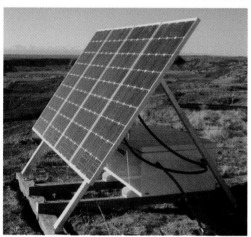

SunWize Technologies

Figure 10-11. *Ground rack mounts are versatile designs that can accommodate both large- and small-scale installations.*

Since rack-mounted arrays on the ground are lower than roof-mounted arrays, shading may be more of a concern due to nearby trees, vegetation, fences, buildings, and other obstructions. Also, arrays installed at ground level may present a safety hazard to the public,

or may be more vulnerable to incidental physical damage or vandalism. If ground-mounted arrays need protection, chain-link fencing is preferred over solid fencing to maintain adequate air circulation around the array.

Pole Mounts. A *pole mount* is a type of array mounting system where modules are installed at an elevation on a pedestal. **See Figure 10-12.** Pole mounts may be fixed tilt, adjustable tilt, or sun tracking, and are typically used for ground mounts or attached to other structures. They are generally not used on buildings. Since pole-mounted modules can be placed at a significant height, they are often protected from harm and positioned above the shadows of nearby obstructions.

Pole Mounts

Figure 10-12. *Pole-mounted arrays can be used in a variety of applications, such as lighting, communications, water pumping, and signage.*

Pole-mounted arrays can be attached to either the top or the side of a pole. Mounting an array at the top of the pole allows more versatility in array orientation.

Pole mounting systems often allow changes to the azimuth orientation by rotating either the array or the pole. Care should be taken to ensure that the pole itself, or any other arrays or equipment located on the pole, will not shadow the array at any critical time, especially when the sun is in the northern part of the sky in morning and afternoon in summer.

Disadvantages for pole-mounted arrays include inherent limitations on the size of array that can be installed on a given pole, and that special lifting equipment is often needed to install, inspect, and maintain the array. Depending on the strength of the pole, the size of the array, and its height above the ground, a pole may be set directly into an auger hole and the soil compacted, or it may be necessary to set the pole base on a concrete foundation, and secure it with fasteners to the steel reinforcement in the concrete.

Pole-mounted PV systems may also contain lights, weather stations, communications equipment, security cameras, or other devices, and these may or may not be powered by the array. For example, a stand-alone PV lighting system may include the array, battery, controller, and light fixture all on one pole. Alternatively, a self-contained unit including a small array, battery, and security camera may be installed on an existing commercial light pole.

> The insolation values for both single-axis and two-axis sun-tracking surfaces can be found in the solar radiation data set for the nearest location. This makes it very easy to compare the potential gain in power output from sun tracking to fixed-orientation surfaces.

Sun-Tracking Systems

Fixed- and adjustable-tilt mounts are the easiest to install and maintain, but they sacrifice energy gain because the array is not always in the optimal orientation to the sun. Sun-tracking arrays can follow the path of the sun, enhancing the amount of energy collected. However, they are complex systems and require a greater investment in expense and installation time.

A *sun-tracking mount* is an array mounting system that automatically orients the array to the position of the sun. This can increase annual solar gain by as much as 40% in some areas when compared to a fixed-tilt mounting design. Sun-tracking mounts are classified according to the number and orientation of the axes rotated to track the sun, and by the means of rotation.

A single-axis tracking mount rotates one axis to approximately follow the position of the sun. The mount may rotate the vertical axis, which changes the azimuth angle, or the north–south axis, which allows the array to follow the sun from east to west. **See Figure 10-13.** Most single-axis tracking mounts can also be manually adjusted for tilt. Single-axis tracking mounts can be mounted atop vertical poles or on horizontal supports in long rows.

To exactly follow the sun's position (solar altitude and azimuth), two-axis tracking is required. A two-axis tracking mount rotates two axes independently to exactly follow the position of the sun. **See Figure 10-14.** Two-axis tracking maximizes the amount of solar energy received. Altitude-azimuth tracking uses the vertical and tilt axes. The vertical axis rotates to follow the sun's azimuth angle and the tilt axis rotates to follow the sun's altitude. Equatorial tracking uses the north–south and tilt axes. The north–south axis rotates to follow the sun from east to west in an arc. The tilt axis needs to be activated to adjust the tilt only periodically throughout the year. Two-axis tracking mounts are generally mounted atop vertical poles to allow sufficient room for movement.

The movement in trackers is produced by either active or passive means. An *active tracking mount* is an array mounting system that uses electric motors and gear drives to automatically direct the array toward the sun. Active tracking mounts may track in either one or two axes. The tracking direction is determined by a computer calculating the sun's position, or with sun-seeking sensors. In order for active tracking mounts to be effective, the solar energy gained by tracking must more than compensate for the higher system cost and additional electrical energy used by active tracking motors.

A *passive tracking mount* is an array mounting system that uses nonelectrical means to automatically direct an array toward the sun. Refrigerants can be used to move the mount, because they vaporize and expand when heated by the sun. The expanding fluid causes the tracker to pivot toward the sun as the weight of the fluid shifts from one side of the tracker to the other. An alternate design uses refrigerant to operate a hydraulic cylinder and linkage arrangement. The control of the tracker direction for both designs uses sunshades to regulate the heating of the fluid by the sun. Passive tracking mounts typically control only one axis.

The value of a tracking array depends on whether the additional energy produced offsets the added cost and complexity of the equipment. The energy gain must also offset the increased maintenance and troubleshooting likely with these systems. The moving parts may require periodic calibration and lubrication.

Utility-scale systems often use tracking systems to maximize array output. Tracking results in the greatest energy enhancement in the summer, when days are long, particularly in high latitudes. Trackers are less beneficial in the winter, when days and sun paths are short.

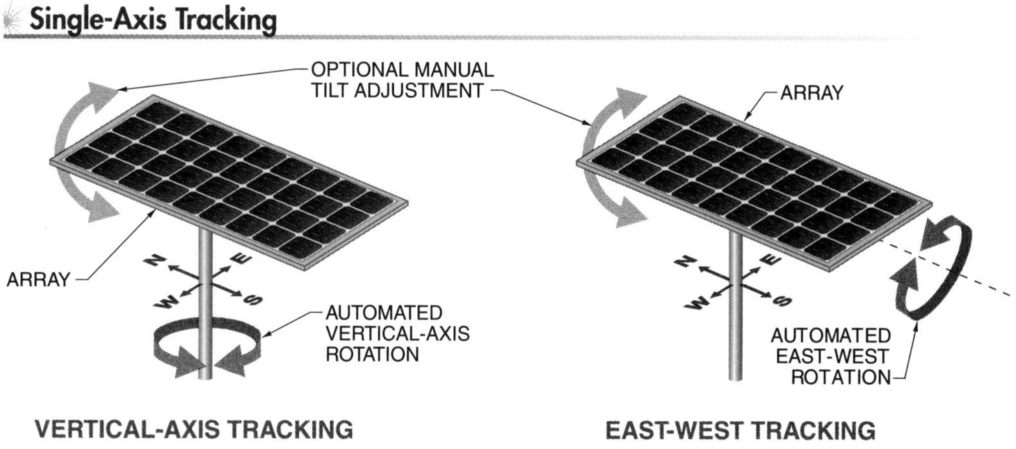

Single-Axis Tracking

VERTICAL-AXIS TRACKING   **EAST-WEST TRACKING**

Figure 10-13. *Single-axis tracking mounts rotate one axis to approximately follow the sun as it moves across the sky.*

Two-Axis Tracking

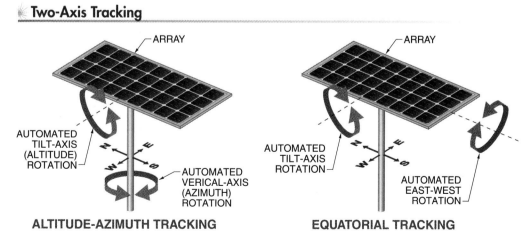

ALTITUDE-AZIMUTH TRACKING   **EQUATORIAL TRACKING**

Figure 10-14. *Two-axis tracking mounts rotate two axes to exactly follow the sun as it moves across the sky.*

MECHANICAL INTEGRATION

Based on the site survey, structural evaluation, customer requirements, and other considerations, a suitable mounting system is chosen for an array. The installer must then consider several factors in its mechanical integration, including selecting the appropriate materials and attachment methods based on the location and an analysis of the structure or foundation.

Materials

Because PV systems are exposed to outdoor elements, materials for structural supports and other system hardware must be chosen to withstand environmental conditions without degrading. Degradation can increase maintenance costs, decrease functionality and serviceability, affect overall appearance, and create an unsafe situation. Any materials used should match the expected 20- to 30-year service lifetime of the overall system under the given site conditions. Wooden exterior structures are usually avoided because they may degrade in less time than PV systems are expected to last. However, wood may be used in interior spaces, such as attics, to help support arrays mounted on the roof, and pressure-treated wood may be used as primary ground support for stand-alone rack-type arrays.

Corrosion can make servicing and adjustments difficult. Wherever possible, corrosion-resistant materials should be used in all parts of the array structure. Corrosion is most prevalent in hot, humid, and marine climates, and corrosion rates may be 400 times higher in southern coastal areas than in the arid desert locations. Stainless steel alloys 316 and 403 are recommended for most fasteners, particularly in humid and marine climates, while hot-dipped galvanized or coated steel fasteners and structural members are acceptable in drier climates. Aluminum structural alloys 6061 and 6063 are commonly used in array designs because they are particularly corrosion resistant, lightweight, and relatively inexpensive.

Galvanic corrosion results from direct contact of dissimilar metals. *Galvanic corrosion* is an electrochemical process that causes electrical current to flow between two dissimilar metals, which eventually corrodes one of the materials (the anode). **See Figure 10-15.** The rate of corrosion depends on the properties of the two metals in contact, as well as temperature and humidity. Aluminum (module frames) and steel (mounting structures) combinations are particularly prone to galvanic corrosion. Galvanic corrosion can be mitigated by electrically insulating the metals with rubber or fiber materials or by adding sacrificial anodes.

A *sacrificial anode* is a metal part, usually zinc or magnesium, that is more susceptible to galvanic corrosion than the metal structure it is attached to, so that it corrodes, rather than the structure. **See Figure 10-16.** Therefore, sacrificial anodes must be replaced periodically. However, they are typically only needed in the most extreme marine or coastal environments.

Galvanic Corrosion

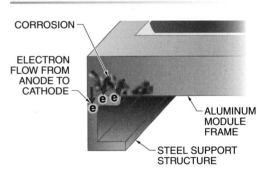

Figure 10-15. *Galvanic corrosion can occur when two dissimilar metals are in contact with each other.*

Sacrificial Anodes

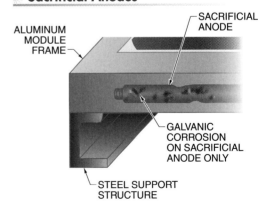

Figure 10-16. *Sacrificial anodes are more prone to galvanic corrosion than the metal they protect, so they corrode first.*

Certain types of rubber compounds may degrade quickly under UV exposure, although EPDM or butyl rubber materials offer good performance and can be used to isolate dissimilar metals. Antiseize compounds applied to threaded fasteners retard corrosion and make the fasteners easier to assemble and remove later. Also, any electrical conductors and other electrical components exposed to sunlight should be explicitly labeled for UV-resistance.

Structural Loads

Arrays, modules, mounting systems, fasteners, and buildings must withstand the maximum forces expected from several types of structural loads. Structural loads on arrays are determined by the size, weight, and orientation of the modules, prevailing site conditions, and other application-specific issues. The principal types of structural loads are dead loads, live loads, wind loads, snow loads, and vibration loads. These loads are either static (constant) or dynamic (changing).

Depending on the location, some loads may be more significant than others. In Florida, for example, wind loads are most important because the area is prone to hurricanes, while snow loads are nonexistent. In the Midwest, however, snow loads may be significant.

Most structural loads are based on unit area, and are typically represented in pounds per square foot (psf or lb/ft^2). For noncritical applications, the load is generally assumed to act evenly over the entire area. **See Figure 10-17.** When the load pressure must be applied to a set number of points, such as the attachment points of an array, the load is divided between the points.

Arrays and their attachment points must be designed and installed to withstand the forces from a combination of structural loads. A *design load* is a calculated structural load used to evaluate the strength of a structure to failure. Each type of structural load has a design load, which is a calculated estimate of the load the structure must withstand. Design loads are estimated because the actual load can sometimes only be measured after the structure

is built. Some types of design loads, such as wind load, are much greater than the actual average load because the source of the load is highly variable and structures must withstand the worst-case scenarios. Other types of loads, such as dead loads, can be calculated from the actual weight of building materials, which is constant, so the design dead loads are usually only a little higher than actual dead loads.

Structural Loads

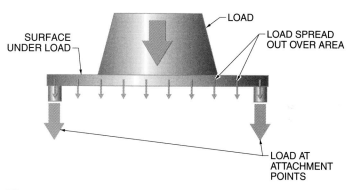

Figure 10-17. *Most structural loads are specified as a force per area. When the area attaches to other structures at certain points, the load is divided between the points.*

Local building codes dictate the requirements for addressing structural loads, though most use the same building standards and have similar requirements. The American Society of Civil Engineers (ASCE) publishes the governing standard on structural loads used in most building codes throughout the United States, ASCE 7-05 *Minimum Design Loads for Buildings and Other Structures.*

Many installations take advantage of pre-engineered mounting systems or complete PV system packages, which have already been analyzed for potential structural loads. If installed according to the manufacturer's recommendations, these structures are designed to withstand all types of structural loads under common conditions. However, custom or altered designs may require independent engineering or testing to certify compliance to local building codes. In either case, the installer might not be responsible for a detailed structural analysis, but should be familiar with the factors affecting loads and understand when to consult with engineering professionals.

The ratio between an expected service load and the corresponding actual failure load is the safety factor. Many formulas, particularly those having to do with structural loads and occupied buildings, include safety factors in the calculations. A safety factor in structural calculations is a multiplier used to increase the design load far above normally expected loads. This means that a structure built to withstand high design loads will easily withstand smaller loads. This also gives the structural design a large margin of safety that should cover worst-case scenarios.

The safety factor may be explicitly given in the formula, such as with a variable, or may be embedded in the formula in a way that is not easily discernable. For example, the allowable withdrawal load for lag screws includes a safety factor of 4.5. That is, the stated allowable load for the screws is less than one-quarter of the actual failure load. For example, if a lag screw has an allowable withdrawal load of 100 lb/in., it can actually handle loads up to 450 lb/in. before the attachment fails. Safety factors in structural design are always equal to at least 4, and may be higher for applications with more significant consequences of failure.

SPG Solar, Inc.

Low tilt angles reduce power output at high latitudes, but also reduce wind loads.

Dead Loads. A *dead load* is a static structural load due to the weight of permanent building members, supported structure, and attachments. **See Figure 10-18.** In the case of PV systems, the dead load is equal to the combined weight of the modules, mounting structure, and BOS components, divided by the area of the array. Less weight or greater area decreases the dead load. Since it is based on weight, a dead load acts only downward. Dead loads affect the mounting structure of the array and the building or foundation to which it is attached. Dead loads are often the smallest of the structural loads. Based on typical weights of modules and support structures and their installation density, dead loads of PV systems are about 5 psf to 10 psf.

Dead Loads and Live Loads

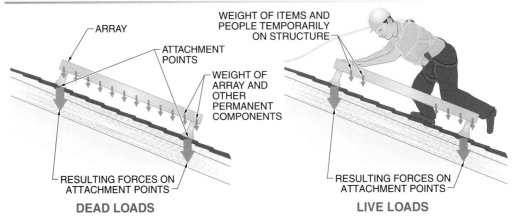

Figure 10-18. *Dead loads result from the weight of arrays and permanent components. Live loads are caused by the weight of people and/or items that are temporarily on the structure.*

Live Loads. A *live load* is a dynamic structural load due to the weight of temporary items and people using or occupying the structure. The live load is equal to the combined weight of all temporary objects, divided by the area of the array. Less weight or greater area decreases the live load. A live load also acts only downward. Generally, live loads on arrays are infrequent and minimal, such as from maintenance equipment or a person briefly leaning on a module. Many live loads to mounting systems can be avoided by using scaffolding or bucket lifts to access the array, or installing modules with enough space between them for access. When unavoidable, live loads can be reduced by distributing a weight over a greater area by using temporary platforms across modules. With these practices, live loads are often as low as 5 psf to 10 psf.

Wind Loads. A *wind load* is a dynamic structural load due to wind, resulting in downward, lateral, or lifting forces. **See Figure 10-19.** Wind loads are typically the most significant of all the types of structural loads, and can range from 25 psf to over 50 psf. Wind loads can be especially large in coastal areas. Local building codes may have a minimum design wind load, often 10 psf, even if calculations indicate lower loads.

Calculating design wind loads involves complex formulas including many variables, some of which are not easily quantified. Mounting system manufacturers usually provide guidance on these calculations with respect to their mounting type and common PV modules. However, for most PV system applications, three primary factors influence wind loads: wind speed, exposure, and array tilt.

The *basic wind speed* is the maximum value of a 3 sec gust at 33' (10 m) elevation, which is used in wind load calculations. The local building code authority may provide this information, or it can be determined from a map. **See Figure 10-20.** Low wind speeds are typical for interior regions, while wind speeds up to 150 mph are used in calculations for coastal areas prone to hurricanes. Special wind regions in certain mountain areas and other areas prone to high winds require more information from the local AHJ. Higher wind speeds increase the wind load on surfaces.

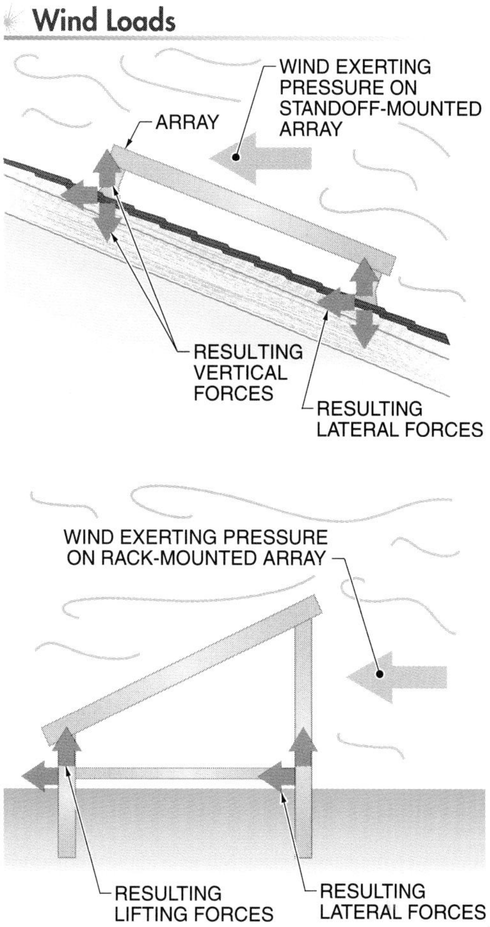

Wind Loads

Figure 10-19. *The wind-load forces at attachment points can be downward, lifting, or lateral forces, depending on wind direction and the orientation of the array.*

Exposure is a wind load factor that accounts for the array height and the characteristics of the surrounding terrain. For example, urban or suburban areas, wooded areas, or other areas with closely spaced obstructions reduce the wind forces on surfaces. Conversely, flat unobstructed areas, such as grasslands and water surfaces, increase potential wind loads.

The tilt angle of a surface greatly affects potential wind loads. Wind loads are small for array tilt angles around 20°, and increase with larger tilt angles up to 90° (vertical). Large wind loads also occur at low tilt angles of 10° to 15°. Even though the array may tilt away from the wind direction (like a wedge), an array at low angles may act like an airfoil and experience lifting forces from the wind.

Basic Wind Speeds

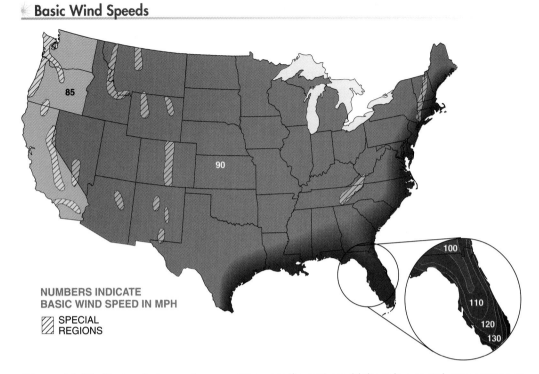

NUMBERS INDICATE
BASIC WIND SPEED IN MPH

SPECIAL
REGIONS

Figure 10-20. *Basic wind speeds are region-specific and are highest in coastal areas prone to hurricanes.*

The location of the array, and therefore the basic wind speed, cannot easily be changed to decrease wind loading. However, several other methods can be used to reduce wind loads. Minimizing the profile height for rack-mounted arrays, perhaps by using multiple short racks instead of a single tall rack, and lowering the overall altitude of the array reduces the exposure to the wind. Avoiding tilt angles greater than the roof pitch and locating arrays near the center of roofs, away from the edges, may also help. Arrays with the longer dimension oriented along the wind direction, as opposed to across it, experience lower wind loads. Keeping standoff heights low and undersides clear of obstructions are additional means of minimizing wind loads.

Wind load calculations include a topographic factor that accounts for the acceleration of wind over steeply sloping terrain, resulting in increased wind loads. This factor may affect PV installations near sharp changes in ground elevation, such as ridges or escarpments.

Snow Loads. A *snow load* is a static structural load due to the weight of accumulated snow. Snow loads cause forces similar to dead loads, but the magnitude can vary greatly according to amount of snow.

Significant snowfall is expected only at high latitudes, where arrays typically have large tilt angles to maximize solar energy gain, and snow is less likely to accumulate on steeply sloped arrays. However, when the tilt angle is smaller because of roof geometry or for maximum summer energy gain, the weight of snow on an array can be significant. Ground-mounted arrays are also vulnerable because snow can accumulate or drift high enough to load the structure.

Some sunlight can pass through light snow into the modules, warming the cells, which usually melts the snow or causes it to slide off the array. If snow does not melt off in a reasonable amount of time, it should be promptly removed. Besides the structural loads from the weight, snow cover shades the array, severely limiting the electrical output.

Snow loads vary greatly by region. Snow load maps show the approximate loads for various regions of the United States. **See Figure 10-21.** However, in some areas snowfall is too variable to assign specific snow load values. Snow loads in such areas must be evaluated on a case-by-case basis to determine the most appropriate value for structural calculations.

Vibration Loads. A *vibration load* is a dynamic structural load due to periodic motion. Vibration loads can produce oscillations of varying magnitudes and frequencies, but the forces are most severe at the resonant frequency. *Resonance* is the condition when a vibration frequency matches the fundamental frequency of the structure. Structures can sometimes be designed to avoid naturally occurring resonant frequencies. Special shock-absorbing structural members can also be used to dampen vibration.

Vibration loads can be caused by nearby construction or heavy equipment, and, in certain regions, seismic activity. For most installations on permanent structures outside of seismically active regions, vibration loads will be minimal. However, portable arrays mounted onto trailers or vehicles are especially prone to vibration loads. In these cases, mounting systems may require special engineering.

Attachment Methods

A variety of attachment methods may be used to install arrays and other equipment on buildings and other structures. The type of mounting system, mounting surface, and anticipated loads determine the best type of attachment method. Structural attachments made directly to a building or other foundation must be able to withstand the expected mechanical loads that work to pull the mount from its foundation. Most designs use conventional threaded fasteners and screws, while others use special clamping fasteners to hold modules in position on support structures. Some arrays require no attachments at all to a building rooftop or the ground.

> Most modules are designed to be rigid, though loads will deflect the glass and frames slightly. For each module model, module manufacturers specify the maximum allowable deflection that does not result in damage. Mounting system manufacturers use this information to recommend the amount of spacing between attachment points to avoid damage from deflection. For a given load, spacing the attachment points closer together leaves less unsupported module area, which decreases deflection.

Snow Loads

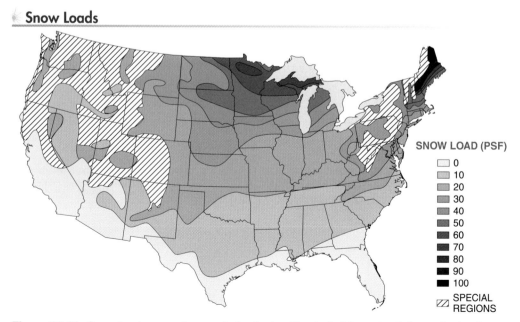

SNOW LOAD (PSF)

- 0
- 10
- 20
- 30
- 40
- 50
- 60
- 70
- 80
- 90
- 100
- SPECIAL REGIONS

Figure 10-21. *Snow loads cause forces similar to dead loads, but the potential magnitude of a snow load varies greatly among geographic regions.*

Individual modules or panels may be installed first on an intermediate support. The connection methods for intermediate supports are manufacturer-specific and those from reputable companies are well engineered for common structural loads. However, a primary connection between a bracket, plate, or structural angle and a building or foundation must be made. This connection method will often be chosen by the installer, though manufacturers may provide recommendations.

Mounting system attachments should be made through the roof cladding and into building structural members, such as rafters. Most mounting systems allow direct attachments from the exterior roof surface, though some methods or situations require access to attics or other under-roof spaces. This may occur when the locations of support brackets are limited and new structural members must be added to accommodate the attachment points. With rare exceptions, arrays should not be attached to plywood decking or other roof covering without also engaging structural members.

Lag Screws. Most mount brackets can be fastened into roof rafters directly from the exterior roof surface. Lag screws are the most common fasteners for attaching array mounts to rooftops since many roofs do not have access to the underside for other mounting methods. **See Figure 10-22.** Lag screws look like very large wood screws with a hex-shaped head. Wrenches are used to drive lag screws into predrilled pilot holes.

Each lag screw must be properly secured into the roof structure. Most rafters are only 1½″ wide, and determining the exact center of the rafters from above a roof can be difficult, particularly when rafters are not exactly parallel. The approximate locations can be found by hitting the roof surface with a hammer and listening for a change in sound. The centers of the rafters can usually then be located with a quality stud finder. If a lag screw continues to turn after seating a bracket, it is likely not embedded into the structural member and only sticking through the plywood decking. In this case, the hole must be sealed and the attachment point redone.

Lag screws are generally rated by their allowable withdrawal load. The *allowable withdrawal load* is the force required to remove a screw from a material by tensile (pulling) force only. Since the required force depends on how deeply the screw is embedded, the allowable withdrawal load is represented in pounds of force per inch of penetration depth. It also increases with the diameter of the screw and the density of the material. **See Figure 10-23.** The penetration depth is the length of threaded portion of fastener embedded in a structural member. Common lag screw sizes are ¼″, ⁵⁄₁₆″, and ⅜″ in diameter and 3″ to 5″ long. Pilot holes are typically 60% to 75% of the screw nominal shank diameter. Larger pilot holes are required for harder woods like common truss lumber, than for softer woods like framing lumber. Tables provide allowable withdrawal loads for various woods and screw sizes.

Lag Screws

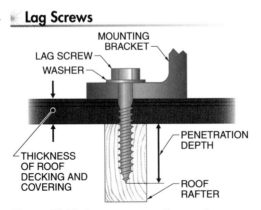

Figure 10-22. *Lag screws are the most common type of fastener used to attach array mounting systems to wood structures, usually residential roofs.*

Allowable Withdrawal Loads*

LAG SCREW DIAMETER†	WOOD TYPE		
	Southern Yellow Pine	White Spruce	Douglas Fir
¼	281	192	167
⁵⁄₁₆	332	227	198
⅜	381	260	226
⁷⁄₁₆	428	292	254
½	473	323	281

* in lb/in.
† in in.

Figure 10-23. *Allowable withdrawal loads for lag screws are greater with larger screw diameter, deeper thread penetration, and higher-density lumber.*

Calculating Allowable Withdrawal Load

For wood types not listed in tables, the allowable withdrawal load can be calculated from the specific gravity of the wood and the nominal screw diameter. Specific gravity is similar to density and relates to the wood's strength and resistance. The specific gravities of various types of woods can be found in engineering reference tables. Allowable withdrawal load can be calculated with the following formula:

$$p = 1800\,\gamma^{3/2}D^{3/4}$$

where

p = allowable withdrawal load (in lb/in.)

γ = specific gravity of wood

D = nominal screw diameter (in in.)

When a lag screw is used to mount a bracket to a roof, it must pass through the combined thickness of the bracket, shingles or other roof covering, roof membrane, and decking, which can add up to more than 1″. This means that roughly 1″ of the length of the screw is not embedded in the rafter. The thickness of these materials must be accounted for in the length of the screw and the thread penetration depth.

Bolts. Bolts are often used to secure modules to mounts, and mounting system manufacturers specify (or sometimes include) the best sizes and types of bolts for use with their products. Bolts or threaded rods can also be used to attach the mounts to a building or structure. The sizes and types of fasteners for this part of the installation may be left for the installer to choose.

Common bolts have a hex head and are secured by a hex nut. Threaded rods are used when bolts are not long enough. Threaded rods are cut to the required length from long pieces, and fastened at both ends. Flat washers spread the load from a hex head or nut over a larger area, and lock washers or special lock nuts keep the fastener from loosening. Antiseize compounds are highly recommended for stainless steel lock nuts to prevent galling and cross-threading, and can facilitate servicing and disassembly as required. Bolts and threaded rods are classified by the mechanical loads they can handle. Typical bolt diameters include ¼″, ⁵⁄₁₆″, and ⅜″.

Attaching mounts to a roof with bolts produces a stronger attachment than with lag screws, but requires access to spaces under the roof. Bolts cannot be fastened into roof trusses or rafters like lag screws, but also should not be fastened to the roof decking alone. Blocking and spanning are techniques used to strengthen bolt attachments.

Blocking is the addition of lumber under a roof surface and between trusses or rafters as supplemental structural support. **See Figure 10-24.** Boards for blocking are short pieces of 2 × 4 or 2 × 6 lumber that are nailed or screwed into the rafters with at least two fasteners on each side. By stacking two blocking boards, lag screws or bolts can be supported when required attachment points do not match rafter locations.

Spanning is the addition of lumber under a roof surface and across trusses or rafters as supplemental structural support. **See Figure 10-25.** Spanners are only used for bolt-type fasteners, usually long threaded rods, and require intermediate blocking between the spanner and underside of the roof deck to keep the spanner from bowing under applied loads.

Hand tools are recommended for tightening fasteners since power tools can strip threads. Mounting system manufacturers may also recommend tightening fasteners to certain torque specifications.

If the mounting system requires future adjustment or disassembly, antiseize paste should be used on the fasteners. This paste also helps during initial assembly.

Bolts with Blocking

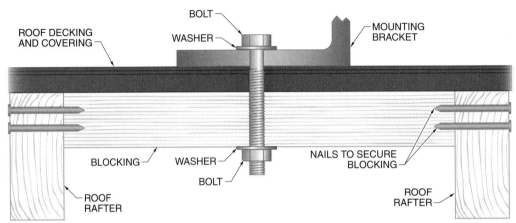

Figure 10-24. *Blocking is used to provide a structural member between roof rafters.*

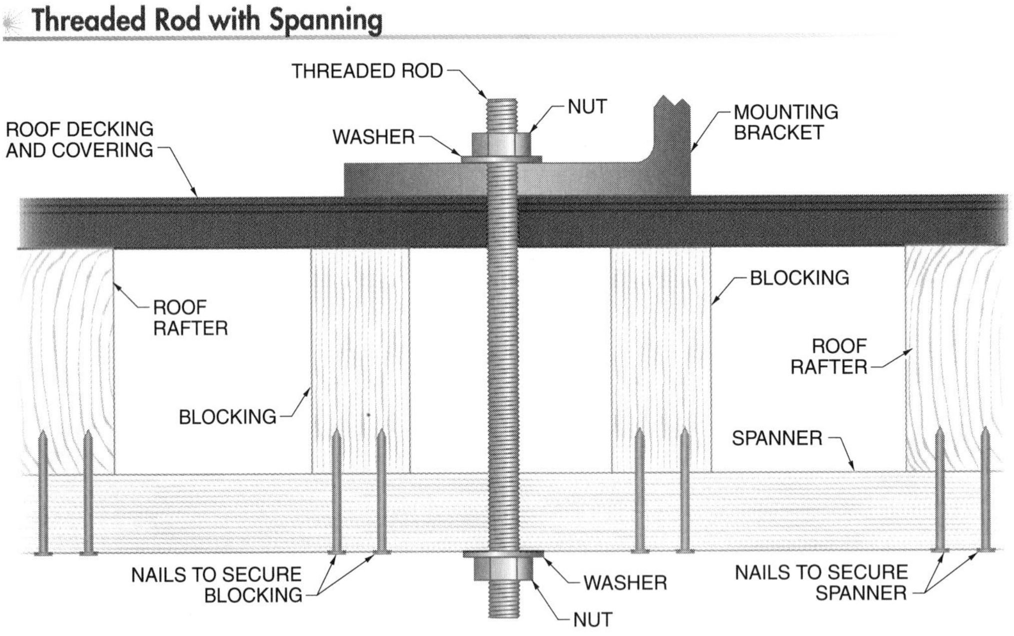

Threaded Rod with Spanning

Figure 10-25. *Spanning is used to provide a structural member across roof rafters. Blocking boards are required to support the spanner.*

J-Bolts. While not as common as lag screws and bolts, J-bolts may be used to attach mounting brackets. A *J-bolt* is a fastener that hooks around a secure support structure and has a threaded end that is used with a nut to secure items. **See Figure 10-26.** The use of J-bolts requires access to the underside of a roof, and they must be carefully sized to fit around the roof rafters or trusses. J-bolts are also less flexible in the placement of attachment points because the attachments must be precisely next to the truss or rafter.

Self-Ballasting. *Self-ballasting* is an attachment method that relies on the weight of the array, support structure, and ballasting material to hold the array in position. Self-ballasted systems do not require direct structural connections with fasteners to a building or foundation. A major advantage of self-ballasting systems is that there are no penetrations into the building surfaces, eliminating concerns about weather sealing of attachment points. They are also installed very quickly and use fewer fasteners. However, self-ballasted systems must be installed on level surfaces and may be limited to regions without high basic wind speeds.

J-Bolts

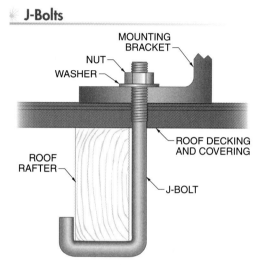

Figure 10-26. *J-bolts secure attachment points by hooking around structural members.*

Typical self-ballasted systems include containers or trays built into the bottom of the mount to hold a large amount of water, concrete, or sand. A simple yet effective mounting system consists of custom brackets that connect rows of modules and provide spaces in between for ballasting material. **See Figure 10-27.**

Ground Foundations. Ground-mounted arrays, unless they use a self-ballasted mounting system, require a foundation for support. Foundation designs vary widely, but typically require concrete footers or bases. For ground-mounted racks, they are arranged similarly to roof attachment points, with the rack being secured at the corners and at set intervals along the edges. The concrete foundations may be embedded into the ground, or may rest on the ground surface like a self-ballasted system. **See Figure 10-28.** Wooden structures may be incorporated into ground rack foundations, typically as part of the aboveground support structure.

Rack Mount Ground Foundations

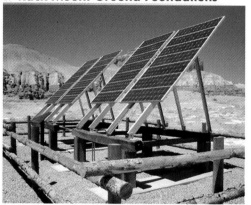

DOE/NREL, Utah Office of Energy Services

Figure 10-28. *Ground foundations for rack mounts typically include concrete footers and may use wood as part of the aboveground rack structure.*

Self-Ballasting

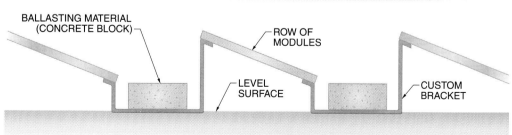

Figure 10-27. *Self-ballasting systems rely on the weight of the array, support structure, and ballasting material to secure the array without making roof penetrations.*

IBEW Local 363 Training School and Union Hall

Location: Harriman, NY (41.3°N, 74.1°W)
Type of System: Utility-interactive with custom roof mounting system
Peak Array Power: 60 kW DC
Date of Installation: September 2005
Installers: Apprentices and journeymen
Purposes: Supplemental electrical power, training, and demonstration

Members of the IBEW Local 363 chose to implement a utility-interactive PV system at their training school and union hall. In addition to augmenting their utility electrical service, the local wanted to use the installation as an opportunity to introduce their apprentices and journeymen to PV-technology systems.

Local 363 teamed up with PV industry organizations to design the mounting system and submit the grant application to the New York State Energy Research and Development Authority (NYSERDA), which provided nearly 60% of the cost of the system.

IBEW Local 363

Apprentices and journeymen build a section of the array indoors to test the installation procedure and examine ways of improving the mounting design.

IBEW Local 363

Local 363's PV system produces about one-third of the electricity used by the building and provides continuing opportunities for PV training and experience.

Roof penetration was a concern, so it was decided that the system should be self-ballasted, relying on the weight of the array itself and heavy ballasting materials to keep the modules on the roof. This type of system also simplifies the installation procedure because the attachment points consist of only a rubber roof pad, a special bracket, and the ballast. In this case, the ballast materials were concrete blocks.

The mounting system is similar to shallow rack mounts, but consists of specially designed brackets to connect rows of modules. The brackets are a custom design that evolved through trial-and-error by the local and NPCP. The apprentices and journeymen would test each design by mocking up a section of the array indoors. Accessibility between module rows and ease of assembly (and disassembly, if needed) were the primary considerations. The first two designs proved to be too labor intensive. Suggestions for improvement were implemented into successive designs until the result was a simple, easy-to-install system that provided enough space between rows to allow personnel to access all parts of the array.

The indoor installations also allowed the apprentices and journeymen to learn the procedures for laying out the brackets and installing the modules before going up on the roof, which saved time during the final roof installation.

The final design utilizes a 12° tilt angle for the modules. While this angle is low for the New York region, it helps the system maximize summer energy production and has some other advantages. The low profile is considered aesthetically pleasing, avoids shading between rows, and reduces the possibility of glare from the modules affecting neighbors.

The system produces an estimated 66,000 kWh of electricity annually, providing about one-third of the facility's electricity demand. The PV system continues to provide training and experience opportunities and increases the knowledge and interest about PV applications among the apprentices and journeymen.

Pole-mounted arrays are secured at only one location, which usually requires a deeper foundation in order to resist the twisting and bending forces from the wind and the weight of the array. The pole is usually embedded in concrete, but in some areas compacted soil alone can support and secure the mount. Another option may be to use screw piles, which are twisted into the ground like giant screws. **See Figure 10-29.**

The type of soil and its strength determine which methods can be used. Local building requirements should be consulted in order to design adequate ground foundations for local soil types.

Weather Sealing

Weather sealing of structural attachments and penetrations is a major concern for arrays mounted on buildings. One small water leak can do considerable damage to a building, both structurally and aesthetically, as well as to consumer confidence. The weather sealing of attachments and penetrations through building surfaces should use accepted roofing industry practices and materials that meet or exceed the lifetime expectations for the PV system.

The number of roof penetrations should be minimized, while still meeting structural support requirements. This saves time, reduces costs, and minimizes the potential for leaks after installation. For example, using fewer strong attachments is generally better than using a greater number of weak attachments.

There are a variety of materials and techniques for weather sealing, and methods of application vary among the types of mounting systems. Caulking or gaskets are often used to seal under and around direct attachments of mounting brackets to building surfaces. Weather sealants must remain flexible over a range of temperatures, maintain adhesion, resist degradation from long-term exposure to UV radiation, dispense easily, and cure in a reasonably short amount of time. Polyurethane, elastomeric, butyl rubber, and asphalt-based compounds are some of the more popular sealants. Alternatively, asphalt or cork gasket tape can be used for sealing. Basic latex, acrylic, or silicon caulks are generally unacceptable, due to their tendency to degrade and lose adhesion to roofing materials.

Pole Foundations

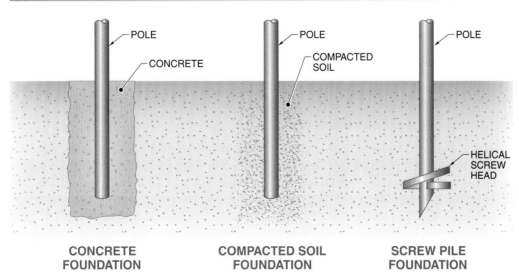

Figure 10-29. *Pole foundations may be encased in concrete or compacted soil, depending on local building requirements and the type of soil.*

Many mounting systems use simple angle brackets, plates, or other footings that attach the array to a rooftop or other building surface. For example, a bracket may be placed on top of a shingle and secured through the roof cladding to the structural members. Weather-sealing material is applied between the bracket and the roof surface, around the fastener, and in the pilot hole. **See Figure 10-30.** While this is a quick, simple, and low-cost approach, this method is not the best roofing practice and may require occasional inspection and resealing to maintain reliable, long-term protection.

Weather Sealing with Caulking

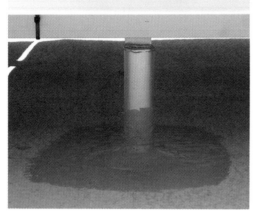

Sharp Electronics Corp.

Figure 10-30. *To weather-seal roof penetrations, caulking material is applied liberally around the entire attachment area to form a continuous seal.*

Direct contact between aluminum and concrete causes aluminum to chemically erode because of the alkaline properties of concrete. Certain weather-sealing techniques, such as the use of rubber gaskets, can be used to isolate the two materials to prevent this type of corrosion.

A better approach uses flashings and rubber boots to seal around roof penetrations. The flashings are installed underneath or in place of shingles, or on top of other roofing systems, and a rubber boot is installed around the footing to provide a weathertight seal. While this method can be used to retrofit existing roofs, it is easiest during initial or reroofing work. **See Figure 10-31.**

Weather Sealing with Flashing

Sharp Electronics Corp.

Figure 10-31. *Flashings and rubber boots provide the highest-quality weather seal for attachment penetrations.*

Electrical conduits routed through exterior building surfaces are also potential sources of water intrusion. Conductors from an array generally enter the building through the roof, eaves, or walls via conduits or junction boxes. These penetrations require weather sealing, usually with caulking, gaskets, or flashings, though in some cases an inverted entrance conduit may be used to route conductors into the building while preventing water intrusion.

Careful attention should be paid to brackets or other roof-mounted equipment that may dam water and debris above or behind it, which may degrade weather sealing and cause leaks. Small, angled flashings installed above the equipment can deflect water and debris away from a weather-sealed penetration.

Structural Analysis

After evaluating the design loads, the strength of the modules, the mounting system, the structure to which the array is mounted, and the attachment points should be considered. Each part must be able to withstand either the largest load (typically the wind load), or a combination of loads (such as the sum of the dead, live, and snow loads). The local building code will determine how loads must be evaluated.

Module manufacturers will often specify allowable loads and acceptable deflections under various support positions. Mounting system manufacturers will often recommend different configurations, attachment methods, or attachment-point spacing depending on the structural loading.

Evaluating attachment points requires a comparison of the total force at each point to the withdrawal strength of the fastener. Lag screw attachments are relatively simple to evaluate for withdrawal strength against lifting forces from wind loads. For example, consider a lifting wind load of 45 psf acting on a 200 ft^2 array that attaches to the roof with 24 support brackets. Each bracket is fastened through the 1″-thick roof surface into a southern yellow pine truss with a single 5⁄16″-diameter lag screw. The total force from the wind load on the entire array is 9000 lb (200 ft^2 × 45 psf = 9000 lb). Therefore, each attachment point must resist 375 lb of force (9000 lb ÷ 24 attachment points = 375 lb per attachment point).

Since 5⁄16″ lag screws have withdrawal resistance of 332 lb/in. in southern yellow pine, the thread penetration depth must be at least 1.13″ (375 lb ÷ 332 lb/in. = 1.13″). Since the roofing thickness is about 1″, the lag screw must be at least 2.13″ long, 2¼″ being the nearest common size.

Refer to Quick Quiz® on CD-ROM

Summary

- Existing structures should be evaluated for structural soundness before installing an array.

- Good access to the array mounting location makes installation safer, easier, faster, and less expensive.

- Only passive cooling means are practical for flat-plate modules, such as mounting in a way to allow air circulation around the modules.

- Arrays mounted directly on building surfaces can increase heat transfer into conditioned spaces.

- Arrays mounted above building surfaces have little effect on building heat gain and may even help reduce interior building temperatures.

- Architectural principles that can improve appearance and public perception can be applied to PV system design and installation without significantly affecting cost or performance.

- Array support structures, hardware, electrical wiring, conduit, and junction boxes should be as inconspicuous as possible and neatly gathered or concealed beneath the array.

- Reducing installation time through careful preparation can significantly reduce costs.

- The simplest and most common type of array mount is the fixed-tilt type.

- Rack mounts are simple to install and flexible in the types and sizes of modules they can accommodate.

- Standoff-mounted arrays are by far the most common, preferred, and least-expensive method for installing arrays as a retrofit to existing rooftops, as well as for new construction.

- BIPV arrays can be integrated into roofing, windows, skylights, curtain wall sections, or complementary architectural features such as awnings, facades, or entranceways.

- Ground-mounted arrays are generally more flexible in location and placement, operate at lower temperatures than most building mounts, and are more accessible.

- Sun-tracking arrays can follow the sun, enhancing the amount of energy collected, but are also complex and require more expense and installation time.

- Wood construction is generally avoided in PV system installations.

- Aluminum, stainless steel, and galvanized steel parts and fasteners are commonly used in array designs because they are corrosion resistant, lightweight, and relatively inexpensive.

- Arrays, modules, mounting systems, fasteners, and buildings must withstand the maximum forces expected from structural loads.

- The principal types of structural loads are dead loads, live loads, wind loads, snow loads, and vibration loads.

- Most attachment methods use conventional threaded fasteners and screws.

- Blocking and spanning techniques can be used to support bolted attachment points.

- Self-ballasted systems do not require direct structural connections to a building or foundation.

- Poles are usually embedded in concrete or compacted soil or are twisted into the ground like giant screws.

- Weather sealing of structural attachments and roof penetrations is a major concern for building-mounted arrays.

- The number of roof penetrations should be minimized.

Definitions

- The *installed nominal operating cell temperature (INOCT)* is the estimated temperature of a module operating in a specific mounting system design.

- A *fixed-tilt mounting system* is an array mounting system that permanently secures modules in a nonmovable position at a specific tilt angle.

- An *adjustable mounting system* is a variation of a fixed-tilt array mounting system that permits manual adjustment of the tilt and/or azimuth angles to increase the array output.

- A *direct mount* is a type of fixed-tilt array mounting system where modules are affixed directly to an existing finished rooftop or other building surface, with little or no space between a module and the surface.

- A *rack mount* is a type of fixed- or adjustable-tilt array mounting system with a triangular-shaped structure to increase the tilt angle of the array.

- A *standoff mount* is a type of fixed-tilt array mounting system where modules are supported by a structure parallel to and slightly above the roof surface.

- A *building-integrated photovoltaic (BIPV) array* is a fixed array that replaces conventional building materials with specially designed modules that perform an architectural function in addition to producing power.

- A *pole mount* is a type of array mounting system where modules are installed at an elevation on a pedestal.

- A *sun-tracking mount* is an array mounting system that automatically orients the array to the position of the sun.

- An *active tracking mount* is an array mounting system that uses electric motors and gear drives to automatically direct the array toward the sun.

- A *passive tracking mount* is an array mounting system that uses nonelectrical means to automatically direct an array toward the sun.

- *Galvanic corrosion* is an electrochemical process that causes electrical current to flow between two dissimilar metals, which eventually corrodes one of the materials (the anode).

- A *sacrificial anode* is a metal part, usually zinc or magnesium, that is more susceptible to galvanic corrosion than the metal structure it is attached to, so that it corrodes, rather than the structure.

- A *design load* is a calculated structural load used to evaluate the strength of a structure to failure.

- A *dead load* is a static structural load due to the weight of permanent building members, supported structure, and attachments.

- A *live load* is a dynamic structural load due to the weight of temporary items and people using or occupying the structure.

- A *wind load* is a dynamic structural load due to wind, resulting in downward, lateral, or lifting forces.

- The *basic wind speed* is the maximum value of a 3 sec gust at 33′ (10 m) elevation, which is used in wind load calculations.

- *Exposure* is a wind load factor that accounts for the array height and the characteristics of the surrounding terrain.

- A *snow load* is a static structural load due to the weight of accumulated snow.

- A *vibration load* is a dynamic structural load due to periodic motion.

- *Resonance* is the condition when a vibration frequency matches the fundamental frequency of the structure.

- The *allowable withdrawal load* is the force required to remove a screw from a material by tensile (pulling) force only.

- *Blocking* is the addition of lumber under a roof surface and between trusses or rafters as supplemental structural support.

- *Spanning* is the addition of lumber under a roof surface and across trusses or rafters as supplemental structural support.

- A *J-bolt* is a fastener that hooks around a secure support structure and has a threaded end that is used with a nut to secure items.

- *Self-ballasting* is an attachment method that relies on the weight of the array, support structure, and ballasting material to hold the array in position.

Review Questions

1. How can mechanical integration strategies help keep arrays cool?

2. How can direct-mounted and standoff-mounted arrays affect the temperatures inside a building?

3. What are some of the aesthetic considerations of mechanical integration?

4. How does the distance between modules and a building surface affect the module's installed nominal operating cell temperature (INOCT) and temperature-rise coefficient?

5. Explain how building-integrated PV (BIPV) arrays offset some building-material costs.

6. How do sun-tracking systems improve the power output of a PV system?

7. What factors influence the added value of a sun-tracking mounting system?

8. Why do the aluminum materials common in module frames present a corrosion problem when installed on steel structures?

9. Explain the difference between static and dynamic structural loads and classify the principal types of structural loads as static or dynamic.

10. Explain the three primary factors that influence wind loads for most PV applications.

11. How can lag-screw attachment points be supported when they do not match rafter locations?

12. Describe the advantages and disadvantages of self-ballasted systems.

11

Electrical Integration

PV systems are subject to all of the same general requirements as most electrical systems, such as overcurrent protection and grounding. However, PV systems are also subject to additional requirements that are specific to PV systems. PV installers must be familiar with all installation codes associated with electrical systems and PV systems in their jurisdiction. These codes, primarily the National Electrical Code®, detail the sizing, specifications, and installation of the electrical balance-of-system (BOS) components required to complete a safe, reliable, and easily maintained PV electrical system.

Chapter Objectives

- Identify the electrical codes, regulations, and practices applicable to PV systems.
- Calculate the voltage and current limits for various circuits of a PV system.
- Determine appropriate conductor ampacities and overcurrent protection ratings for various circuits.
- Identify the appropriate types of conductors for PV system circuits based on application and environment.
- Describe the required types of disconnects and their locations.
- Identify acceptable PV system grounding methods.
- Describe the functions and requirements of electrical balance-of-system (BOS) components.

NATIONAL ELECTRICAL CODE®

Electrical integration is arguably the most important, and misunderstood, part of installing a PV system. Since there are many different ways to configure a PV system, the electrical integration requirements can be very different between systems. These requirements involve calculating circuit parameters, such as voltage and current, and using this information to size, specify, and locate various components. The majority of the regulations governing electrical installations, including PV systems, are found in NFPA 70: National Electrical Code® (NEC®).

The National Electrical Code® is a nationally recognized standard on safe electrical installation practice and is used as the governing electrical code in most jurisdictions in the United States. The primary intent of the NEC® is to safeguard persons and property from electrical hazards. It is not intended as an instruction manual for untrained or unqualified persons.

> Most PV electrical installations must be completed by a licensed electrician or contractor and inspected by the authority having jurisdiction (AHJ) for code compliance. Code compliance can be ensured through the plans review, permitting, inspection, and approval process of electrical installations.

Electrical integration includes the installation of all wiring, overcurrent protection, disconnects, grounding, and other electrical equipment in a PV system.

Nearly all types of electrical installations are covered by the NEC®, including PV systems installed on public and private premises, commercial and residential buildings, as well as mobile homes and RVs. It also applies to the installation of conductors and equipment that connect to an electricity supply, such as a PV system or any other distributed power system.

Unfortunately, the evolving nature of PV systems is such that it sometimes fosters improperly installed systems. Because PV systems are still relatively uncommon, prospective PV system operators may not fully understand all that is involved with installing a PV system, believing that it is a simple do-it-yourself project or that there is no danger in working with DC power. It may also be difficult to source DC-rated components that are adequately tested and listed, leading some installers to use inappropriate or unsafe equipment.

Improper electrical integration can have significant consequences, including poor performance, component damage, and electrical shock hazards. PV installers must understand all applicable electrical codes, regulations, and recommendations, use only listed and appropriate equipment, follow safe working and installation practices, and educate their customers and others about the importance of code-compliant electrical integration.

Article 690

Article 690, "Solar Photovoltaic Systems," appeared for the first time in the 1984 NEC®. Since then, major revisions and additions have been made to this article based on industry input and developments. Article 690 addresses requirements for all PV installations covered under the scope of the NEC®. Article 690 is divided into nine parts: General, Circuit Requirements, Disconnecting Means, Wiring Methods, Grounding, Marking, Connection to Other Sources, Storage Batteries, and Systems Over 600 Volts. Each covers a certain aspect of PV systems.

Many other articles of the NEC® are referenced in Article 690 or otherwise apply to

PV installations. Applicable articles depend on the type and configuration of the system. **See Figure 11-1.** Whenever the requirements of Article 690 and other articles differ, the requirements of Article 690 apply.

Selected Applicable NEC® Articles

110*	Requirements for Electrical Installations
200	Use and Identification of Grounded Conductors
210*	Branch Circuits
220	Branch-Circuit, Feeder, and Service Calculations
230*	Services
240*	Overcurrent Protection
250*	Grounding and Bonding
280	Surge Arrestors, Over 1 kV
285	Surge-Protective Devices (SPDs), 1 kV or Less
300	Wiring Methods
310*	Conductors for General Wiring
334	Nonmetallic-Sheathed Cable: Types NM, NMC, and NMS
338	Service-Entrance Cable: Types SE and USE
400*	Flexible Cords and Cables
422	Appliances
445	Generators
450*	Transformers and Transformer Vaults
480*	Storage Batteries
490*	Equipment, Over 600 Volts, Nominal
690	**Solar Photovoltaic Systems**
702	Optional Standby Systems
705*	Interconnected Electric Power Production Sources
720	Circuits and Equipment Operating at Less Than 50 Volts

* Articles directly referenced in Article 690

Figure 11-1. *Many articles in the NEC® are applicable to the electrical integration of a PV system, particularly Article 690.*

The beginning of Article 690 defines specific terminology for PV system components and circuits. Most are easily distinguishable, though a few may require special attention when referencing the code. For example, the NEC® differentiates between a PV power source and a PV source circuit. **See Figure 11-2.**

A *PV power source* is an array or collection of arrays that generates DC power. A *PV source circuit* is the circuit connecting a group of modules together and to the common connection point of the DC system. PV source circuits are usually (but not always) a string of series-connected modules. For small systems, the PV power source may be only one source circuit. For larger systems, the PV power source is usually composed of several paralleled PV source circuits. A *PV output circuit* is the circuit connecting the PV power source to the rest of the system. Distinctions such as these are very important for understanding code requirements.

Other circuits defined in Article 690 are named logically as an input or output of a system component. However, this can cause confusion with some configurations. For example, in an interactive system, the PV output circuit and the inverter input circuit are essentially the same circuit. Since a disconnect is required in this circuit, some may adopt the convention of naming the array side of the disconnect as the PV output circuit and the inverter side as the inverter input circuit.

VOLTAGE AND CURRENT REQUIREMENTS

Article 690, Part II, addresses circuit requirements for PV systems, such as maximum system voltages and currents. These values must be determined because they affect the sizing and ratings for conductors, overcurrent protection devices, disconnects, and grounding equipment.

Maximum PV Circuit Voltage

According to Section 690.7, the maximum DC voltage of a PV source circuit or output circuit must be less than the maximum voltage limits of all components on the DC side of the system, including the modules, inverter, charge controller, disconnects, and conductors. Also, for one- and two-family homes, the maximum output-circuit voltage cannot be greater than 600 V. Systems over 600 V are allowed for commercial or utility-scale PV systems, but must follow different code regulations, particularly those in Article 490, "Equipment, Over 600 V, Nominal."

Electrical Integration

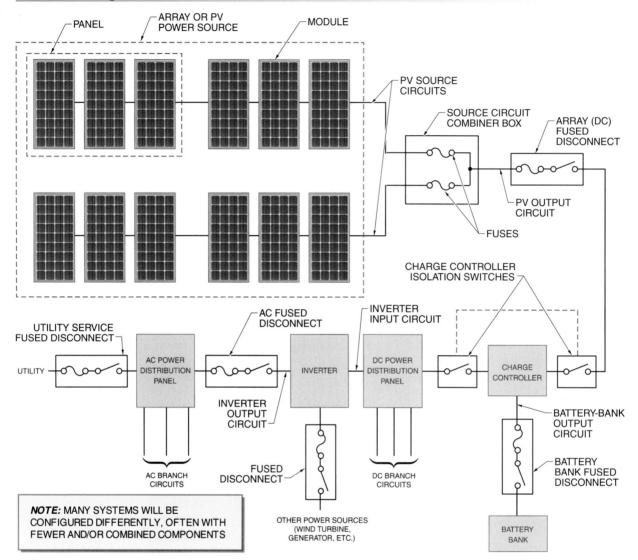

Figure 11-2. *The NEC® defines the various circuits and components in PV systems and specifies their requirements.*

Since temperature affects the voltage output of PV devices, the maximum possible voltage is the array's open-circuit voltage at the lowest expected temperature for the location. Low temperatures for certain locations are included in the solar radiation data sets and other sources. The voltage is then calculated using the module manufacturer's temperature coefficient for voltage.

If temperature coefficients are not provided by the manufacturer, the NEC® provides a temperature correction factor calculation for estimating the maximum voltage.

See Figure 11-3. The maximum PV source-circuit or output-circuit voltage is estimated with the following formula:

$$V_{max} = V_{oc} \times n_m \times C_T$$

where

V_{max} = maximum PV circuit voltage (in V)

V_{oc} = module rated open-circuit voltage at 25°C (in V)

n_m = number of series-connected modules

C_T = low-temperature voltage correction factor

Voltage Correction Factors for Low Temperatures

AMBIENT TEMPERATURE*	VOLTAGE CORRECTION FACTOR
24 to 10	1.02
19 to 15	1.04
14 to 10	1.06
9 to 5	1.08
4 to 0	1.10
−1 to −5	1.12
−6 to −10	1.14
−11 to −15	1.16
−16 to −20	1.18
−21 to −25	1.20
−26 to −30	1.21
−31 to −35	1.23
−36 to −40	1.25

* in °C

NEC® Table 690.7. Reprinted with permission from NFPA 70-2008, the National Electrical Code® Copyright® 2007, National Fire Protection Association, Quincy, MA 02169. This reprinted material is not the official position of the NFPA on the referenced subject which is represented solely by the standard in its entirety.

Figure 11-3. *Array open-circuit voltage is corrected for low temperatures to yield the maximum possible PV circuit voltage.*

For example, consider an array of 22 series-connected modules, each with an open-circuit voltage of 21.7 V. The lowest expected temperature of the array is −25°C, which corresponds to a low-temperature correction factor of 1.20. What is the maximum PV circuit voltage?

$$V_{max} = V_{oc} \times n_m \times C_T$$
$$V_{max} = 21.7 \times 22 \times 1.20$$
$$V_{max} = \textbf{573 V}$$

Since the maximum possible voltage is less than 600 V, this output-circuit voltage is acceptable, provided that it does not exceed the rating of any connected component.

Maximum PV Circuit Currents

The NEC®, Section 690.8, distinguishes between the maximum DC currents of the PV source circuits and the PV output circuit. Each has specific requirements that affect the sizing of conductors and components in the circuit separately.

For PV source circuits, the maximum current is 125% of the sum of the short-circuit current ratings of parallel-connected modules. This 125% factor accounts for the fact that PV devices can deliver currents higher than their rated short-circuit current under enhanced irradiance. Since a source circuit usually consists of a single series string of modules, the maximum current is simply 125% of the module's short-circuit current.

For example, a source circuit consisting of one string of series-connected modules, each with a short-circuit current of 4.8 A, has a maximum current of 6 A, regardless of the number of modules in the series string. However, if a source circuit includes multiple strings in parallel, the maximum current is 125% of the module short-circuit current multiplied by the number of parallel strings.

For PV output circuits, the maximum current is the sum of the maximum currents of the parallel-connected source circuits. For example, a PV output circuit combining three parallel strings of modules, each with a maximum source circuit current of 6 A, has a maximum PV output circuit current of 18 A (3×6 A = 18 A).

These maximum current numbers are involved in further calculations when determining the appropriate conductor size and required overcurrent protection ratings. This applies to PV source circuits and the PV output circuit.

System voltage and current calculations, along with the application and environment, affect the wiring methods and conductors chosen for PV systems.

Maximum Inverter Input Current

For an interactive inverter with the PV output circuit connected directly to the inverter, the inverter input circuit is the same as the PV output circuit and, therefore, has the same maximum current.

For stand-alone systems with batteries, the inverter input current depends on battery voltage. As battery voltage decreases, the inverter input current increases to provide the same power input. At low battery voltages and peak power output, this current can be considerably higher than the inverter input current rating at nominal battery voltages. Thus, the highest possible input current is associated with the lowest inverter operating voltage. Maximum inverter input current is calculated with the following formula:

$$I_{max} = \frac{P_{AC}}{V_{min} \times \eta_{inv}}$$

where

I_{max} = maximum inverter input current (in A)

P_{AC} = rated inverter maximum AC power output (in W)

V_{min} = minimum inverter operating voltage (in V)

η_{inv} = inverter efficiency

For example, consider a 5500 W stand-alone inverter that is 85% efficient and can operate at input voltages down to 44 V. What is the maximum inverter input current?

$$I_{max} = \frac{P_{AC}}{V_{min} \times \eta_{inv}}$$

$$I_{max} = \frac{5500}{44 \times 0.85}$$

$$I_{max} = \frac{5500}{37.4}$$

$$I_{max} = \textbf{147 A}$$

Maximum Inverter Output Current

Since inverters are limited-power devices, their AC output circuits are sized based on the maximum inverter output rather than load calculations. The maximum current for the inverter output circuit is equal to the inverter's continuous output current rating. This information is provided on inverter labeling and specifications. For example, a 2500 W inverter outputting 240 VAC will have a listed continuous output rating of approximately 10.4 A (2500 W ÷ 240 VAC = 10.4 A).

CONDUCTORS AND WIRING METHODS

Any conductors and wiring methods allowed by NEC® Chapter 2, "Wiring and Protection," and Chapter 3, "Wiring Methods and Materials," may be used with PV systems, in addition to any wiring equipment specifically identified for use with PV systems. Part IV of Article 690 further covers the applications and conditions for conductors used in PV systems, and special equipment and practices allowed.

Generally, only copper conductors of various types and sizes are used in PV systems. They may be exposed to a variety of conditions, including extreme temperatures, mechanical stress, UV exposure, and moisture. These factors determine the type of conductor permitted by the NEC® for the application, while the maximum circuit current dictates the appropriate conductor size. For any application, the conditions of use must be well understood to ensure that proper types and sizes of conductors are specified and installed.

Conductor Size

Conductor sizes used in most electrical systems are expressed in American Wire Gauge (AWG) numbers. **See Figure 11-4.** Larger diameter conductors have smaller AWG numbers. Larger conductors have greater current-carrying capacity and less resistance. However, solid (single wire) conductors can be stiff and difficult to work with. Therefore, conductors are also available stranded (made up of multiple smaller wires), which makes them more flexible. Solid and stranded conductors of the same AWG size have the same cross-sectional area, though stranding makes the diameter slighter larger. At size 6 AWG and larger, conductors are generally only available in stranded versions.

Conductor Sizes*

AWG	SOLID CONDUCTOR			STRANDED CONDUCTOR			
	Diameter†	Area‡	Resistance**	Strands	Diameter†	Area‡	Resistance**
18	0.040	1.620	7.77	7	0.046	1.620	7.95
16	0.051	2.580	4.89	7	0.058	2.580	4.99
14	0.064	4.110	3.07	7	0.073	4.110	3.14
12	0.081	6.530	1.93	7	0.092	6.530	1.98
10	0.102	10.38	1.21	7	0.116	10.38	1.24
8	0.128	16.51	0.764	7	0.146	16.51	0.778
6	—	—	—	7	0.184	26.24	0.491
4	—	—	—	7	0.232	41.74	0.308
3	—	—	—	7	0.260	52.62	0.245
2	—	—	—	7	0.292	66.36	0.194
1	—	—	—	19	0.332	83.69	0.154
0 (1/0)	—	—	—	19	0.372	105.6	0.122
00 (2/0)	—	—	—	19	0.418	133.1	0.0967
000 (3/0)	—	—	—	19	0.470	167.8	0.0766
0000 (4/0)	—	—	—	19	0.528	211.6	0.0608

* conductor alone, not accounting for insulation
† in in.
‡ in kcmil
** in Ω/kft at 75°C (167°F)

Excerpted from NEC® Chapter 9, Table 8. Reprinted with permission from NFPA 70-2008, the National Electrical Code® Copyright© 2007, National Fire Protection Association, Quincy, MA 02169. This reprinted material is not the official position of the NFPA on the referenced subject which is represented solely by the standard in its entirety.

Figure 11-4. *Conductor sizes typically used in PV systems range from 18 AWG to 4/0 AWG. Conductors may be solid or stranded. Larger conductors have lower resistance for a given length.*

Conductor size is chosen to be the smallest size that safely conducts the maximum continuous current of a circuit with the appropriate margin of safety. For PV source circuits, conductors must be sized to handle at least 125% of the maximum current. *Note:* This 125% is in addition to the 125% used to estimate the maximum current under high irradiance. For example, if a string of series-connected modules has a short-circuit current of 4.8 A, the maximum current is 6 A (4.8 A × 125% = 6 A) and the minimum ampacity for conductor sizing is 7.5 A (6 A × 125% = 7.5 A).

Conductor Ampacity. Conductor sizing is based on a conductor's ampacity. *Ampacity* is the current that a conductor can carry continuously under the conditions of use without exceeding its temperature rating. Nominal conductor ampacity at 30°C is determined by the conductor material (copper or aluminum), size, insulation type, and application (direct burial, conduit, or free air). **See Figure 11-5.**

Since temperature affects a conductor's ampacity, this nominal ampacity is derated (reduced) for ambient application temperatures higher than the nominal 30°C. The derating of conductor ampacities is covered in NEC® Section 310.15.

Large conductors are nearly always stranded while small conductors may be solid or stranded.

Nominal Ampacities of Insulated Copper Conductors*

	TYPE OF INSULATION	TW, UF	RHW, THHW, THW, THWN, XHHW, USE, ZW	TBS, SA, SIS, FEP, FEPB, MI, RHH, RHW-2, THHN, THHW, THW-2, THWN-2, USE-2, XHH, XHHW-2, ZW-2
	AWG	60°C Rated	75°C Rated	90°C Rated
CONDUCTORS IN A RACEWAY, CABLE, CONDUIT, OR EARTH (DIRECTLY BURIED)	18	—	—	14
	16	—	—	18
	14	20	20	25
	12	25	25	30
	10	30	35	40
	8	40	50	55
	6	55	65	75
	4	70	85	95
	3	85	100	110
	2	95	115	130
	1	110	130	150
	0 (1/0)	125	150	170
	00 (2/0)	145	175	195
	000 (3/0)	165	200	225
	0000 (4/0)	195	230	260
CONDUCTOR IN FREE AIR	18	—	—	18
	16	—	—	24
	14	25	30	35
	12	30	35	40
	10	40	50	55
	8	60	70	80
	6	80	95	105
	4	105	125	140
	3	120	145	165
	2	140	170	190
	1	165	195	220
	0 (1/0)	195	230	260
	0 (2/0)	225	265	300
	000 (3/0)	260	310	350
	0000 (4/0)	300	360	405

* Based on ambient temperature of 30°C (86°F) and not more than three current-carrying conductors when in a raceway, cable, or earth (directly buried). Excerpted from NEC® Table 310.16 and Table 310.17. Reprinted with permission from NFPA 70-2008, the National Electrical Code® Copyright© 2007, National Fire Protection Association, Quincy, MA 02169. This reprinted material is not the official position of the NFPA on the referenced subject which is represented solely by the standard in its entirety.

Figure 11-5. *Ampacity is the current-carrying capacity of a conductor, which depends on the conductor's type, size, and application.*

Ambient-temperature-based ampacity correction factors are given in a table. **See Figure 11-6.** The correction factor is multiplied by the nominal ampacity to calculate the derated ampacity. Therefore, for a certain current-carrying capacity rating, the size of the required conductor must be increased to account for deratings.

Since conduits exposed to direct sunlight on or above rooftops experience higher temperatures, an extra value is added to the ambient temperature before determining the temperature-based correction factor. **See Figure 11-7.** This adder increases the effective ambient temperature by an amount dictated by the distance of the conduit from the rooftop.

For example, 10 AWG USE-2 conductors in a conduit have a 90°C rating and nominal ampacity of 40 A. However, this nominal ampacity is for 30°C. Since the conductors will experience higher temperatures, their ampacity must be derated. The highest expected ambient temperature for the location is 43°C. Since the conduit will be installed about 1″ above the roof surface, an additional 22°C is added to the ambient temperature. The correction factor for an adjusted ambient temperature of 65°C (43°C + 22°C = 65°C) is 0.58. Therefore, their ampacity is reduced to 23.2 A (40 A × 0.58 = 23.2 A) for operating under those temperatures.

When more than three current-carrying conductors are installed together in a conduit or raceway longer than 24′, conductor ampacities must be further derated. This situation can occur in PV systems with arrays having multiple source circuits. Because bundling several current-carrying conductors together affects their ability to dissipate heat, an additional correction factor is applied to the temperature-corrected ampacity. **See Figure 11-8.** For example, if USE-2 conductors are used for three source circuits run through a conduit, each with positive and negative conductors, the total is six current-carrying conductors. The correction factor is 0.80 and the ampacity of each conductor is further reduced to 18.6 A (23.2 A × 0.80 = 18.6 A).

More often, however, these calculations are done together and in reverse. In such a case, a minimum conductor size must be determined from a maximum circuit current, taking into account both types of ampacity deratings. The resulting nominal ampacity is calculated with the following formula:

$$I_{nom} = \frac{I_{max}}{CF_{temp} \times CF_{conduit}}$$

where

I_{nom} = conductor nominal ampacity (in A)

I_{max} = maximum circuit current (in A)

CF_{temp} = correction factor for temperature

$CF_{conduit}$ = correction factor for number of current-carrying conductors in a conduit or cable

Ampacity Correction Factors for High Temperatures

AMBIENT TEMPERATURE*	CONDUCTOR TEMPERATURE RATING		
	60°C Rated	75°C Rated	90°C Rated
21 to 25	1.08	1.05	1.04
26 to 30	1.00	1.00	1.00
31 to 35	0.91	0.94	0.96
36 to 40	0.82	0.88	0.91
41 to 45	0.71	0.82	0.87
46 to 50	0.58	0.75	0.82
51 to 55	0.41	0.67	0.76
56 to 60	—	0.58	0.71
61 to 70	—	0.33	0.58
71 to 80	—	—	0.41

* in °C

Excerpted from NEC® Table 310.16. Reprinted with permission from NFPA 70-2008, the National Electrical Code® Copyright© 2007, National Fire Protection Association, Quincy, MA 02169. This reprinted material is not the official position of the NFPA on the referenced subject which is represented solely by the standard in its entirety.

Figure 11-6. *Conductor ampacity must be derated for high temperatures.*

Ambient Temperature Adjustments for Conduits Exposed to Sunlight On or Above Rooftops

DISTANCE FROM ROOF TO BOTTOM OF CONDUIT*	TEMPERATURE ADDER†
0 to ½	33
½ to 3½	22
3½ to 12	17
12 to 36	14

* in in.
† in °C

NEC® Table 310.15(B)(2)(c). Reprinted with permission from NFPA 70-2008, the National Electrical Code® Copyright© 2007, National Fire Protection Association, Quincy, MA 02169. This reprinted material is not the official position of the NFPA on the referenced subject which is represented solely by the standard in its entirety.

Figure 11-7. *For conduits installed on rooftops, an extra temperature adder is needed to account for the extreme ambient temperatures of the environment. The adjusted ambient temperature is then used to determine the temperature-based ampacity correction factor.*

Ampacity Correction Factors for Number of Conductors

NUMBER OF CURRENT-CARRYING CONDUCTORS	CORRECTION FACTOR
4 to 6	0.80
7 to 9	0.70
10 to 20	0.50
21 to 30	0.45
31 to 40	0.40
Over 40	0.35

NEC® Table 310.15(B)(2)(a). Reprinted with permission from NFPA 70-2008, the National Electrical Code® Copyright© 2007, National Fire Protection Association, Quincy, MA 02169. This reprinted material is not the official position of the NFPA on the referenced subject which is represented solely by the standard in its entirety.

Figure 11-8. *Conductor ampacity must be derated for more than three current-carrying conductors together in a conduit or cable.*

For example, four source-circuit conductors must each carry 15 A of current at 40°C ambient. The conductors are installed together in a conduit that is secured 4″ above a rooftop, which adds another 17°C to the ambient temperature. If an XHHW-2 conductor is used, the temperature correction factor for 57°C is 0.71. Since there are four current-carrying conductors, the multiple-conductor correction factor is 0.80. What is the minimum nominal ampacity, and therefore minimum size, for an XHHW-2 conductor in this situation?

$$I_{nom} = \frac{I_{max}}{CF_{temp} \times CF_{conduit}}$$

$$I_{nom} = \frac{15}{0.71 \times 0.80}$$

$$I_{nom} = \frac{15}{0.568}$$

$$I_{nom} = \mathbf{26.4\,A}$$

Therefore, the nominal conductor ampacity must be at least 26.4 A in order to have sufficient ampacity after derating for the installation conditions. From the table of nominal ampacities, this means that the 12 AWG conductor is of the appropriate minimum size.

These circumstances illustrate the significant effect temperature has on conductor ampacity. Over 40% of this conductor's nominal ampacity cannot be utilized in this application.

Voltage Drop. Voltage drop can be an issue in any electrical system, but it is particularly important to address in PV systems. The PV output circuit may have a relatively low voltage and the conductor runs can be long, increasing voltage drop. Excessive voltage drop can also affect charge controllers, batteries, inverters, loads, and other devices that require certain voltages to operate properly. Because voltage drop is not considered a safety concern, the NEC® does not establish specific requirements. However, it does recommend a maximum voltage drop of 3%.

Voltage drop and the associated percentage of voltage drop of conductors are calculated with the following formulas:

$$V_{drop} = I_{op} \times R_C \times L$$

$$V_{drop\%} = \frac{V_{drop}}{V_{op}}$$

where

V_{drop} = voltage drop (in V)

I_{op} = operating current (in A)

R_C = conductor resistance (in Ω/kft)

L = total conductor length (in kft)

$V_{drop\%}$ = voltage drop

V_{op} = operating voltage (in V)

Conductor resistance is determined by the conductor material, size, and ambient temperature. Also, the total conductor length is the total (round-trip) distance that current travels in a circuit. Therefore, the total length used in calculations is twice the length of the conductor run.

For example, consider the voltage drop in an 80 V output circuit from the array to a disconnect, a distance of 100′. The circuit operating current is 15 A and uses 10 AWG stranded copper conductors. The DC resistance of this conductor is 1.24 Ω/kft and the total length is 200′ (0.2 kft). What is the voltage drop in this circuit?

$$V_{drop} = I_{op} \times R_C \times L$$

$$V_{drop} = 15 \times 1.24 \times 0.2$$

$$V_{drop} = \mathbf{3.7\,V}$$

$$V_{drop\%} = \frac{V_{drop}}{V_{op}}$$

$$V_{drop\%} = \frac{3.7}{80}$$

$$V_{drop\%} = \mathbf{0.046 \text{ or } 4.6\%}$$

This voltage drop is higher than the recommended maximum, but there are a number of potential options to reduce the voltage drop in the circuit.

First, shortening the conductor length will reduce the voltage drop, but this may not be feasible. The distance between the array and the rest of the system is usually not changeable, since the array location is chosen based on factors that are more significant.

More commonly, circuit resistance is reduced by increasing the conductor size, as larger conductors have lower resistance than smaller conductors. (This will likely be beyond the minimum size determined from ampacity calculations.) If the example array used larger 8 AWG conductors instead, the voltage drop would be an acceptable 2.9%. Also, since conductor connections can affect circuit resistance, and therefore voltage drop, circuits should be designed with as few splices and connections as possible.

Conductor Insulation

Insulation protects a bare conductor from coming into contact with personnel or equipment. The insulating material defines three critical properties for a conductor: its maximum operating temperature, its application and environmental resistance (such as to sunlight, oil, or moisture), and its permissible installation locations (such as direct burial, in conduit, or exposed). This information is marked on the outer insulation or jacket of the conductor, sometimes in abbreviations or codes, along with other information, such as maximum voltage, manufacturer, and size. **See Figure 11-9.**

As current flows through a conductor, the resistance of the material, though very small, causes power loss in the form of heat. The amount of heat is proportional to the square of the current, as in the formula $P = I^2R$. An ampacity rating is determined by how much heat the conductor can withstand without damage, which depends on the insulation material and how well the heat can be dissipated.

Conductor Insulation Codes

Conductor insulation is typically identified and marked with letter codes. Some codes are only abbreviations of conductor names, such as "SE" for service-entrance conductors. Others give information about the actual material type and ratings of the insulation. Each letter or pair of letters describes something about the conductor type or ratings. Common insulation codes include the following:

Insulation Materials:
    E = elastomer
    R = thermoset
    S = silicone
    T = thermoplastic
    X = cross-linked synthetic polymer

Outer Covering:
    N = nylon jacket

Heat Ratings:
    H = 75°C
    HH = 90°C

Special Conditions:
    O = oil resistant jacket
    OO = oil resistant jacket and conductors
    U = underground (direct burial) applications
    W = moisture resistant (usually for 60°C)
    W-A = outdoor applications
    -2 = high temperature (90°C) and moisture resistance

For example, a conductor marked "THWN" has a thermoplastic insulation material covered by a nylon jacket, which offers 75°C heat resistance and moisture resistance.

Insulation letter codes are intended as a reminder only and can easily be misinterpreted. For example, the "S" in "SE" does not stand for silicone. In this case, the abbreviation refers to the conductor's application (service entrance) rather than specific materials or ratings. However, "USE" does indicate a service-entrance conductor that is suitable for underground installation Therefore, if a conductor type is unfamiliar from previous experience, its specifications should be checked to ensure that it has the proper insulation and ratings for the intended application. Some information may be printed separately on the conductor more explicitly, such as "90°C," or the information may be found in manufacturer literature. Also, tables in NEC® Article 310 give more information about the ratings and uses of various conductor and insulation types.

Conductor Insulation Markings

Figure 11-9. *Size, insulation type, resistances, and other information are printed on the outer jacket of conductors.*

Conductor insulation types used in PV systems must be compatible with the environmental conditions and ratings of the associated equipment, connectors, or terminals. PV systems can include many of the common conductor types used in buildings, but they must be chosen separately for different parts of the system. **See Figure 11-10.**

PV Source-Circuit Wiring. PV module electrical connections are usually installed with full exposure to the elements, including temperature extremes, sunlight (UV) exposure, and precipitation. Consequently, any conductors used for these circuits may need to be rated for outdoor applications with high temperature, moisture,

and sunlight resistance. **See Figure 11-11.** Single-conductor USE-2 is widely used because it has high temperature, moisture-resistance, and sunlight-resistance ratings, and is readily available. Also, single-conductor cable listed and labeled as "photovoltaic wire" or "PV wire" was added in the 2008 NEC® as approved wiring.

Source-Circuit Wiring

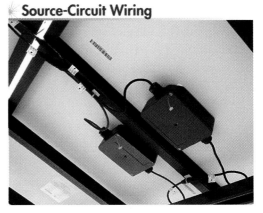

Figure 11-11. *Source-circuit conductors are permitted to be exposed if the conductor insulation has the required environmental resistances.*

PV modules and arrays operating on hot days under full sunlight can reach very high temperatures. Any conductors, conduits, junction boxes, connectors, or terminals located close to arrays or installed on rooftops in direct sunlight may experience these high temperatures as well. For this reason, conductors and other components used for PV source circuits may need to be rated for at least 90°C (194°F).

Recommended Insulation Types for PV Systems

APPLICATION	REQUIRED RESISTANCES				NUMBER OF CABLE CONDUCTORS		INSTALLATION		RECOMMENDED INSULATION TYPE
	Moisture	Sunlight	≥ 90°C	Fire	One	Multiple	Exposed	Conduit	
PV source-circuit wiring	✓	✓	✓		✓		✓	✓	USE-2, PV WIRE
PV output-circuit wiring	✓		✓		✓			✓	USE-2, XHHW-2, RHW-2, THWN-2
Interior wiring			✓		✓			✓	THHN, THW, RHW, XHHW, RH
			✓			✓	✓*		NM, NMB, UF
Battery wiring	✓				✓		✓		USE, RHW, THW

* may not be permitted in local jurisdiction

Figure 11-10. *Conductors in different parts of a PV system have different application requirements.*

Conductors in array tracking systems have special considerations. These conductors must be especially flexible to withstand repeated movement, which can weaken or crack regular insulation and break conductor strands. The cables must be rated as "hard-service" flexible cables, in addition to being listed for outdoor use. Requirements and ampacity derating regulations are detailed in Section 690.31 and Article 400, "Flexible Cords and Cables."

PV Output-Circuit Wiring. The conductors in the PV source circuits do not need to be the same as the conductors for the PV output circuit. Since USE-2 and PV wire are more expensive than most other types of conductors, it may be advisable to transition to a different conductor type for the rest of the exterior wiring. This transition is usually done at the source-circuit combiner box.

Conductors between the combiner box and the DC disconnect are usually enclosed in conduit for protection, so they do not need to be sunlight resistant, but they must still be rated for at least 90°C and wet conditions. (Even though the conductors are protected within conduit, the conduit is exposed and may trap water inside. Therefore, the conductors require a wet rating.) Single-conductor RHW-2, THW-2, THWN-2, and XHHW-2 types are commonly used for the PV output circuit.

Interior Wiring. When DC conductors enter a building, they no longer require high temperature ratings (except when routed through attics), sunlight resistance, or moisture resistance. However, they must be contained within metal conduit or enclosures (up to the first readily accessible disconnect) and rated for fire resistance. Single conductors in conduit, or multiconductor sheathed cables may be used, except where the cables are subject to damage from exposure or where otherwise required to be in conduit. For residential applications, types NM and NM-B (both known as "Romex®") and UF are common acceptable multiconductor sheathed cables. USE-2 conductors that also include RHH or RHW-2 designations may be used inside buildings, but only when installed in conduit.

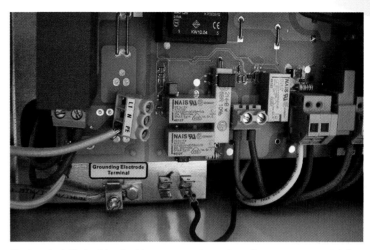

Factory-installed wiring inside components such as inverters may not follow North American color codes for field wiring. These conductors should not require modification, but the manufacturer's installation instructions should be carefully consulted to avoid any confusion with this wiring.

Battery Wiring. Battery conductors include those between batteries or cells within the battery bank, and those connecting the battery bank to the nearest circuit device, such as a junction box or disconnect. Such conductors must be listed for hard-service use and moisture resistance. It is recommended that the conductors be flexible, as identified in Article 400, since they may be sized at 2/0 or larger. Flexible cables put less stress on battery terminal connections. Finely stranded building-wiring conductors such as USE, RHW, and THW are commonly used. The connectors permitted on these types of conductors must be specifically identified for such use, but certain types of ring lugs are generally compatible. Conductors intended for welding, automotive, or marine applications are not permitted.

Conductor Insulation Color Codes

The color of a conductor's insulation is used to identify the purpose of the conductor in the circuit and to match the ends of a conductor. The insulation may be a solid color or may consist of mostly one color with one or two stripes of another color. Most conductors are available in many colors. For conductors larger than 6 AWG, which are usually only black, colored tape is wrapped around the ends for identification.

There are separate color designations for grounded, grounding, and ungrounded conductors and they apply to both AC and DC circuits. Grounded conductors are white or gray. Grounding conductors can be bare or can have insulation that must be either solid green or green with yellow stripes. Ungrounded conductors may be of various colors, but must be distinguishable from the grounded or grounding conductors, so they cannot be white, gray, or green.

The common convention in AC systems is that the first three ungrounded conductors are colored black, red, and blue. The ungrounded conductor for DC circuits (usually the positive conductor) may be black, red, or any other color except green, white, or gray. These color conventions do not always apply to factory-installed wiring inside of components.

Wiring Connections

Quality wiring connections are critical to the safe and reliable operation of PV systems. Poor connections add circuit resistance, increasing voltage drop, and power loss. In the worst cases, loose or corroded connections cause hot spots and thermal expansion at the connection, leading to failure of the connection or the entire system. To minimize the chances of wiring failure, installers should use high-quality connectors and be careful to install them properly. In humid environments, terminations should be coated with an antioxidation compound to retard corrosion. It is also important to use as few connections as possible.

A variety of connectors, terminals, splices, and lugs may be used for field-installed PV system wiring, as long as they are appropriate and rated for the application, such as for conductor material, size, and environment. All wiring connections other than module connections must be made within junction boxes or equipment enclosures using approved methods and listed devices. This protects the connections from the environment but allows access for inspections and service.

Module Connectors. Modules are typically connected together in PV source circuits with external, exposed connectors. **See Figure 11-12.** These connectors must be listed for the conditions of use and have ratings at least equal to those of the rest of the wiring system.

Module Connectors

Figure 11-12. *Modules are typically connected together in PV source circuits with external, exposed connectors.*

Additionally, these connectors must meet the following five requirements from Section 690.33:

- The connectors must be polarized and noninterchangeable with other electrical equipment on the premises. This prevents accidentally reversing the array connections or connecting the incorrect circuits.
- The connectors must protect the live parts from accidental contact with persons during assembly and disassembly.
- The connectors must have a latching or locking mechanism to prevent accidental disconnection. When readily accessible and used in circuits over 30 V, a tool is required for opening.
- When multiple connections are made, the grounding conductor must be the first conductor to make contact and the last to break contact.
- The connectors must be either capable of safely interrupting the circuit current or require a tool to open and are marked "Do Not Disconnect Under Load" or "Not for Circuit Interrupting."

Connectors meeting these requirements facilitate quick, yet secure, safe, and reliable array connections. Some modules may not include the conductors and/or connectors, and require special crimping and installation tools to attach the connectors. However, many module manufacturers now factory-install the conductors and connectors, eliminating many preassembly requirements. The module junction boxes are often then sealed since the installer no longer needs access.

Screw Terminals. Screw terminals use the compressive force of a screw to secure a conductor to a terminal and are highly reliable. **See Figure 11-13.** A screwdriver or Allen wrench is used to tighten the screw terminal to a certain torque. These mechanical connections are common on many types of electrical components, including disconnects, overcurrent protection devices, PV modules, charge controllers, and inverters.

Screw Terminals

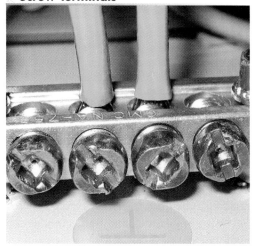

Figure 11-13. *When tightened properly, screw terminals produce secure and low-resistance connections.*

Special terminals can be used as junction points combining several smaller conductors in parallel to connect with a single larger conductor. Screw terminals should not be used with finely stranded or multiple conductors unless listed for such use.

Lugs. Lugs are used to terminate conductors with special connectors. In most cases, screw terminals are preferable to lugs because they are easier to install correctly and involve one fewer connection. Poorly installed lugs are problematic in electrical installations. However, some PV equipment may require lugs. Lugs are specified by the conductor size, number of conductors, temperature rating, insulation, lug type, and lug size. **See Figure 11-14.**

Lugs

Figure 11-14. *Lugs are crimped conductor terminations in ring, fork, spade, or pin shapes.*

Lug types include fork, ring, and disconnect (spade or pin) lugs. Fork and ring lugs are used for connections made by threaded screws or nuts and the lug is sized according to screw diameter designations, such as 1/4″ and #8. Fork terminals can be easily removed by loosening the terminal screw or nut, while ring terminals require complete removal of the fastener. For this reason, ring lugs are usually used for large conductors, so they cannot accidentally slip out from under the screw or nut due to their weight or movement. Spade- or pin-type lugs have both male and female elements that connect only to each other and can be disconnected without tools.

Some electrical circuits use aluminum or copper-clad aluminum conductors. These conductors have lower ampacity than all-copper conductors for the same sizes but are significantly lighter. However, they require special connectors and terminating methods and are not generally used in PV systems.

Most types of lugs require crimping. Crimping compresses the conductive lug material around the exposed end of a conductor, making secure mechanical and electrical connections. Heavy-duty crimpers are required for conductor sizes greater than 8 AWG. Secure crimping is vital, so only quality crimping tools should be used. Every connection should be tugged and checked after crimping to ensure that a quality connection has been made. Some battery cables are available with preinstalled lugs that are both crimped and soldered for extra strength.

Lugs may or may not include insulation around the crimped portion. To ensure quality crimping without interference from the plastic insulation, uninsulated lugs are preferred. If insulation is required, it can be added with tape or heat-shrink tubing after the crimp connection is made and checked.

Splices. Splices are used in PV systems to connect or extend conductors, parallel array source circuits, or tap service-entrance conductors for supply-side interconnections. The splice and any exposed areas on the conductor must be insulated (covered with appropriate tape or heat-shrink tubing) with a level of protection and rating equivalent to the conductor insulation.

Splicing devices include screw-terminal, split-bolt, crimped-barrel, and twist-on connectors. **See Figure 11-15.** Splicing devices for direct burial must be listed for that use. With the exception of direct-burial splices, all splices must be made in an approved junction box or enclosure.

Twist-on splicing devices, called "wire nuts," are commonly used in many electrical applications. Some types are rated for wet locations or direct burial and are prefilled with antioxidation compounds. However, these connectors can easily produce poor splices and are not commonly used in PV systems.

DC Plugs and Receptacles. Branch circuits installed for powering DC loads may require the use of plugs and receptacles for making quick or temporary connections. The National Electrical Manufacturers Association (NEMA) standardizes the configurations of dozens of plug and receptacle designs. The NEC® does not require a specific design for use in DC branch circuits, but allows any that conform to three requirements: the plugs and receptacles must be listed for DC power with a rating of at least 15 A, must have a separate equipment-grounding terminal, and must be different from any other receptacles used on the premises.

Screw Terminal Block Splices

Figure 11-15. *Splicing devices, such as screw terminal blocks, are used in PV systems to connect or extend conductors, parallel array source circuits, or tap service-entrance conductors for supply-side interconnections.*

Only one NEMA plug-and-receptacle configuration is specifically designed for DC only, and a few two-pole, three-wire designs are specifically for AC only. Most others can be used for either type of power. For most residential and light commercial applications, several NEMA styles satisfy the other two requirements, since they have separate grounding terminals and are not commonly used otherwise in these locations. **See Figure 11-16.**

Two-conductor plugs and receptacles, such as automotive "cigarette lighter" styles, cannot be used. It is also not acceptable to use the third grounding conductor of a three-conductor plug or receptacle to carry common negative return currents on a combined 12/24 V system.

Junction Boxes

A *junction box* is a protective enclosure used to terminate, combine, and connect various circuits or components together. Junction boxes may be empty enclosures, within which conductors are routed or spliced together, or they may contain terminal strips for making connections.

Junction boxes can be either a box permanently attached to the outside of a major component for making connections to terminals of that component, or a separate box for combining circuits. Junction boxes located behind modules or panels must be accessible by the removal of module or support fasteners. Thus, modules may not be permanently fastened or welded to a support structure.

The common types of junction boxes used in PV systems are module junction boxes and source-circuit combiner boxes. Separate junction boxes must be listed for the application.

Module Junction Boxes. Most PV modules include a plastic junction box attached to the back surface. The junction box contains the module's electrical terminals and often the bypass diodes. **See Figure 11-17.** Some module junction boxes have multiple knockouts sized for conduit or individual conductors. For the junction box to be used with conduit, the module nameplate must include a CR rating.

Exposed conductors must use a method of mechanically securing the conductor to reduce any pulling forces on the connection. Junction boxes may include strain-relief posts for this purpose, or the installer may use strain-relief clamps at the entrance to the junction box.

Some junction box terminals allow limited reconfiguration of the cell circuit arrangements within the module. For example, by changing terminal connections, a module consisting of two series strings of cells in parallel can be reconfigured into one long series string.

Permanently sealed module junction boxes are becoming increasingly common. The module's primary electrical connections are encapsulated in a sealed enclosure on the back of the module. Short lengths of conductors extend from the box and terminate with positive and negative connectors. This makes field installation faster, safer, and more reliable.

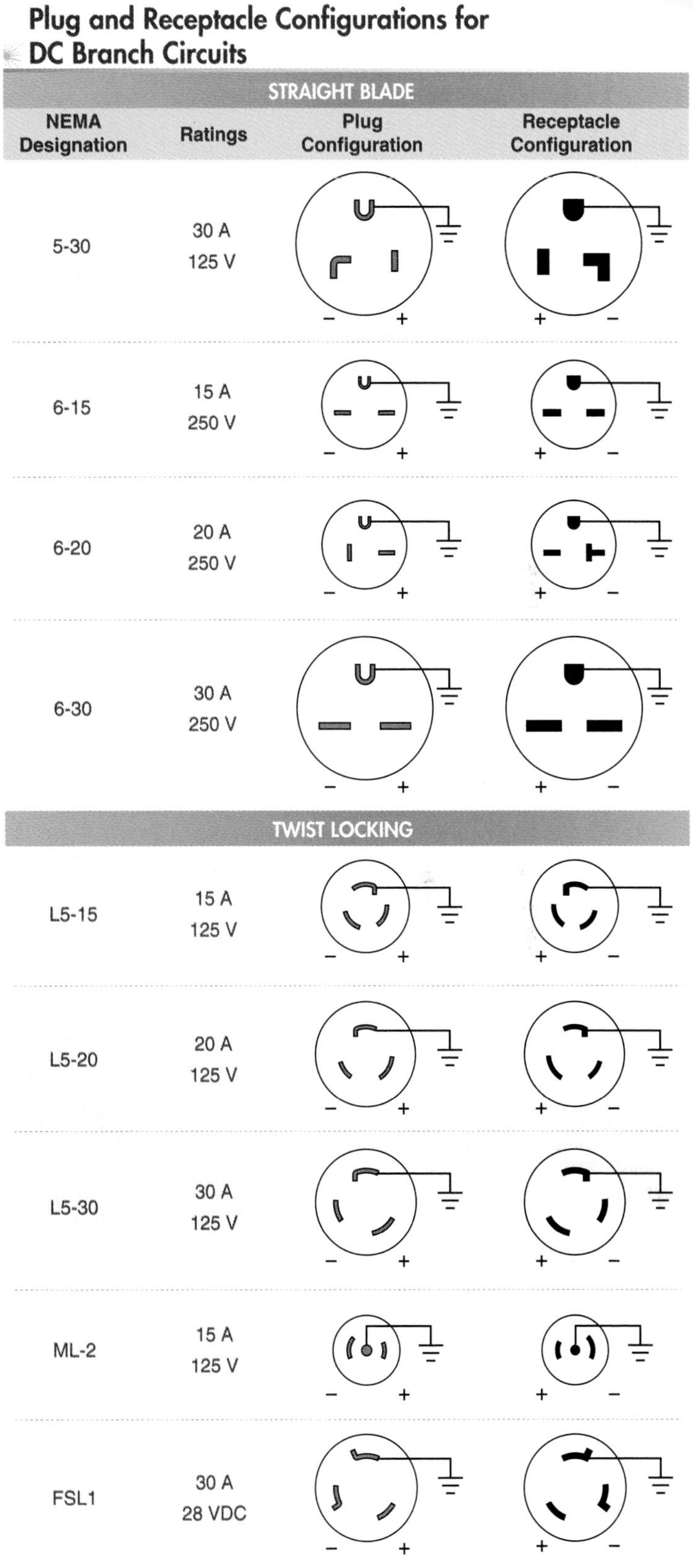

Plug and Receptacle Configurations for DC Branch Circuits

STRAIGHT BLADE			
NEMA Designation	Ratings	Plug Configuration	Receptacle Configuration
5-30	30 A 125 V		
6-15	15 A 250 V		
6-20	20 A 250 V		
6-30	30 A 250 V		
TWIST LOCKING			
L5-15	15 A 125 V		
L5-20	20 A 125 V		
L5-30	30 A 125 V		
ML-2	15 A 125 V		
FSL1	30 A 28 VDC		

Figure 11-16. *Several NEMA plug-and-receptacle configurations are acceptable for use with DC branch circuits.*

Module Junction Boxes

Figure 11-17. *Module junction boxes contain and protect the module terminal connections and diodes in the source circuit. Some are field-accessible.*

PV Source-Circuit Combiner Boxes. A *combiner box* is a junction box used as the parallel connection point for two or more circuits. Combiner boxes are used to combine parallel array source circuits into the PV output circuit. These boxes typically include terminal blocks, fuses, and fuse holders.

A source-circuit combiner box facilitates and protects array wiring connections and provides a central point to test and troubleshoot arrays. **See Figure 11-18.** It also allows one source circuit to be disconnected without interrupting the other source circuits. Combiner boxes may be located indoors or outdoors, near or inside an inverter, or as a separate enclosure near the array.

Combiner boxes often have circuit boards that include built-in provisions for PV source-circuit fuses and wiring connections.

Source-Circuit Combiner Boxes

AMtec Solar

Figure 11-18. *Multiple PV source circuits are combined into the PV output circuit within the combiner box.*

Protection Diodes

In PV systems, diodes are used in the array circuits as either blocking diodes or bypass diodes. These diodes are typically built into the PV modules, installed inside the module junction boxes, or built into components that interface with the array output circuit. When field-installed, these diodes must be properly sized for the maximum expected voltages and currents.

Blocking Diodes. A *blocking diode* is a diode used in PV source circuits to prevent reverse current flow. For example, in a simple self-regulated PV system, a fully charged battery may begin to discharge through the array when the array voltage falls below the battery voltage at nighttime. A blocking diode is used to prevent this undesirable discharge. Similarly, in an array with multiple source circuits, a blocking diode prevents one source circuit from feeding another source circuit, such as when the second circuit is shaded and consumes power. A blocking diode is installed in the ungrounded conductor in series with the module string to prevent these types of power losses. **See Figure 11-19.**

Protection Diodes

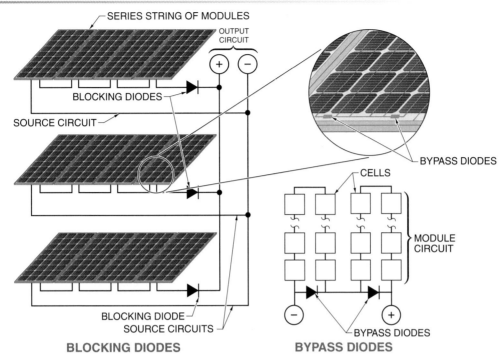

SERIES STRING OF MODULES

OUTPUT CIRCUIT

BLOCKING DIODES

SOURCE CIRCUIT

BYPASS DIODES

CELLS

MODULE CIRCUIT

BLOCKING DIODE

SOURCE CIRCUITS

BYPASS DIODES

BLOCKING DIODES   **BYPASS DIODES**

Figure 11-19. *Blocking diodes are installed in the source circuit and bypass diodes are installed within a module or its junction box. These diodes prevent power loss due to reverse current or high-resistance conditions.*

Blocking diodes are rarely installed today as separate components. If used separately, they are usually installed in the PV source-circuit combiner box or the module junction box at the head of each series string. More commonly, they are incorporated into components that control battery charge and discharge current, such as charge controllers and bimodal inverters. In shunt-type charge controllers, a blocking diode also prevents the controller from shorting the battery bank during charge regulation.

Bypass Diodes. A *bypass diode* is a diode used to pass current around, rather than through, a group of PV cells. In contrast to blocking diodes, bypass diodes are installed in parallel with modules or groups of cells. Each module may include one to three bypass diode(s).

Under normal operating conditions, current does not pass through bypass diodes. However, if cells develop an open-circuit or high-resistance condition because of shading or module damage, bypass diodes minimize power

output losses by routing current around the strings of affected cells.

Bypass diodes can be factory-installed by the module manufacturer or field-installed by the installer. They may be accessible in module junction boxes, or they may be encapsulated in a junction box or module laminate. **See Figure 11-20.**

Bypass Diodes

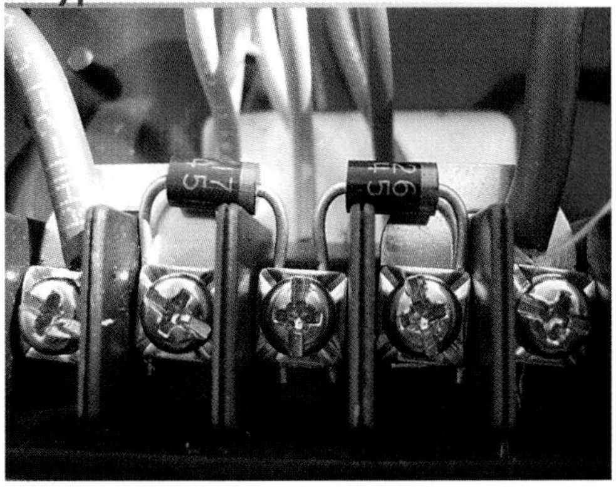

Figure 11-20. *Bypass diodes may be field-installed in the module junction box.*

Conduit Selection

Most PV system conductors are installed in conduit. However, source-circuit and output-circuit conductors are required to be in conduit only when installed in a readily accessible location and their maximum circuit voltages are greater than 30 V. Otherwise, the extra lengths of exposed conductors should be neatly gathered and secured to the mounting structure to prevent abrasion damage from wind or ice.

Conduit options for throughout PV systems may include electrical metallic tubing (EMT), rigid nonmetallic conduit (electrical PVC, schedule 40 or 80), and electrical nonmetallic tubing (ENT), assuming their use does not exceed their ratings. Rigid metal conduit (RMC) and intermediate metal conduit (IMC) may be used when extra protection is needed. **See Figure 11-21.** Like outdoor conductors, outdoor conduit must be rated for high temperatures, sunlight resistance, and moisture resistance.

Source-Circuit Conduit

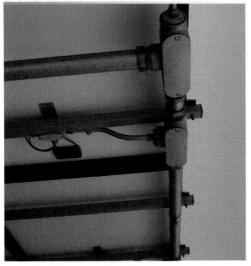

Figure 11-21. *A number of different types of conduit may be used in PV systems if they have the necessary resistances, such as moisture and high temperature resistance for source circuits.*

When DC conductors from a PV power source are run inside a building, they must be contained in metallic conduits or enclosures up to the point of the first readily accessible disconnect. In the case of building-integrated arrays, the entire array circuit falls under this requirement. Conduit provides physical protection for these conductors. Metallic conduit also provides fire resistance and a ground-fault path for ground-fault protection devices.

The conductors of PV source and output circuits must not be contained in the same conduit, junction box, or raceway with conductors of any other electrical system. However, DC circuits can be run together with AC circuits if they are both directly related to a specific PV system.

OVERCURRENT PROTECTION

An *overcurrent protection device* is a device that prevents conductors or devices from reaching excessively high temperatures due to very high currents by opening the circuit. High temperatures can damage components and conductor insulation, causing electrical shock and fire hazards. The overcurrent protection device opens the circuit before the overcurrent damage can occur. **See Figure 11-22.** An overcurrent condition can be the result of an overload, ground fault, or short circuit.

Overcurrent protection devices are classified by how quickly they activate. A non-current-limiting device operates slowly, allowing damaging short-circuit currents to build up to full values before opening. A current-limiting device opens the circuit in less than one-quarter cycle of short-circuit current, before the current reaches its highest value, limiting the amount of destructive energy allowed into the circuit. Current-limiting devices are required in some parts of PV systems.

Overcurrent Protection Devices

Overcurrent protection devices include fuses and circuit breakers. **See Figure 11-23.** A *fuse* is a metallic link that melts when heated by current greater than its rating, opening the circuit and providing overcurrent protection. Requirements for overcurrent protection in PV systems are covered in Section 690.9. Overcurrent protection devices must be listed and specifically rated for their intended use, such as for DC. Automotive fuses may not be used in PV systems.

Overcurrent Protection

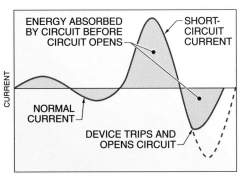

NONCURRENT-LIMITING
OVERCURRENT PROTECTION DEVICE

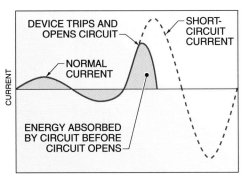

CURRENT-LIMITING OVERCURRENT
PROTECTION DEVICE

Figure 11-22. *Current-limiting overcurrent protection devices open a short circuit before current reaches its highest value.*

Overcurrent Protection Devices

Figure 11-23. *Overcurrent protection devices include fuses and circuit breakers of various types and ratings.*

Fuses are either non-time-delay or time-delay types. A non-time-delay fuse detects any overcurrent and opens the circuit almost instantly. This provides maximum protection, but can cause nuisance tripping for short, harmless surge currents. A time-delay fuse detects and opens a short circuit almost instantly, but allows small overloads to exist for a short time. Time-delay fuses may be used for protecting circuits with momentary current surges, such as circuits involving transformers or motors.

A *circuit breaker* is an electrical switch that automatically opens as a means of overcurrent protection, and that can be manually opened as a disconnecting means. The advantages of circuit breakers include being resettable after an overcurrent trip and that they can be used as disconnects for installation and maintenance. A circuit breaker can be used to satisfy both the overcurrent protection and disconnect requirements in some parts of a PV system.

Current Ratings

The current rating of an overcurrent protection device depends on the ampacity of the circuit conductors. The overcurrent protection device rating required for PV circuits is 125% of the calculated maximum circuit current. (Note that this is in addition to the 125% required to calculate the maximum circuit current for PV source-circuit conductors from the module short-circuit current.) This is the same calculation as for the required conductor ampacity. Since overcurrent protection devices are available in only certain ratings, the nearest higher standard rating may be used.

For example, consider a PV output circuit composed of three source circuits, each with a rated short-circuit current of 7 A. The maximum current of the output circuit is 26.25 A (7 A × 3 × 125% = 26.25 A). The required overcurrent protection device rating is 32.8 A (26.25 A × 125% = 32.8 A). The overcurrent protection device rating should be the next higher available standard size, or 35 A.

Fuses must be safely isolated from all sources of power prior to replacement. There must not be any voltage on either end of a fuse before contact is made by a person. This requirement can be satisfied by installing a switch on both sides, but this complicates and potentially adds extra costs and resistance to the system. Manufacturers have developed fuse-holding systems that simplify compliance with this requirement.

Replaceable fuses in components such as inverters can be disconnected from the circuit and removed by unscrewing a nonconductive end cap in the side of the enclosure. When in the circuit, the cap holds the fuse up against the electrical contacts. When removed, the completely disconnected fuse can be safely handled. The circuit contacts are still potentially live but completely enclosed within the component and inaccessible.

Alternatively, "finger-safe" fuse holders are commonly used for field-installed fuses, such as array source-circuit supplementary fuses. These holders either pull the fuse out using a nonconductive tool or include a mechanism to eject a fuse from its contacts while keeping the contacts entirely within a nonconductive enclosure.

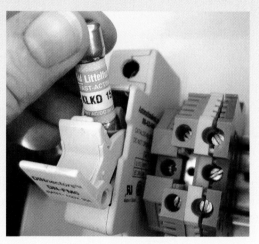

Finger-safe fuse holders eject a fuse from its contacts with the touch of a finger on the nonconductive enclosure.

Interrupting Rating

The *interrupting rating* is the maximum current that an overcurrent protection device is able to stop without being destroyed or causing an electric arc. This protects devices from high fault currents and short circuits. All listed current-limiting overcurrent protection devices have interrupting ratings, which are usually marked with "AIR" for amperes interrupting rating. Overcurrent protection devices without adequate interrupting ratings can rupture, burn, or become welded together by a short circuit.

For PV source circuits, the interrupting rating is not critical, since the circuit is current-limited by nature. However, battery circuits can deliver short-circuit currents of several thousands of amperes, which can cause battery explosions or fires. Therefore, overcurrent protection devices for battery circuits must have a very high interrupting rating. Some fuses are available with interrupting ratings up to 200,000 A. DC-rated magnetic circuit breakers are available with interrupting ratings of only a few thousand amperes, which is generally not considered sufficient protection.

PV System Overcurrent Protection

General information about overcurrent protection is covered in Article 240, "Overcurrent Protection," and specific applications for PV systems are covered in Article 690. Nearly all PV system circuits require overcurrent protection, including the PV source circuits, PV output circuit, inverter output circuit, and any battery or load circuits. Circuits must be protected from every source of power. For the purposes of sizing overcurrent protection devices, all PV system currents are considered to be continuous. In most systems, every ungrounded conductor must be protected. There are only a few exceptions, which primarily apply to PV source and output circuits.

Array Overcurrent Protection. Generally, all ungrounded array conductors must include overcurrent protection. The grounded conductors must not normally include overcurrent protection (or a disconnect), since opening this circuit will disconnect part of the system from ground. All array overcurrent protection devices must be listed for DC applications and have appropriate voltage and current ratings.

In addition to overcurrent protection on the PV output circuit, overcurrent protection is usually required in each PV source circuit. Requirements for source-circuit overcurrent protection must consider possible back-feed currents from other PV source circuits, the inverter, or battery circuits. Parallel power sources may inadvertently back-feed each other in a way that could exceed the conductor ampacity and damage the module. Also, PV source-circuit overcurrent devices must not exceed the maximum overcurrent device rating marked on the modules.

This source-circuit overcurrent protection may be provided by supplementary overcurrent protection devices. **See Figure 11-24.** A *supplementary overcurrent protection device* is an overcurrent protection device intended to protect an individual component and is used in addition to a current-limiting branch circuit overcurrent protection device. In this case, a series string of modules is treated as a single component. The use of such devices enables the source-circuit protection to be closer to the specified ratings required on module labels. These devices must be accessible, but are not required to be readily

accessible, so they may be installed in junction boxes or other enclosures on rooftops or in attics.

Source-Circuit Fuses

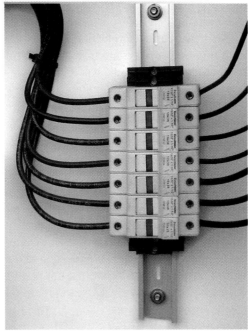

Figure 11-24. *Array source circuits are fused individually within the source-circuit combiner box.*

Overcurrent Protection Exceptions

Conductors and equipment require overcurrent protection when connected to power sources that can potentially produce higher-than-normal current. Since PV devices are inherently current-limited, they cannot be a source of overcurrent, provided the conductors are properly sized for the device's short-circuit current. Therefore, conductors that are series-connected to only PV power sources do not require overcurrent protection. For example, single source circuits in direct-coupled systems do not require overcurrent protection because the conductors should already be sized for the greatest amount of current that the source circuit is capable of producing.

Likewise, an array with only two source circuits does not require overcurrent protection. Each source circuit is sized to handle its own maximum current, which is the same current that could come from the other source circuit if one circuit back-feeds the other. This could occur if one circuit was shaded or damaged. Conductors sized normally (according to the current from one source circuit) and without overcurrent protection will not be adequate, however, for arrays with three or more source circuits. Two circuits could potentially back-feed the third, overloading the third circuit's conductors with twice the current they were sized for.

Regardless of the number of source circuits, overcurrent protection is required to protect the array conductors from potential back-feed currents from other power sources, particularly a battery bank or inverter. Battery banks are always a potential source of high back-feed currents, although some inverters may be designed to avoid this risk. Whether an inverter is non-back-feeding can be investigated in the product listing or by independent test results provided by the manufacturer.

Inverter-Output Overcurrent Protection. Overcurrent protection on the inverter output circuit conductors is sized based on the maximum continuous output rating of the inverter. This protects the conductors on the AC side of the system from overcurrent from the inverter. If the PV system is interactive, current-limiting overcurrent protection devices are used to protect the system from overcurrent from utility-service faults, which usually requires a higher interrupting rating.

AC output overcurrent protection for interactive systems depends on the location of the utility interconnection. **See Figure 11-25.** Some systems may use a separate fused disconnect. Alternatively, the inverter output may be connected to the AC distribution panel through a back-fed circuit breaker. A *back-fed circuit breaker* is a circuit breaker that allows current flow in either direction. The back-fed circuit breaker provides overcurrent protection of the branch circuits from the inverter, and the panel's main service circuit breaker provides protection for the service conductors.

Battery Overcurrent Protection. Battery banks can produce very high currents when short-circuited. This can affect the DC branch circuits and other connected equipment, including the array, charge controller, and inverter connected directly to the battery system. The battery-bank output circuit must include current-limiting overcurrent protection with an especially high interrupting rating.

Branch-Circuit Overcurrent Protection. PV systems may include DC branch circuits, AC branch circuits, or both. In either an AC or DC system, the conductors in each branch circuit must be protected with an appropriate and properly sized overcurrent protection device. Circuit breakers are commonly used since they also function as disconnects.

Inverter-Output Overcurrent Protection and Disconnects

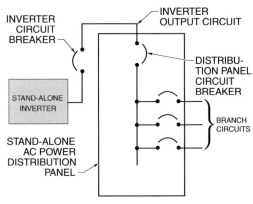

STAND-ALONE SYSTEM

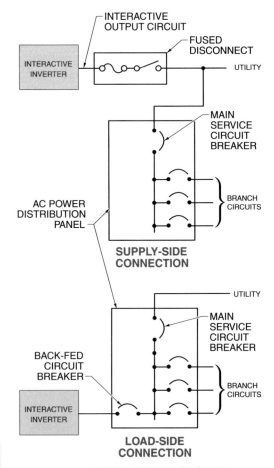

INTERACTIVE SYSTEM

Figure 11-25. *Overcurrent protection for the inverter output circuit depends on the system or utility interconnection type. Overcurrent protection and disconnecting means for this circuit may also be combined by using circuit breakers or fused disconnects.*

A particular overcurrent problem arises when one stand-alone inverter with a 120 V output supplies a 120/240 V distribution panel. If the system includes a multiwire branch circuit with two 120 V loads or one 240 V load, the single grounded (neutral) conductor can become dangerously overloaded. **See Figure 11-26.** Since current on the two ungrounded conductors will be in-phase, instead of out-of-phase as with a normal 240 V split-phase supply, currents in a multiwire branch circuit will add when they return on the shared grounded conductor. Therefore, the grounded conductor may carry twice its rated circuit current. (A similar problem can occur with interactive systems when PV systems are added to buildings that are already wired for standard 120/240 V service.)

There are a number of ways to solve this problem. A transformer can be used to convert the 120 V inverter output to 240 V split-phase. The multiwire branch circuit can be rewired into two separate circuits, each with its own neutral conductor. Or, two 120 V inverters can be installed to supply split-phase 240 V.

Neutral Loading

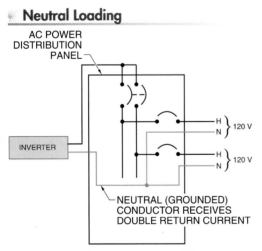

Figure 11-26. *Connecting a 120 V inverter to a 120/240 V system with multiwire branch circuits causes dangerous overloading in the grounded (neutral) conductor and must be avoided.*

Transformer Overcurrent Protection. Transformers are used in PV systems to convert an inverter output voltage to a different level, for example from 208 V to 240 V or from 120 V to 240 V. For interactive inverters, the transformer

is energized from two sources, the PV system and utility grid. In this case, both windings must be considered primaries, which normally requires overcurrent protection.

An exception is made when the transformer has a current rating equal to or higher than the maximum short-circuit current rating of the inverter output. In such a case, overcurrent protection is not required on that side of the transformer.

DISCONNECTS

A *disconnect* is a device used to isolate equipment and conductors from sources of electricity for the purpose of installation, maintenance, or service. Article 690, Part III, addresses disconnect requirements for PV systems. Disconnects are required for both the DC and AC sides of a PV system. Disconnects are also required to isolate other power sources, such as batteries, and may be included at additional points to facilitate system testing or maintenance.

A disconnect must be manually operable and may be either a switch or a circuit breaker. Either type of device, when used as a disconnect, must be located where readily accessible, be externally operable and guarded, plainly indicate whether it is in the open or closed position, and have an appropriate interrupting rating for the circuit fault conditions.

Array Disconnects

A disconnect must be provided to isolate all current-carrying conductors of a PV power source from all other conductors in a building or structure. **See Figure 11-27.** This disconnect is also known as the DC disconnect or PV disconnect. Disconnects used in the PV output circuit must be rated for DC and identified as such. Equipment such as PV source-circuit isolating switches, overcurrent protection devices, and blocking diodes are permitted on the array side of the array disconnect.

All current-carrying conductors in the PV output circuit must include a switch, circuit breaker, or some other disconnecting means. This includes the grounded conductor, unless the disconnecting means would leave that conductor ungrounded and energized. The array disconnect means must be installed at a readily accessible location either on the outside of the building or inside nearest the point of entrance of the PV output-circuit conductors.

Since PV arrays cannot be turned OFF with a disconnect, special care must be taken to disable an array for installation or service to minimize electrical shock hazards. This can be accomplished by open-circuiting or short-circuiting the array, or by covering the modules' front surfaces with an opaque material. For large systems, the array may be divided into small sections for individual disablement.

Array Disconnects

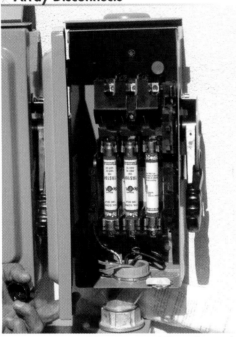

Direct Power and Water Corporation

Figure 11-27. *The array disconnect opens all current-carrying conductors in the PV output circuit.*

AC Disconnects

A disconnect must also be installed on the AC side of a PV system to isolate the system from the rest of a building's electrical system. In an interactive system, the PV system's AC disconnect is often located near the main utility disconnect for convenience. **See Figure 11-28.** Although this is not an NEC® requirement, this location can satisfy utility interconnection requirements for an accessible, visible-break, lockable PV system disconnect.

When an electrical system includes multiple power sources, such as a PV array, battery bank, engine generator, wind turbine, or utility power, there must be a disconnect for each power source. These disconnects should be grouped together and should require no more than six hand motions to completely remove any one power source from the electrical system.

AC Disconnects

Figure 11-28. *The AC disconnect of an interactive PV system may be located close to the main utility service disconnect, which can satisfy utility requirements for an external, visible-break, and lockable PV system disconnect.*

Equipment Disconnects

Disconnects must be provided to open all ungrounded conductors to every power source and each piece of PV system equipment, including inverters, battery banks, charge controllers, and other major components. If equipment is connected to more than one power source, each power source must have disconnecting means. **See Figure 11-29.** These disconnects must be circuit breakers or switches, though circuit breakers are more common because they are less expensive and easier to source for DC circuits.

Some equipment, particularly charge controllers, may operate erratically if left connected on one side and not the other. The solution is to install a pair of circuit breakers, one for the component's input circuit and one for the output circuit, and mechanically connect the switches together.

Equipment Disconnects

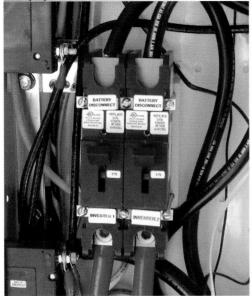

Direct Power and Water Corporation

Figure 11-29. *Switches or circuit breakers are required to isolate and disconnect all major components in a PV system from all ungrounded conductors of all power sources.*

GROUNDING

Grounding provides a path for fault currents and lightning-induced surges to dissipate safely, protecting people and equipment from hazards or damage. It also helps suppress electromagnetic interference. The NEC® requires most electrical systems to be grounded. Article 250, "Grounding and Bonding," establishes general requirements for grounding. Article 690, Part V, covers particular grounding requirements for PV systems.

A PV power source operating over 50 V must have one grounded conductor, with the exception of ungrounded systems meeting other special requirements. This voltage is the temperature-corrected maximum open-circuit voltage of the system, not the nominal voltage. Therefore, nominal 12 V stand-alone PV systems do not require a grounded conductor, but nominal 24 V systems may, since the actual array open-circuit voltage may reach higher than 50 V under extreme conditions. Typically, the negative DC conductor is the grounded conductor for two-wire PV arrays.

Grounding Electrode System

A *grounding electrode* is a conductive rod, plate, or wire buried in the ground to provide a low-resistance connection to the earth. Article 250, Part III, covers the materials, shapes, sizes, and depths of grounding electrodes for adequate grounding. Section 690.47 establishes requirements for the grounding electrode system used in PV systems. These requirements cover three specific grounding cases: AC systems, DC systems, or a combination of both.

AC Grounding. For AC systems, the grounding electrode system requirements are established in Article 250, Part III. These are the requirements for most common residential and commercial electrical systems, and are familiar to most electricians. The grounding connection is typically made at the main electrical distribution panel, where the grounded conductor and the grounding electrode conductor are connected to the same bus bar. In PV systems, this section applies to AC-only systems, such as systems with AC modules and no field-installed or accessible DC circuits, or to the AC side only of a PV system with both AC and DC circuits.

DC Grounding. For DC-only systems, such as small stand-alone PV systems supplying only DC loads, the grounding electrode system must meet the requirements of Article 250, Part VIII. The DC grounding electrode conductor must be at least the size of the largest conductor supplied by the PV system, or 8 AWG copper, whichever is greater. However, when the DC grounding electrode conductor is the only connection to a rod, pipe, or plate electrode, the grounding electrode conductor is not required to be larger than 6 AWG copper. This avoids the use of very large grounding electrode conductors when the sole connection to the grounding electrode is made at the battery or inverter DC terminals.

The NEC® allows a number of different types of grounding electrodes. Existing metal parts of a structure, such as buried water piping, structural steel, or concrete-encased rebar, are ideal. If these are unavailable, electrodes made of buried metal rods, pipes, plates, or rings are acceptable.

Grounding Terminology

Terms for the conductors and parts in a grounding system can be confusing but must be well understood to properly wire an electrical system. This terminology is the same for all electrical systems covered by the NEC®, including PV systems.

Grounded is the condition of something that is connected to the earth or to a conductive material that is connected to the earth. A *grounding electrode* is a conductive rod, plate, or wire buried in the ground to provide a low-resistance connection to the earth. A *grounding electrode conductor (GEC)* is a conductor connecting the grounding electrode to the rest of the electrical grounding system. This conductor carries current only during ground faults.

A *grounded conductor* is a current-carrying conductor that is intentionally grounded. In an AC system, this conductor is often called the neutral conductor. In a DC system, this conductor is usually the negative conductor.

The *equipment grounding conductor (EGC)* is a conductor connecting exposed metallic equipment, which might inadvertently become energized, to the grounding electrode conductor. This conductor does not normally carry current. Grounding conductors from various branch circuits and equipment are bonded together on a common busbar. This busbar also includes the grounding electrode conductor and the main bonding jumper, which connects the grounding system to the grounded (neutral) conductor.

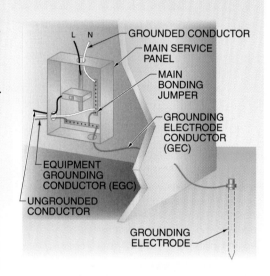

An *ungrounded conductor* is a current-carrying conductor that has no connection to ground. These conductors are also referred to as "hot" conductors.

The DC grounding connection may be made at any single point on the PV output circuit. Multiple DC grounding connections must be avoided, as they provide an alternate path for current to flow and may cause some ground-fault protection devices to operate improperly. Since the extent of the PV output circuit is difficult to define, connections on either side of a DC disconnect or charge controller are generally considered acceptable. Grounding connections close to the PV power source provide the system with better protection against lightning surges, but may be more complex to install. Common grounding points in the DC circuit are at the source-circuit combiner box and at the inverter.

AC and DC Grounding. Most PV systems involve both AC and DC, which involves the incorporation of the grounding requirements

for these portions of the PV system. Since the DC grounded conductor is not directly connected to the AC grounded conductor, they are considered separate systems. These two grounding systems must be bonded together. **See Figure 11-30.** There are two acceptable methods of meeting this requirement.

In the first method, separate AC and DC grounding electrodes are connected with the bonding conductor. The bonding conductor must be no smaller than the larger of the two grounding electrode conductors, in either the AC or DC grounding system.

In the second method, the DC grounding system is bonded to the AC grounding system, which is then connected to a single grounding electrode through an AC grounding electrode conductor. This method usually uses the existing grounding electrode for the building's AC power system. Both the bonding conductor and the grounding electrode conductor must be sized to meet the requirements of both the AC and DC grounding systems. This method is more common for grounding interactive PV systems.

Array Grounding. The 2008 NEC® added a new and separate array grounding requirement to the grounding electrode system. Arrays may be mounted some distance away from the building that holds the majority of the

system components, including the grounding electrode conductor. In order to maintain uniform grounding throughout the entire system, an additional and separate grounding electrode is required at the array location, if the array grounding electrode would be more than 6′ from the premises wiring electrode. This electrode does not need to be directly bonded to the other electrodes, though the grounding systems are effectively connected through the equipment grounding conductors.

AC and DC Grounding Methods

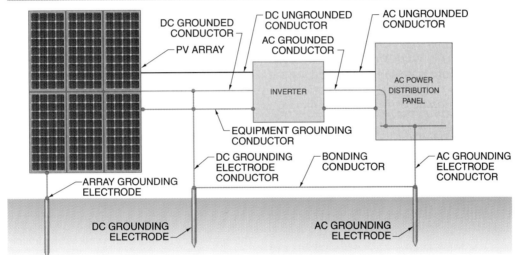

SEPARATE GROUNDING ELECTRODES

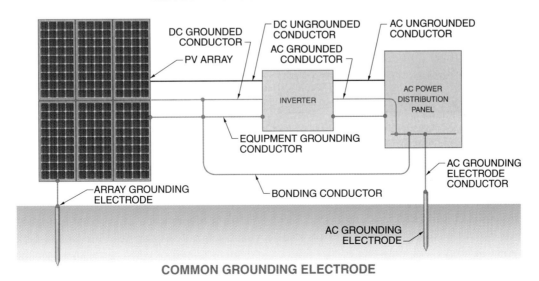

COMMON GROUNDING ELECTRODE

Figure 11-30. *The DC grounding system and the AC grounding system must be connected together with a bonding conductor. The array may also require a separate grounding electrode system.*

Ground-Fault Protection

A *ground fault* is the undesirable condition of current flowing through the grounding conductor. Ground faults are typically caused by damage to the protective insulation of normally current-carrying conductors. The copper material may then contact and energize metallic equipment such as enclosures, conduit, structures, and bare grounding conductors. Ground faults are a significant shock hazard. A ground fault current can also be a fire hazard as nearby combustible materials can be ignited by arcing or the bare metal heated by fault-current flow.

Ground-fault protection is the automatic opening of conductors involved in a ground fault. This stops the fault-current flow and disables the system until it can be inspected and repaired for conductor insulation damage or other causes of ground faults. Two types of ground-fault protection devices are used in PV systems. Although both types of devices provide ground-fault protection, they are very different in purpose and operation.

Ground-fault detection and interruption (GFDI) circuits in inverters sense the loss of a grounding connection from a ground fault and quickly shut down the inverter.

Array Ground-Fault Detection and Interruption. Most arrays are required to include ground-fault protection, as described in Section 690.5. Exceptions are ground- or pole-mounted arrays with only one or two source circuits and all DC circuits isolated from buildings, and arrays at other than residences with appropriately sized equipment grounding conductors. Arrays mounted on residential roofs must include ground-fault protection in the DC circuit. This is because a roof is considered a more serious fire risk, with greater potential for loss of life and property than, for example, a ground-mounted array.

An array ground-fault protection device must detect a ground fault in the PV output circuit, interrupt the flow of fault current, and provide an indication of the fault. Therefore, these devices are sometimes called ground-fault detection and interruption (GFDI) devices. The faulted circuits are isolated by either automatically disconnecting the ungrounded conductors or causing the inverter or charge controller to automatically cease supplying power to output circuits. If the grounded conductor is opened as well, all conductors of the faulted circuit must then be opened, and it must be automatic and simultaneous.

All ground faults must flow through the grounding electrode conductor bonding connection, and there must be only one grounding connection in the DC circuit. Therefore, the array ground-fault protection device must be located at this point. Depending on the system configuration, this may be at the combiner box, array disconnect, or inverter.

The grounding connection must pass through the ground-fault protection device. Disconnecting this grounding connection, with either a fuse or circuit breaker, effectively interrupts the ground fault. The fuse or circuit breaker is activated when the ground fault exceeds a certain amount, normally 1 A.

Ground-fault protection is often built into the inverter, which includes a serviceable fuse. **See Figure 11-31.** Inverters are designed to immediately shut down and disconnect the ungrounded conductor if the fuse is opened.

For low-voltage PV systems, a pair of circuit breakers can be used to provide array ground-fault protection. **See Figure 11-32.** A lower-rated circuit breaker is mechanically tied to a higher-rated array circuit breaker, which acts only as a switch. The ground-fault circuit breaker trips when current between the grounded and grounding conductors exceeds its rating and forces the other circuit breaker to open the ungrounded conductor.

Array Ground-Fault Protection with Inverter Fuse

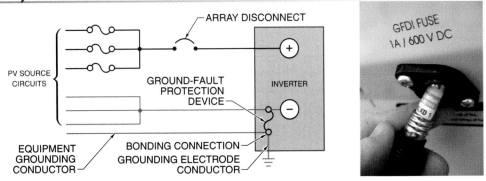

Figure 11-31. *Some inverters include fuses as array ground-fault protection in their DC input circuits.*

Array Ground-Fault Protection with Circuit Breakers

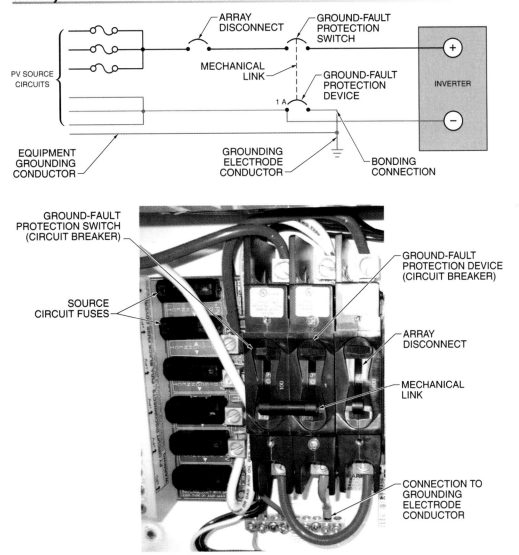

Figure 11-32. *Circuit breakers can be used for array ground-fault protection when the inverter does not already provide this protection.*

Ground-Fault Circuit Interrupters. A *ground-fault circuit interrupter (GFCI)* is a device that opens the ungrounded and grounded conductors when a ground fault exceeds a certain amount, typically 4 mA to 6 mA. It does this by sensing a difference between the current flowing out through the ungrounded conductor and returning through the grounded conductor. **See Figure 11-33.** For GFCI devices to function properly, the grounded conductor must be properly bonded to the equipment grounding conductor, typically at the service equipment.

Ground-Fault Circuit Interrupter

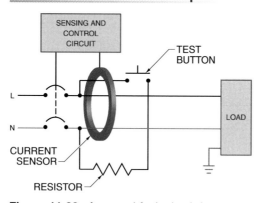

Figure 11-33. *A ground-fault circuit interrupter (GFCI) senses differences between the current in the grounded and ungrounded conductors, indicating a ground fault, and opens the circuit in response.*

A GFCI should not be confused with array ground-fault protection. A GFCI device is used in AC branch circuits to protect persons from electrical shock. GFCI protection is often included in receptacles and is required in wet environments, such as bathrooms, with a greater potential for ground faults. Article 210, "Branch Circuits," provides details and required locations for GFCI devices.

Some GFCI devices may not activate at the set fault current when used with modified square wave inverters, due to the way they sense fault currents. While some GFCI devices may work with these inverters, a sine wave inverter ensures the most reliable operation of GFCI devices.

Equipment Grounding

Equipment grounding protects personnel from the shock hazard of equipment enclosures and parts that may become energized under fault conditions. An equipment grounding conductor connects all exposed equipment, raceways, and enclosures to the grounding electrode conductor at the main service enclosure.

Equipment grounding requirements for PV systems are covered in Sections 690.43 and 690.45. All exposed non-current-carrying metal parts of module frames, support structures, enclosures, or other equipment in PV systems must be grounded. Equipment grounding is required regardless of system voltage, even for small 12 V or 24 V systems not otherwise required to have a grounded current-carrying conductor. Damaged PV arrays can energize their metal frames or support structures, but effective equipment grounding diverts this fault current safely to the ground.

The integrity of the electrical contact between the module frames and a grounded mounting structure cannot always be assured with typical fasteners. This is because the thin anodized layer of aluminum frames and structures, or the corrosion of inappropriate materials, may prevent a good electrical connection. Specially listed and identified devices that provide a secure electrical connection can be used to bond module frames to grounded mounting structures or other module frames. Alternatively, equipment grounding can be accomplished with continuous runs of bare conductor that are secured to each module with a special connector. **See Figure 11-34.**

When ground-fault protection is used, PV circuit equipment grounding conductors are sized in accordance with Article 250. The article establishes the minimum size for equipment grounding conductors based on the overcurrent protection rating for the circuit (regardless of whether an overcurrent protection device is actually used). **See Figure 11-35.** For example, if the PV output circuit overcurrent protection device is 60 A, then a 10 AWG equipment grounding conductor is required. The size of the equipment grounding conductor may need to be increased for voltage drop considerations.

Module Frame Grounding

Wiley Electronics LLC
BONDING TO GROUNDED STRUCTURE

CONTINUOUS CONDUCTOR

Figure 11-34. *Modules should be connected to each other and the mounting structure with grounding conductors to ensure a continuous grounding connection.*

Grounding Conductor Sizing

RATING OF OVERCURRENT PROTECTION DEVICE IN CIRCUIT*	CONDUCTOR SIZE†
15	14
20	12
30	10
40	10
60	10
100	8
200	6

* in A
† in AWG for copper conductors

Excerpted from NEC® Table 250.122. Reprinted with permission from NFPA 70-2008, the National Electrical Code® Copyright® 2007, National Fire Protection Association, Quincy, MA 02169. This reprinted material is not the official position of the NFPA on the referenced subject which is represented solely by the standard in its entirety.

Figure 11-35. *Equipment grounding conductors are sized based on the rating of the overcurrent protection device in the circuit.*

When ground-fault protection is not used, the equipment grounding conductor is sized for twice the temperature- and conduit-fill-derated ampacity of the circuit's current-carrying conductors. For example, a source circuit with maximum current of 8 A requires conductors that can safely carry 8 A of current when derated. Therefore, this circuit would require an equipment grounding conductor sized for at least 16 A (8 A × 2 = 16 A).

Although this ampacity could be met with a smaller conductor, a minimum of 14 AWG conductor is required for adequate mechanical strength. Similarly, an output circuit containing four of these strings in parallel would require an equipment grounding conductor rated for at least 64 A (8 A × 4 × 2 = 64 A).

Grounding Continuity

The continuity between all grounding conductors and the grounding electrode must be maintained. When components are removed for service or replacement, other equipment or components may become disconnected from the grounding electrode conductor, causing a safety hazard. A bonding jumper must be installed to maintain grounding continuity to the entire system while equipment is removed. Both the equipment grounding system and the grounded conductor system must be maintained with this method. This requirement also applies to modules and panels removed from an array for access or replacement.

Ungrounded PV Systems

In an effort to harmonize requirements in the United States with those in Europe, which currently has more experience with ungrounded PV systems, the NEC® Section 690.35 permits PV arrays to have ungrounded source and output circuits, but only when certain conditions apply. This allowance is in addition to permitting PV systems operating below 50 V to be ungrounded. However, neither allowance exempts a system from equipment grounding requirements.

In ungrounded systems, all source and output circuit conductors must be in sheathed multiconductor cables, installed in conduit, or be listed and identified as PV wire. Both ungrounded array conductors (positive and negative) must include disconnecting means, overcurrent protection, and ground-fault protection. Inverters and charge controllers used with ungrounded PV source circuits must be listed for ungrounded systems. Because of the additional disconnects, overcurrent protection, and other equipment required for ungrounded systems, grounded systems are usually less expensive and easier to install.

The most common type of grounding electrode is a long, metal rod that is driven into the ground near the building.

Lightning Protection Systems

Lightning strikes can cause dangerous and damaging voltage transients. Because PV arrays are often mounted on elevated structures, such as rooftops, many PV systems must be protected from potential lightning damage. Lightning protection is especially important in the southeastern United States, which experiences the highest rates of lighting strikes in the country. **See Figure 11-36.** Lightning protection system requirements are covered briefly in Article 250 and more extensively in NFPA 780, *Standard for the Installation of Lightning Protection Systems.*

Lightning protection systems consist of a low-impedance network of air terminals (lightning rods) connected to a special grounding electrode system. This does not violate the rule mandating only one ground connection for the DC system since the lightning grounding electrode system is not connected to the grounded conductor. The system conducts any surges induced by direct or indirect lightning strikes to ground, safely away from the building and equipment. **See Figure 11-37.**

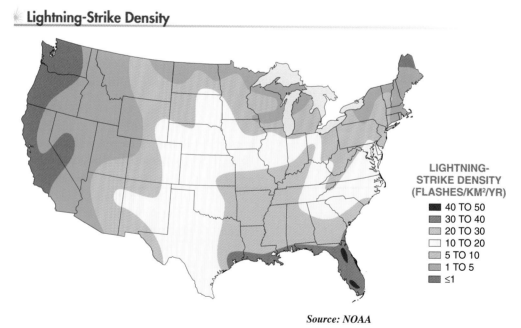

Lightning-Strike Density

LIGHTNING-
STRIKE DENSITY
(FLASHES/KM²/YR)

- 40 TO 50
- 30 TO 40
- 20 TO 30
- 10 TO 20
- 5 TO 10
- 1 TO 5
- ≤1

Source: NOAA

Figure 11-36. *Lightning protection is especially important in the southeastern states, which have the highest lightning-strike density in the United States.*

Lightning Protection System

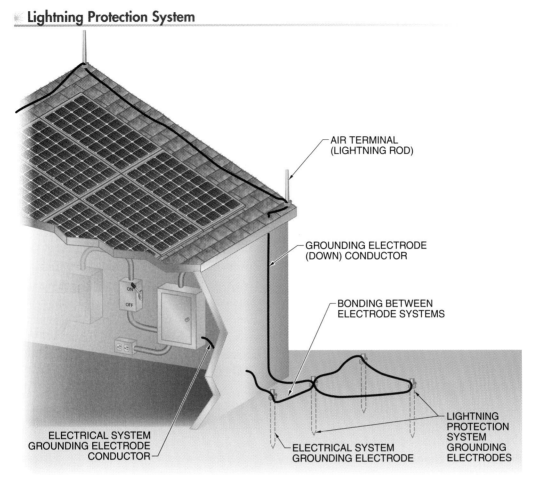

Figure 11-37. *A lightning protection system includes a network of air terminals, a grounding electrode (down) conductor, and a set of grounding electrodes.*

The grounding electrodes of the lightning protection system must also be bonded to the electrical-service grounding electrode system. At points other than the grounding electrode, the effectiveness of a lightning protection system depends on the separation distance between its conductors and other conductors within the electrical system.

Surge Arrestors

A *surge arrestor* is a device that protects electrical devices from transients (voltage spikes). It does this by either limiting or shorting to ground all voltage above a certain threshold. A surge arrestor is also known as a surge suppressor or surge protector. The *clamping voltage* is the voltage at which a surge arrestor initiates its transient protection. For PV systems, the clamping voltage should be greater than the maximum expected open-circuit voltage so that the device will activate only in the event of an extreme voltage surge, but still low enough for adequate protection.

The maximum current that the device can handle should be at least as great as any potential fault currents from the PV system or the utility system. The energy rating of a surge arrestor is the maximum energy-dissipating capability of the device. This rating is given in joules. Surge arrestors with higher ratings provide more protection.

Article 280, "Surge Arrestors, Over 1 kV," covers installation requirements for surge arrestors. Surge arrestors may be used on the AC or DC sides of PV systems, or both. When used in PV systems, surge arrestors are typically connected between the positive and negative leads of each source circuit or between the output circuit and ground.

Surge arrestors must be listed and marked with their ratings, and because they rely on a connection to ground to dissipate surges, they cannot be used in ungrounded systems. For maximum protection, surge arrestors should be located as close to the protected equipment as possible. Some components may already include surge arrestors, but they can be field-installed as separate devices. **See Figure 11-38.**

Surge Arrestors

Figure 11-38. *Surge arrestors may be incorporated into equipment or can be installed on circuits as separate devices.*

Varistors. A *varistor* is a solid-state device that has a high resistance at low voltages and a low resistance at high voltages. The term "varistor" is a combination of the words "variable resistor." Varistors are used as surge arrestors in many electrical circuits, including in inverters and charge controllers, to protect from surges induced on PV output circuits. When a voltage above the clamping voltage is applied, the resistance of the varistor falls, allowing large currents to flow to ground.

A varistor can handle repeated surges, but if a surge exceeds its energy rating, such as from a direct lightning strike, the device may be damaged or destroyed. Common types of varistors include the metal-oxide varistor (MOV) and the silicon-oxide varistor (SOV). MOVs may be included inside components, but are not available as separate listed devices so they are generally

not field-serviceable. SOVs are available as listed or recognized components. They may be installed separately or inside electrical equipment such as combiner boxes and inverters.

Transient Voltage Surge Suppressors. A transient voltage surge suppressor is another type of surge arrestor. A *transient voltage surge suppressor (TVSS)* is a surge-protective device that limits transient voltages by diverting or limiting surge current. This device is also known as a surge protective device (SPD), following the terminology of the NEC® and the UL standard that covers their testing.

A TVSS is represented by two opposing Zener diodes. Under normal conditions, the TVSS maintains a high resistance. When a transient voltage is encountered, the Zener diodes exhibit a controlled breakdown that allows excess current to flow to ground. TVSS devices are used similarly to varistors in electrical circuits, including PV systems. Article 285, "Surge-Protective Devices (SPDs), 1 kV or Less," addresses requirements for TVSS installation.

BATTERY SYSTEMS

General information about the installation of battery systems is included in Article 480, "Storage Batteries." Requirements for the interconnection of batteries and charge control in PV systems are covered in Article 690, Part VIII. Many requirements depend on the size of the battery bank, which is quantified by its nominal operating voltage.

Battery Banks Less Than 50 V

Residential PV-system battery banks are limited to less than 50 V nominal, or no more than 24 series-connected, 2 V lead-acid cells (which are actually 48 V nominal). All live battery parts must be guarded in accordance with Article 110, "Requirements for Electrical Installations." Whenever the available short-circuit current in a battery system exceeds the ratings of other equipment in that circuit, current-limiting overcurrent protection devices are required.

Battery Banks Greater Than 50 V

Battery systems of greater than 48 V must include a disconnect to divide the system into segments of 48 V or less for maintenance. There must also be a main disconnect for the entire battery bank that is accessible to qualified persons only. This disconnect must open the grounded conductor in the battery circuit, but not any other grounded conductors from other parts of the system. A bolted or plug-in nonload break switch may be used for this purpose. **See Figure 11-39.**

Battery Bank Disconnects

Figure 11-39. *Connectors may be used for disconnecting high-voltage battery banks for servicing.*

Battery systems greater than 48 V may operate with ungrounded conductors, provided the PV source and output circuits are grounded, all AC and DC load circuits are grounded, the ungrounded battery conductors have overcurrent protection and disconnect means, and a ground-fault detector and indicator is installed to monitor battery bank faults.

Flooded, lead-acid batteries in these battery banks may not have conductive cases or be installed on metal racks that come within 6″ of the tops of the cases. This is because ground faults have been attributed to battery systems with higher voltages when an electrolyte film accumulates on the tops of flooded batteries and creates short circuits between the battery terminals and grounded cases or racks. This requirement does not apply to any type of sealed battery.

Charge Control

PV systems with batteries must also include some method of controlling the charge applied to the batteries. In the case of self-regulated systems, charge control is accomplished through careful sizing and matching of the array to the battery bank. Most systems, however, which have a potential charging current greater than C/33 (3% of the battery capacity per hour), must include an active means of charge control. Charge controllers may be a separate component, or charge control features may be included with inverters in power conditioning units.

Diversion loads are commonly used for charge control and must be appropriately rated for this application. The load's voltage rating must be greater than the maximum battery voltage, the current rating must be less than or equal to the charge controller's current rating, and the power rating must be at least 150% of the array's power rating. The conductors and overcurrent protection devices in the diversion-load circuit must be sized for 150% of the maximum current rating of the charge controller.

Even when diversionary charge controllers are used, an additional independent means of charge control must be provided as a backup. This is because the 100% availability of the diversion load cannot be guaranteed and the batteries must always be protected from potential overcharge.

Summary

- The majority of the regulations governing electrical installations, including PV systems, are found in NFPA 70: National Electrical Code® (NEC®)

- Improper electrical integration can have significant consequences, including poor performance, component damage, and electrical shock hazards.

- Article 690 addresses requirements for all PV installations covered under the scope of the NEC®.

- Maximum system voltages and currents must be calculated because they affect the sizing and ratings for conductors, overcurrent protection devices, disconnects, and grounding equipment.

- Since temperature affects the voltage output of PV devices, the maximum possible voltage is the array's open-circuit voltage at the lowest expected temperature for the location.

- For PV source circuits, the maximum current is 125% of the sum of the short-circuit current ratings of parallel-connected modules.

- Conductor size is chosen to be the smallest size that safely conducts the maximum continuous current of a circuit with the appropriate margin of safety. Voltage drop may further increase required conductor size.

- Nominal conductor ampacity at 30°C is determined by the conductor material (copper or aluminum), size, insulation type, and application (direct burial, conduit, or free air).

- The insulating material defines three critical properties for a conductor: its maximum operating temperature; its application and environmental resistance such as to sunlight, oil, or moisture; and its permissible installation locations such as direct burial, in conduit, or exposed.

- A variety of connectors, terminals, splices, and lugs may be used for field-installed PV system wiring, as long as they are appropriate and rated for the application, such as for conductor material, size, and environment.

- The NEC® does not require a specific design for use in DC branch circuits, but allows any that conform to three requirements: the plugs and receptacles must be listed for DC power with a rating of at least 15 A, must have a separate equipment-grounding terminal, and must be different from any other receptacles used on the premises.

- Most PV modules include a plastic junction box attached to the back surface. The junction box contains the module's electrical terminals and often the bypass diodes.

- Combiner boxes are used to combine parallel array source circuits into the PV output circuit. These boxes typically include terminal blocks, fuses, and fuse holders.

- Most PV system conductors are installed in conduit. However, source-circuit and output-circuit conductors are required to be in conduit only when installed in a readily accessible location and their maximum circuit voltages are greater than 30 V.

- A circuit breaker can be used to satisfy both the overcurrent protection and disconnect requirements in some parts of a PV system.

- The current rating of an overcurrent protection device depends on the ampacity of the circuit conductors. The overcurrent protection device rating required for PV circuits is 125% of the calculated maximum circuit current.

- In most systems, every ungrounded conductor must include overcurrent protection. There are only a few exceptions, which primarily apply to PV source and output circuits.

- Disconnects are required for both the AC and DC sides of a PV system. Disconnects are also required to isolate other power sources, such as batteries, and may be included at additional points to facilitate system testing or maintenance.

- A PV power source operating over 50 V must have one grounded conductor, with the exception of ungrounded systems meeting other special requirements.

- Grounding electrode system requirements for PV systems cover three specific grounding cases: AC systems, DC systems, or a combination of both.

- Since the DC grounded conductor is not directly connected to the AC grounded conductor, they are considered separate systems. These two grounding systems must be bonded together.

- In order to maintain uniform grounding throughout the entire system, an additional and separate grounding electrode is required at the array location if the array grounding electrode would be more than 6′ from the premises wiring electrode.

- Most arrays are required to include ground-fault protection, which must detect a ground fault in the PV output circuit, interrupt the flow of fault current, and provide an indication of the fault.

- Equipment grounding protects personnel from the shock hazard of equipment enclosures and parts that may become energized under fault conditions.

- Specially listed and identified devices that provide a secure electrical connection can be used to bond module frames to grounded mounting structures or other module frames.

- Because PV arrays are often mounted on elevated structures, such as rooftops, many PV systems must be protected from potential lightning damage.

- Residential PV-system battery banks are limited to less than 50 V nominal, or no more than 24 series-connected, 2 V lead-acid cells (which are actually 48 V nominal).

- PV systems with batteries must also include some method of controlling the charge applied to the batteries.

Definitions

- A *PV power source* is an array or collection of arrays that generates DC power.

- A *PV source circuit* is the circuit connecting a group of modules together and to the common connection point of the DC system.

- A *PV output circuit* is the circuit connecting the PV power source to the rest of the system.

- *Ampacity* is the current that a conductor can carry continuously under the conditions of use without exceeding its temperature rating.

- A *junction box* is a protective enclosure used to terminate, combine, and connect various circuits or components together.

- A *combiner box* is a junction box used as the parallel connection point for two or more circuits.

- A *blocking diode* is a diode used in PV source circuits to prevent reverse current flow.

- A *bypass diode* is a diode used to pass current around, rather than through, a group of PV cells.

- An *overcurrent protection device* is a device that prevents conductors or devices from reaching excessively high temperatures due to very high currents by opening the circuit.

- A *fuse* is a metallic link that melts when heated by current greater than its rating, opening the circuit and providing overcurrent protection.

- A *circuit breaker* is an electrical switch that automatically opens as a means of overcurrent protection, and that can be manually opened as a disconnecting means.

- The *interrupting rating* is the maximum current that an overcurrent protection device is able to stop without being destroyed or causing an electric arc.

- A *supplementary overcurrent protection device* is an overcurrent protection device intended to protect an individual component and is used in addition to a current-limiting branch circuit overcurrent protection device.

- A *back-fed circuit breaker* is a circuit breaker that allows current flow in either direction.

- A *disconnect* is a device used to isolate equipment and conductors from sources of electricity for the purpose of installation, maintenance, or service.

- *Grounded* is the condition of something that is connected to the earth or to a conductive material that is connected to the earth.

- A *grounding electrode* is a conductive rod, plate, or wire buried in the ground to provide a low-resistance connection to the earth.

- A *grounding electrode conductor (GEC)* is a conductor connecting the grounding electrode to the rest of the electrical grounding system.

- A *grounded conductor* is a current-carrying conductor that is intentionally grounded.

- The *equipment grounding conductor (EGC)* is a conductor connecting exposed metallic equipment, which might inadvertently become energized, to the grounding electrode conductor.

- An *ungrounded conductor* is a current-carrying conductor that has no connection to ground.

- A *ground fault* is the undesirable condition of current flowing through the grounding conductor.

- *Ground-fault protection* is the automatic opening of conductors involved in a ground fault.

- A *ground-fault circuit interrupter (GFCI)* is a device that opens the ungrounded and grounded conductors when a ground fault exceeds a certain amount, typically 4 mA to 6 mA.

- A *surge arrestor* is a device that protects electrical devices from transients (voltage spikes).

- The *clamping voltage* is the voltage at which a surge arrestor initiates its transient protection.

- A *varistor* is a solid-state device that has a high resistance at low voltages and a low resistance at high voltages.

- A *transient voltage surge suppressor (TVSS)* is a surge-protective device that limits transient voltages by diverting or limiting surge current.

1. How do Article 690 and other NEC® articles apply to the planning and installing of a PV system?

2. What maximum voltage and current calculations are required by the NEC® and how are they used?

3. What factors affect conductor ampacity?

4. List ways in which voltage drop in the PV output circuit can be reduced.

5. What are the requirements for conductors used in source-current wiring?

6. What are the requirements for plugs and receptacles used for DC branch circuits?

7. Explain the two most common applications for junction boxes in PV systems.

8. What are the differences between blocking diodes and bypass diodes?

9. How are conductors and overcurrent protection in array circuits sized in relation to short-circuit currents?

10. Which PV-system circuits require overcurrent protection and which conductors must be protected?

11. How are supplementary overcurrent protection devices used in PV circuits?

12. Which PV-system circuits require disconnects?

13. Describe the two acceptable methods of grounding a PV system that includes both AC and DC circuits.

14. Under what circumstances may the DC circuits in a PV system be ungrounded?

15. When does a PV system with batteries require active means of charge control?

12

Utility Interconnection

Interconnection is the technical and procedural process of connecting and operating PV and other distributed generation systems with the electric utility system. However, since an interconnected PV system may affect the grid network and the safety of electrical workers, the system must adhere to certain requirements, and utilities have the right to approve equipment and installations. These interconnections are then governed by contractual agreements between the utility and the PV system owner.

Chapter Objectives

- *Compare the differences in the interconnection of rotating generators and electronic inverters.*
- *Identify the applicable codes and standards for utility interconnection.*
- *Describe how interconnected PV systems can affect utility operations.*
- *Differentiate between load-side and supply-side interconnections and identify the code and installation requirements for each type.*
- *Compare the technical and policy issues between net metering and dual metering.*
- *Identify the common issues addressed in interconnection agreements for PV systems.*

DISTRIBUTED GENERATION

Distributed generation is a system in which many smaller power-generating systems create electrical power near the point of consumption. Operating these systems in parallel with the utility's distribution grid makes these systems interactive. **See Figure 12-1.**

SolarWorld Industries America

Interconnected PV systems provide supplemental power and some can be used as back-up power in the event of a utility outage.

Distributed generation is an increasingly common supplement to traditional central power generation. This arrangement increases the diversity and security of the electrical energy supply and benefits both customers and electric utilities. For customers, these systems provide power to on-site loads and some provide back-up in the event of a utility outage. For utilities, the additional power sources supplying their excess power to the grid increase the utility's capacity to serve customers without building new power plants.

Accordingly, a great deal of attention is focused on interconnecting these power resources. However, there are technical, procedural, and contractual requirements to making interconnections that are safe and reliable and that do not adversely impact the rest of the electricity distribution system.

Many distributed-generation technologies can be interconnected to the utility grid, including engine generators, fuel cells, wind turbines, micro-hydroelectric turbines, and, of course, PV systems. Distributed-generation technologies are also classified by the way AC power is generated, either with rotating generators or with electronic inverters. This has important implications for connecting these systems to the electric utility grid.

Distributed Generation

Figure 12-1. *With distributed generation, utility customers are served by both the centralized power plants and the power exported from interconnected distributed generators.*

Generators

Most electrical power is produced by rotating generators that are mechanically driven. The mechanical energy is produced by engines or by turbines driven by steam, wind, or water.

Current is induced in a conductor as it moves through a magnetic field. If the movement is from rotation, the current varies sinusoidally, producing AC power. If a set of coils rotates through a stationary magnetic field, the design is known as a generator. With three sets of equally-spaced coils, three-phase power is produced. **See Figure 12-2.** Alternatively, a design with a magnetic field rotating around a set of stationary coils produces the same output. The resulting voltage is proportional to the strength of the magnetic field, and the frequency is proportional to the rotational speed.

Generators

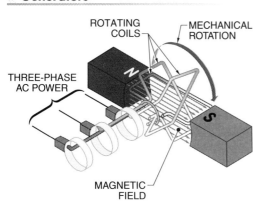

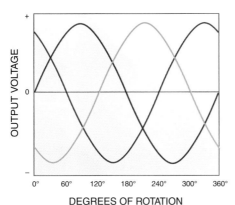

Figure 12-2. *Most power is generated by rotating generators, which require precise interconnection procedures to avoid damage to the equipment and the utility.*

The electric utility system consists of hundreds or thousands of interconnected generators operating in parallel. These generators must all be synchronized with one another to prevent damaging themselves or the interconnection interfaces. *Synchronizing* is the process of connecting a generator to an energized electrical system. Synchronizing is extremely critical and involves precisely matching the phase sequence, frequency, and voltage of the generator to the rest of the electrical system before they are connected.

The electric utility system must also maintain a critical balance between load and generation. If generation exceeds the load on the utility system, the frequency of the system increases. Conversely, if the load exceeds the generation, frequency decreases. Frequency is one of the most important utility parameters because it establishes the speed for devices using motors, including some clocks. In North America, it must be maintained at exactly 60 Hz. Frequency control is established by varying power generation to match the load.

Other parameters are also constantly monitored for variations, which requires external sensors and controls. If the interface parameters go beyond acceptable limits, the controls automatically disconnect the generator from the utility system.

Inverters

Solid-state electronic inverters are substantially different from generators in the way they operate, which has important implications for connecting them to the utility system. Solid-state inverters use electronic switching elements to produce AC power, so they have no moving parts.

Generators act like a voltage source, while interactive inverters act like a current source. Generators can operate independently of the grid and produce high fault currents. Inverters feed much less current into a fault and are less capable of supporting an islanded electrical power system. However, unlike generators, inverters cannot act as loads and consume power from other generators or inverters.

Inverters are also loaded differently than generators. Since the sun cannot be turned ON and OFF, most interactive PV inverters initially load the array at its open-circuit voltage where it produces no power. Once loaded, the inverter decreases the array operating voltage to the maximum power voltage. Maximum power point tracking (MPPT) electronics then continually make adjustments to the array operating voltage to maximize output. Power output is limited by the capacity of the PV array, the inverter output rating, and temperature.

For interactive inverters, synchronizing functions are performed automatically and internally. Since inverters have the ability to monitor and regulate system output directly through microprocessor controls, synchronizing can be done cost-effectively and directly. Inverters can also incorporate additional protective and safety features that may otherwise be required as external equipment on generators. Since the interactive inverter is the primary utility interface device, it must meet all requirements for utility interconnection and be listed and identified for use in interactive PV systems. **See Figure 12-3.**

Interactive Inverter Labeling

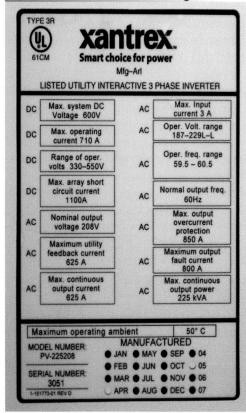

Figure 12-3. *Inverters must be identified as interactive and listed to required standards before being interconnected.*

Inverters can be stand-alone, interactive, or bimodal types. There is no special type for use in hybrid systems, as it is not possible for most inverters to interconnect multiple sources. Instead, multiple separate inverters are used when necessary.

Interactive-Only Systems. The most common type of interactive PV system is one that does not use energy storage. The array is connected to the DC input of an interactive-only inverter. The inverter output interfaces with the utility, typically at a site distribution panel or electrical service entrance. **See Figure 12-4.**

Power can flow in both directions at the point of service, making the PV system a supplementary generation source operating on the electric utility network. When on-site power demand exceeds the supply from the PV system, supplementary power is drawn from the utility. If the PV array produces more power than is needed on-site, the excess power is fed to the grid.

Bimodal Systems. Bimodal systems are interactive systems that include battery storage, so they can operate in either interactive or stand-alone mode, providing a back-up power supply for critical loads such as computers, refrigeration, water pumps, or lighting. Unlike interactive-only inverters, the DC input of the inverter is connected to the battery bank, not the array. The array charges the battery bank, through the charge controller. **See Figure 12-5.**

Under normal circumstances, bimodal systems operate in interactive mode. The inverters serve the on-site loads or send excess power back to the grid while the array keeps the battery bank fully charged. If the grid de-energizes, control circuitry in the inverter opens the connection with the utility and draws power from the batteries to supply an isolated subpanel, typically for critical loads.

Interactive-Only Systems

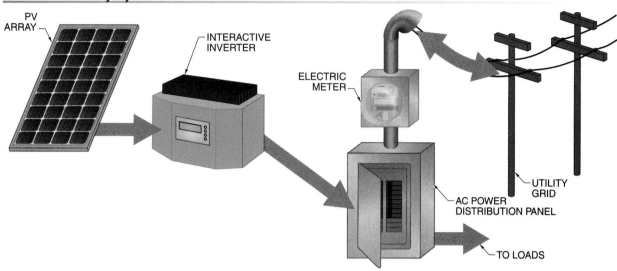

Figure 12-4. *Interactive-only systems are the simplest way to interconnect a PV system with the utility.*

Bimodal Systems

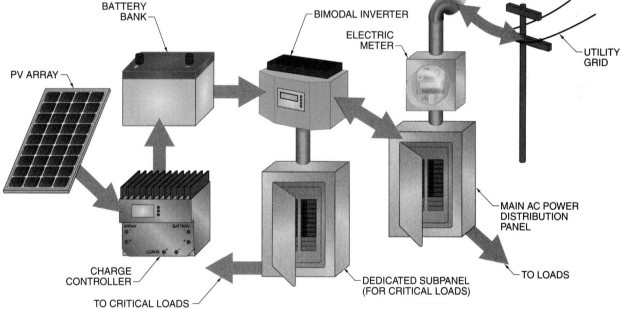

Figure 12-5. *The inverters in bimodal systems can continue to supply power to certain loads in the event of a utility outage.*

For servicing the system without interrupting the load operation, a manual transfer switch and bypass circuit can isolate the PV array, battery, and inverter from the system and directly connect the subpanel loads to the utility supply.

Bimodal systems with battery storage can also be employed to manage and optimize utility energy use by utility customers that are billed using time-of-use electric rates, or those that incur demand charges for peak power use. Bimodal inverters are programmed to supply electrical loads with energy from the battery and PV generation during peak times, minimizing the use of high-priced utility energy. During off-peak times, less-expensive utility energy powers system loads and charges the battery system.

Interconnection Codes and Standards

In the United States, the technical requirements for PV systems are established through national codes and standards published by the Institute of Electrical and Electronics Engineers (IEEE), Underwriters Laboratories (UL), and the National Fire Protection Association (NFPA). These organizations work collectively to ensure electrical equipment and installations are safe, through the combination of standards, equipment testing and certification, and enforceable codes. As a PV system is an electrical system, any general electrical codes and standards apply, such as the National Electrical Code®. Additional standards dealing specifically with the interconnection of PV systems also apply. **See Figure 12-6.**

IEEE 929. ANSI/IEEE 929-2000, *Recommended Practice for Utility Interface of Photovoltaic (PV) Systems* provided uniform interconnection requirements widely accepted by many utilities and jurisdictions, and recommended no additional requirements for small PV systems of 10 kW and less. IEEE 929 is no longer an active standard, but its requirements have been incorporated into the current standard ANSI/IEEE 1547.

IEEE 929 was important to the early history of PV systems because it established requirements for power quality and safety features for interactive inverters. Power quality requirements included specifications for service voltage, frequency, harmonic distortion, and power factor. Safety features included anti-islanding, reconnection to the utility service after an interruption, and responses to abnormal utility conditions, such as voltage and frequency disturbances. It also provided guidance on DC isolation, grounding, and manual disconnects. These and other requirements have been incorporated into IEEE 1547.

Interconnection Codes and Standards

DOCUMENT NUMBER	TITLE
IEEE 929*	Recommended Practice for Utility Interface of Photovoltaic (PV) Systems
IEEE 1547	Standard for Interconnecting Distributed Resources with Electric Power Systems
IEEE 1547.1	Standard Conformance Test Procedures for Equipment Interconnecting Distributed Resources with Electric Power Systems
IEEE 1547.2	Application Guide for IEEE Standard 1547, Standard for Interconnecting Distributed Resources with Electric Power Systems
IEEE 1547.3	Guide for Monitoring, Information Exchange, and Control of Distributed Resources Interconnected with Electric Power Systems
IEEE P1547.4	Draft Guide for Design, Operation, and Integration of Distributed Resource Island Systems with Electric Power Systems
IEEE P1547.5	Draft Technical Guidelines for Interconnection of Electric Power Sources Greater than 10 MVA to the Power Transmission Grid
IEEE P1547.6	Draft Recommended Practice for Interconnecting Distributed Resources with Electric Power Systems Distribution Secondary Networks
IEEE P1547.7	Draft Guide to Conducting Distribution Impact Studies for Distributed Resource Interconnection
UL 1741	Inverters, Converters, Controllers, and Interconnection System Equipment for Use with Distributed Energy Resources
NEC® Article 690	Solar Photovoltaic Systems

* IEEE 929 is no longer active and has been replaced by IEEE 1547.

Figure 12-6. *Several codes and standards address specific interconnection issues with PV systems.*

IEEE 1547. ANSI/IEEE 1547-2003, *Standard for Interconnecting Distributed Resources with Electric Power Systems,* is a broader interconnection standard addressing requirements for all types of distributed power sources, including PV systems, fuel cells, wind turbines, engine generators, and large combustion turbines. It establishes requirements for testing, performance, maintenance, and safety of interconnections, as well as responses to abnormal events, islanding, and power quality.

The focus of IEEE 1547 is on distributed-generation resources with capacities of less than 10 MW that are interconnected to the electrical utility system at typical primary or secondary distribution voltages. The standard provides universal requirements to help ensure safe and technically sound interconnections. It does not address limitations or impacts on the utility system in terms of energy supply, nor does it deal with procedural or contractual issues associated with the interconnection.

IEEE 1547 is actually a family of standards, guides, and recommended practices. While IEEE 1547 addresses core issues regarding interconnection, specific technical issues are addressed in the IEEE 1547.X series.

IEEE 1547.1-2005, *Standard Conformance Test Procedures for Equipment Interconnecting Distributed Resources with Electric Power Systems,* specifies commissioning tests that shall be performed to demonstrate that the equipment and operation of distributed power sources conforms to IEEE 1547. In particular, emphasis is placed on anti-islanding protection.

IEEE 1547.2-2008, *Application Guide for IEEE Standard 1547, Standard for Interconnecting Distributed Resources with Electric Power Systems,* provides technical background and application details to support the understanding of IEEE 1547. It characterizes the various distributed-generation technologies and their associated interconnection issues and discusses the background and rationale of the technical requirements.

IEEE 1547.3-2007, *Guide for Monitoring, Information Exchange, and Control of Distributed Resources Interconnected with Electric Power Systems,* facilitates the interoperability of one or more distributed-generation resources interconnected with electric power systems.

IEEE P1547.4, *Draft Guide for Design, Operation, and Integration of Distributed Resource Island Systems with Electric Power Systems,* provides alternative approaches and practices for the interconnection of distributed-generation systems that also provide critical load backup (bimodal systems). It includes provisions to separate from and reconnect to an area electric power system while providing power to the islanded local power system.

IEEE P1547.5, *Draft Technical Guidelines for Interconnection of Electric Power Sources Greater than 10 MVA to the Power Transmission Grid,* provides guidelines for interconnecting utility-scale power sources to a bulk power transmission grid.

IEEE P1547.6, *Draft Recommended Practice for Interconnecting Distributed Resources with Electric Power Systems Distribution Secondary Networks,* provides guidance on technical issues associated with the interconnection of distributed-generation systems, including recommendations for the performance, operation, testing, safety considerations, and maintenance of the interconnection. Consideration is given to the needs of the local utility to be able to provide enhanced service to site loads as well as to other loads served by the grid.

IEEE P1547.7, *Draft Guide to Conducting Distribution Impact Studies for Distributed Resource Interconnection,* provides criteria and scope for engineering studies regarding the impact of interconnected distributed power sources to an area electric power distribution system. This guide facilitates a methodology for when such impact studies are appropriate, what data is required, how the studies are performed, and how the study results are evaluated.

> The IEEE 1547 base standard, and the related 1547.1, 1547.2, and 1547.3 standards, are approved and published documents. The 1547.4 through 1547.7 standards are still in development, designated by the letter "P" before their numbers and the word "Draft" in their names.

UL 1741. UL 1741-2005, *Inverters, Converters, Controllers, and Interconnection System Equipment for Use with Distributed Energy Resources,* addresses requirements for distributed-generation equipment, including inverters and charge controllers, and for the utility interface. The standard is intended to supplement IEEE 1547 and IEEE 1547.1. The products covered by these requirements are intended to be installed in accordance with the NEC® and NFPA 70E.

NEC® Article 690. The NEC® Article 690, "Solar Photovoltaic Systems" covers installation requirements for all types of PV systems. In particular, Part VII of Article 690 addresses the requirements for connecting interactive PV systems to other power sources, such as the electric utility grid. Many of these requirements are based on equipment standards and listing requirements for interactive inverters under UL 1741.

Article 690 includes several fundamental requirements for interactive inverters. First, inverters must be listed and identified for interactive operation. This information must be included on the inverter nameplate or label. Also, inverters must be anti-islanding, or capable of disconnecting from the utility grid when grid voltage is lost. The PV system must remain disconnected until grid voltage is restored.

> Large-scale systems owned and operated by the electric utility and connected to the utility's own distribution substations often are not always subject to the same requirements for installation and interconnection as residential and commercial interactive systems.

Interconnection Concerns

Electric utilities have legitimate concerns about connecting distributed-generation equipment to their grid system. For one, the electric utility distribution system was not designed to handle many distributed power sources. For example, many utility service meters are not designed to monitor two-way power flow. Also, because utilities do not operate, control, or maintain customer-owned distributed-generation equipment connected to their system, there are concerns about the impact it may have on safety and the reliability of utility service to other customers.

Islanding. *Islanding* is the undesirable condition where a distributed-generation power source, such as a PV system, continues to transfer power to the utility grid during a utility outage. **See Figure 12-7.** Islanding is a serious safety hazard for utility lineworkers working to restore power after an outage. Since the workers expect the grid to be de-energized, they may be shocked by the power present in the system from an islanding inverter. Also, islanding can damage utility equipment by interfering with the utility's normal procedures for restoring service following an outage, primarily because the islanded electrical system is no longer in-phase with the utility system.

All utility-interactive inverters must be able to detect power outages and discontinue transfer of power until the utility system returns to normal operation. This precaution is evaluated under UL 1741 as part of the equipment listing. Inverters meeting this standard are often referred to as "non-islanding" or "anti-islanding" interactive inverters. As an added precaution against islanding, utilities may also require easily accessible outdoor switches to physically disconnect interactive PV systems. Bimodal systems may continue operating in stand-alone mode if completely disconnected from the utility.

Power Quality. Power quality encompasses the voltage, frequency, harmonic distortion, power factor, DC injection, voltage flicker, and noise on the utility system. Power quality is influenced by the performance of the electrical generation and distribution equipment, as well as electrical loads operating on the system.

Power quality is among a utility's chief concerns for interconnected distributed-generation equipment. Because electrical loads are designed to operate at prescribed conditions, the electric utility system must be maintained within stringent power quality limits. Otherwise, poor performance or even damage to electrical loads and utility system equipment may result.

Islanding

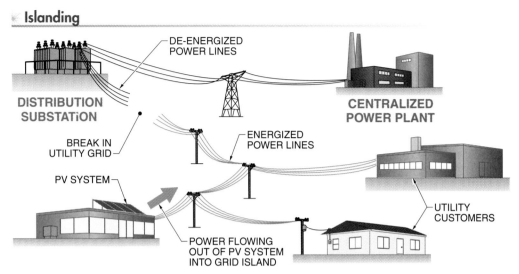

Figure 12-7. *During a utility outage, an islanding inverter can energize the utility lines around the PV system, potentially damaging equipment and creating a serious safety hazard.*

Utilities routinely test and monitor their generation, transmission, and distribution equipment for performance and power quality. Correspondingly, utilities also require assurances that interactive distributed-generation systems are operating within these limits and do not adversely affect the quality of utility service to other customers.

Neutral Loading. Inverter and utility service types must be matched to prevent overloading. If a two-wire, single-phase inverter is connected to the neutral and one of the ungrounded conductors of a split-phase 120/240 V service or a three-phase, wye-connected service, the return current on the neutral (grounded) conductor will not balance and may overload the conductor. **See Figure 12-8.** Therefore, the neutral conductor must be oversized. NEC® Section 690.62 requires that the sum of the maximum load between the neutral and ungrounded conductor and the inverter's current output rating must not exceed the ampacity of the neutral conductor. Some inverters avoid this problem with two ungrounded outputs. Alternatively, multiple inverters can be installed, one for each ungrounded conductor.

Unbalanced Phases. Section 690.63 does not allow the connection of single-phase interactive PV inverters to three-phase power systems

unless the interconnection can be designed to minimize unbalanced voltages between the phases. **See Figure 12-9.** One option would be to use three small inverters and connect each inverter to a different phase. A better solution would be to use a single three-phase inverter. With a three-phase inverter, all phases must automatically de-energize when voltage becomes unbalanced or any phase is lost completely.

Neutral Loading

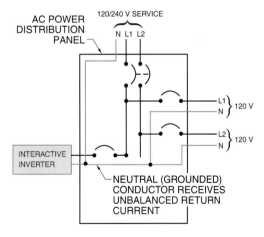

Figure 12-8. *When a single-phase inverter is added to a system with more than one ungrounded (hot) conductor, the neutral conductor can become overloaded.*

Unbalanced Phases

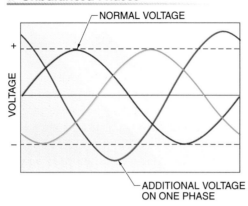

Figure 12-9. *Adding single-phase output from an interactive inverter to a three-phase power system can result in unbalanced voltages between the phases.*

Point of Connection

The *point of connection* is the location at which an interactive distributed-generation system makes its interconnection with the electric utility system. Power flow between the utility and distribution system may occur in both directions at the service connection, so all the service equipment must be sized and rated to allow this.

Section 690.64 permits the output of interactive PV inverters to be connected to either the load side (customer side) or supply side (utility side) of the service disconnect. **See Figure 12-10.** For many smaller systems, the point of connection is usually made on the load side, usually at a circuit breaker in the distribution panel. When the requirements for a load-side connection are not possible (such as due to the size of the PV system), interactive systems may be connected to the supply side. In cases of very large PV installations, existing service conductor ampacity may not be sufficient and separate services may need to be installed.

Load-Side Interconnections. Many small interactive PV systems are interconnected on the load side of the main service disconnect at a customer's facility. The NEC® permits load-side connections at any distribution equipment on the premises, providing the following seven conditions are satisfied:

Point of Connection

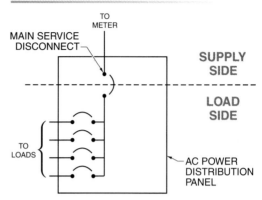

Figure 12-10. *Interactive inverters can be connected to either the load side or the supply side of the main service disconnect.*

1. Each source interconnection must be made at a dedicated circuit breaker or fused disconnect. **See Figure 12-11.** Multiple inverters are considered multiple sources. Each requires a dedicated interconnection device, unless their outputs are first combined at a subpanel.

2. The sum of the ratings of overcurrent protection devices in all circuits supplying power to a busbar or conductor must not exceed 120% of the rating of the busbar or conductor. This prevents potential overload conditions from occurring at the point of connection.

Load-Side Connection

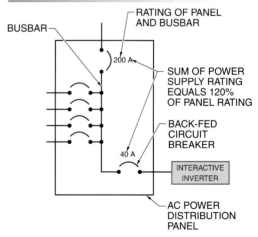

Figure 12-11. *Interconnection on the load side of the service disconnect is done through back-fed circuit breakers.*

This limit applies only to the breakers for power sources, which are the main utility-fed circuit breakers and any back-fed circuit breakers from PV systems that supply power to the busbar. It does not include load circuit breakers.

For example, consider a 200 A distribution panel with a main circuit breaker and busbar rated at 200 A. The sum of the ratings for devices supplying power is allowed to be 120% of the busbar rating, or 240 A (200 A × 120% = 240 A). Since the panel busbar is already fed with a 200 A circuit breaker, additional circuit breakers with ampere ratings totaling up to 40 A are permitted to supply power from a PV system. Alternatively, if the busbar rating were 225 A and the main circuit breaker rating were 200 A, then the PV system's back-fed breakers could have a total rating of up to 70 A.

3. The interconnection point must be on the supply side of all ground-fault protection equipment. This requirement prevents potential damage to, or improper operation of, ground-fault protection equipment. An exception is allowed for connection to the load side of ground-fault protection equipment that is protected from all ground-fault current sources.

4. All panels with more than one source of power must be marked showing all sources of power. This labeling is required to alert maintenance and service personnel to the presence of multiple power supply sources to distribution equipment. It is not required for equipment with power supplied from a single source. This requirement can be met by installing placards on all back-fed panelboards and fused disconnects indicating the rated output current and nominal line voltage of all connected inverters and utility services.

5. Any back-fed circuit breaker used for an interconnection must be identified for such operation. Any circuit breaker not marked with "line" and "load" designations are considered suitable for back-feeding.

6. Dedicated circuit breakers back-fed from listed utility-interactive inverters are not required to be individually clamped to the panelboard busbars with additional fasteners.

7. A back-fed circuit breaker in a panelboard shall be positioned at the opposite (load) end from the main circuit location. **See Figure 12-12.** A permanent warning label must be applied by the back-fed breaker with the following or equivalent marking: "WARNING: INVERTER OUTPUT CONNECTION, DO NOT RELOCATE THIS OVERCURRENT DEVICE."

Bimodal systems present special issues in meeting these requirements because many of these inverters are capable of delivering 60 A continuously. Although the inverter may be rated for 60 A continuous, the external conductors and circuit breaker must be able to handle 125% of this current continuously. This requires an 80 A circuit breaker and conductors sized to handle 75 A. (Reducing the inverter output current to 48 A may allow the use of equipment rated at 60 A, but this would also reduce the peak output capability of the inverter.)

Back-Fed Circuit Breakers

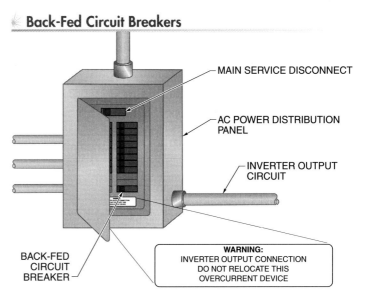

MAIN SERVICE DISCONNECT

AC POWER DISTRIBUTION PANEL

INVERTER OUTPUT CIRCUIT

BACK-FED CIRCUIT BREAKER

WARNING:
INVERTER OUTPUT CONNECTION
DO NOT RELOCATE THIS
OVERCURRENT DEVICE

Figure 12-12. *Back-fed circuit breakers are circuit breakers on the load side of the main service disconnect that supply PV power to the busbar.*

The capacity of existing service and distribution equipment limits the size of PV systems that can be interconnected to the load side of a utility service without modifications. For example, a 2500 W, 240 V inverter has a rated current output of 10.4 A. Overcurrent protection must be sized for 125% of the inverter's rated output, yielding 13 A. This value is rounded up to the next available overcurrent protection device size of 15 A. Since a common panel with a 200 A busbar and 200 A main circuit breaker is limited to 40 A of additional back-fed circuit breakers, the interconnection at this panel is limited to only two of these inverters.

Inverter Subpanels

To increase the allowable PV system capacity, a subpanel can be installed that feeds into a single 40 A circuit breaker on the main panel. The subpanel can accommodate three inverters with three 15 A circuit breakers. The system is sized correctly, because the overcurrent protection device rating for each inverter is actually 13 A, and the total is 39 A (3×13 A = 39 A). An additional 40 A circuit breaker protects the entire subpanel.

The subpanel rating is sized based on the total ampacity of circuit breakers supplying this panel and the application, which is 85 A ([15 A $\times$ 3] + 40 A = 85 A). Since panels are permitted to be supplied with circuit breaker ampacities at 120% of the busbar rating, this panel would need to be sized for at least 71 A (85 A $\div$ 120% = 71 A). This would still require a standard 100 A panel.

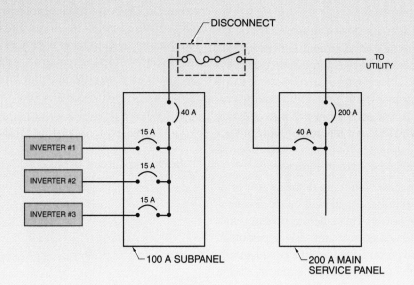

Panel Ratings

Another, though often impractical, method of increasing the limit for additional back-fed circuit breakers is to decrease the rating of the main panel utility service circuit breaker. For example, reducing the main circuit breaker rating from 200 A to 175 A would allow for an additional 25 A of back-fed circuit breakers. However, this approach limits the utility supply to the panel and may cause nuisance circuit breaker trips or violate the original panel load calculations.

A better solution would be to install a panel with a higher-rated busbar and a downsized main circuit breaker. For example, if a 320 A rated panel (busbar) with a 200 A main circuit breaker were installed, the allowable additional back-fed circuit breakers would be 184 A ([320 A $\times$ 120%] − 200 A = 184 A). When panel modifications become impractical to meet this requirement, a supply-side connection may be used.

Also, a 60 A back-feed circuit breaker requires a panelboard rated to at least 300 A to meet the second requirement. Residential panels rated above 300 A are available but uncommon. For these reasons, a supply-side interconnection might be a more practical alternative.

Supply-Side Interconnections. When PV systems are too large to interconnect on the load side, due to the capacity of the distribution equipment, a supply side interconnection must be used. The supply-side connection of interactive PV systems requires adding another service in parallel with an existing service disconnect. Accordingly, the rules for service entrance equipment apply to this connection. The equipment must be rated as service equipment and the existing service conductors must be rated for the additional PV system output. The interconnection requires tapping the service entrance conductors, which is usually done between the existing main service disconnect and meter socket. Service entrance equipment is available that splits the incoming service into individual services, which can be used for this purpose. Alternatively, the connection can be made at a new meter, which would constitute a new service.

The NEC® requires this new service to have disconnects and overcurrent protection as described in Article 230, "Services." **See Figure 12-13.** A service-rated circuit breaker or fused disconnect meets these requirements and may also satisfy a utility's requirement for an accessible, lockable safety disconnect that clearly indicates open status. Because a service disconnect with at least a 60 A rating must be used (per Article 230), even for lower-output-rated PV systems, smaller fuses and adapters may be required. Because services and taps are unprotected on the line side up to the fuse on the primary side of the distribution transformer, this new service disconnect must also have appropriate interrupting ratings consistent with potential utility fault currents and other service equipment. Fused disconnects with 200,000 A interrupting ratings are available to meet this requirement.

Supply-Side Connection

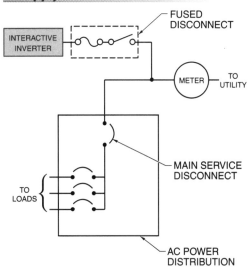

Figure 12-13. *Interconnection on the supply side of the service disconnect must include a separate service-rated fused disconnect or circuit breaker.*

The load-side interconnection requirements in the 2005 NEC®, particularly those regarding the allowable ratings of back-fed circuit breakers, differentiated between commerical and residential installations. This distinction was removed in the 2008 edition, making compliance and inspections easier.

Interconnection standards provide guidelines on the placement, connection, and operation of utility-interactive equipment.

The service conductors must be sized for at least 125% of the continuous load current as required in Article 230, based on the PV system inverter maximum current output rating. However, the conductors between the service tap and disconnect must be rated at no less than the disconnect rating, which is at least 60 A. Consequently, smaller systems may require larger service tap conductors than the PV system output would otherwise dictate. Best practice suggests using tap conductors as large as the service equipment terminals will permit. The best location for the tap will vary depending on the application and equipment. This connection may be done at the main distribution panel prior to the existing main service disconnect, or it may be done on the load side of the meter socket.

Regardless of the point of connection, NEC® Article 705, "Interconnected Electric Power Production Sources," requires that a permanent directory be placed at each service location showing all power sources for a building. **See Figure 12-14.** If the PV system is connected on the load side, this labeling may be in addition to labeling at the panelboard. If the point of connection is on the supply side, this labeling may serve the purposes of a supply-side label near the service disconnects.

Metering

Electricity is metered to determine the amount of energy delivered to (or from) a customer's facility for billing purposes. Metering used for billing is the responsibility of electric utilities and is often called "revenue metering." Revenue meters are installed at the service entrance and establish the transition between utility and customer-owned equipment. The type of metering installed in facilities with interactive systems is determined by the type and size of the facility and distributed-generation equipment and the applicable interconnection policies and rules.

Net Metering. Net metering uses one meter that can operate in both directions, effectively subtracting exported electricity from imported electricity. **See Figure 12-15.** This assigns them the same, full retail value, which is most advantageous to the customer. Sometimes a customer's existing meter is capable of operating backwards without any modifications. If a different meter must be installed, it is the responsibility of the utility to do this, though it may charge the customer a fee.

Interconnection Labeling

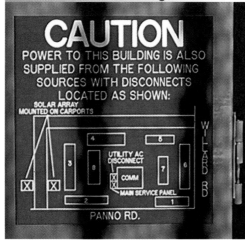

Sharp Electronics Corp.

Figure 12-14. *Interconnected PV systems must include labeling that clearly identifies the disconnects and point of interconnection.*

Net Metering

POWER FLOWS IN BOTH DIRECTIONS

BIDIRECTIONAL METER

AC POWER DISTRIBUTION PANEL

POWER FROM INTERACTIVE INVERTER

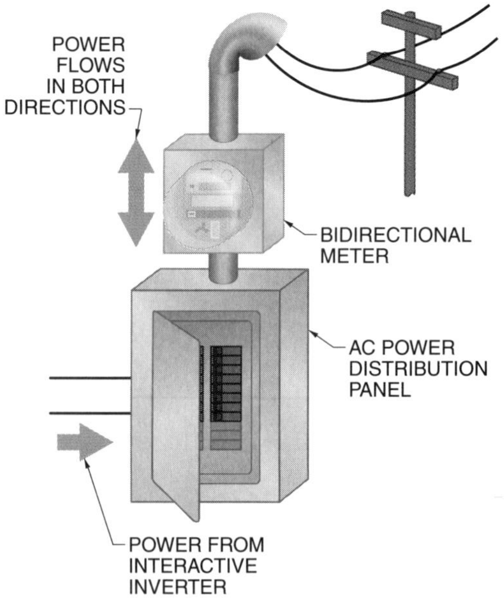

Figure 12-15. *With a net metering arrangement, exported power makes the electric meter run backward, crediting the PV system owner for power supplied to the utility grid at the retail rate.*

Metering Basics

For billing purposes, electricity consumption is metered as power enters a customer's building. For residential and small commercial buildings, standard watt-hour meters accumulate total energy passing through the meter, and the utility charges the customer a certain rate per kilowatt-hour. For larger commercial facilities, metering typically also includes peak power demand in kilowatts (kW), and customers are billed for both peak power demand and energy consumption.

Electromechanical induction watt-hour meters are conceptually similar to motors. Electricity passing through coils creates magnetic fields, which rotate a metal disk. The rate of disk revolution is proportional to the power passing through the meter, so each revolution accounts for a certain amount of energy transfer. The watt-hour constant (designated "K_h") is the number of watt-hours per disk revolution. The number of disk revolutions is counted mechanically and the corresponding energy use is displayed on the register in digits or dials.

Watt-hour meters rotate in the positive direction when the meter is fed from the overhead service entrances. However, electromechanical meters can operate with power passing through in the opposite direction, which simply reverses the direction of disk rotation, and the meter register counts backward instead of forward. This is the basis for net metering.

Today, many electromechanical watt-hour meters are being replaced by electronic meters using current and voltage transformers and microprocessors to measure, process, and record data. Some can record other electrical service information, including peak power demand, power factor, reactive energy, time-of-use consumption, or, in the case of distributed generation, exported energy. Many electronic meters can also be used for automated meter reading (AMR) applications, allowing the meter data to be read remotely by either infrared, radio frequency, telephone, wide-area network, or power line carrier signals.

Under most state rules, residential, commercial, and industrial customers with small to medium-sized PV systems are eligible for net metering. In other areas, eligibility can vary by location, customer type, utility, and technology (such as PV array, wind turbine, or engine generator). Net metering is also a low-cost and easily administered way of promoting direct customer investment in renewable energy, though at the utility's expense.

Dual Metering. Two-meter arrangements are more common for larger independent power producers, though they are also used for a variety of PV systems in states that have not yet mandated net metering rules. Two unidirectional meters, or a single multi-register meter, record exported and imported energy separately. **See Figure 12-16.**

Because they are measured separately, the energy sent to or supplied from the utility can be assigned different values. Usually the utility pays a lower rate for electrical energy exported from a customer, which is less advantageous to the customer than net metering. Under the most common dual-metering arrangement,

referred to as "net purchase and sale," excess electrical energy produced by a customer can be purchased by the utility at a different rate than it charges to sell electricity to the customer. Under this arrangement, the utility usually pays its avoided cost when purchasing electricity. There is generally a significant difference in the retail rate and the avoided cost. For instance, a retail rate might be about 9¢/kWh, while the avoided-cost rate may be only 3¢/kWh.

Electromechanical induction watt-hour meters are the most common type of meters installed by electric utility companies.

Dual Metering

Direct Power and Water Corporation

TWO METERS

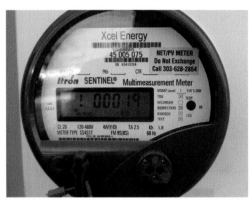

MULTIREGISTER METER

Figure 12-16. *Dual metering can be accomplished with two separate meters or with one meter that can measure and record energy flow in both directions separately*

UTILITY INTERCONNECTION POLICIES

Utility interconnection policies and practices have been a major barrier to the expanded adoption of PV and other distributed power systems, though procedures are gradually becoming more consumer friendly in many areas. Some policies dealing with the interconnection of these systems are legislated by federal, state, and local governments. Where governmental policies are absent, interconnection policies are established by local utility companies.

Public Utilities Regulatory Policy Act (PURPA)

Over the past 30 years, a number of policies developed at the state and federal levels have impacted the utility interconnection of privately owned power generation systems. The first significant legislation was the Public Utilities Regulatory Policy Act (PURPA) of 1978, passed by the U.S. Congress during the energy crises of the 1970s. This law was intended to decrease U.S. dependence on foreign oil by increasing energy conservation and efficiency and by encouraging the use of renewable energy and cogeneration resources. PURPA established the first opportunities for non-utility power producers by eliminating barriers that previously hindered their entry into a market controlled by electric utilities. The most significant part of PURPA was that it required electric utilities to purchase power from independent power producers (IPPs) and establish the technical and procedural requirements for their interconnection to the utility system, subject to state regulatory approval.

Qualifying Facility. PURPA defines a class of IPPs known as qualifying facilities. **See Figure 12-17.** A *qualifying facility (QF)* is a non-utility large-scale power producer that meets the technical and procedural requirements for interconnection to the utility system. PURPA mandates that utilities purchase power from QFs at the utility's avoided cost. *Avoided cost* is the cost that a utility would normally incur to generate a given amount of power, often synonymous with the wholesale market value of electricity. When purchasing this energy, the utility "avoids the cost" of generating it themselves.

Public Utilities Regulatory Policy Act (PURPA)

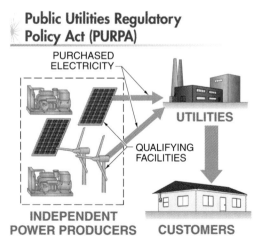

Figure 12-17. *PURPA defines the entities that can contribute to the collective energy supply.*

Tri-City JATC (Latham, NY)

Location: Latham, NY (42.8°N, 73.8°W)
Type of System: Utility-interactive
Peak Array Power: 28.1 kW DC
Date of Installation: October 2004
Installers: Apprentices and journeymen
Purposes: Training, supplemental electrical power

In an effort to make PV systems more affordable and encourage installer training, the New York State Energy Research and Development Authority (NYSERDA) granted $148,000 toward the installation of a large PV system at the Tri-City JATC in Latham, NY. This covered more than 60% of the system costs. The local worked with PV organizations to design the system, and the local apprentices and journeymen gained PV system experience by installing the system themselves.

The annual output of the PV system is about 28,500 kWh, which offsets a large proportion of the electricity demand of the building and significantly lowers the utility bill. The system is also utility-interactive, so any excess electricity when demand is low is transferred back to the utility grid. However, the local utility company supports net metering for residential customers only. For commercial customers like the training center, the utility installs a second meter to record any exported energy. The billing for energy used and credits for energy exported are processed separately. The wholesale rate the utility pays for exported energy varies daily. At the end of each month, the training center receives both a low bill and a refund check.

The modules are set on a standard rack mounting system designed for ground applications. However, instead of using typical concrete foundations for the main pole structures, the training center worked with the rack mount manufacturer to adapt the design for screw pile foundations. This is a unique

Tri-City JATC

The Tri-City JATC PV system consists of 152 modules arranged in groups to supply 15 inverters with DC power.

application of screw pile foundations. Screw piles are strong metal posts with an auger tip that is drilled into the ground under pressure. Screw piles are usually used to support failing foundations or structures in difficult soil conditions. They are more than strong enough for the relatively light array mounting structure, with a real advantage in ease of installation. This innovation turned out to be a considerable improvement.

Tri-City JATC

One meter measures energy used from the utility and another meter measures energy exported back to the utility.

Tri-City JATC

The ground mounting system was adapted to use screw piles as a foundation.

It is important to note that avoided costs are only part of the retail price for electricity supplied to the consumer. Retail price also includes the utility's administration costs and the costs of building, operating, and maintaining the electricity transmission and distribution system, and bringing services to consumers. These are services that a QF often uses but does not own or operate. Typically, avoided costs are less than half of retail electricity prices.

The Federal Energy Regulatory Commission (FERC) is responsible for overseeing the electric utility industry in the United States, including the implementation of PURPA. FERC's responsibilities include the regulation of the wholesale and interstate utility markets, power exchange transactions, and any rates, terms, or conditions established by state public utility commissions. Under PURPA, public utilities are required to submit their QF rates and billing structures to FERC for approval. The rates are an important part of interconnection agreements between utilities and QFs and in large part determine the economic value of distributed power generation.

SolarWorld Industries America
Large-scale PV systems may be subject to special interconnection requirements.

Qualifying-Facility Agreements. A *qualifying-facility agreement* is a contract between a utility and a qualifying facility that establishes the terms and conditions for interconnection and the rates or tariffs that apply. Traditionally, QF agreements are targeted to IPPs with generation levels of about 10 MW and greater. These agreements include contractual commitments regarding prices and expected levels of generation.

In the past, with the absence of interconnection agreements written specifically for small PV systems, utilities often used their general QF agreements for PV interconnection requests. These agreements are intended for large-scale power producers and include many requirements that are unnecessary for small, customer-owned systems. In addition, insurance requirements under QF agreements often exceed the coverage that most homeowners and small businesses carry. These complex legal documents therefore became financial, technical, and regulatory barriers to interconnecting small PV systems. This has prompted most states to pass legislation and adopt rules for a streamlined utility interconnection process and agreements for PV and other small distributed-generation systems.

Interconnection Agreements

An *interconnection agreement* is a contract between a distributed power producer and an electric utility that establishes the terms and conditions for the interconnection. Many utilities have simple interconnection agreements for the installation of small PV systems at residential and commercial facilities.

Permission of the local utility distribution provider is required to interconnect PV systems with the utility grid. Approvals for PV system interconnections are granted by electric utilities in cooperation with the local authority having jurisdiction (AHJ).

The interconnection process begins with completing the system design and interconnection plan, and submitting applications for plans review and permits with the AHJ. Concurrently, applications are made to the local utility for interconnection. Once permits are received, the installation is completed and a request for inspection is made with the AHJ. After completing inspection, the utility is notified, inspects the system (if necessary), approves the interconnection agreement, and grants approval to interconnect the system. Usually, electric utilities permit installers to test a PV system with pre-approval. Finally, the system may be interconnected, commissioned, and operated.

Utility interconnection agreements are governed by state utility commissions and local utility boards, so they vary somewhat by region. However, most have common requirements based on national codes and standards.

Size Restrictions. Interconnection agreements typically limit the size of the distributed-generation system covered under the agreement, though the maximum size is often far larger than is feasible for most residential and commercial systems. Limits vary by state and range from 10 kW peak to a few MW peak. A larger system may still be interconnected, but will likely fall into the category of a QF, which involves more stringent technical requirements and legal issues. The size limitation allows for a simplified contract, making interconnection more accessible to homeowners and small businesses.

Liability Insurance. Because of safety concerns, liability has always been a consideration for interconnecting distributed power sources. Liability insurance is required by most utilities that have interconnection standards, as a way to protect themselves and their employees should there be any accidents due to the operation of the customer's generating system.

Liability insurance in the amount of $100,000 is considered adequate for small PV systems by most utilities, and is generally covered in most home or property owner's policies. If not, the customer must obtain a policy rider or an additional policy to cover the system. Utilities may require the PV system owner to indemnify the utility for any potential damages as a result of operation of the PV system, which may also be covered under a liability policy.

Inspection and Monitoring. When interconnecting a PV system to the electric utility, the utility company assumes some of the risks and responsibilities of the system. It must ensure that the system is safe for the customer, their neighbors, and the utility lineworkers who may work on or near the system. The utility must also be sure that the PV system will not adversely affect the operation of the electric utility system.

Therefore, most utilities require verification, inspections, and sometimes even testing of PV systems to ensure they are operating within the specified voltage and frequency limits. Inspection rights allow the utility to check for listed equipment and proper installation before interconnection.

System Maintenance. After the utility inspections are completed and the PV system is interconnected with the utility system, it is the customer's responsibility to properly maintain their PV equipment and to promptly contact the utility if there are any problems with the interconnection. The customer is also responsible for the protection of the PV system from the utility system, during both normal and abnormal operation, including installing and maintaining all protection devices.

Disconnects. Utilities may require that PV systems have disconnects that are outside the building and accessible by utility personnel. **See Figure 12-18.** The utility may retain the right to disconnect the system, without prior notice to the customer, if work is necessary on the utility's part of the system or the customer fails to comply with the interconnection agreement.

Utility Accessible Disconnects

SolarWorld Industries America

Figure 12-18. *Utility interconnection agreements commonly require outside disconnects for PV systems so that the system can be isolated in the event of an outage or emergency.*

When a PV system includes an anti-islanding inverter, utilities may relax some requirements for the utility-accessible disconnect, such as its location. Also, in accordance with Article 690, PV inverters must already have a manual means of isolation from the grid, which should satisfy the utility requirement if installed outdoors.

Additional metering may be installed by a customer to record energy between portions of a PV system, such as between the inverter output and a critical load subpanel. This is called submetering.

SPG Solar, Inc.
Some utilities install electronic meters because they are easier to read and may include additional metering features.

Refer to Quick Quiz® on CD-ROM

Interconnection Fees. To offset the additional costs of inspecting, monitoring, billing, and completing paperwork for interconnecting PV systems, utilities may impose interconnection application fees. The fees, if applicable, may be flat rates or may be based on the size of the system. The fee structures vary by state or utility, but range from about $20 to $800 for most small PV systems. State interconnection rules may limit the amount utilities are allowed to charge for interconnection applications.

Additionally, the interconnection agreement may also include a schedule of fees for other services and equipment needed to interface with the PV system, especially for larger systems. For example, a new meter may need to be installed or an additional inspection may be needed to remedy code compliance issues. However, the interconnection agreement should also include assurances to customers that paid services or equipment installations will be completed in a reasonable timeframe.

Metering and Billing. Interconnection agreements must establish some means to credit the customer for excess power supplied to the utility, usually via either net metering or dual metering. The interconnection agreement defines each party's responsibilities regarding metering, billing, and rates. The agreement should also specify whether net metering, dual metering, or some other metering arrangement will be used, as well as the party responsible for installing and paying for the metering devices, which is usually the utility.

If credits will be issued for exported electricity, customers are generally not monetarily compensated, but are allowed to carry energy credits over from month to month. Usually credits will be used quickly, during times when the customer's electricity demand exceeds the PV electricity supply, but if credits begin to accumulate and carry over, the utility may specify an expiration date. Credits normally expire after one year, at which time the utility may purchase them at a special rate (usually the wholesale rate) or the credits may expire without compensation. Therefore, it is not advantageous to oversize a PV system without special agreements in place to purchase the excess energy.

If the utility is in a state that mandates metering requirements, the utility must abide by those rules. In 2009, policies on net metering exist in 44 states and the District of Columbia. **See Figure 12-19.** However, not every state has established rules regarding interconnections and metering. In states without established policies, utilities may choose to offer metering programs, though the implementation and requirements will vary between utility companies.

States that mandate that utilities allow net metering still have varying rules on other requirements, such as how much electricity a customer can export. Customers should also be careful to note in the interconnection agreement whether the utility will claim the renewable energy certificates (RECs) produced by the PV system, since these can be a valuable financial resource for the customer.

Net Metering Rules

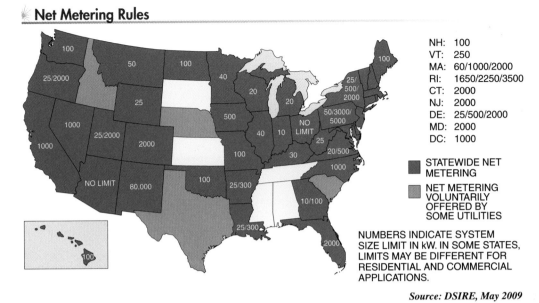

NH: 100
VT: 250
MA: 60/1000/2000
RI: 1650/2250/3500
CT: 2000
NJ: 2000
DE: 25/500/2000
MD: 2000
DC: 1000

■ STATEWIDE NET METERING

▨ NET METERING VOLUNTARILY OFFERED BY SOME UTILITIES

NUMBERS INDICATE SYSTEM SIZE LIMIT IN kW. IN SOME STATES, LIMITS MAY BE DIFFERENT FOR RESIDENTIAL AND COMMERCIAL APPLICATIONS.

Source: DSIRE, May 2009

Figure 12-19. *Net metering policies vary by state and sometimes also by utility.*

Summary

- Distributed-generation systems provide power to on-site loads and back up the normal utility service in the event of an outage.

- Rotating generators produce most of the power on the utility grid, but require precise synchronization procedures before interconnection.

- Solid-state inverters include automatic and sophisticated synchronizing functions and have no moving parts.

- Interactive-only PV systems connect the array directly to the inverter.

- Bimodal systems connect the battery bank to the inverter, and the batteries are charged by the array through the charge controller.

- In the event of a utility outage, bimodal inverters can power certain loads by drawing power from the batteries.

- In addition to the NEC®, standards apply that deal specifically with the interconnection of PV systems.

- Islanding results from an interactive inverter that continues to transfer power to the utility grid that would otherwise be de-energized from an outage.

- Islanding is a serious safety hazard and can damage utility equipment.

- Power quality is among a utility's chief concerns for interconnected distributed-generation equipment.

- The output from an interactive inverter may be connected to either the load side or the supply side of the service disconnect.

- Interconnected systems should be clearly labeled in a central location, and other applicable locations, with all the building's power sources and their disconnects indicated.

- Net metering uses one meter that operates in both directions to subtract the amount of energy exported from the amount of energy purchased.

- Dual metering uses two meters to measure exported energy and purchased energy separately.

- Most utilities have simple interconnection agreements for the installation of small PV systems at residential and commercial facilities.

- Approvals for PV system interconnections are granted by electric utilities in cooperation with the local authority having jurisdiction (AHJ).

Definitions

- *Distributed generation* is a system in which many smaller power-generating systems create electrical power near the point of consumption.

- *Synchronizing* is the process of connecting a generator to an energized electrical system.

- *Islanding* is the undesirable condition where a distributed-generation power source, such as a PV system, continues to transfer power to the utility grid during a utility outage.

- The *point of connection* is the location at which an interactive distributed-generation system makes its interconnection with the electric utility system.

- A *qualifying facility (QF)* is a non-utility large-scale power producer that meets the technical and procedural requirements for interconnection to the utility system.

- *Avoided cost* is the cost that a utility would normally incur to generate a given amount of power, often synonymous with the wholesale market value of electricity.

- A *qualifying-facility agreement* is a contract between a utility and a qualifying facility that establishes the terms and conditions for interconnection and the rates or tariffs that apply.

- An *interconnection agreement* is a contract between a distributed power producer and an electric utility that establishes the terms and conditions for the interconnection.

Review Questions

1. Explain how the interconnection of solid-state inverters is different from synchronizing rotating generators.

2. Why is the DC input connection of an inverter different between interactive-only and bimodal systems?

3. How is the standard ANSI/IEEE 1547 related to interconnection?

4. Why is islanding a safety hazard to utility lineworkers?

5. What are the two points of interconnection allowed by the NEC®?

6. How do net metering and dual metering affect the billing and crediting of power exchanged with the utility?

7. What types of interconnection-related labeling are potentially required for an interactive PV system?

8. Why are qualifying-facility agreements not appropriate for small residential and commercial PV systems?

9. Why are insurance requirements, inspection rights, and interconnection fees commonly included in interconnection agreements?

BUILDING PERMIT

NO. _____

Has Been Issued For Construction
Upon These Premises in Accordance With

Permit Expiration Date _____

Permit Issued Date _____

No _____, Street _____

Subdivision _____

Owner of Premises _____

General Contractor _____

PROJECT _____

Inspections Required

The following inspections you MUST call for on this project:

• Footings — before pouring concrete
• Foundations — (for basement walls only) after damp-proofing and drain tile before backfilling.
• Electric Service
• Framing — after plumbing and electric is roughed in, BEFORE INSULATION
• Insulation
• Flatwork — Before pouring any concrete — slabs, decks, or support post holes.
• Before back___

13

Permitting and Inspection

The requirements for PV system installations are governed by building codes adopted by local jurisdictions. PV installation approvals are granted by local jurisdictions through the permitting, plans review, and field inspection processes. There are many steps and requirements for proper permitting and inspection, but the process ensures safe and quality systems for installers and property owners. An inspection checklist is a useful guide to common PV system issues for inspectors, system designers, installers, and operators.

Chapter Objectives

• Understand the role of building codes and code enforcement in electrical installations.
• Describe the common requirements for permit applications.
• Identify the applicable articles of the NEC® for both general electrical system requirements and PV-specific requirements.
• Describe the labeling requirements for PV systems and components.
• Understand the function of an inspection checklist in checking a PV system for common installation code compliance issues.

BUILDING CODES AND REGULATIONS

Each community has a building code that applies to the installation of electrical equipment on residential and commercial properties. A *building code* is a set of regulations that prescribes the materials, standards, and methods to be used in the construction, maintenance, and repair of buildings and other structures. The purposes of building codes are to protect public safety and ensure quality construction projects. Building codes cover all construction trades, including electrical, mechanical, plumbing, structural, fire and life safety, and energy efficiency.

Building codes are applicable to the construction and installation aspects of PV systems, just as for any other electrical system. These regulations may dictate who can install PV systems and the requirements for different system types and sizes. Any designer, contractor, or code official involved with integrating, installing, and inspecting PV systems should be familiar with these requirements.

Building code requirements are both technical and administrative in scope. They affect manufacturers and suppliers of equipment and materials, as well as installers, contractors, code officials, and end users. Building codes are enforced by inspections, the issuance of building permits and certificates of approval or occupancy, and the imposition of fines for violations.

Product safety standards, installation codes, and code enforcement are separate but related components of the electrical safety system. This system depends on a close working relationship among the organizations responsible for the development of product standards and installation codes and the electrical inspection community. All three components must be in place for the electrical safety system to be effective. **See Figure 13-1.**

Each major city in the United States once had its own building code. Since building codes are very expensive to develop and maintain, most cities eventually adopted model codes. Only a few cities, such as Chicago, continue to maintain local codes.

Electrical Safety System

Figure 13-1. *Product standards, installation codes, and code enforcement are separate but related functions that result in higher-quality and safer installations.*

Code Adoption

A building code must be adopted by the state or local jurisdiction as law to apply and be enforceable. The jurisdiction may choose to develop its own building code, but this is uncommon. Most adopt model building codes, sometimes making small modifications to suit their locale. A *model building code* is a building code that is developed and revised by a standards organization independently of the adopting jurisdictions. The National Electrical Code® (NEC®) is an example of a model building code.

Building codes may contain guidance on the administration and enforcement of the code, such as requirements for electrical boards, the responsibilities of code officials, and the processes for plans review, permitting, and inspections. Article 80 in Annex H of the NEC®

is such a guide, though it applies legally only if it has been formally adopted along with the rest of the code by a local jurisdiction. Even when Article 80 is not officially adopted, most jurisdictions follow a similar approach.

Authority Having Jurisdiction

An *authority having jurisdiction (AHJ)* is an organization, office, or individual designated by local government with legal powers to administer, interpret, and enforce building codes. The AHJ approves equipment, materials, installations, and procedures by issuing permits, conducting inspections, and granting certificates of approval or occupancy.

With respect to electrical construction, the basic responsibility of the AHJ is to verify that electrical installations, including PV systems, comply with the NEC®. The AHJ is permitted to waive, in writing, specific requirements of the NEC® or to permit alternative methods where it is assured that equivalent safety objectives are maintained. The AHJ also has the authority to disconnect electrical systems when hazardous conditions exist and until such problems are corrected.

The AHJ delegates powers to qualified plans examiners and electrical inspectors. These electrical officials are typically licensed and certified at the state or local level, and have demonstrated a thorough knowledge of the applicable codes and standard materials and methods used in electrical installations. Responsibilities and requirements for electrical inspectors and plans examiners are established in Article 80 of the NEC®, as well as in local building codes or by licensing boards.

Plans Examiners. A *plans examiner* is a local official qualified to review construction plans and documentation for compliance with applicable codes and standards. Plans examiners evaluate drawings, schematics, specifications, instructions, and equipment manuals, and verify the suitability and proper application of equipment. A plans examiner may specialize in the electrical, mechanical, or structural aspects of construction plans.

Electrical Inspectors. An *electrical inspector* is a local official qualified to evaluate electrical installations in the field for compliance with applicable codes and standards. Electrical inspectors visually evaluate the safety of electrical installations in accordance with the approved plans and local codes. They identify the suitability and proper installation of equipment, conductors, overcurrent protection devices, disconnects, and bonding and grounding. Other building inspectors may evaluate PV installations according to mechanical and structural requirements.

DOE/NREL, Utah Office of Energy Services
The review of a new PV system is completed by official building inspectors.

IAEI

The International Association of Electrical Inspectors® (IAEI®) is a trade organization for the electrical code official community and has thousands of members in over 30 countries. The IAEI is a valuable resource for electrical inspectors and promotes uniform standards and a common interpretation of the NEC® and other codes. The IAEI seeks to promote cooperation between inspectors, the electrical industry, and the public, and to collect and disseminate information related to the safe use of electricity.

The IAEI regularly publishes articles on PV systems and the NEC® in the *IAEI News*. These articles address specific areas of PV system installations, their associated code requirements, and solutions to common problems. These articles are a valuable resource for any contractor, code official, or utility representative involved with the installation of PV systems.

Approved Equipment

The use of approved equipment is required by the NEC® and facilitates the plans review, permitting, and field inspection processes. Approvals are based on examinations of the equipment for safety, installation, and conditions of use. In most cases, equipment is approved for its intended application by a recognized testing laboratory. A *nationally recognized testing laboratory (NRTL)* is an OSHA-recognized, accredited safety testing organization that certifies equipment or materials to meet applicable standards.

This independent approval process allows AHJs to avoid the time-consuming and potentially inconsistent process of conducting their own equipment evaluations, while assuring AHJs, contractors, and consumers that electrical equipment is safe and appropriate for a given application. Approved equipment is identified with product listing marks on its label. *Listing* is the process used by an NRTL for certifying that equipment or materials meet applicable standards.

To obtain a product listing, a manufacturer produces a product in accordance with the applicable standards and submits prototypes to an NRTL for evaluation and approval. The NRTL tests the prototypes for compliance with the standards and, if the standards are met, authorizes the manufacturer to use its product certification mark. The manufacturer's production facilities are then periodically inspected by the NRTL to ensure continuing compliance with the standards and conditions of listing.

PV modules, inverters, charge controllers, and other electrical equipment are typically listed by Underwriters Laboratories, though it is common to see other types of product certification marks on PV components.

Underwriters Laboratories (UL). Underwriters Laboratories is perhaps the most well-known NRTL and provides safety certification services for a variety of products used throughout the world. There are several UL marks, each with specific meaning. These marks may only be used on or in connection with products certified by UL, under the terms of a written agreement between UL and the manufacturer.

The UL Listing Mark is one of the most common UL marks. **See Figure 13-2.** It signifies that UL has tested representative samples of the product in accordance with UL's safety standards and that the manufacturer's production facilities are routinely inspected by UL for compliance with the standards. Products may be listed to either Canadian or U.S. safety standards, or both. Canadian listing marks appear with a "C" on the left side of the mark, and U.S. listing marks appear with a "US" on the right side. The UL Listing Mark is accepted by the NEC® and the Canadian Electrical Code as evidence of approved equipment. The UL Listing Mark includes the UL symbol with the word "LISTED," a control number, and a product name such as "Photovoltaic Module" or "Utility-Interactive Inverter."

UL Listing Mark

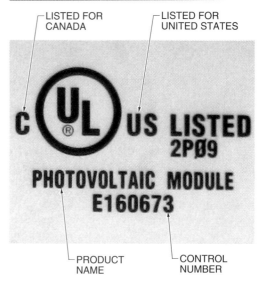

Figure 13-2. *The UL Listing Mark indicates if the product is listed for both the United States and Canada.*

Listing requirements for PV modules are established by UL 1703, *Standard for Flat-Plate Photovoltaic Modules and Panels*. Listing requirements for inverters and charge controllers are governed by UL 1741, *Inverters, Converters, Controllers, and Interconnection System Equipment for Use with Distributed Energy Resources*.

The UL Classification Mark is used on products that UL has evaluated only with respect to specific properties, a limited range of hazards, or suitability for use under limited or special conditions. Classified products may be tested to the same standards as listed products and may be of equal quality. The difference is that listed products are evaluated for all reasonably foreseeable situations applicable to the standard, while classified products are evaluated for only a certain set of circumstances. Products classified by UL include building materials, protective gear, and fire safety equipment. Roofing products and PV modules may also be classified for fire rating.

Some products may be both listed and classified and use a combination listing/classification mark with the text "Also Classified in Accordance with" and the applicable classification standard. Listed PV modules can be classified in accordance with IEC 61215 or IEC 61646 design qualification standards. These classifications are indicators of the quality of module design and reliability. Design classifications are not required for code compliance, but may be specified when the most reliable PV modules are desired.

The UL Recognized Component Mark indicates that a component is certified to meet the requirements for a limited, specified use as part of a larger product or system and is not intended for separate installation in the field. When products submitted to UL include recognized components, the evaluation and approval process of the overall product is faster. However, equipment composed of recognized components does not automatically constitute approval or listing for the complete equipment. UL Recognized Component Marks are found on a wide range of electrical products, such as switches, power supplies, motors, terminal blocks, and batteries.

Equipment or products that are modified or do not bear a third-party certification mark may be evaluated by UL in the field at the request of a building owner, manufacturer, or regulator. Products that meet appropriate safety requirements are labeled with a tamper-resistant UL Field Evaluated Product Mark.

Canadian Standards Association. The Canadian Standards Association (CSA) is an OSHA-accredited NRTL that tests using similar processes and to the same industry standards as other NRTLs. The CSA also publishes the Canadian Electrical Code. CSA Marks are recognized and accepted throughout the United States by federal, state, and local authorities. A CSA Mark with the indicator "US" or "NRTL" means that the product is certified for the U.S. market to the applicable U.S. standards. **See Figure 13-3.**

Listing and Conformity Marks

CSA MARK

ETL LISTING MARK

CE MARK

Figure 13-3. *PV equipment may bear the marks of listing and certification organizations other than Underwriters Laboratories.*

Fire-Rating Classifications

Building codes establish fire-resistance requirements for construction materials, including roofing products. These requirements apply to PV modules because they can be installed on rooftops and in building-integrated applications. Most PV module manufacturers have their products tested by UL and classified for fire-resistance rating. Modules that have not been classified with respect to fire exposure are marked "Not Fire Rated."

The fire-resistance rating is the level of resistance to external fire that a roof covering provides to the roof decking below. Classes A, B, and C represent different levels of fire resistance. All classified products will not spread fire to the roof deck, slip from position, or produce flying debris under the fire conditions of their classification. The differences are in the severity of fire exposure that the products can withstand.

MAX. SYST.
OPEN CKT. VOLTAGE

600 V

FIRE RATING

CLASS C

FIELD WIRING

COPPER ONLY, 14 AWG MIN.
INSULATED FOR 90° C MIN.

Module fire ratings are included on module labels.

These classes are defined by ASTM E108, *Standard Test Methods for Fire Tests of Roof Coverings,* and ANSI/UL 790, *Standard Test Methods for Fire Tests of Roof Covering Materials.* Class A roofing products are effective against severe fire exposures. Most building codes recognize brick, masonry, slate, clay or concrete tile, exposed concrete, and copper shingles as Class A roofing materials without being tested. Class A roofing products are generally only required for buildings where the roof falls under interior fire rating requirements or for construction in special fire districts. PV modules used for building-integrated installations may be required to have a Class A fire rating.

Class B roof coverings are effective against moderate fire exposure. Building codes recognize aluminum roofing as Class B without testing. Class C roof coverings are effective against light fire exposure. Class C fire ratings are acceptable for PV modules used in most residential and commercial rooftop applications.

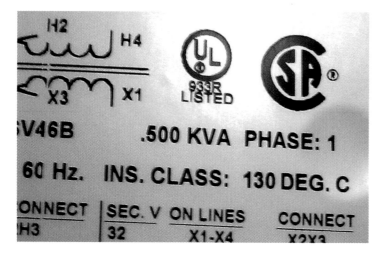

Many components are certified for more than one market and include multiple listing or certification marks on their labels.

Intertek ETL. Intertek is an OSHA-recognized NRTL and provides testing, inspection, and certification of electrical products for manufacturers and retailers around the world. Among their certification and conformity marks are the ETL Listed Mark for North America and the CE Mark for Europe. ETL Listed Marks are found on some PV inverters, which are tested in accordance with the standard UL 1741.

CE Mark. The CE Mark is a European marking of conformity that indicates that a product complies with the requirements of standards with respect to safety, health, environment, and consumer protection. It is used on products to facilitate trade between member countries of the European Union (EU). Unlike the UL

Mark, the CE Mark is not a safety certification. It is based on a self-declaration by the manufacturer, as opposed to a third-party independent certification. By itself, it does not indicate compliance to North American safety standards or installation codes. However, quality products may bear the CE Mark in addition to listing marks.

Other Marks. Other listing and certification marks may appear on PV equipment, especially on equipment produced for markets outside of North America. For example, PV products produced for European markets are often certified by TÜV Rheinland and are labeled with corresponding TÜV Marks. The European Solar Test Installation (ESTI) is another certification body recognized by the European Commission for PV module testing and certification capabilities.

Contractor Licensing

All electrical work should be performed by qualified persons. A *qualified person* is a person with skills and knowledge of the construction and operation of electrical equipment and installations and is trained in the safety hazards involved. In most cases, local and state contracting laws and regulations require an electrical contractor to be licensed in order to apply for permits and perform electrical work, including work on PV systems. Licenses are granted by state or local governments and are different from trade certifications, which are conferred by nongovernmental organizations such as NABCEP.

Licensing ensures that the contractor understands and complies with the applicable construction laws and regulations. Licensed contractors should complete all installations in a timely manner and to the owner's satisfaction. If there is a problem with a contractor's work, licensing also provides a means of addressing the issue, by filing a complaint with the license-issuing authority.

Local electrical contractor licenses that are offered by counties and municipalities limit work to a relatively small area. However, many states have statewide licensing programs that

are accepted by every AHJ in the state. Several states also have reciprocity agreements with neighboring states with similar contractor licensing requirements. This means that contractors licensed in any one of the states in the group may also work in any of the other states in the group through a simplified application process.

Solar Contractors. A few states, including Florida, California, Oregon, and Nevada, have a solar contractor license classification. The scope of work for this licensure may include both PV and solar thermal systems. However, these licenses are controversial in that they generally do not require rigorous electrical systems training. The solar contractor may be limited to performing only incidental electrical work and may need to hire an electrical subcontractor to install any premise wiring or make connections to the utility grid.

A qualified electrical contractor has extensive knowledge and experience and the license does not limit the scope of the contractor's electrical work on PV systems in any way.

⚠ **Qualified Person**

A qualified person is trained to recognize potential electrical hazards and can administer first aid and CPR in an emergency. Training must include the use and inspection of personal protective equipment (PPE) and use of insulated tools and test equipment. Persons working on or near exposed energized conductors operating at 50 V or above must be able to identify exposed live parts and their voltage, assess the risks for the type of work to be performed, and determine the appropriate PPE and other safety precautions required. Persons may be considered qualified for certain equipment or methods and still be unqualified for others. Apprentices may be considered qualified if they have demonstrated the applicable knowledge and skills and are under the supervision of a qualified person. OSHA regulations and NFPA 70E, *Standard for Electrical Safety in the Workplace,* include more information and descriptions of job safety requirements.

Owners wishing to act as their own contractor face a lot of responsibility. They must carefully study the applicable building codes and be prepared to address any questions the inspector may have. The owner-contractor must schedule inspections and remedy any compliance issues that the inspector finds. If they fail an inspection, they may also be charged a fee to cover additional inspection costs.

Owner-Contractors. In some jurisdictions, property owners are allowed to apply for permits and act as their own contractor without having a license. These owner-contractor exemptions allow owners to install electrical wiring and equipment to their own residence or commercial building, for projects costing up to a certain dollar amount. The owner may be required to occupy the building, and may be restricted from selling or leasing the property for one year after the work is completed. Owner-contractors are not relieved from any normal contractor responsibilities or the requirements of building codes and other regulations. Owner-contractors are also prohibited from hiring unlicensed persons to act as their electrical contractor or employing any unlicensed individuals to perform the work. These provisions are in line with contracting law and are intended to protect public safety.

PV systems should be installed by licensed contractors or qualified persons.

Construction Bonds. AHJs may require evidence of financial responsibility on the part of the contractor as a condition of licensure. Commonly, this requirement is satisfied by a construction bond. A *construction bond* is a contract in which a surety company assures that a contractor will complete their work in accordance with contracting laws. If a contractor does not meet their obligations, a claim can be filed against their bond by property owners, employees, subcontractors, or suppliers. As opposed to an insurance policy, a contractor is obligated to reimburse the surety company for any payments from the bond.

Construction bonds represent the amount of money available for claims for all jobs a contractor performs over the life of the bond. Minimum bond amounts for residential contractors are usually $10,000, with higher amounts required for commercial work or if the contractor has been subjected to licensing disciplinary action. Most state laws require the surety company to report claims on a contractor's bond to licensing boards. The surety may also cancel the bond and the licensing board may suspend the contractor's license until the surety is reimbursed for any claims. Contractors can avoid claims against their bond by using written contracts and communicating openly with their suppliers, customers, and employees to resolve any problems.

Building Restrictions

Building restrictions include easements, covenants, and ordinances, and are imposed by state or local governments or homeowner's associations. These limitations may restrict permanent construction on certain parts of a property, which can affect PV systems. Easements restrict construction within a certain distance of the property lines to allow utility access. Homeowner's association covenants may regulate solar installations for aesthetic reasons. Community ordinances may ban structures and antennas above a certain height, which may affect PV arrays installed on tall racks. Although these concerns are rare for most PV installations, the owner or installer should investigate potential issues in their jurisdiction.

Some regulations, however, benefit a property owner seeking to install a PV system. For example, some regulations protect a property owner's right to solar resources by prohibiting the erection of structures or planting of trees on adjacent properties that blocks the solar resource to a property. Also, some states make it illegal for covenants or deed restrictions to prohibit the installation of renewable energy systems on homes. More states are expected to follow in this trend.

PERMITTING

Permits are issued by city or county building departments under the authority of the AHJ. A *permit* is permission from the AHJ that authorizes construction work to begin and establishes the inspection requirements, but does not represent an approval of compliance with codes and standards. Permits are generally required for all electrical work, including the installation of PV systems. Few exceptions apply.

The basic recommendations for electrical permits are given in the NEC®, but some jurisdictions may require a special permitting process for PV systems. Differences in requirements may depend on the location and size of the system to be installed. Where an AHJ governs both code compliance and utility systems, permitting and utility interconnection applications may be combined. In other cases, separate AHJ permits and inspections are required for utility interconnection approval.

Proper documentation for PV system installations is essential to working with the AHJ and facilitates the permitting, plans review, inspection, and approval processes. Since PV systems are relatively new and many code officials are not familiar with their requirements, comprehensive documentation is particularly important for permitting and inspections.

Permit Applications

Permits are applied for by either the property owner or the owner's contractor. Permit applications require certain information about the scope and specifications of the work, typically including the construction tasks, location, permit applicant, expiration date, and inspection requirements. In most cases, three copies of all construction plans are submitted for review.

Permit applications for PV systems should contain site drawings, electrical diagrams, specifications for major components and equipment, and array mounting information. This documentation is the basis for the plans review and subsequent on-site inspections.

Site Drawings. A simple drawing of the site layout should indicate the locations of major PV system components including the array, inverter, and disconnects, and their relationships to electrical services, property lines, streets, and other features. **See Figure 13-4.** A property survey or plot plan can be used as a starting point for marking the locations of PV equipment. Drawings prepared during the site survey and planning stages of the project may be used if they include all required information and the equipment locations are up to date.

Electrical Diagram. At a minimum, a PV permit application should include a one-line electrical diagram. **See Figure 13-5.** Three-line electrical diagrams provide additional detail. These drawings should show all major system components and their interconnections with existing electrical equipment. The types, sizes, and ratings of the conductors, overcurrent protection devices, disconnects, and grounding equipment used throughout the PV system should also be provided on the diagrams or included with other permit submittals.

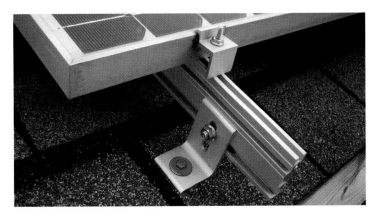

Permit applications will typically require descriptions and drawings of the array mounting system.

Site Drawings

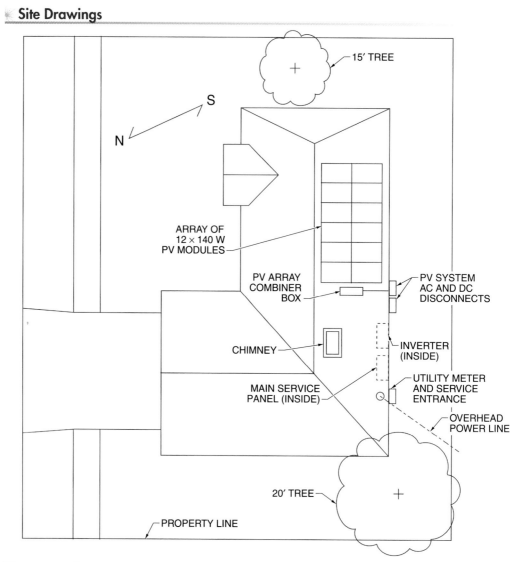

Figure 13-4. *Site drawings should be provided with permit applications indicating locations and providing descriptions of major components.*

For most large commercial installations, complete construction drawings with an engineer's stamp are required. Many packaged PV systems offered by major integrators include detailed diagrams and equipment specifications, eliminating the installation contractor's burden to produce such materials for permit applications.

Equipment Specifications. Permit applications should include specifications for all major components, including the manufacturer, ratings, operating parameters, and listing information. This includes PV modules, inverters, charge controllers, and batteries, as applicable.

This information is needed to determine if the equipment is appropriate and if conductors, overcurrent protection devices, and disconnects have been adequately sized.

Equipment specifications are available from the manufacturer's equipment manuals, labels, or downloadable information sheets. The permit application should be accompanied by either a comprehensive list of equipment specifications or a collection of product literature. **See Figure 13-6.** Summarized and key equipment specifications may also be included on the electrical diagrams in addition to the full set of specifications.

Electrical Diagrams

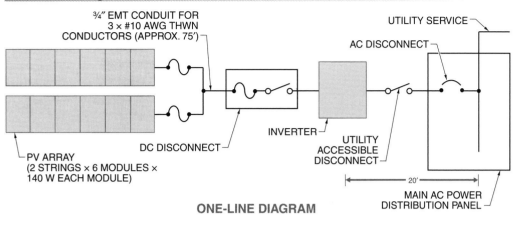

3⁄4″ EMT CONDUIT FOR
3 × #10 AWG THWN
CONDUCTORS (APPROX. 75′)

UTILITY SERVICE

AC DISCONNECT

INVERTER

DC DISCONNECT

UTILITY
ACCESSIBLE
DISCONNECT

PV ARRAY
(2 STRINGS × 6 MODULES ×
140 W EACH MODULE)

20′

MAIN AC POWER
DISTRIBUTION PANEL

ONE-LINE DIAGRAM

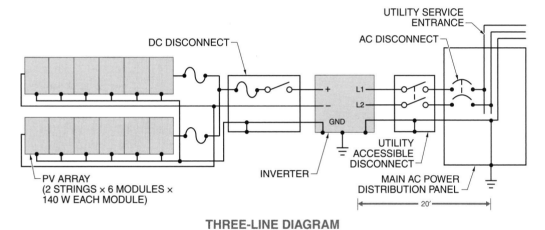

DC DISCONNECT

UTILITY SERVICE
ENTRANCE

AC DISCONNECT

+
−
L1
L2

GND

INVERTER

UTILITY
ACCESSIBLE
DISCONNECT

PV ARRAY
(2 STRINGS × 6 MODULES ×
140 W EACH MODULE)

MAIN AC POWER
DISTRIBUTION PANEL

20′

THREE-LINE DIAGRAM

Figure 13-5. *Permit applications usually require either a one-line or a three-line electrical diagram. Additional electrical information may be included as a separate document.*

Array Mounting Design. Most permit applications for PV systems require descriptions and/or drawings of the array mounting design and materials to ensure that the roof or structure can support the additional weight of the PV array and that the array will be well secured. **See Figure 13-7.** Roof information includes the age, composition, covering, pitch, and the size and spacing of structural members. Array information includes details about weight, attachment points, and weather sealing. Some AHJs require complete engineering reviews to verify that the structure and the attachment points have sufficient strength for the expected loads, particularly for large installations, unusual mounting schemes, or regions with high winds.

Equipment Specifications

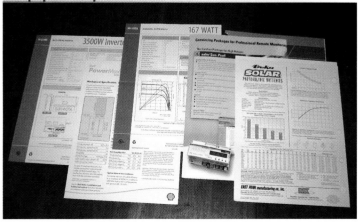

Figure 13-6. *Specifications for PV modules, inverters, and other equipment are required for code calculations and compliance.*

Array Mounting Design Detail

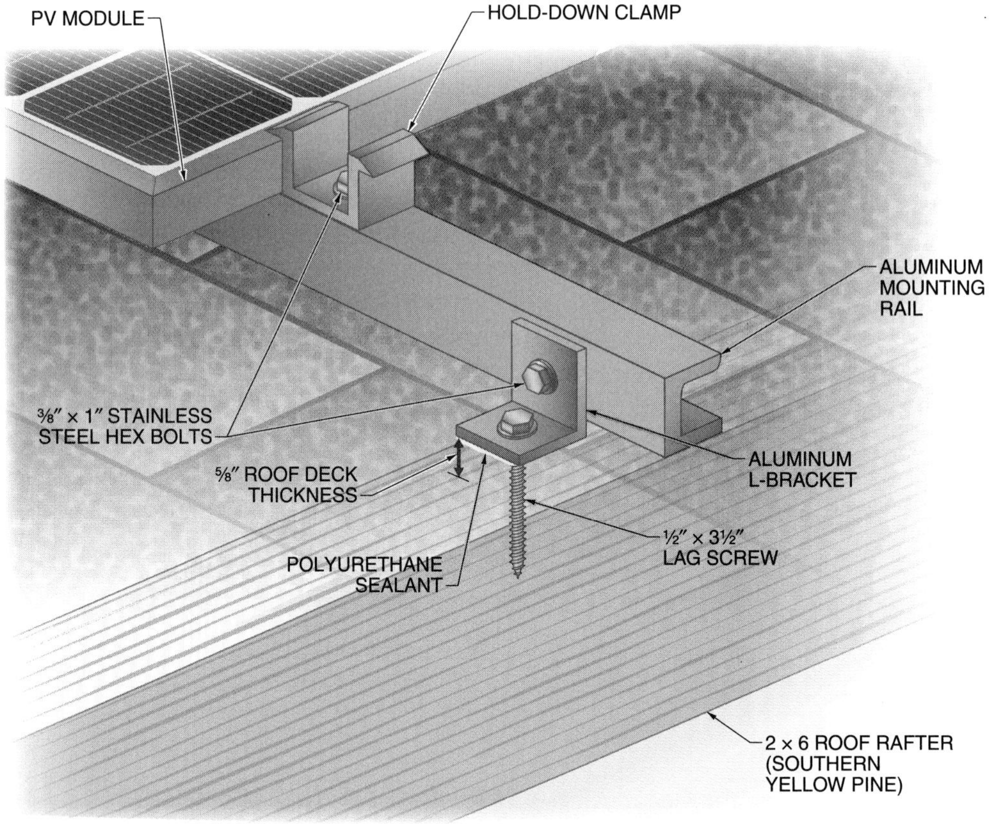

Figure 13-7. *Descriptions and drawings of the array mounting design and materials are used to analyze the structural integrity of the structure and the array.*

Permit Fees

Fees are assessed for permit applications to offset the costs of administering the code enforcement system. Permit fees for PV systems may be determined in a variety of ways. Valuation-based fees are calculated according to the total installed cost of a system. Some AHJs assess fees based on the number of PV modules or the required engineering review and inspection services. The flat-fee method applies the same permit fee for a wide range of system sizes or costs. These factors result in significant differences in permit fees from one AHJ to another. Permit fees for a medium-sized interactive residential PV system can range from $50 to several hundred dollars. Low permit fees for PV systems are associated with progressive AHJs with PV-knowledgeable inspectors and plans examiners.

Plans Review

A *plans review* is an evaluation of system-design documentation as part of the permitting process. The AHJ's plans reviewer is responsible for reviewing construction documents and drawings for any new construction or alterations to electrical systems. They must conduct this review in a reasonable timeframe and, if plans are not approved, provide reasons for nonacceptance. Corrected plans can then be resubmitted.

Permit applications list materials and documentation required for plans review. The applicant has the responsibility of providing applicable construction documents in accordance with governing codes and regulations. The plans review process by the AHJ does not relieve the applicant's obligations and responsibilities for code compliance.

Permit Issuance

When the permit application is complete, the fees are paid, and the plans review (including any amendments to the design) is complete, the permit is issued. The permit may be in the form of a separate document, sometimes a brightly colored card, or may simply be a signed copy of the permit application form. **See Figure 13-8.** The permit will often include information about each required inspection, such as the timing, sequencing, and elements covered, and a place for the inspector's signature. The permit must then be posted in a conspicuous location on the job site premises for the duration of the project.

Building Permits

Figure 13-8. *Building permits include information about the construction project and inspections, and must be posted in a conspicuous location on the job site.*

INSPECTION

Following installation, and sometimes at intervals during installation, the contractor or owner must contact the AHJ to schedule an inspection. An inspector evaluates the installation's conformity to application documents and applicable codes, laws, and regulations. Upon successful completion of all necessary inspections, the AHJ issues a certificate of approval or completion, authorizing the operation and connection of the system to other electrical systems, as applicable. Copies of certificates should be held by the system owner for documentation and insurance purposes, as well as for arranging interconnection approval with the electric utility in the case of interactive systems.

When inspections find a system to be non-compliant, the AHJ provides written notice of the discrepancies. Final approval is then contingent on rectifying the problems and passing a follow-up inspection. The AHJ may grant extensions to the permit expiration date upon request.

To facilitate the inspection process, the installer should have copies of all documentation submitted for the permit application on hand, including drawings, equipment specifications, and manuals. Any changes from the original design should be indicated on as-built drawings or as modifications to originals. All equipment should be accessible to inspectors, particularly equipment containing field-installed wiring.

All aspects of the electrical installation governed by codes and standards are subject to inspection, including conductor sizes, wiring methods, overcurrent protection, and grounding. For PV systems, most inspectors will pay particular attention to general electrical workmanship and safety, and the proper application of information and warning labels on system equipment.

Workmanship

The NEC® requires electrical equipment to be installed in a "neat and workmanlike manner." The installation should include structurally adequate supports, plumb and level mountings, proper conduit bends, and other quality matters not otherwise covered by NEC® safety requirements. **See Figure 13-9.** Since AHJs may interpret this requirement differently, the National Electrical Contractors Association (NECA) and industry partners publish standards to help define performance and workmanship expectations for electrical construction. One example is ANSI/NECA 1, *Standard Practice of Good Workmanship in Electrical Contracting.* These standards are often referenced in contract documents to establish installation quality expected by engineers, contractors, and customers.

> The AHJ has the responsibility of investigating the causes and circumstances of any fires, explosions, or other hazardous conditions affecting public health, safety, or welfare.

Quality Workmanship

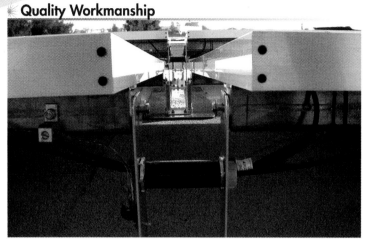

Schott Solar

Figure 13-9. *Quality workmanship results in a neat and efficient electrical installation.*

Working Space

PV system inspections also ensure that there is adequate access and safe working space around all electrical equipment. Meeting these requirements can be challenging when installing PV system components in buildings with existing electrical systems. Building-integrated PV systems can also present special challenges. Article 110 requires safe access and working space be maintained for electrical equipment operating at less than 600 V, which covers the majority of PV systems.

Dedicated space is the clear space reserved around electrical equipment for the existing equipment and potential future additions. For indoor installations, dedicated space for electrical equipment must be equal to the width and depth of the equipment and must extend from the floor to a height of 6′ above the equipment or to the structural ceiling, whichever is lower. No piping, ductwork, or other equipment not related to the electrical installation shall be located in this space. **See Figure 13-10.**

Working space is the clear space reserved around electrical equipment so that workers can inspect, operate, and maintain the equipment safely and efficiently. The depth of working space is the minimum clear space in front of electrical equipment with exposed live parts. This distance depends on the nominal voltage of the equipment and on whether similar equipment with live parts is located on the other side of the space.

Dedicated Space

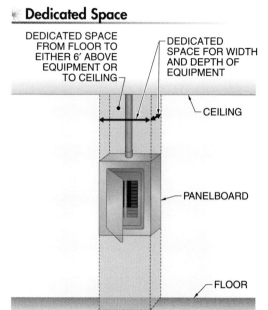

Figure 13-10. *Dedicated space is the clear space reserved around electrical equipment for the existing equipment and potential future additions.*

For less than 150 V, the clear working space must be at least 3′ deep. For equipment operating between 151 V and 600 V, 3′ of depth is required if exposed live parts are on only one side, 3.5′ of depth is required if grounded parts are located on the opposite side of the space, and 4′ of depth is required if live parts exist on both sides of the space. **See Figure 13-11.** With special permission from the AHJ, smaller working spaces may be permitted for equipment operating at no greater than 30 VAC or 60 VDC. This provision may apply, for example, to small stand-alone PV lighting systems with 12 VDC or 24 VDC loads. When equipment is not permitted to be serviced in an energized state, these requirements may not apply.

The width of the working space in front of any electrical equipment shall be at least 30″ or the width of the equipment, whichever is greater. This space may overlap the width of working space for adjacent equipment. The working space must also allow at least 90° openings for any doors or hinged panels on equipment. Working heights must be at least 6.5′ or the height of equipment, whichever is higher. At least one entrance of sufficient area must be provided for access to working spaces. In all cases, storage of materials must not block or interfere with safe access to equipment.

Working Space

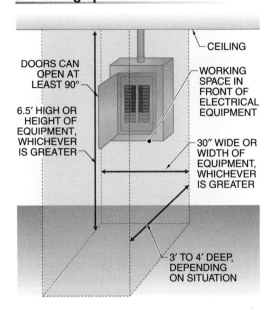

DOORS CAN OPEN AT LEAST 90°

6.5' HIGH OR HEIGHT OF EQUIPMENT, WHICHEVER IS GREATER

CEILING

WORKING SPACE IN FRONT OF ELECTRICAL EQUIPMENT

30" WIDE OR WIDTH OF EQUIPMENT, WHICHEVER IS GREATER

3' TO 4' DEEP, DEPENDING ON SITUATION

Figure 13-11. *Working space is the clear space reserved around electrical equipment so that workers can install, inspect, operate, and maintain the equipment safely and efficiently.*

Accessibility

Accessibility is explicitly covered throughout the NEC® and evaluated during electrical inspections. There are different meanings for accessibility, depending on context. Accessibility with regard to equipment means that close approach is possible and the equipment is not protected by means such as locked doors, fencing, or elevation. Accessibility with regard to wiring means that the conductors are capable of being removed or exposed and are not concealed by a building structure or its finish.

When parts of an electrical installation are to be concealed at the completion of construction, interim inspections may be scheduled to approve this work before covering. For example, pre-drywall inspections are required when any electrical or mechanical work will become permanently concealed behind walls. While this is not typically a concern for retrofit

PV installations, it may impact new construction when PV systems are installed at the same time as the building electrical system.

When the completed installation restricts access to any equipment, wiring systems, or equipment labels, the installer should make provisions to allow access to these areas for inspections. For example, certain PV modules may be attached in a way that allows them to be easily removed from the array for inspection of labels, junction boxes, and grounding connections behind them.

Live Parts

Live parts are energized conductors and terminals. Live parts must be protected from accidental contact for any electrical equipment operating at 50 V or more. This can be accomplished by using approved enclosures and conduit, elevating the equipment 8' or more from the floor or work surface, or locating the equipment in a special room. When live parts are installed in protected locations, warning signs are required at the entrance prohibiting access to unqualified persons.

For one- and two-family dwellings, access to energized PV circuits over 150 V is prohibited to other than qualified persons. This protection can be in the form of enclosures that are lockable or require tools for access. Panel covers that are readily accessible and can be easily opened by unqualified persons do not meet these requirements.

Battery terminals are exposed live parts that must be adequately protected to prevent accidental contact. This protection includes battery or terminal covers, or installing batteries in enclosures. **See Figure 13-12.**

Labels and Marking

All electrical equipment must have durable labels identifying the manufacturer and any listing marks or certifications. Article 690, Part VI addresses additional marking requirements for PV systems. Labels are required for individual components and are typically applied by the manufacturer. For example, PV modules, inverters, and charge controllers must include voltage, current, and power ratings.

Battery Terminal Covers

Figure 13-12. *Rubber or plastic terminal covers are used to prevent shorts across battery terminals during maintenance.*

Labels are also required at certain points in the electrical system to identify the locations, purposes, or interfaces of various components. These labels are prepared by the installer. The labels identify power sources, utility interconnection information, disconnects, and safety hazards. All labels must be visible and accessible during inspections.

PV Module Labels. Module labels must include the terminal polarity, the maximum overcurrent protection device rating, and six module performance parameters: open-circuit voltage, maximum power (operating) voltage, maximum system voltage, maximum power (operating) current, short-circuit current, and maximum power. **See Figure 13-13.** Additional, but not required, information may include allowable conductor types and sizes, temperature ratings, and terminal torque specifications.

AC module labels must identify the terminals and ratings for nominal operating AC voltage and frequency, maximum AC power and current, and maximum overcurrent protection device rating.

Array-Disconnect Labels. The array (DC) disconnect must be clearly identified on a permanent and easily visible label. Since the array is a power source, its operating parameters must also be included on the disconnect labeling. The label must include the maximum power (operating) current, maximum power (operating) voltage, maximum system voltage, short-circuit current, and rated output for charge controllers (if installed). **See Figure 13-14.**

PV Module Labels

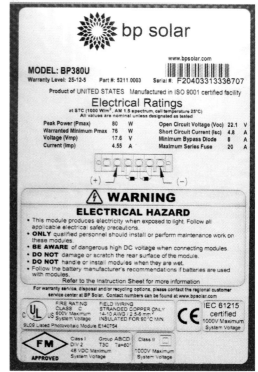

Figure 13-13. *PV module labels must include terminal polarity information and ratings for voltage, current, and power.*

AC-Disconnect Labels. AC disconnects must be clearly identified and include critical circuit information. A visible label must include the maximum AC output current and the operating AC voltage. **See Figure 13-15.** This information is obtained from inverter specifications.

Additionally, when all terminals of a disconnect device may remain energized in the open position, the device must include a warning sign. The sign shall be clearly legible and have the following words or equivalent: "WARNING—ELECTRIC SHOCK HAZARD. DO NOT TOUCH TERMINALS. TERMINALS ON BOTH THE LINE AND LOAD SIDES MAY BE ENERGIZED IN THE OPEN POSITION." This is especially important for AC disconnects in interactive systems and for battery bank disconnects because there are power sources on both sides of the disconnects.

Array-Disconnect Labels

*Southwest Technology Development Institute/
New Mexico State University*

Figure 13-14. *A label indicating the operating current, operating voltage, maximum system voltage, and short-circuit current must be displayed at the DC disconnect of a PV array.*

Point-of-Connection Labels. For interactive systems, a label must be located at the point of interconnection and must identify the system as interconnected with utility service. This point-of-connection identification will likely be in addition to other AC disconnect label requirements. The point-of-connection label may be at a back-fed circuit breaker in the main power distribution panel or at an AC disconnect switch.

Battery-System Labels. For PV systems with batteries or other forms of energy storage, system labels must indicate the maximum operating voltage, equalization voltage (if applicable), and the polarity of the grounded circuit conductor (typically the negative conductor). Additional information such as the nominal voltage or capacity may be included on this label. **See Figure 13-16.** Additional warning labels about chemical safety are recommended.

AC-Disconnect Labels

*Southwest Technology Development Institute/
New Mexico State University*

Figure 13-15. *Labels identifying the PV system as a power source and including its maximum output operating current must be posted at the AC disconnect of a PV system.*

Battery-System Labels

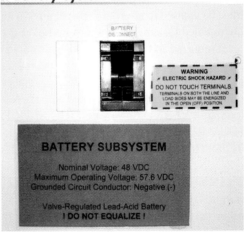

Figure 13-16. *PV systems with batteries must have labels indicating the battery maximum operating voltage, equalization voltage, and polarity of the grounded circuit conductor.*

PV systems may include optional watt-hour meters in addition to the utility service meter. These may be used for monitoring, production incentive metering, or other purposes. It is best practice to label each meter with its purpose to avoid confusion with required metering devices.

Power-Source Identification. For stand-alone PV systems, a permanent plaque or directory must be installed on the exterior of the structure identifying the presence of a stand-alone electrical system and indicating the location of the system disconnects.

For interactive systems, a permanent directory must indicate the locations of the service disconnect and the PV-system disconnect, if they are not located together. This safety requirement is in place to alert emergency and maintenance personnel to the presence of other power sources when utility service is disconnected.

Ground-Fault-Indicator Labels. Ground-fault indicators require labels for PV arrays mounted to roofs. These labels should be located near the ground-fault device and must state the following: "WARNING—ELECTRIC SHOCK HAZARD. IF A GROUND FAULT IS INDICATED, NORMALLY GROUNDED CONDUCTORS MAY BE UNGROUNDED AND ENERGIZED."

Ungrounded-System Labels. In addition to any other required labels, special warnings must be applied to ungrounded systems. In an ungrounded system, leakage currents below the level required to trip the ground-fault device may be present on the conductors, presenting an electrical shock hazard. For this reason, ungrounded systems must be labeled with the following: "WARNING—ELECTRIC SHOCK HAZARD. THE DC CONDUCTORS OF THIS PHOTOVOLTAIC SYSTEM ARE UNGROUNDED AND MAY BE ENERGIZED." This warning must be displayed at each junction box, combiner box, disconnect, and any other device where ungrounded conductors may be exposed during service.

Single 120 V Supply Labels. Stand-alone inverters with 120 V output are permitted to supply 120/240 V distribution panels, but no 240 V loads or multiwire branch circuits may be connected. The panelboard must be marked with the following label or equivalent: "WARNING—SINGLE 120-VOLT SUPPLY. DO NOT CONNECT MULTIWIRE BRANCH CIRCUITS." **See Figure 13-17.**

Single 120 V Supply Labels

Figure 13-17. *Single 120 V supply panelboards must be marked to prohibit connection of multiwire or 240 V branch circuit loads.*

Operating Parameters. Additional recommended labels include operating, maintenance, or extra safety information. **See Figure 13-18.** Examples include charge controller and inverter setpoints and maintenance schedules, and the name and contact information of the system installer.

Maintenance Labels

Figure 13-18. *Maintenance or operation labels are recommended for detailing nominal equipment settings and adjustment procedures.*

Inspection Checklist

An inspection checklist provides an extensive list of code references for PV systems based on the NEC® and industry standards. NEC® section citations are included for reference.

However, since code numbers can change with cycle revisions, the citations may need to be updated. A general inspection checklist is not intended to include every applicable requirement, but can be a useful guide to common PV system issues for inspectors, as well as PV system designers, installers, operators, and owners.

Also, since many component installation practices are governed by equipment manufacturer's instructions, the inspector should reference these materials as well as other documentation (schematics and equipment specifications) as part of a plans review or an in-field inspection.

Refer to Quick Quiz® on CD-ROM

Summary

- Building codes are applicable to the construction and electrical aspects of PV systems, just as for any other project.

- A building code must be adopted by the local jurisdiction to apply and be enforceable.

- The authority having jurisdiction (AHJ) approves equipment, materials, installations, and procedures by issuing permits, conducting inspections, and granting certificates of approval or occupancy.

- The use of approved equipment is required by the NEC® and facilitates the plans review, permitting, and field inspection processes.

- Underwriters Laboratories is the best-known NRTL and provides several types of certification marks, each with specific meaning.

- All electrical work should be performed by qualified persons.

- In some states, it is illegal for covenants or neighbors to restrict the solar resource to a person's property or to prohibit the installation of renewable energy systems.

- Building permits issued by city or county building departments authorize construction work to begin and establish the inspection requirements.

- Permit applications require certain information about the scope and specifications of the work, typically including the construction tasks, location, permit applicant, expiration date, and inspection requirements.

- If any portion of an electrical design is determined to be noncompliant with the applicable standards during a plans review, corrected designs must be resubmitted.

- The permit must be posted in a conspicuous location on the job site premises for the duration of the project.

- Following installation, and sometimes at intervals during installation, the contractor or owner must notify the AHJ to schedule an inspection.

- An inspector evaluates the conformity of an installation to applicable codes, laws, and regulations.

- PV system inspections ensure that there is adequate access and safe working space around all electrical equipment.

- All electrical equipment must have labels identifying the manufacturer and any listing marks or certifications.

- Labels are required at certain points in the electrical system to identify the locations, purposes, or interfaces of various components, and are prepared by the installer.

- An inspection checklist provides an extensive list of code references for PV-systems based on the NEC® and industry standards.

- A *building code* is a set of regulations that prescribes the materials, standards, and methods to be used in the construction, maintenance, and repair of buildings and other structures.

- A *model building code* is a building code that is developed and revised by a standards organization independently of the adopting jurisdictions.

- An *authority having jurisdiction (AHJ)* is an organization, office, or individual designated by local government with legal powers to administer, interpret, and enforce building codes.

- A *plans examiner* is a local official qualified to review construction plans and documentation for compliance with applicable codes and standards.

- An *electrical inspector* is a local official qualified to evaluate electrical installations in the field for compliance with applicable codes and standards.

- A *nationally recognized testing laboratory (NRTL)* is an OSHA-recognized, accredited safety testing organization that certifies equipment or materials to meet applicable standards.

- *Listing* is the process used by an NTRL for certifying that equipment or materials meet applicable standards.

- A *qualified person* is a person with skills and knowledge of the construction and operation of electrical equipment and installations and is trained in the safety hazards involved.

- A *construction bond* is a contract in which a surety company assures that a contractor will complete their work in accordance with contracting laws.

- A *permit* is permission from the AHJ that authorizes construction work to begin and establishes the inspection requirements, but does not represent an approval of compliance with codes and standards.

- A *plans review* is an evaluation of system-design documentation as part of the permitting process.

- *Dedicated space* is the clear space reserved around electrical equipment for the existing equipment and potential future additions.

- *Working space* is the clear space reserved around electrical equipment so that workers can inspect, operate, and maintain the equipment safely and efficiently.

1. Why are applicable building codes sometimes different between local jurisdictions?

2. Who is responsible for ensuring that building codes are followed properly?

3. Explain the difference between products bearing the UL Listing Mark and those bearing the UL Classified Mark.

4. How does contractor licensing improve the likelihood of quality installations?

5. What types of documents are commonly required for permit applications?

6. Explain the difference between dedicated space and working space.

7. How can inspection checklists be used by installers prior to the PV system-inspection by the AHJ?

14

Commissioning, Maintenance, and Troubleshooting

Although most PV systems require minimal maintenance, regular inspections and maintenance are important to maximizing the output and reliability of the system. Maintenance requirements vary according to the system configuration, installation type, and location. Qualified PV service technicians should have a complete understanding of the system design, equipment, and performance specifications in order to effectively conduct maintenance and troubleshooting activities. In all activities, appropriate safety measures must be followed.

Chapter Objectives

- *Describe the steps involved with commissioning a new PV system.*
- *Identify the maintenance tasks involved with maximizing array output, battery health, and other equipment operation.*
- *Develop a maintenance plan based on system configuration, installation, and location.*
- *Compare the various methods of monitoring system parameters for performance verification and troubleshooting.*
- *Troubleshoot PV systems based on a logical and efficient process.*

COMMISSIONING

Once a PV system has been installed, inspected, and approved, it can then be commissioned. *Commissioning* is the starting and operation of a PV system for the first time. Commissioning is also the time when the installer formally transfers responsibility for the system to the owner or operator, and marks the completion of the installation contract. Any warranties or service agreements should be finalized at commissioning.

The requirements for commissioning vary depending on the complexity of the system, though some general guidelines apply to most situations.

Final Checkout

Before any PV system is fully operated for the first time, the system should be checked thoroughly to ensure that the installation is complete and that the system is safe and ready for operation. A final checklist includes many of the same items as on maintenance, testing, and inspection checklists. **See Figure 14-1.** All disconnects should be in the open (OFF) position during the final checkout.

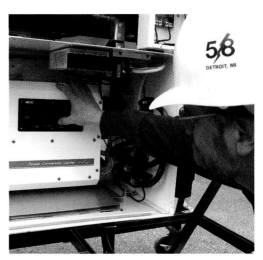

Commissioning, maintenance, and troubleshooting tasks include setting and adjusting parameters on inverters or charge controllers.

Final checkout should include attention to details associated with a neat and professional installation. All tools and debris should be removed from the site and the area cleaned. The customer's property should be restored to its original condition or better. Any property belonging to the customer that is damaged during the installation must be repaired or replaced at the installer's expense.

Commissioning Checklist

☐ Installation conforms to system design documents

☐ Conductors are of appropriate types and sizes

☐ Wiring is correct, protected, and secure

☐ Terminal connections are tight and properly identified

☐ Equipment is securely mounted

☐ Array mounting system is secure

☐ Roof penetrations are properly weather sealed

☐ Safety features are installed and operational

☐ Applicable warning and operational labels are posted

☐ Job site is clean, neat, and orderly

☐ Documentation package is complete

Figure 14-1. *A commissioning checklist should be reviewed before the initial startup of any PV system.*

Initial Startup

The initial startup is the first time that all the system components are energized. A general start-up procedure begins at the array and ends at the loads. If there are any installation problems, this reduces the chances of causing safety hazards and damaging equipment. One at a time, each disconnect is closed and each component is switched ON, and voltage and current are compared with expected values. **See Figure 14-2.** (Array and battery bank voltages should be confirmed before closing their disconnects.) Meters, indicators, and displays on charge controllers and inverters are used for measuring and verifying system parameters. If at any step the system parameters are outside of the acceptable range, the startup should be aborted and troubleshooting procedures should be used to find the cause of the problem.

⚡ General Start-Up Procedure

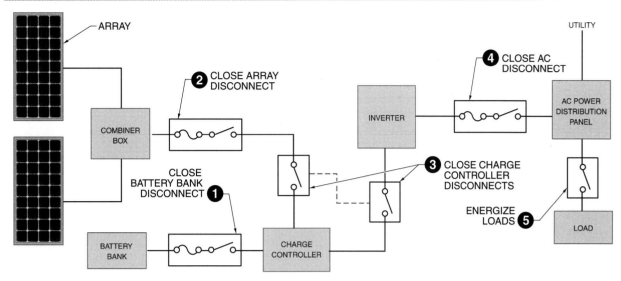

Figure 14-2. *A general start-up procedure begins at the array and ends at the loads, though exact procedures may differ.*

It is critically important to review all equipment documentation for any special start-up procedures. Improper startup may cause safety hazards, component damage, or impaired performance. Some component manufacturers specify variations on the general start-up procedure, which should always be followed. For example, some interactive inverters require that connections be made in a certain sequence to properly synchronize with utility-supplied power. The manual details each step and typically includes troubleshooting procedures for out-of-range results.

The voltage and current measurements taken during the initial startup should be recorded and kept with the other system documentation. These measurements serve as a baseline that will be helpful for subsequent maintenance and troubleshooting tasks.

System Documentation

A complete documentation package should be assembled and given to the customer at the time of commissioning. The package should include system and equipment manuals, permitting documents, and any warranty or service information.

Comprehensive manuals for package systems, such as those pre-engineered by an integrator, may be available. System manuals generally include safety requirements, electrical and mechanical drawings, parts lists with sources, system warranty information, and procedures for operation, maintenance, and troubleshooting. A documentation package should include the manuals for all major components, including PV modules, inverters, and charge controllers. Copies of permits, certificates of inspection, and utility interconnection agreements should also be included. These documents, in addition to installation contracts, may be required for insurance or financing purposes.

Original documentation should be kept in a safe place where it will not be easily damaged, but will always be available. If it is desired that operating or servicing instructions be posted, copies rather than original documents should be used in order to avoid loss or damage of the original documents. The installer should also retain copies of all documentation provided to the owner to aid with any warranty claims or follow-up service.

Proper safety precautions must be taken during all aspects of PV-system installation, operation, maintenance, and troubleshooting. These tasks can expose personnel to electrical, chemical, explosion, fire, exposure, and ergonomic hazards. Qualified persons have a thorough understanding of the equipment and procedures involved, and the skills and knowledge to recognize and mitigate the associated safety hazards. Specific safety regulations and standards required for construction and maintenance activities are covered in OSHA regulations 1926 and 1934 and NFPA 70E, *Standard for Electrical Safety in the Workplace.*

Depending on the system and the task, special tools and equipment, fall protection, and personal protective equipment (PPE) may be required. PPE includes face shields, safety glasses, hoods, helmets, harnesses, footwear, and clothing appropriate for the tasks to be conducted. Additional safety equipment may be permanently installed with the PV system, such as fire extinguishers and smoke detectors.

All sources of power should be disconnected, de-energized, or disabled before conducting any maintenance on electrical equipment. If tasks require working on energized equipment, special PPE must be used. It is highly recommended to work with a partner, especially around hazards.

User Training

Installers should provide operation and some maintenance training to the PV-system owner. The appropriate level and amount of training will vary depending on the type of system, the capabilities of the owner/operator, and the owner's/operator's expected involvement with the system. At a minimum, the owner should understand the overall functions of system components, principles of operation and safety, and maintenance requirements. Means of disconnecting the system should be demonstrated, and the owner should be given clear instructions about what to do or who to contact if service is required.

For relatively simple residential interactive systems, user training may require only a short walkthrough. **See Figure 14-3.** If the customer observes much of the installation and commissioning processes, additional training time may not be required.

For stand-alone and hybrid systems, additional training on safety and maintenance requirements is recommended, particularly regarding batteries, charge controllers, engine generators, dedicated electrical loads, and other equipment as applicable. Energy-management training is especially important for stand-alone systems. The customer must be aware that excessive load usage may result in overdischarged batteries and failure to meet system load requirements.

Customer Walkthrough

DOE/NREL, Rich Chartier

Figure 14-3. *A detailed walkthrough allows the installer to explain the basic operation of the PV system and the maintenance requirements to the owner.*

In addition to required system labeling, posting line diagrams, operating procedures, and programming instructions near key components can be very helpful when explaining system operation during user training, and for maintenance and troubleshooting tasks afterward. The installer should also review the contents and key sections of system documentation with the customer. Quality documentation greatly facilitates user training and allows the owner to continue learning about the system after the commissioning.

System Warranties

Major components such as PV modules, inverters, charge controllers, and batteries typically include warranties from the manufacturer. Some integrators and installers offer additional warranties on selected subsystems or the entire system. Terms vary widely regarding what is covered by a warranty and the length of time a system or component is covered. Installers may also extend warranties by offering a service contract. Local contractors may work under an agreement with a large national integrator that backs warranty claims for systems and major components installed by its recognized installers.

The installer should discuss and clarify all warranties with the system owner at the time of commissioning. This includes the method and manner in which the warranties are exercised, such as through the installer or dealer or directly with the product manufacturer.

MAINTENANCE

Most PV systems require relatively little maintenance, especially when designed appropriately for the application and installed according to best practices and with quality components. However, a modest investment in periodic inspections and maintenance ensures safety and the best possible performance. Routine maintenance also helps identify problems that require corrective service.

The degree and frequency of maintenance required depends on the system configuration, installation type, and location. Interactive systems require the least maintenance, while stand-alone and hybrid systems, because of the inclusion of batteries or additional power sources, require maintenance that is more extensive and frequent. Manufacturers may provide maintenance guidance or procedures in their component instructions.

Simple and nonhazardous maintenance can sometimes be performed by the system owner. Advanced maintenance, including troubleshooting and equipment replacement, may require the experience and expertise of a qualified installer or service technician familiar with PV systems, components, and proper safety procedures.

SPG Solar, Inc.
Array maintenance includes cleaning performance-reducing dust and dirt from the module surface.

Operating manuals, electrical diagrams, programming instructions, and troubleshooting procedures for PV modules, inverters, charge controllers, batteries, and other major system components are usually available on-line from the manufacturers' web sites.

Array Maintenance

Common maintenance tasks for arrays include module inspections, shading control, debris removal, array mount inspections, and tilt adjustments. For activities involving working around or touching modules, the array should be disabled by covering the modules or opening the array disconnect.

Module Inspection. Modules should be visually inspected for signs of any physical damage, including bent frames or broken glass. **See Figure 14-4.** Fractured or damaged modules should be replaced, even if they are still functioning electrically. If a damaged module is left in service, moisture may enter the module, causing dangerous shorts or ground faults. Most modules use tempered glass, which shatters into small pieces when broken from stress or impact.

Delamination is the separation of the bonded layers of glass and/or plastic encasing the PV cells of a module. Delamination allows moisture intrusion and corrosion within modules, particularly near the edges. This is visible as discolorations or bubbles in the laminate.

Module Inspection

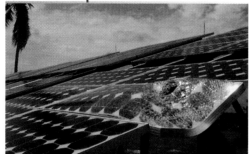

PHYSICAL DAMAGE

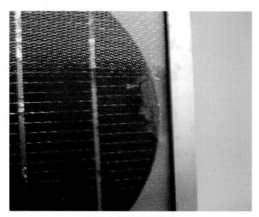

DELAMINATION

BURNED CONNECTIONS

Figure 14-4. *Visual module inspections involve checking for damage from physical impacts, delamination, burned internal connections, and other problems.*

In rare cases, the internal solder connections between cells can degrade. The increased resistance causes hot spots that can burn through the back of the module and result in module failure. Problems internal to a module, such as delamination or degradation of cell connections, are typically covered under the manufacturer's warranty.

When module junction boxes have removable covers, a few modules should be randomly chosen at each inspection and checked for secure wiring and moisure intrusion into the junction box. Any exposed conductors should have proper strain relief and should be neatly tied and concealed beneath the array. Conduit and fittings should be inspected for damage.

Equipment-grounding connections at each module should be inspected for corrosion. **See Figure 14-5.** This type of corrosion usually results from the contact of incompatible metals, so any connections exhibiting corrosion should be replaced with the proper stainless-steel fasteners.

Corroded Grounding Connections

Figure 14-5. *Corroded grounding connections usually result from the contact of incompatible materials or extreme environments.*

Shading Control. Even a relatively small amount of shading on the array can significantly reduce electrical output. The shading analysis conducted during the site survey establishes the optimal location with the least shading. However, some conditions can change over time, resulting in increased array shading. **See Figure 14-6.** Routine maintenance may be required to control excessive shading.

The growth of nearby trees and vegetation is an ongoing shading concern. Ground-mounted arrays may also be susceptible to shading from shrubs or long grass. These plants should be trimmed or mowed on a regular basis. The planting of trees or erection of new buildings or structures on adjacent properties may also result in additional array shading. A few municipalities and states have enacted legislation that protects the solar rights of property owners against excessive shading from adjacent properties, but in most areas there are few remedies to this situation.

Shading Control

TREES

SOILING

Figure 14-6. *Periodic shading control involves trimming vegetation and cleaning a soiled array.*

Excessive soiling can also cause shading. *Soiling* is the accumulation of dust and dirt on an array surface that shades the array and reduces electrical output. Soiling deposits may result from bird droppings, nearby industrial emissions, smoke, dust, dirt, and other atmospheric particles that settle on the array surface. Extensive soiling can reduce array output by 10% or more.

The amount of soiling on an array depends on the location. Arrays mounted near dirt roads or in dusty and windy areas are more likely to become soiled. Arrays mounted next to busy roadways or airports may become soiled with hydrocarbon emissions and grime. However, even where conditions are likely to cause soiling, frequent heavy rainfall and/or a high array tilt will tend to keep modules cleaner.

Soiled arrays can be cleaned by spraying the modules with a garden hose or pressure washer (with relatively low pressure). If material is stuck on a module, a soft brush and/or a light detergent may also be needed. A long-handled nonconductive brush makes it easier to reach all areas of the array surface. Direct contact with energized modules while washing must be avoided.

Debris Removal. Leaves, trash, or other debris should not be permitted to collect around arrays or any other electrical equipment. When dry, these materials are a fire hazard. When damp, they attract insects and can contribute to mold or mildew problems. **See Figure 14-7.** In humid climates, mildew can develop in the shaded portions of a roof under standoff-mounted arrays.

Mold

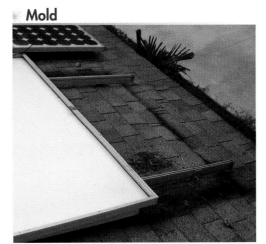

Figure 14-7. *Damp leaf debris trapped under arrays can cause mold and mildew problems.*

Possible contact between modules and flammable materials, such as dry leaves or building materials, is the reason that ground-fault protection is required for arrays mounted on buildings.

Debris can be removed by hand or with a garden hose when cleaning soiled arrays. A mildew cleaner or weak bleach solution can be used to control mold and mildew. When using bleach or any sort of cleaning solution, only the recommended concentrations should be used, and any components or plants exposed to the runoff should be protected or thoroughly rinsed afterward.

Array Mount Inspection. Array mounts should be inspected for any signs of corrosion or weakness. The inspection should include both the attachment of the modules to the mounting structure and the attachment of the mounting structure to the buildings or ground.

Special attention should be paid to the weather sealing of roof penetrations from mounting attachment points and conduits. Sealant cracks or shrinkage, broken gaskets, and corroded metal flashings are all signs of weather sealing degradation, which can quickly result in water leaks. **See Figure 14-8.** Inspecting the underside of the roof from the attic (if accessible) may reveal minor water leaks not previously discovered. Water stains, soft or rotten wood, constant dampness, or insect infestation indicate possible roof leaks. Even if leaks are not found, any degradation of weather sealing should be promptly corrected by removing the old sealant materials completely and applying new sealant.

Degraded Weather Sealing

Figure 14-8. *Cracked or deteriorated weather sealing around attachment-point penetrations can quickly develop water leaks.*

Tilt Adjustment. Some mounting systems allow manual tilt adjustments to optimize array orientation. If adjustments are to be made seasonally, they can coincide with the biannual inspection and maintenance activities. Tilt adjustments may involve removing and replacing pins, fasteners, or clamps. The array must be properly supported and secured during and after adjustments.

Battery Maintenance

Batteries are usually the most maintenance-intensive components in any PV system. Regular maintenance is important to maximizing battery life and minimizing hazardous conditions, though requirements vary significantly depending on the battery design and application. Open-vent batteries are the most maintenance-intensive type of batteries because they need periodic water additions and cleaning. Sealed batteries require much less maintenance.

Battery maintenance tasks include cleaning, tightening terminals, watering, and checking battery health and performance. Performance checks include specific gravity measurements, load tests, and capacity tests.

Battery manufacturers usually provide detailed maintenance recommendations for their batteries. For more information, IEEE 450, *Recommended Practice for Maintenance, Testing, and Replacement of Vented Lead-Acid Batteries for Stationary Applications* is a useful guide for servicing batteries.

Safety is of the utmost importance when performing battery maintenance. The battery bank should be isolated from the remainder of the system by opening the battery-bank disconnect. For battery banks greater than 48 V nominal, special disconnects are required to electrically divide the battery banks into lower-voltage sections for maintenance. Special insulated tools must be used when battery terminals are exposed during maintenance.

Battery Enclosure Inspection. The battery enclosure should be visually inspected for signs of electrolyte leakage, corrosion, or damage. Racks should be checked for adequate structural support, as well as proper placement of any straps or rails that hold batteries in place. **See Figure 14-9.** Trays should be checked for proper positioning beneath batteries. Also, the battery enclosure must have adequate ventilation. Obstructions that prevent airflow or otherwise clutter areas around the battery bank should be removed.

Batteries contain toxic materials and must be disposed of properly. Fortunately, nearly all of the materials in a lead-acid battery can be recycled. Batteries are typically accepted for recycling by battery manufacturers and retailers, recycling centers, and hazardous waste collectors.

Battery Enclosures

Figure 14-9. *Battery enclosures should be inspected for strength, cleanliness, and adequate ventilation.*

Individual battery cases should be inspected for any cracks or distortion, as well as the cleanliness of the top surface near the terminals and vents. If battery racks, trays, or cases need cleaning, a mild soap and water solution may be used with a soft brush or rag.

Terminal Inspection. Since battery terminals are made of soft lead alloys, the connections at the terminals can become loose over time. Loose connections increase the electrical resistance and voltage drop within the battery bank, resulting in unequal charge and discharge currents between individual batteries or cells. The high resistance also creates heat, which is a fire hazard, and in severe cases can overheat the battery-terminal connection until it deforms or melts.

Regular battery maintenance should include checking all terminal connections for looseness or corrosion. Any connections that move when pulled on vertically must be tightened. (Terminals should not be pulled laterally or twisted because internal battery components could be damaged.) Lock washers can be used with bolted battery terminals to prevent future loosening. The crimped lugs on battery cables should also be checked.

Any corrosion on battery terminals or connectors should be cleaned with a wire brush. **See Figure 14-10.** A weak solution of baking soda and water may be used to wipe down the terminals and top surfaces of open-vent lead-acid batteries as needed. Special care should be taken to ensure cleaning solutions do not enter battery cells. Terminals can be coated with petroleum jelly, grease, or special battery-terminal corrosion inhibitors as required. Also, check for adequate protection of live battery terminals, including boots or battery covers.

Battery-Terminal Cleaning

Figure 14-10. *Battery terminals are particularly susceptible to corrosion and may require frequent cleaning.*

Routine battery maintenance should include inspection of the safety and auxiliary equipment, including personal protective equipment (PPE), fire extinguishers, smoke detectors, ventilation equipment, and battery-testing tools and instruments.

Watering. The electrolyte in flooded, open-vent batteries must be maintained at proper levels by periodically adding water. This replaces water lost through battery gassing during charging. Water additions are required for both flooded lead-acid and flooded nickel-cadmium batteries. Pure, distilled water is recommended. Any salts and minerals dissolved in the water, even at relatively low concentrations such as in hard water from a domestic water supply, will slowly degrade a battery. Under no circumstances should acid (or base, for nickel-cadmium batteries) be added to refill a battery.

The level of electrolyte must not be allowed to fall below the tops of the battery plates. Exposed portions of the plates will oxidize, leading to premature capacity loss and battery failure. If the plates are exposed, water should be added to cover the plates. Because the electrolyte expands and the level rises slightly as the battery charges, batteries should only be completely filled or topped off after they are fully charged. Otherwise, the battery may overflow electrolyte from the cell vents when charged.

The frequency of watering required depends on a number of factors. The rate of water loss increases with battery age, higher charge rates, higher regulation voltage, and higher operating temperature. Watering intervals may be extended when batteries have reserve electrolyte capacity. Advanced multistage charge control and temperature compensation reduces water loss. Excessive water loss may be due to unnecessarily frequent charging cycles caused by a faulty charge controller, failed temperature compensation, or an improper regulation setpoint. Comparatively low water loss in individual cells indicates a weak or failing cell. The cell may need either an equalization charge or to be replaced.

The electrolyte level is checked by removing the cell vent caps and comparing the level

with respect to an established mark. A small flashlight is helpful to see inside the battery. Alternatively, if the battery case is partially transparent, it may be possible to see the electrolyte level from the outside. **See Figure 14-11.** Some batteries include an electrolyte level gauge that does not require cap removal. This gauge is a tiny captured float in contact with the electrolyte surface. When the float is visible from above through a clear window in the cap, the electrolyte level is adequate. When the electrolyte falls enough that the float cannot be seen in the window, the electrolyte is too low.

Checking Electrolyte Levels

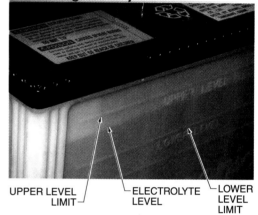

UPPER LEVEL LIMIT — ELECTROLYTE LEVEL — LOWER LEVEL LIMIT

Figure 14-11. *Battery maintenance includes checking for an adequate level of electrolyte.*

Water is added directly to cells through the vent caps. Care should be taken not to overfill the cells. A funnel or large syringe can be used to add water to cells with small openings. **See Figure 14-12.** A special watering can with a valve on the spout to prevent against overfilling may be used. Large battery banks may use automated watering systems to administer water to each cell through a network of plastic hoses. The date and amount of water added to each cell should be recorded in maintenance records.

Specific Gravity Measurement. The specific gravity of the electrolyte in open-vent batteries should be checked as part of a regular maintenance schedule, and more often if problems are suspected. Specific gravity can be used to estimate battery state of charge in lead-acid batteries (though not in nickel-cadmium batteries). Open-circuit voltage can also be measured and used in conjunction with specific gravity to estimate battery state of charge. **See Figure 14-13.** With either method, state of charge is most accurate when the battery has been at steady-state (neither charging nor discharging) for 5 min to 10 min. After watering, specific gravity should not be measured until a charging cycle has mixed the electrolyte.

Watering

Figure 14-12. *Battery watering replaces water lost from gassing during charging.*

Battery State of Charge Measurements*

STATE OF CHARGE	ELECTROLYTE SPECIFIC GRAVITY	OPEN-CIRCUIT VOLTAGE
100%	1.265	12.6
75%	1.225	12.4
50%	1.190	12.2
25%	1.155	12.0
0%	1.120	11.8

* for lead-acid batteries at 25°C

Figure 14-13. *Battery state of charge can be related to both specific gravity and voltage.*

When taking specific gravity readings, the variations between cells are as important as the overall average of the readings. Significantly low specific gravity of individual cells indicates cell failure or shorts, and likely requires battery or cell replacement. When variations between individual cells are small (within ±0.004), an equalization charge may be required.

A *hydrometer* is an instrument used to measure the specific gravity of a liquid. Two types of hydrometers used with battery electrolyte are the Archimedes hydrometer and the refractive index hydrometer. **See Figure 14-14.**

An Archimedes hydrometer is a bulb-type syringe that extracts electrolyte from the battery cell into a chamber. A float in the chamber experiences a buoyant force equivalent to the weight of the displaced electrolyte. Consequently, the float is more buoyant at higher specific gravities and floats at a higher level. The float may be a glass tube, plastic ball, or a lever attached to the side of the chamber. Marks on the side of the float or chamber are calibrated to indicate the specific gravity directly.

Hydrometers

ARCHIMEDES HYDROMETER

REFRACTIVE INDEX HYDROMETER

Figure 14-14. *Either one of two types of hydrometers can be used to measure the specific gravity of battery electrolyte.*

A refractive index hydrometer, also called a refractometer, uses a prism to measure the refractive index of the electrolyte, which is related to specific gravity. The refractive index is the amount that a substance bends light that passes through it. A small drop of electrolyte is placed on the prism, and light refracts at an angle related to the density and specific gravity of the electrolyte. The user then observes the refracted light though a viewfinder on a scale calibrated in specific gravity units. **See Figure 14-15.** The main advantage of a refractive index hydrometer is that only a very small amount of electrolyte is required.

Refractive Index Hydrometer Scale

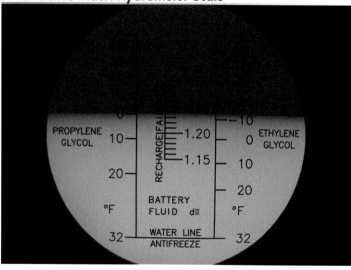

Figure 14-15. *When viewing through a refractive index hydrometer, the specific gravity of the tested fluid is measured against a calibrated scale.*

Hydrometer markings are calibrated to a specific temperature, typically 80°F (27°C). When specific gravity is measured at temperatures lower or higher than the reference temperature, a correction factor must be applied. First, the temperature of the electrolyte must be accurately measured. The adjustment of the specific gravity reading is based on the difference between the measured temperature and the reference temperature. A standard correction factor of 0.004 specific gravity units, often referred to as "points," is applied for every 10°F (5.6°C) difference. This correction factor applies regardless of the reference temperature of the hydrometer. Points are added for temperatures above the reference temperature and subtracted for temperatures below the reference temperature. For example, at 90°F (32°C), a hydrometer reading of 1.250 would be corrected to 1.254. Conversely, at 70°F (21°C), a hydrometer reading of 1.250 would be corrected to 1.246.

Load Testing. The ability of a battery to maintain voltage while under load is another indication of battery condition. While a degraded battery may accept a charge, it may not perform well under load. A *battery load tester* is a test instrument that indicates battery health by drawing a high discharge current from a battery for a short period. The load is typically drawn at a rate of C/1 or greater for no more than 15 sec. At the same time, battery voltage is measured and recorded. A minimum voltage of 9.6 V for a nominal 12 V lead-acid battery is considered acceptable for this test. Results should be recorded in a log, which helps identify long-term trends.

Capacity Testing. Available battery capacity can be measured by load testing at rates comparable to those during normal system operation. Starting with a fully charged battery bank, the array is disconnected, but the system load is left connected. At regular intervals, such as every hour, the battery voltage and load current are measured and recorded. The load remains connected until the battery reaches the cutoff voltage. The total capacity is equal to the average load current multiplied by the discharge time. For example, a battery that is discharged at an average of 10 A for 8 hr delivers 80 Ah. After a capacity test, the load should be left disconnected and the array or an auxiliary charging source connected until the battery regains full state of charge.

Electrical Equipment Maintenance

Routine maintenance should include visual inspections of inverters, chargers, charge controllers, transformers, and any other electrical equipment in the PV system. This equipment requires adequate surrounding space for accessibility and airflow, which allows for heat dissipation. Therefore, equipment temperature should be measured. Infrared (IR) noncontact thermometers are ideal for this task. **See Figure 14-16.** Equipment at higher-than-normal temperatures requires immediate attention to address possible overloading conditions or poor airflow.

All wiring should be inspected, including conductors, terminations, conduit, and junction boxes. Disconnects, fuses, and circuit breakers should be checked for proper operation. Any exposed conductors should be checked for insulation damage, clean and secure terminals, adequate strain relief, and properly connected and supported conduits.

Infrared (IR) Thermometers

Fluke Corporation

Figure 14-16. *An infrared (IR) thermometer can measure the temperature of electrical equipment, including PV system components.*

Tools and Testing Equipment

So that maintenance and troubleshooting can be conducted in a prompt and efficient manner, PV system service personnel should have certain basic tools and testing equipment available at all times.

Tools

A standard PV maintenance tool kit should include screwdrivers, pliers, wire cutters, wire strippers, crimpers, wrenches and socket sets, utility knives, and flashlights. PV-specific tools include a handheld pyranometer, compass, caulking gun, and a variety of battery-maintenance equipment, as required.

A variety of spare parts and materials should be readily available, including fasteners, fuses, connectors, lugs, conductors, weather sealant, lubricant, corrosion inhibitor, electrical tape, wire ties, and distilled water.

Besides basic PPE, safety equipment should include a first aid kit, electrical gloves, fire-resistant clothing, hats, sunscreen, rubber gloves, and baking soda.

Testing Equipment

A digital multimeter is the most important and versatile test instrument for maintenance, monitoring, and troubleshooting of electrical systems. Most models can measure voltage, current, and resistance. Other parameters, such as temperature, can also be measured, either as a built-in function or with add-on accessories. This test instrument is vital to maintaining any PV system.

Other test instruments are not always required, but can be very useful. A clamp-on ammeter makes current measurements easier and safer, since the circuit does not need to be opened to take measurements.

Watt-hour meters record cumulative energy and can be useful in many different ways. At the utility connection, a pair of watt-hour meters can record net energy exporting and importing. Another watt-hour meter is recommended for the inverter output, if the inverter does not already monitor energy output. On the array output circuit and the battery-bank charging circuit, ampere-hour meters are used in a manner similar to watt-hour meters. In the battery system, ampere-hour meters can also show the net energy that is in a battery, and therefore its state of charge.

Maintenance Plans

PV-system maintenance should be carefully planned and scheduled to ensure that all necessary tasks are being performed and to minimize the time and expense required. A *maintenance plan* is a checklist of all required regular maintenance tasks and their recommended intervals. **See Figure 14-17.** Maintenance plans are developed during the design and commissioning process based on the typical maintenance requirements for the system configuration, installation type, and location. Much of this information is found in manufacturer's recommendations.

However, maintenance plans can evolve as the needs for the particular system are determined or changing conditions necessitate different maintenance requirements. For example, an array should normally be inspected every six to twelve months. However, if inspections reveal that soiling conditions are severe and are causing significant shading, the array may require inspection and cleaning every three months. The maintenance plan should then be changed to reflect the new requirement.

It is highly recommended that records of all maintenance tasks be kept in a maintenance log. A *maintenance log* is a collection of past maintenance records. The log should include the maintenance tasks completed, the date and time, results, problems encountered, and any recommendations for future maintenance, such as changes to the maintenance plan or tips for making tasks easier or more effective.

Most rooftop maintenance tasks require some form of fall protection, such as warning lines.

Maintenance Plan*

TASK	RECOMMENDED INTERVAL		
	As Required	**Monthly**	**Semiannually**
Inspect modules for damage			✓
Address array shading issues	✓		
Remove debris around array	✓		✓
Inspect array mounting system			✓
Adjust array tilt	✓		
Check inverter and/or charge controller for correct settings		✓	
Inspect battery enclosure		✓	
Inspect battery terminals and connections		✓	
Equalize batteries	✓	✓	
Water batteries	✓	✓	
Measure specific gravity of each battery cell	✓	✓	
Load-test batteries			✓
Capacity-test batteries			✓
Inspect and clean all electrical equipment			✓
Monitor system for voltage and current	✓	✓	

* maintenance plan tasks and recommended intervals will vary between systems

Figure 14-17. *A maintenance plan includes all the necessary maintenance tasks and their respective schedules.*

MONITORING

Monitoring is the repeated measurement of electrical, environmental, or battery parameters at certain intervals. PV-system monitoring can serve several purposes. Primarily, it provides information about system status and alerts users to possible problems. Owners and installers also monitor system performance to ensure that energy production meets load requirements and justifies the system cost. Monitoring also includes short-term measurements for troubleshooting. Production-based financial incentives and renewable energy certificate (REC) programs rely on accurate system output measurements to appropriately award incentives. Utilities monitor interactive PV systems to determine the energy exchanged between the customer and the utility grid for billing purposes.

A current transformer consists of a toroidal coil in which a proportionally smaller current is induced from the higher current in the conductors passing through it. These devices are used to monitor current flow in various applications.

Measured Parameters

A variety of parameters may be measured by monitoring equipment in PV systems. The number and types of parameters depend on the system configuration and size and the intended use of the information.

Electrical Parameters. Measuring electrical parameters is the most important type of monitoring. At a minimum, electrical power and energy are measured, though various voltage and current measurements are also common. These parameters are the basic indicators of array and inverter performance and are extensively used to troubleshoot system problems. Performance is based on comparisons with expected output (based on sizing calculations) and baseline measurements taken during commissioning.

Energy is also often measured (in kilowatt-hours) on daily, monthly, and total cumulative bases. These measurements are particularly useful for the inverter output circuit, but may be used elsewhere. In interactive systems, the total load requirements are the energy supplied by the PV system plus any energy imported from the grid. In stand-alone systems, energy flow into and out of batteries is measured with ampere-hour meters.

As part of a service contract, a technician may monitor a PV system remotely. The monitoring system may be configured to trigger a service call for certain events or if the measured parameters fall outside the normal range.

There are three primary locations in a PV system for measuring electrical parameters: the array output circuit, the inverter output circuit, and the battery-bank output circuit (if applicable). **See Figure 14-18.** Since the inverter is connected to all three circuits, most modern inverters include integrated monitoring functions as standard features. If this is not the case, external monitoring equipment can be added to measure electrical parameters. For temporary use, such as for troubleshooting, handheld test instruments are used. Sytem designers and equipment manufacturers sometimes incorporate test points into circuits for safe and easy access by test instruments. For some installations, permanently installed meters are used.

Depending on the system configuration, electrical parameters may also be measured at other parts of the system, such as PV source circuits, the charge-controller output circuit, and within the power-distribution system.

Primary Monitoring Points

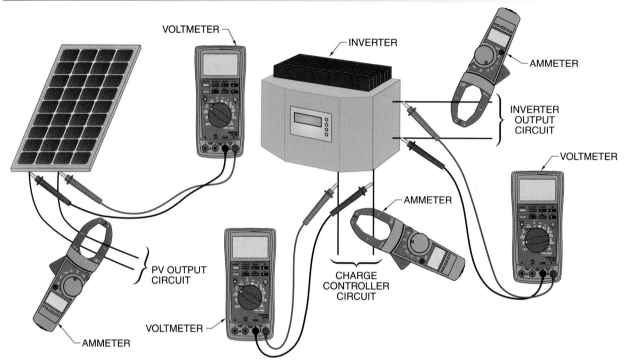

Figure 14-18. *The three most important points for measuring voltage and current information are the array output circuit, inverter output circuit, and charge controller circuit (if applicable).*

Environmental Parameters. Including environmental parameters in monitoring allows system output to be correlated with weather conditions. These parameters usually include irradiance and ambient temperature. **See Figure 14-19.** For example, current output from a PV device increases proportionally with irradiance. However, if the current measurements do not correlate with the irradiance measurements, there may be a problem with the array, such as excessive shading, soiling, or blown source circuit fuses. For an accurate representation of the array's environmental conditions, all sensors should be placed as close to the array as possible and in the same plane.

By adding a special temperature sensor to the back surface of a module, cell temperature can also be measured. This information can then be used to determine the temperature-rise coefficient for this installation.

Battery Parameters. Sophisticated monitoring systems may have the capability to automatically measure battery specific gravity and water levels. This feature is usually employed for only the largest battery-based systems, where the large number of batteries makes manual battery tests impractical. The battery bank may require manual maintenance only when the monitoring system indicates an out-of-range measurement.

Environmental Sensors

Kipp & Zonen, Inc.

Figure 14-19. *Sensors for measuring irradiance and other parameters allow technicians to correlate PV system output with weather conditions.*

Monitoring Output

In order to be useful, measurements must be displayed in some form. Monitoring of a PV system can be as simple as LED status indicators on a component or as complex as a separate data network measuring and recording multiple parameters.

Indicators. The simplest type of monitoring employs small indicator lights to show the status of components and their functions. These are usually LEDs of different colors. For example, an LED may indicate that the battery bank is fully charged, even if the charge controller also displays the actual battery-bank voltage. **See Figure 14-20.** Indicators on major components may indicate normal operation, ground faults, low-voltage load disconnect, battery charging, battery discharging, battery equalizing, or component malfunctions. For efficiency of space on the front panel, indicator LEDs may have multiple meanings depending on their colors (for bicolor LEDs) and the rate at which they blink.

Status Indicators

Morningstar Corp.

Figure 14-20. *Equipment status indicators may have multiple functions.*

Display Screens. Most inverters and charge controllers include small display screens. This is often in addition to a few indicator lights. The display typically cycles through the various measurements and displays the present values. Some inverter manufacturers provide the option of adding a remote display unit to view these parameters from a distant location.

The connection to remote units may be wired or wireless.

Data-Acquisition Systems. The most sophisticated monitoring systems include data acquisition. *Data acquisition* is the recording and processing of data from a monitoring system. The same types of parameters may be measured as with other monitoring systems, but the data is recorded at regular time intervals and stored for future analysis. Data acquisition generates a wealth of information and provides a much more complete picture of system performance than measurements at isolated moments in time.

As the primary electrical interface device in most PV systems, the inverter is the core of most data-acquisition systems. Some inverters include data acquisition as an on-board feature and others rely on separate computer software to read the data output at intervals and save the data to a storage device.

Some data-acquisition systems employ separate units to collect and process the data from multiple sources. In addition to taking measurements from the inverter, these systems may allow a variety of other sensors to be added, such as irradiance and temperature. Such a system requires the added time and expense of separate data wiring that connects each sensor and the data-acquisition unit, but provides much more information.

Using computer software, this data can then be plotted to illustrate the changes of values over time. **See Figure 14-21.** This facilitates spotting trends and correlations between parameters. When troubleshooting, data-acquisition systems are invaluable tools. They allow the technician to review past records and identify slow trends over time or short-term events that may have impaired system performance. By connecting the computer or data-acquisition unit to the Internet, the raw data and plots can also be made available to users through web sites.

> Several leading inverter manufacturers and independent companies offer data-acquisition equipment and sensors for interfacing directly with system components.

Monitoring Data Plots

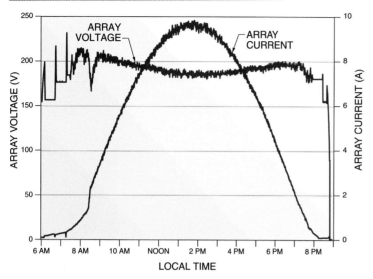

Figure 14-21. *Data-acquisition systems gather, record, and process information from many sources that can be used to observe trends or problems in PV system operation.*

TROUBLESHOOTING

Troubleshooting is a systematic method of investigating the cause of system problems and determining the best solution. Problem diagnoses begin with observations and measurements of the overall system. Through a process of confirming or eliminating factors, the investigation narrows to subsystems or specific components until the cause of a problem is found.

The focus of troubleshooting is on remedying the cause, not the symptoms, of a problem. While identifying symptoms is a critical first step in determining the specific cause of a problem, there may be several contributing factors. For example, a dead battery is a common symptom. While the individual battery may fail solely due to the end of its useful life, aggravating circumstances could also be preventing the battery from charging. This includes factors that disrupt the energy balance of the system, such as excessive loads or a reduction in array output. If the battery is replaced without eliminating contributing

factors, the problem of undercharge and premature failure may affect other batteries. Many possibilities must be investigated to prevent a misdiagnosis or partial diagnosis of the actual problem, as failure to correct the underlying problem results in wasted time and money.

Troubleshooting Levels

Troubleshooting procedures are designed to identify malfunctions by examining the system at progressively narrower levels. A *troubleshooting level* is the depth of examination into the equipment or processes that compose a system. Troubleshooting levels are very similar for all electrical systems, even though they may be named differently. For PV systems, the appropriate levels are the system level, subsystem level, component level, and element level. **See Figure 14-22.**

System Level. The system level encompasses the entire PV and power-distribution system, sometimes including the on-site loads. Systems may be isolated or may interface with other related systems. For example, a stand-alone PV system is a self-contained electrical system. A utility-interactive system, however, interacts with the utility grid. Even though the utility grid is much larger in scale, it is considered a parallel system.

Subsystem Level. The system is broken down into functional areas called subsystems. Subsystems may include many components, but they are all related to a particular function within the system, such as producing and controlling DC power. The subsystem level includes the energy production, storage, conditioning, distribution, and consumption functions. These functions correspond to the array, battery bank, inverter (or power conditioning unit), distribution panel, and loads, respectively. **See Figure 14-23.**

Each subsystem also includes all the auxiliary equipment related to the safe and efficient operation of that subsystem. For example, besides the batteries, the energy-storage subsystem includes the charge controller, disconnects, fuses, conductors, electrolyte containment, and ventilation devices.

Troubleshooting Levels

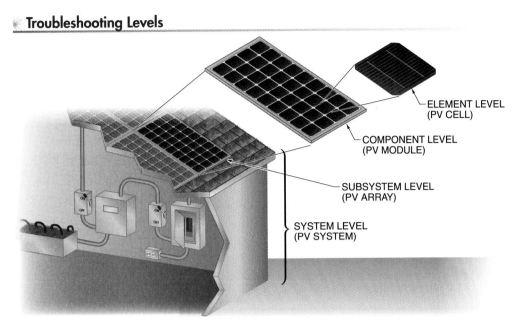

Figure 14-22. *The system level includes all the components of a PV system.*

PV Subsystems

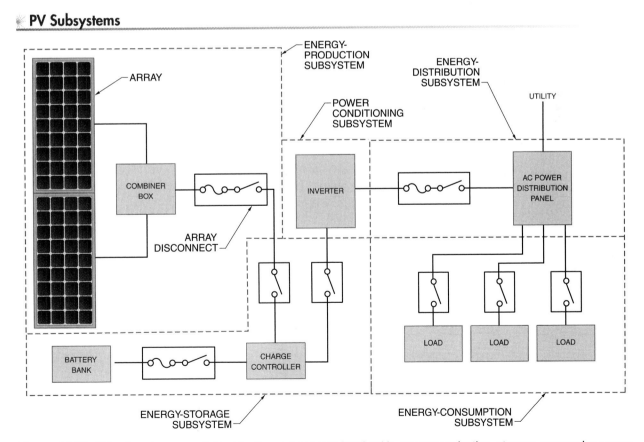

Figure 14-23. *PV subsystems are divided by the components involved in energy production, storage, processing or conditioning, distribution, and consumption.*

The number and types of subsystems will depend on the system configuration. For example, interactive-only systems do not include an energy-storage subsystem. Also, some systems may include multiple subsystems of the same type. For example, a very large system may have multiple arrays, each feeding a separate inverter. Or, a system may include both a DC power-distribution subsystem for DC loads and an AC power-distribution subsystem for AC loads.

Component Level. The component level includes the discrete pieces of equipment that are acquired and installed individually to make up subsystems. This includes modules, disconnects, inverters, charge controllers, batteries, and loads. Each component performs a specific task by itself, such as opening a circuit, processing power, or storing a certain amount of energy.

Element Level. Within the components are smaller elements. Individually, these elements are too small or too specialized to be useful in any way other than as a part of a component. For example, a PV cell is an element within a module component. A PV cell is not easily used by itself because it is extremely fragile and a system would require hundreds or thousands of cells to produce an appreciable amount of power. A PV cell must be combined with other cells and additional elements (such as conductors, glass, and frames) to form a module component, the smallest practical PV unit for building a PV system.

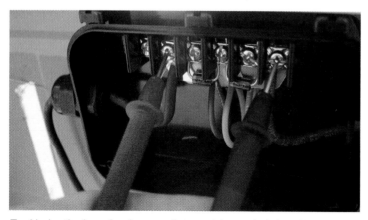

Troubleshooting loss of performance from modules can include making electrical measurements at the junction-box terminals.

The distinction between the component and element levels can sometimes be unclear. For example, battery cells are considered components if they are large individual units, but they are considered elements if they are inseparable parts of a battery, which is then a component of the battery-bank subsystem. Fuses can sometimes be considered components and other times be considered elements, especially when they operate from within other components such as inverters.

Some elements, such as fuses, circuit breakers, surge suppressors, and diodes, are easily field-serviceable, but many are not. When elements are not reasonably serviceable, the entire component may need to be replaced, even if only one small element is faulty. For this reason, field troubleshooting does not always reach the element level. For example, if an impact damages a module, it is not worthwhile to disassemble the module and identify the individual broken cells since the module will still need to be replaced.

For only certain components, such as inverters, is it cost effective to troubleshoot at the element level, and troubleshooting of this sort is not usually performed by field technicians. When found to be faulty during component-level troubleshooting, these components are usually sent to the manufacturer for service. Depending on the design of the component, subassemblies or circuit boards with a faulty element may be replaced to remedy the problem.

Troubleshooting Procedures

While troubleshooting at progressively more detailed levels, a systematic procedure is followed to determine the cause of the problem. While specific troubleshooting procedures will vary according to the type of system and equipment involved, a general strategy guides the overall troubleshooting process in a logical and efficient manner.

Observation. The first step in troubleshooting is to identify the symptoms of a problem. Some problems may be obvious, while others may be noticed only during inspections and maintenance activities or when comparing the system performance with past experience of its normal operation. Therefore, some problems may be discernable only by the system owner/operator, who is most familiar with the system. For example, system output will normally fall during cloudy weather, but if output falls more than past experience predicts or fails to return to higher output later, a problem exists.

Observation involves determining which part of the system is not working properly and observing the conditions that may have contributed to the cause. This includes obvious problems such as broken or burned equipment, water intrusion, and blown fuses. All equipment with monitoring, such as inverters and charge controllers, should be checked for error or fault indicators. Some indicators will provide very detailed information. Observation also includes attention to more subtle clues, such as any recent changes in the load requirements, maintenance plan, or weather patterns.

If a problem is creating a hazard to persons or equipment, it may be necessary to immediately shut down the entire system. However, as some troubleshooting tasks, such as taking electrical measurements, require certain components or the entire system to be operating, extreme caution must be exercised when restarting the system. As with maintenance and testing tasks, all proper safety precautions should be followed.

Research. Research involves gathering all the relevant system documentation together, particularly equipment manuals, and finding information about normal operating parameters, specifications, compatibilities, maintenance requirements, precautions, and error codes. Some manuals include troubleshooting instructions or flow charts that may help narrow the cause of the problem. **See Figure 14-24.**

If the system includes a data-acquisition system, past records can be used to look for unusual events or patterns in the measured parameters. Researching the maintenance log and troubleshooting reports for any previous symptoms or related problems can also help uncover the cause or at least suggest a line of investigation.

Troubleshooting Levels

The troubleshooting levels are important at the observation step in the troubleshooting process. The problem may be system-wide with no obvious cause, such as an unexplained low system output, which requires troubleshooting to begin at the system level. Further steps in the troubleshooting process will help narrow the cause of the problem down to the subsystems or individual components causing the problem.

Alternatively, the problem may be relatively localized and the troubleshooter can begin observing the symptoms at the more focused levels. However, it may also be necessary to return to the overall system level if the cause of the problem cannot be determined. The chosen focus may have been an incorrect assumption.

Ultimately, the troubleshooter makes a judgment about where to begin troubleshooting in order to find the cause of the problem quickly and efficiently.

Investigation. Through observation and research, the problem typically becomes well understood. The aim of investigation is to determine the root cause of the problem. For example, a blown fuse is easily fixed by replacing the fuse. However, an underlying problem caused the fuse to blow. Unless the cause is identified and addressed, the new fuse will likely blow relatively quickly. An investigation would seek to determine the cause of the overcurrent, so that the underlying problem could be remedied.

Investigation is conducted primarily through inspection and testing. The causes that are the most likely and require the simplest remedies should be investigated first. These include damage from physical impacts, worn insulation, excessive heat or cold, electrical arcing, smoke, or fire. Portable test instruments are used to take electrical measurements at various points in the circuits to trace the locations of voltage drops, phantom loads, short circuits, and open circuits. Temperature measurements check for overheating equipment.

Troubleshooting Flow Charts

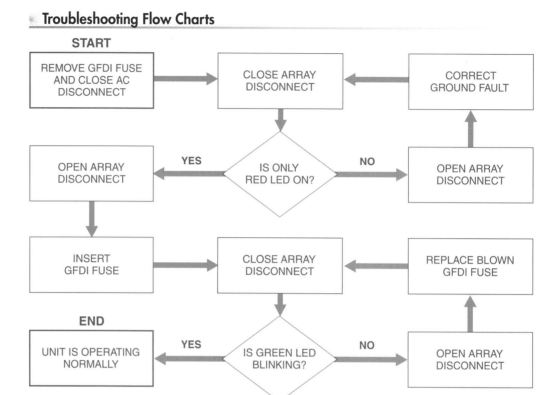

Figure 14-24. *Most equipment manuals include troubleshooting flow charts or procedures, which narrow the possible causes of a problem by following a specific set of instructions.*

Refer to Quick Quiz® on CD-ROM

It may be necessary to replicate the conditions associated with the symptoms of the problem. In order to determine which conditions contribute to the problem, conditions should be tested and eliminated, preferably one by one. A systematic investigation should gradually eliminate certain subsystems and components until the cause is isolated.

Remediation. When the cause is discovered, it can be remedied, fixing the problem. This usually involves repairing or replacing components, reterminating or tightening connections, or replacing damaged conductors. If the problem occurs during or soon after the system commissioning, there may be an inherent flaw in the design of the system, requiring more extensive remediation.

Occasionally, the cause of the problem turns out to be minor enough that the remedy is not a high priority. That is, if the problem is not a safety issue and causes only a minor inconvenience, the owner may choose to live with it, especially if the estimated costs to remedy the problem are high. Still, the issue should be monitored and fully documented in case the problem worsens.

Documentation. When troubleshooting has successfully identified the cause of the problem, it is important to document the results. Troubleshooting reports should describe the problem, the steps taken to determine the cause of the problem, and the repair that eliminated or reduced the cause. This includes the symptoms, the equipment manuals consulted, the test instruments used, the measurements taken, and a detailed description of the remedy.

Troubleshooting reports may also include recommendations for preventive maintenance to avoid a recurrence of the problem. These records should be kept with all the other system documentation so that they can be reviewed again as needed during future troubleshooting.

Summary

- Commissioning is the time when the installer demonstrates system operation and formally transfers responsibility for the system to the owner or operator.

- A general start-up procedure begins at the array and ends at the loads, though certain components or system configurations recommend different procedures.

- It is critically important to review all equipment documentation for any special start-up procedures.

- Interactive systems require the least maintenance while stand-alone and hybrid systems, because they include batteries or additional power sources, require maintenance that is more extensive.

- Simple and nonhazardous maintenance may sometimes be performed by the system owner. Advanced maintenance tasks, including troubleshooting and equipment servicing or replacement, usually requires the experience and expertise of a qualified installer or service technician.

- Shading conditions can change over time, so routine maintenance may be required to control excessive shading.

- Sealant cracks or loss, broken gaskets, and bent metal flashings are all signs of weather sealing degradation, which can quickly result in roof leaks.

- Batteries are usually the most maintenance-intensive components in any PV system.

- Battery maintenance tasks include cleaning, tightening terminals, watering, and checking battery health and performance.

- Battery state of charge is estimated by specific gravity or open-circuit voltage measurements.

- PV-system maintenance should be carefully planned and documented to ensure that the necessary tasks are being performed on the required schedule.

- PV-system monitoring provides information about system performance and status and alerts persons to possible problems.

- Power, voltage, and current are the basic indicators of array and inverter performance and these measurements are extensively used to troubleshoot problems.

- Data acquisition facilitates spotting trends and correlations between parameters and is particularly useful for troubleshooting.

- Troubleshooting procedures are designed to identify malfunctions by examining the system at progressively narrower levels.

- When troubleshooting has successfully identified the cause of the problem, it is important to document the results.

Definitions

- *Commissioning* is the starting and operation of a PV system for the first time.
- *Delamination* is the separation of the bonded layers of glass and/or plastic encasing the PV cells of a module.
- *Soiling* is the accumulation of dust and dirt on an array surface that shades the array and reduces electrical output.
- A *hydrometer* is an instrument used to measure the specific gravity of a liquid.

- A *battery load tester* is a test instrument that indicates battery health by drawing a high discharge current from a battery for a short period.

- A *maintenance plan* is a checklist of all required regular maintenance tasks and their recommended intervals.

- A *maintenance log* is a collection of past maintenance records.

- *Monitoring* is the repeated measurement of electrical, environmental, or battery parameters at certain intervals.

- *Data acquisition* is the recording and processing of data from a monitoring system.

- *Troubleshooting* is a systematic method of investigating the cause of system problems and determining the best solution.

- A *troubleshooting level* is the depth of examination into the equipment or processes that compose a system.

Review Questions

1. Why should PV systems be commissioned by using a careful start-up sequence?

2. Why might strat-up procedures vary for different PV systems?

3. What should be included in the system documentation package?

4. What factors are involved in the long-term control of array shading?

5. What tasks are typically involved in battery maintenance?

6. What factors affect the frequency of battery watering?

7. Why is electrolyte specific gravity measured?

8. How is a maintenance plan developed?

9. How are data-acquisition records particularly useful when troubleshooting?

10. Why is it sometimes unnecessary to troubleshoot beyond the component level?

11. What is involved in the observation step when troubleshooting?

15

Economic Analysis

The value of a PV system is often associated with environmental issues and energy conservation. However, PV systems can provide financial benefits as well. A PV system may be able to pay back its initial investment and possibly earn appreciable amounts of money, especially with help from incentive programs. Sometimes a PV system is simply less expensive than alternate energy sources. To determine if a PV system is cost effective, an analysis of the life-cycle cost and potential financial payback is conducted.

Chapter Objectives

- Compare the numerous incentive options based on type, source, availability, and requirements.
- Describe how present and future costs are calculated.
- Understand how economic variables such as discount rate affect the present value of future costs.
- Compare energy-production systems based on total life-cycle costs.
- Differentiate between the ways different incentives affect the life-cycle costs of PV systems.
- Determine whether a PV system can pay back its costs against an alternate energy source.

INCENTIVES

Many people support the idea of PV and other renewable-energy systems, but assume that these systems are too expensive. Even in cases where the system cost is likely to be recovered, the significant initial costs keep some people from investing in a system. For these reasons, government entities, utilities, and nonprofit organizations sponsor incentives to make renewable energy more affordable. **See Figure 15-1.** An *incentive* is a monetary inducement to invest in a certain type of capital improvement, such as an energy-generating system or energy-conservation measure.

SolarWorld Industries America
Many government entities are involved in promoting renewable energy sources.

Incentives

INCENTIVE TYPE	PRIMARY SPONSORING ENTITIES				
	Federal Government	State Governments	Local Governments	Utilities	Nonprofit Organizations
Rebates		✓	✓	✓	
Grants	✓	✓	✓	✓	✓
Loans	✓	✓	✓	✓	
Tax incentives	✓	✓			
Production-based incentives	✓	✓		✓	✓
Renewable energy certificates					✓

Figure 15-1. *Several types of incentives are available for PV and other renewable-energy systems.*

Incentives from Utilities

It may not seem logical for utility companies to support the expansion of renewable-energy systems among small-system operators because it encroaches upon the utilities' business. Indeed, some utilities are mandated through legislation to provide incentives. However, offering incentives can be in the utilities' best interest for several reasons.

First, many states have implemented renewables portfolio standards (RPSs), which mandate that energy producers operating within the state produce a certain percentage of their energy with renewable-energy technologies. Electricity from PV systems sold to the utility through interconnection agreements counts toward their share of the goal.

Also, encouragement of additional utility-connected power producers helps alleviate the need for utilities to build "peaker" power plants that produce extra power for periods of peak power demand. PV systems are especially well suited for "peaking" because they produce the most electricity during the highest-demand period, the middle of the day.

Finally, increased demand for PV systems fosters improvement in module efficiency and manufacturing, lowering the costs for installing new systems. This includes large-scale systems that the utilities may use in the future.

Incentives also stimulate the renewable-energy technology industries, because making systems more affordable increases demand. Higher demand increases production and promotes research into more-efficient systems and manufacturing methods, which both lower system costs. At some point, the costs of renewable energy will be competitive with traditional fossil-fuel sources, so there will no longer be a need for subsidies. It is uncertain when this will occur, perhaps in as short a time as 10 years. Incentive programs are intended as a relatively short-lived effort.

Incentives may take many forms. One-time reimbursement payments include rebates and grants, which offset a portion of the initial investment. Loan programs provide low-interest or no-interest loans for the purchase of renewable-energy systems, lowering the cost of borrowing money. Production incentives award money to system operators for every unit of renewable energy produced. Tax incentives lower the owner's tax liability, reducing future costs.

The availability of incentives varies widely by system size, system technology, state, utility, sector (residential, commercial, educational, etc.), and implementation. Many incentives are also time-limited, so information must be checked frequently for new programs and to ensure that a program has not expired. The on-line Database of State Incentives for Renewables and Efficiency (DSIRE) is a comprehensive resource of up-to-date incentive information.

Rebates

Rebates are the type of incentive most commonly available to owners of renewable-energy systems. A *rebate* is a one-time refund for a portion of an original purchase price. Rebate programs often require proof of purchase and installation to qualify. Most rebates are offered by state agencies and utilities. Eligible sectors usually include residences and businesses, although many rebate programs are available to industry, institutions, and government agencies as well.

The amount of money available through a rebate program may be specified in various

ways. Some programs offer a certain amount per system unit of capacity, such as rated watts for PV systems. Others offer a percentage of the purchase price. All specify an upper limit to rebate amounts. Rebates available per applicant range from $300 to over $1,000,000. Rebate amounts may also be limited by the total amount of money set aside to fund the rebate program each year.

Grants

A *grant* is a one-time monetary payment for certain types of projects. Grants may be awarded by a government entity, utility, or nonprofit organization. Grant programs typically require applications and written proposals, as grants are limited in number and amount. Grant-awarding organizations may target a certain sector, such as small businesses or technology research, in hopes of spurring economic development or technology improvements in that area. Grant programs may also restrict the use of the award to certain renewable-energy technologies and require documentation that the project is fulfilling its objectives. Grants typically range from $500 to $1,000,000 or more.

In addition to incentives for the purchase and installation of renewable energy systems, there are also a large number of incentive programs for improving and maintaining energy efficiency.

Incentives are designed to make many types of renewable energy systems, including PV systems, more affordable.

Loans

Loan programs offer low-interest or no-interest financing for the purchase of renewable-energy systems, including PV systems. Most loan programs are offered by states and utility companies, though there are federal programs, and a few programs are offered by local governments and utilities. These programs are typically available for a broad range of renewable-energy system technologies and applicant sectors, including residential, commercial, industrial, transportation, public, and nonprofit sectors. Interest and repayment terms vary and may be determined on an individual-project basis, though loan lifetimes are commonly 7 to 10 years.

Alternatively, PV and other renewable-energy systems can be financed with conventional bank loans, which may not offer below-normal interest rates but are highly flexible. The purchase and installation of a PV system may be amended to a mortgage loan or addressed separately as a home equity or capital improvement loan.

Tax Incentives

A *tax incentive* is a measure to reduce owed taxes as an inducement for investment. Just as there are different types of taxes, there are different types of tax incentives. Only government bodies can offer tax incentives, because they are the only entities entitled to levy taxes. Most tax incentives are offered by state or local governments, though there are some federal programs.

Tax incentives are typically categorized as deductions or credits. **See Figure 15-2.** Exemptions are fundamentally similar to deductions. When an incentive is offered as a tax deduction, an amount is removed from a taxable value. Since the tax owed is calculated as a percentage, the smaller taxable value results in less tax owed. In contrast, tax credits do not reduce a taxable value. Instead, they cancel a portion of actual taxes owed. Credits may be based on a set dollar amount, a percentage of the cost of the system, or a certain amount per unit of energy production, such as cents per kilowatt-hour.

Tax Incentives

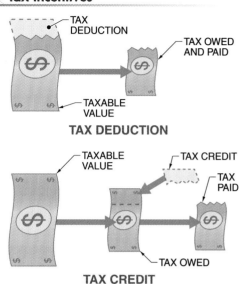

Figure 15-2. *Tax incentives reduce the amount of tax that must be paid, whether they are tax exemptions or tax credits.*

Income Tax. Income tax incentives are available for both personal and corporate income taxes. Income tax credits are offered by several states and the federal government. Personal income tax incentives are aimed at homeowners who install renewable-energy systems on their residences. The incentives are intended to help eliminate the personal income tax that would have been paid on the income used to purchase and install a renewable-energy system for a home.

Corporate income tax incentives allow corporations to receive credits of 10% to 35% of the total cost of the system. Some states allow the tax credit only if the corporation has invested a certain dollar amount into a renewable-energy project.

Property Tax. Property tax incentives are offered as either exemptions or credits. Exemptions function similarly to deductions and make up the majority of property tax incentives. When an exemption is applied, the added value of the renewable-energy system is not included in the valuation of the property for taxation purposes. If the renewable-energy technology, such as solar thermal water heating, replaces a conventional technology, the system is assessed at the value of the conventional system. This avoids additional taxes that would result from the value added to the property due to the renewable-energy system. When a property tax credit is applied, the value of the renewable-energy system is included in the property assessment, and the credit is applied toward the additional taxes.

For example, consider a property with an assessed value of $100,000 that is taxed annually at a rate of 1%, for a tax liability of $1000 ($100,000 × 1% = $1000). The owners then add a PV system worth $10,000 to the property. When a property tax exemption is applied, the assessment is not increased to account for the PV system and the taxes are assessed for the original $100,000 value. When a property tax credit is applied, the taxable value of the PV system is included in the assessment, which totals $110,000. Therefore, the tax liability is $1100 ($110,000 × 1% = $1100). However, if a tax credit of $100 is applied, the credit covers the additional tax liability resulting from the value of the PV system. In this example, the result is that the tax paid ($1000) is equal in both cases. Neither exemptions (deductions) nor credits are automatically better than the other. Their ultimate value to the consumer depends on several factors. It is important to note that the details of tax credit and tax exemption programs vary, and that the full value of taxes resulting from the addition of a PV system may not always be covered completely.

Property taxes are collected locally; some states allow local authorities the option of providing a property tax incentive for renewable-energy systems. As of 2009, most states have such provisions.

Sales Tax. About half of the states offer some type of sales tax incentive. Sales tax incentives typically provide an exemption from the entire amount of state sales tax on the purchase of renewable-energy equipment. Most programs are offered for systems of any size, but a few programs stipulate that the system must be certified to output a certain amount of energy. In these cases, the sales tax is paid at the time of purchase and refunded when the system is complete and operating.

Production Incentives

A *production incentive* is a recurring payment program that provides project owners with cash payments based on electricity production. The incentive is typically a certain dollar amount per kilowatt-hour of electricity. The term of the payments is limited, such as for a certain number of years. One of the largest programs is the Renewable Energy Production Incentive (REPI), which is a federal program that currently offers 1.5¢/kWh (in 1993 dollars and indexed for inflation) for up to ten years of system operation. Numerous other programs are offered by state and local governments, utilities, and nonprofit organizations.

Production incentives are based on the idea that payments based on performance rather than capital investments can be a more effective mechanism for ensuring quality and efficient projects.

SPG Solar, Inc.
Homeowners with PV systems can usually take advantage of property tax incentives to lower the total system costs.

Renewable Energy Certificates

Renewable energy certificates are similar to production incentives. A *renewable energy certificate (REC)* is a tradable commodity that represents a certain amount of electricity generated from renewable resources. RECs are also known as renewable energy credits, "green tags," or tradable renewable certificates (TRCs). RECs are not necessarily sponsored by government entities or utilities. Rather, they are facilitated by organizations that arrange the transactions and keep a small percentage for their services.

An owner of a renewable-energy system, such as a PV system or wind farm, receives one REC for every 1000 kWh of electricity produced and exported to the utility grid. **See Figure 15-3.** A certifying agency ensures that RECs are allocated accurately and that each is assigned a unique serial number. The REC can then be sold on the open market. Several organizations exist that facilitate this trading. The price of RECs floats like shares of stock and varies depending on supply and demand.

However, the market for RECs is unique. In states with renewables portfolio standards (RPSs), which require that utilities produce a minimum amount of their electricity from renewable resources, those utilities that fail to meet their quota can purchase RECs. Purchasing RECs gives buyers the right to claim "green" energy produced elsewhere as their own to fulfill a quota. The price paid for the REC acts as a penalty mechanism. In return for the REC, the renewable-energy producer receives funds that help offset the costs of the initial system investment and/or contribute to future investment in renewable-energy technologies.

However, quotas are not the only force driving the REC market. Strong public opinions about the use of renewable energy sources add to the value of RECs. For example, some companies may advertise that they operate from renewable energy in an effort to attract consumers who highly value energy conservation. Companies may purchase RECs based on the belief that some consumers are willing to pay more for products that support the use of renewable energy.

For example, a product may be marketed with the claim that it was "produced using renewable energy." The actual electricity used in manufacturing the product likely came from a variety of sources, including fossil-fuel-burning power plants, because there is no way to differentiate the sources of power on the grid. However, by purchasing RECs, a buyer is able to take credit for a certain amount of electricity that was produced from renewable sources and added to the grid system. By carefully tracking RECs through their serial numbers, the "green" energy generated is matched to the "green" energy claimed, enabling the REC trading system to operate.

Renewable Energy Certificates

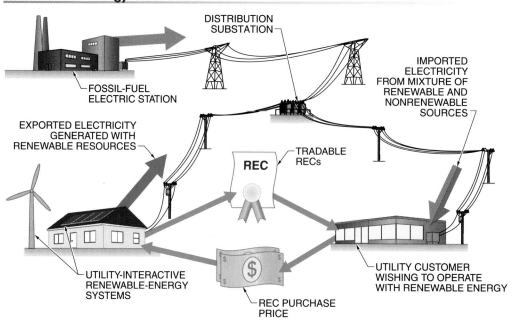

Figure 15-3. *Renewable energy certificates (RECs) allow a utility or its customers to claim renewable energy while providing a financial incentive for renewable-energy producers.*

Carbon Credits

The concept of trading in renewable energy certificates (RECs) is similar to the global market for carbon credits. Under the Kyoto Protocol, an international agreement, countries are allocated maximum levels of carbon dioxide that their industries may release into the atmosphere. The objective is to stabilize and then reduce the release of carbon dioxide, the primary greenhouse gas, into the atmosphere by setting low target levels of emissions. The protocol also sets up a market for carbon credits. If a country's emissions fall below their

target level, they can sell the surplus carbon dioxide allocation to a country that has exceeded its target. One carbon credit equals 1 metric tonne (2205 lb) of carbon dioxide. The price of carbon credits fluctuates on the open market like shares of stock.

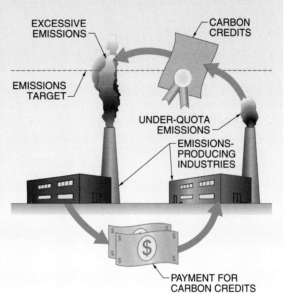

This activity has no net effect on the total amount of carbon dioxide emissions. However, it encourages further improvement in emissions reductions because those who meet their target receive rewards in the form of salable credits. Meanwhile, those who do not meet their goal are penalized by having to pay for credits. This creates an incentive for industries to reduce pollution, in that they may soon find it less expensive to improve their processes or install emissions-control equipment than to pay for expensive carbon credits.

RECs and metering credit for electricity exported to the utility are not mutually exclusive. That is, customers can receive both if both programs are available in their area. However, in some interconnection agreements, the utility retains the right to claim any RECs generated by an interconnected customer. In such a case, the customer cannot receive payment for RECs.

COST ANALYSIS

A *cost analysis* is a comparison of two or more options that considers both the cost and value of each. It is important to note that financial cost is not the only factor. Factors such as lifetime, availability, practicality, and ease of use all contribute to the value of an option, and can be just as important. For example, option A may cost a little more than option B, but if it lasts twice as long, it is likely a better value. A cost analysis takes as many of these factors into

account as possible, though some are difficult to quantify. Of course, this type of analysis is still only a tool, and decisions may also be based on other, qualitative factors.

Some manufacturers are using purchased RECs to market their products as being "made with green energy."

San Diego Electrical Training Center

Location: San Diego, CA (32.4°N, 117.1°W)
Type of System: Utility-interactive
Peak Array Power: 89.4 kW DC
Date of Installation: August 2004
Installers: Journeymen, apprentices
Purposes: Training, supplemental power

The San Diego Electrical Training Center hosts one of the largest PV systems in its area. The array consists of 455 modules installed at a slight angle on a tall standoff mounting system. The system is divided into 34 subarrays that each feed a separate utility-interactive inverter. Complemented by an extensive investment in energy efficiency, the system has virtually eliminated the facility's electric utility bills. As an added benefit, by shading most of the building's roof, the PV system reduces the facility's summer cooling loads. The system cost over a half million dollars to install, but is producing considerable energy-cost savings for the training center.

The system was installed in two phases. The first phase cost $266,000 and was commissioned in September 2002. The second phase cost approximately $288,000 and was commissioned in August 2004. The entire system was subsidized through a rebate incentive program available through the California Energy Commission that covered 50% of the labor, materials, and administration costs. The total rebate exceeded $278,000.

For net metering credit, the utility requires an approved watt-hour meter at the system's main service equipment. Output data is monitored through a web-based data-acquisition system. Every twelve months, system energy production is totaled and any net exported energy is paid for by the utility. With monthly production of 11,000 kWh to 14,000 kWh, the training center produced over $35,000 worth of energy in its first six months of system operation. With the peak array energy production at mid-day matching the peak time-of-use utility rates, the training center is maximizing its net-metering revenue.

The local union's leaders are also involved in promoting renewable energy in the city of San Diego. The city has committed to having 50 MW in electricity-generating capacity from renewable energy sources by 2013, including 35 MW from solar technologies alone. The city has established a committee to promote increased use of renewable-energy technologies. Jim Westfall, the training director at the San Diego Electrical Training Center, is among the committee members working toward this goal.

With local incentives to encourage new PV systems in San Diego and surrounding communities, electrical contractors will need electricians with training in PV-system installation. By involving many of its apprentices and journeymen in the PV-system installation processes, the San Diego Electrical Training Center is adding valuable skills to the electrical workforce.

San Diego Electrical Training Center
The PV system at the San Diego Electrical Training Center meets nearly all of the facility's electrical needs.

San Diego Electrical Training Center
Electricians installed the modules on an elevated rack mount system with a small tilt angle.

When analyzing PV systems against alternate energy sources, a standardized life-cycle cost analysis is used. The *life-cycle cost* is the total cost of all the expenses incurred over the life of an electricity-generating system. Examples of electricity-generating systems include utility connections, engine generators, wind turbines, and PV systems. A life-cycle cost analysis is a comparison of the life-cycle costs of various electricity-supply options. **See Figure 15-4.** For example, a consumer may need to decide between installing a PV-only system with a large array or a hybrid system with a small PV array and a wind turbine. While life-cycle costs are not the only consideration, an analysis comparing the various options will help quantify some of the financial pros and cons that lead to a decision.

Life-Cycle Cost Analysis

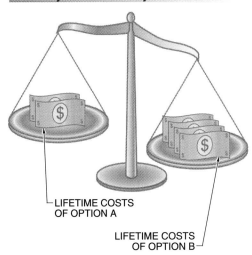

LIFETIME COSTS OF OPTION A

LIFETIME COSTS OF OPTION B

Figure 15-4. *A life-cycle cost analysis compares the life-cycle costs of various electricity-supply options.*

Life-Cycle Costs

A life-cycle cost assigns a monetary value to every aspect of system installation and operation. Some costs are simple to quantify, such as the price of a purchased component. Other costs must be estimated, such as installation labor and maintenance. The most difficult costs or values to quantify are impacts on the environment or the operator's quality of life, such as pollution, noise, aesthetics, ease of use, and peace of mind.

A life-cycle cost analysis is most useful when conducted prior to the start of a project, because it can be used to choose the system type, configuration, and components to maximize value and minimize costs. However, a life-cycle cost analysis can also be conducted after a system is installed to compare actual performance to alternatives. One advantage of a post-installation analysis is that most of the costs have been documented, so the analysis will be more accurate than one done from pre-installation estimates. The information and lessons learned from the analysis may also be useful for future upgrade or equipment-replacement projects.

Since PV arrays have no moving parts and require very little maintenance, they are expected to last significantly longer than 20 years. Many of the other major components, such as inverters, exhibit excellent long-term reliability as well. The life-cycle cost period of 20 years is often chosen, however, as a compromise between the expected lifetime of PV systems and the alternatives.

In order to make a valid comparison, the systems analyzed must meet the same load demands with the same reliability over the same lifetime. This may require adding additional components, such as an energy-storage system, to match system power output to power demand. Components for system redundancy may also be required to make the two systems equivalent in reliability and system availability. For example, if an engine generator requires a periodic engine rebuild that takes one week, another electricity-generating system must be available to meet the load demand with the same reliability during that time. This additional cost must be accounted for in the analysis.

A length of time must also be chosen for use in the analysis. **See Figure 15-5.** The period typically represents the expected lifetime of the system component with the longest life. Again, the shorter-lived system components may require maintenance or replacement to extend their lifetimes to the length of the analysis period, adding costs. Life-cycle cost for PV systems is typically evaluated using a 20- to 30-year period.

Analysis Time Period

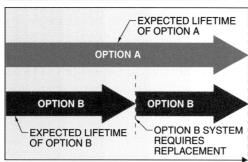

Figure 15-5. *Each system option must meet the same requirements, including the length of time used to calculate life-cycle cost.*

Financial costs included in the total life-cycle cost include initial costs, maintenance costs, energy costs, and repair and replacement costs. These costs vary significantly over the analysis period. **See Figure 15-6.** The salvage value of the system at the end of its life is also considered, and is a credit rather than a cost.

Initial Costs. An *initial cost* is an expense related to the design, engineering, equipment, and installation of a system. Initial costs include both goods and services. These costs may also be known as capital costs

Recurring costs, such as maintenance, energy, repair, and replacement costs, are figured on an annual basis for life-cycle cost analyses. That is, even if the costs occur monthly, quarterly, or semiannually, they are figured based on an entire year.

Prices for equipment can be taken from invoices, catalogs, or dealer quotes. Labor rates for design, engineering, and installation services are acquired from service estimates. However, it is important to ensure that service estimates include the expenses for any additional equipment to be used or consumed during the providing of these services. For example, the cost of renting aerial lift equipment to access an array location is an expense that should be included in the initial costs.

The value of the owner's time is an important factor in system comparisons. Any labor provided by the system owner should be valuated at a reasonable market rate. This ensures a fair comparison between a system to be self-installed and a system that requires professional installation. This also helps to insure the owner against unexpected costs if the owner later decides to hire professional installers.

The total of all these initial costs is considered as a single expense at the beginning of the life of the system.

Maintenance Costs. A *maintenance cost* is an expense related to the operation and maintenance of a system. These costs occur periodically throughout the life of the system. Maintenance costs include inspections, cleanings, insurance, property taxes, and operator salaries, but not fuel or equipment-replacement costs. Maintenance costs may be itemized or they may be covered by a service contract with a set annual fee.

Life-Cycle Costs

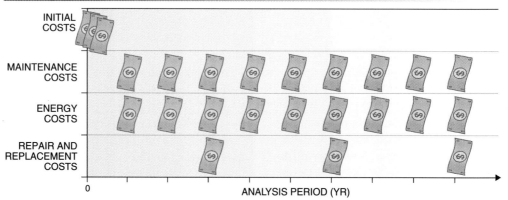

Figure 15-6. *The various types of life-cycle costs occur at different points in the life cycle of a power-generating system.*

Energy Costs. An *energy cost* is an expense related to the energy input of an electricity-generating system. This energy input refers to gasoline or diesel fuel for engine generators, hydrogen for fuel cells, or, for comparison, the electricity from the utility. Wind and solar radiation are, of course, free of charge.

Replacement and Repair Costs. A *replacement cost* is an expense related to the periodic replacement of components with expected lifetimes shorter than the expected lifetime of the entire system. For example, batteries typically have life of only several years, so it is reasonable to assume that a battery bank will need to be replaced a few times during the 20-plus years of system operation.

A *repair cost* is an expense related to the restoration of a component or system to its nominal operating state following damage or wear. It is very difficult to predict which components may require future repairs, so repair costs are often lumped together with a conservative value for future replacement costs. A repair budget might include a modest amount for replacement components such as blown fuses, broken brackets, and a damaged module, and would be added to the replacement costs.

Salvage Value. The *salvage value* is the monetary value of a system at the end of its expected useful lifetime. It is common to assign a value of 20% of the original purchase price for equipment that can be moved. This figure can be increased or decreased based on the condition and obsolescence of the system. Note that the original purchase price is not the same as the initial costs, which also include the costs of installation services.

However, salvage value can be misleading, because it assumes that the system is dismantled at the end of the estimated lifetime period and the components sold individually. Since it is expected that PV systems operate efficiently for appreciably longer than 20 years, the system likely has more value as a complete system than as the sum of the values of its components. In fact, some homeowners find that an operational PV system installed on a residence adds more value to their home when they sell the home than the original amount paid for the system. Studies have shown that a home's value increases by about $20 for every $1 reduction in annual utility bills through energy efficiency and supplemental electricity-producing systems. This is partly because, with lower utility bills, the buyer can afford a larger mortgage.

The salvage value assigned to the system is subjective and difficult to quantify, but must not be ignored since it can be significant. Multiple life-cycle costs with high, medium, and low figures for salvage value may be useful to appreciate how this variable affects the total system cost.

Value of Money

Because the life-cycle cost includes monetary values at the beginning, middle, and end of a system's lifetime, which can span 20 years or more, it is important to consider how the value of money changes over time. The value of a dollar today is greater than the value of a dollar next year and much greater than the value of a dollar 20 years from now. **See Figure 15-7.** This change in the value of money affects the relative values of the various life-cycle costs.

Value of Money

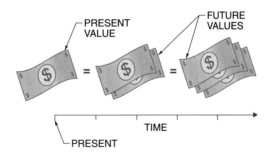

Figure 15-7. *A certain amount of present money is equal to a greater face-value amount of future money. The difference in face value depends on the difference in time.*

When life-cycle costs include fuel, such as diesel or natural gas, all necessary fuel services and equipment should be included in the energy costs. These costs may include taxes, delivery, storage tanks, metering, and safety equipment.

Discount Rate. In order to compare future monetary value to the present value, it is discounted by a certain rate. The *discount rate* is the rate at which the value of future money is reduced to its present value. The present value of a future cost is calculated using the following formula:

$$PV = \frac{FV}{(1+r)^t}$$

where

PV = present value (in $)

FV = future value (in $)

r = discount rate

t = time period (in years)

For example, what is the present value of the 10-year replacement of an inverter with a price of $3000 if the discount rate is 5% per year?

$$PV = \frac{FV}{(1+r)^t}$$

$$PV = \frac{3000}{(1+0.05)^{10}}$$

$$PV = \frac{3000}{(1.05)^{10}}$$

$$PV = \frac{3000}{1.63}$$

$$PV = \mathbf{\$1842}$$

Therefore, $3000 at 10 years from now has the same value as $1842 now. The discount rate can have a significant effect on present values of future costs. **See Figure 15-8.** When assessing the present value of many future costs, small differences in the discount rate can add up to significant cost variations, which render a life-cycle cost analysis ineffective. A change of only ±1% in the discount rate would change this amount to $1675 or $2027. If the discount rate is too low, the value of future costs will be exaggerated, but if the discount rate is too high, the value of future costs will be underestimated. Therefore, it is important to choose the discount rate carefully. Local banks are a good source for estimated discount rate information.

Discount Rates

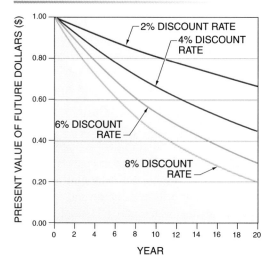

Figure 15-8. *The present value of future money falls more quickly with higher discount rates.*

Discount Rate

The discount rate is the combination of the investment rate and the inflation rate. The investment rate is the annual rate of return expected on an invested sum of money. This rate is difficult to determine because returns on investments vary widely, depending on the type of investment and the timing. For investments in the stock market, the annual rate of increase in the Dow Jones Industrial Average is one measure of an average rate of return. For example, $1000 invested in the stock market may return $1200 in a few years.

However, $1200 in the future may not have the same buying power as $1200 in the present. The inflation rate is the annual rate of increase in prices, which makes future earnings have less value. The consumer price index is a measure of the rate of inflation in the United States. Inflation in the last few years has ranged between 0% and 4%.

If the investment rate is greater than the inflation rate, then it is advantageous to invest. For example, because of inflation, the buying power of the $1200 investment return is equal to about $1100 in the present. Since this is a positive result, investing is a sensible way to increase net worth.

It may be prudent to use more than one discount rate for different types of costs. If fuel prices or utility rates are expected to rise faster than inflation, the energy cost calculations should use a lower discount rate. Also, multiple life-cycle cost calculations based on two or three different discount rates give the final values a reasonable margin of error. For example, with a discount rate of 5% ±1%, $3000 at 10 years from now has a present value of about $1842 ±$176.

Single Present Value Factors. As an alternative to the formula method of determining present value, a simple multiplication factor can be used to determine the present value. The *single present value factor* is a multiplication factor for determining the present value of a single future cost for a given discount rate and period. The factors are arranged in a table of discount rates against time in years. **See Figure 15-9.** The factor for a given rate and period is multiplied by the future value to yield the equivalent present value. For example, the single present value factor for a 5% discount rate at 10 years is 0.614. Multiplying $3000 by 0.614 yields $1842.

Recurring Present Value Factors. Some costs occur in every year, such as fuel costs for generators or the cost of utility electricity. Some types of generating systems may have maintenance costs every year. For example, an engine generator may require periodic oil changes that add up to approximately $100 each year. To determine the present value of these costs with the discount rate formula or the single present value factors would require a separate calculation for each year. The present value of $100 (at a 5% discount rate) in Year 1 is $95, in Year 2 is $91, in Year 3 is $86, in Year 4 is $82, and so on. **See Figure 15-10.** All these individual costs are added together to yield the total present value of the recurring cost. If the life-cycle analysis is calculated for 20 years, this method is inconvenient and prone to mistakes.

The term "discount rate" must be used with care because it has slightly different meanings in different financial and economic contexts.

Single Present Value Factors

YEAR	DISCOUNT RATE							
	1%	2%	3%	4%	5%	6%	7%	8%
1	0.990	0.980	0.971	0.962	0.952	0.943	0.935	0.926
2	0.980	0.961	0.943	0.925	0.907	0.890	0.873	0.857
3	0.971	0.942	0.915	0.889	0.864	0.840	0.816	0.794
4	0.961	0.924	0.888	0.855	0.823	0.792	0.763	0.735
5	0.951	0.906	0.863	0.822	0.784	0.747	0.713	0.681
6	0.942	0.888	0.837	0.790	0.746	0.705	0.666	0.630
7	0.933	0.871	0.813	0.760	0.711	0.665	0.623	0.583
8	0.923	0.853	0.789	0.731	0.677	0.627	0.582	0.540
9	0.914	0.837	0.766	0.703	0.645	0.592	0.544	0.500
10	0.905	0.820	0.744	0.676	0.614	0.558	0.508	0.463
11	0.896	0.804	0.722	0.650	0.585	0.527	0.475	0.429
12	0.887	0.788	0.701	0.625	0.557	0.497	0.444	0.397
13	0.879	0.773	0.681	0.601	0.530	0.469	0.415	0.368
14	0.870	0.758	0.661	0.577	0.505	0.442	0.388	0.340
15	0.861	0.743	0.642	0.555	0.481	0.417	0.362	0.315
16	0.853	0.728	0.623	0.534	0.458	0.394	0.339	0.292
17	0.844	0.714	0.605	0.513	0.436	0.371	0.317	0.270
18	0.836	0.700	0.587	0.494	0.416	0.350	0.296	0.250
19	0.828	0.686	0.570	0.475	0.396	0.331	0.277	0.232
20	0.820	0.673	0.554	0.456	0.377	0.312	0.258	0.215

Figure 15-9. *The single present value factor is used to quickly calculate the present value of a single future value.*

Recurring Costs

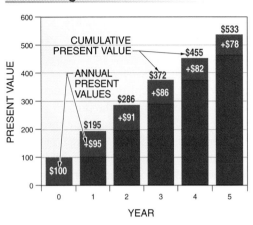

Figure 15-10. *Related costs that occur every year are calculated and added together for each year to determine the total present value.*

Refer to worksheet on CD-ROM

Similar to the single present value factor, a simple multiplication factor can be used to easily calculate this recurring cost. The *recurring present value factor* is a multiplication factor for determining the present value of an annually recurring cost for a given discount rate. These factors are arranged in a table similar to that for the single present value factors and are used in the same way. **See Figure 15-11.** For example, the recurring present value factor for $100 every year for 20 years at a 5% discount rate is 12.5. Therefore, the total present value of this recurring cost is $1250 ($100 × 12.5).

However, there are limitations to this method. The cost must occur every year between the first year and the final year and must be the same each year. This method will not work for costs beginning in years other than the first year or occurring in multiples of years, such as every five years. If a recurring cost does not fit these requirements, the individual costs calculations must be done separately. Fortunately, it is reasonably easy to set up a computer spreadsheet to complete or double-check the calculations automatically.

The present value factors can be calculated with formulas for any discount rate and any year. The single present value factor is equal to $(1 + r)^{-t}$ and the recurring present value factor is equal to $[1 - (1 + r)^{-t}] \div r$.

DOE/NREL, NC Solar Center
For remote or isolated loads, PV systems can be a cost-effective energy source.

Life-Cycle Cost Analysis

The life-cycle cost for any energy system is the sum of the present values of the various types of costs over its expected lifetime, minus the present value of the system at the end of its lifetime. This is represented in the following formula:

$$LCC = I + M_{PV} + E_{PV} + R_{PV} - S_{PV}$$

where

LCC = life-cycle cost (in $)

I = initial costs (in $)

M_{PV} = present value of all maintenance costs (in $)

E_{PV} = present value of all energy costs (in $)

R_{PV} = present value of all repair and replacement costs (in $)

S_{PV} = present value of salvage value (in $)

From this calculation, the analysis is straightforward. The system with the lower life-cycle cost is the better long-term value, assuming all other factors, such as reliability or environmental impact, are equal. **See Figure 15-12.** The most common types of life-cycle cost analyses are comparisons between a PV system and either an engine generator or utility electricity.

Recurring Present Value Factors

YEAR	DISCOUNT RATE							
	1%	2%	3%	4%	5%	6%	7%	8%
1	0.99	0.98	0.97	0.96	0.95	0.94	0.94	0.93
2	1.97	1.94	1.91	1.89	1.86	1.83	1.81	1.78
3	2.94	2.88	2.83	2.78	2.72	2.67	2.62	2.58
4	3.90	3.81	3.72	3.63	3.55	3.47	3.39	3.31
5	4.85	4.71	4.58	4.45	4.33	4.21	4.10	3.99
6	5.80	5.60	5.42	5.24	5.08	4.92	4.77	4.62
7	6.73	6.47	6.23	6.00	5.79	5.58	5.39	5.21
8	7.65	7.33	7.02	6.73	6.46	6.21	5.97	5.75
9	8.57	8.16	7.79	7.44	7.11	6.80	6.52	6.25
10	9.47	8.98	8.53	8.11	7.72	7.36	7.02	6.71
11	10.37	9.79	9.25	8.76	8.31	7.89	7.50	7.14
12	11.26	10.58	9.95	9.39	8.86	8.38	7.94	7.54
13	12.13	11.35	10.63	9.99	9.39	8.85	8.36	7.90
14	13.00	12.11	11.30	10.56	9.90	9.29	8.75	8.24
15	13.87	12.85	11.94	11.12	10.38	9.71	9.11	8.56
16	14.72	13.58	12.56	11.65	10.84	10.11	9.45	8.85
17	15.56	14.29	13.17	12.17	11.27	10.48	9.76	9.12
18	16.40	14.99	13.75	12.66	11.69	10.83	10.06	9.37
19	17.23	15.68	14.32	13.13	12.09	11.16	10.34	9.60
20	18.05	16.35	14.88	13.59	12.46	11.47	10.59	9.82

Figure 15-11. *The recurring present value factor is used to quickly calculate the present value of a future value recurring for a number of years.*

Life-Cycle Cost Analysis

LIFE-CYCLE COSTS

System:

General Discount Rate:
Energy Costs Discount Rate:

Cost Description	Cost/Year	Single Cost Year	Recurring Cost Years	Present Value Factor	Present Value
Initial Costs					
Maintenance Costs					
Energy Costs					
Repair and Replacement Costs					
Salvage Value					
					Total Life Cycle Costs

Figure 15-12. *A life-cycle cost analysis totals the present values of all the life-cycle costs over the length of the analysis period.*

For example, consider a newly constructed home in a rural area. The home features many energy conservation measures to minimize electrical requirements. The owner is considering three options to supply electricity to the home: a utility connection, an engine generator, and a PV system. In order to help determine the most cost-effective system, the customer performs a life-cycle cost analysis of all three options. The peak power demand is estimated at 5 kW, the yearly energy consumption is estimated at 7800 kWh, and the discount rate is estimated at 4%.

Engine generators have relatively low initial costs but require significant maintenance and fuel expenses that result in high life-cycle costs.

Utility Connection Costs. The life-cycle cost of a utility connection is the simplest to calculate. There are no maintenance, repair, or replacement costs to consider. There is also no real salvage value, though in certain cases, a utility connection may add value to a rural home when it is up for sale. The only potential sources of costs for a utility connection are initial connection costs and annual energy costs. **See Figure 15-13.**

The initial costs are the utility's charges for connecting to their electricity grid, which depends on the customer's location. If the customer is located in a relatively developed area, within a certain distance of existing electric service (such as 250′), the utility will make the connection at no cost to the customer. However, if the connection requires a substantial extension of the utility lines to

reach the customer, the utility will charge the customer for a portion of this improvement. These charges may also depend on the expected annual revenue from the customer. Depending on the terrain, line extensions can cost $10,000 to over $50,000 per mile. In the example situation, since the home is in a rural location and requires extension of service, the utility quotes the customer a connection fee of $25,000.

The major component of the life-cycle cost of a utility connection is the energy cost, which is an annually recurring cost. At the average electricity rate of $0.10/kWh, the first year's energy cost is $780 (7800 kWh × $0.10/kWh = $780). A recurring present value calculation is required to determine the present value of 20 years' worth of utility electricity. Since the cost of electricity is expected to rise faster than most costs, a discount rate of 3% is used. The lower discount rate will increase the present value of future costs. The recurring present value factor is 14.88, so the present value of the energy is $11,606 ($780 × 14.88 = $11,606).

The total life-cycle cost of the utility connection is calculated as follows:

$$LCC = I + M_{PV} + E_{PV} + R_{PV} - S_{PV}$$
$$LCC = 25,000 + 0 + 11,606 + 0 - 0$$
$$LCC = \mathbf{\$36,606}$$

Engine-Generator System Costs. The life-cycle cost of an engine-generator system is more complicated and includes each type of cost. **See Figure 15-14.** First, the initial costs include the purchase and installation of a 5 kW prime power diesel generator with an auxiliary fuel tank, a 500 Ah battery bank, and a 5 kW inverter. The battery system stores excess energy when the generator is running, which means that the generator runs less often and at a more efficient full load. The total initial costs are $8500.

Engine generators require more maintenance than any other type of residential power-generating system. The engine must be inspected and the oil changed regularly. The annual cost of this maintenance is estimated at $200. Therefore, the present value of 20 years of maintenance is $2718 ($200 × 13.59 = $2718).

Life-Cycle Costs for Utility Connection

LIFE-CYCLE COSTS

General Discount Rate: 0.04
System: Utility Connection   Energy Costs Discount Rate: 0.03

Cost Description	Cost/Year	Single Cost Year	Recurring Cost Years	Present Value Factor	Present Value
Initial Costs					
Utility Line Extension	$25,000	0		1.000	$25,000
Maintenance Costs					
Energy Costs					
Electricity	$780		20	14.88	$11,606
Repair and Replacement Costs					
Salvage Value					
				Total Life Cycle Costs	**$36,606**

Figure 15-13. *The life-cycle cost analysis of a utility connection includes the initial connection costs and annual energy costs.*

Life-Cycle Costs for Engine Generator

LIFE-CYCLE COSTS

General Discount Rate: 0.04
System: Engine Generator   Energy Costs Discount Rate: 0.03

Cost Description	Cost/Year	Single Cost Year	Recurring Cost Years	Present Value Factor	Present Value
Initial Costs					
System Purchase and Installation	$8,500	0		1.000	$8,500
Maintenance Costs					
Engine Inspection and Oil Changes	$200		20	13.59	$2,718
Energy Costs					
Diesel Fuel	$2,145		20	14.88	$31,918
Repair and Replacement Costs					
Engine Rebuild	$1,000	5		0.822	$822
Engine Rebuild	$1,000	10		0.676	$676
Engine Rebuild	$1,000	15		0.555	$555
Inverter Replacement	$2,000	10		0.676	$1,352
Battery Bank Replacement	$1,500	8		0.731	$1,097
Battery Bank Replacement	$1,500	16		0.534	$801
Salvage Value					
Salvage	$1,500	20		0.456	-$684
				Total Life Cycle Costs	**$47,755**

Figure 15-14. *The life-cycle cost analysis of an engine-generator system includes each type of cost.*

The cost of fuel over 20 years is the most difficult cost to estimate, since the price of fuel is highly volatile. The generator will run for 1560 hr at full load over the course of each year to supply 7800 kWh of electricity (7800 kWh ÷ 5 kW = 1560 hr). Since the engine consumes 0.55 gal/hr at full load, 858 gal of diesel fuel will be needed each year (1560 hr × 0.55 gal/hr = 858 gal). At an average price of $2.50/gal, the annual fuel cost is estimated at $2145. At a 3% discount rate (for rising and volatile fuel costs), the total cost over 20 years is estimated at $31,918 ($2145 × 14.88 = $31,918).

Repair and replacement costs include engine rebuilds and replacements of the inverter and battery bank. Engine rebuilds cost approximately $1000 and are expected to be needed every 5 years. Rebuild costs total $2053 ($822 at Year 5, $676 at Year 10, and $555 at Year 15). A $2000 replacement inverter at Year 10 costs $1352. Battery banks cost $1500 and are expected at Year 8 ($1097) and Year 16 ($801), totaling $1898. Therefore, the total repair and replacement costs are $5303.

Salvage value is estimated at $1500 at Year 20 ($684). Therefore, the total life-cycle cost of this system is calculated as:

$$LCC = I + M_{PV} + E_{PV} + R_{PV} - S_{PV}$$
$$LCC = 8500 + 2718 + 31,918 + 5303 - 684$$
$$LCC = \textbf{\$47,754}$$

Clearly, an engine-generator system can be extremely expensive in the long-term, especially when compared to a utility connection. When there is no possibility of a utility connection due to remoteness, these costs are considered more acceptable. However, when located in a region with only moderate insolation, a well-designed PV system can meet an equal load profile for a significantly lower life-cycle cost.

PV System Costs. The life-cycle cost of a PV system is significantly simpler to determine than for most other systems. There are no energy costs, and maintenance, repair, and replacement costs are minimal. The most significant factor is the initial cost. **See Figure 15-15.** A 5 kW PV array with two 2.5 kW inverters and a 1000 Ah battery bank is estimated to cost $30,000, including installation.

Maintenance costs include yearly inspection and battery maintenance, expected to cost $100 each year. The present value of this cost is $1359 ($100 × 13.59 = $1359). Replacement costs include a battery bank and perhaps a damaged module for $2200 at Year 8 ($1608) and Year 16 ($1175), and a $2000 inverter at Year 10 ($1352), for a total present value of $4135. The salvage value is conservatively estimated at $5000 ($2280).

$$LCC = I + M_{PV} + E_{PV} + R_{PV} - S_{PV}$$
$$LCC = 30,000 + 1359 + 0 + 4135 - 2280$$
$$LCC = \textbf{\$33,214}$$

The PV system is roughly two-thirds of the cost of an equivalent engine-generator system. In addition to being a better financial value, a PV system has many other advantages over an engine generator. With a PV system, there is no noise, no pollution, no regular deliveries of fuel, no storage of large quantities of toxic liquids, very little maintenance, and the system is nearly effortless to operate.

Annualized Costs

Another way to compare system options is to compare their annualized costs. The *annualized cost* is a portion of the total life-cycle cost attributed equally to each year of operation. One might assume that, to calculate annualized cost, the total life-cycle cost of the system is divided by the number of operating years. However, there are two possible problems with this method.

First, simply dividing the total life-cycle cost by the number of operating years assumes that costs are spread evenly over the operating period, which is rarely the case. For example, PV systems require large up-front costs, but then require few expenses for the rest of the operating life of the system. The actual cost for a year is high in years with major expenses and low in years with few expenses. Annualizing the life-cycle cost spreads all the expenses equally over the years. From a cash-flow perspective, this method of annualizing inaccurately profiles the life-cycle cost. However, the method is still commonly used for overall cost comparisons.

Life-Cycle Costs for PV System

LIFE-CYCLE COSTS				General Discount Rate: 0.04	
System: PV System				Energy Costs Discount Rate: 0.03	
Cost Description	Cost/Year	Single Cost Year	Recurring Cost Years	Present Value Factor	Present Value
Initial Costs					
System Purchase and Installation	$30,000	0		1.000	$30,000
Maintenance Costs					
Inspections	$100		20	13.59	$1,359
Energy Costs					
Repair and Replacement Costs					
Inverter Replacement	$2,000	10		0.676	$1,352
Battery Bank & Module Replacements	$2,200	8		0.731	$1,608
Battery Bank & Module Replacements	$2,200	16		0.534	$1,175
Salvage Value					
Salvage	$5,000	20		0.456	-$2,280
				Total Life Cycle Costs	**$33,214**

Figure 15-15. *The most significant costs in the life cycle of a PV system are the initial costs.*

Also, the simple division assumes that the costs are numerically equal for every year, which is also not the case. The annualized cost method assigns an equal present value for each year, but because of the time value of money, the numerical values for future costs will vary. To determine the annualized cost, the recurring present value factor is used in a manner opposite to how it was used in previous calculations. The calculation asks for an annually recurring cost with equal present values over 20 years that has a cumulative and final present value equal to the life-cycle cost. Therefore, the life-cycle cost is divided by the recurring present value factor, such as in the following formula:

$$ALLC = \frac{LLC}{F_{RPV}}$$

where

$ALLC$ = annualized cost (in $)

LLC = life-cycle cost (in $)

F_{RPV} = recurring present value factor

For example, the PV system has a life-cycle cost of $33,214 over 20 years at a 4% discount rate. The recurring present value factor for this scenario is 13.59. What is the annualized cost?

$$ALLC = \frac{LLC}{F_{RPV}}$$

$$ALLC = \frac{33,214}{13.59}$$

$$ALLC = \textbf{\$2444}$$

This means that an annually recurring cost of $2444 over 20 years has the same total present value as the total life-cycle cost. This can be illustrated in a plot of cumulative annualized costs over the operating period. **See Figure 15-16.** In contrast to the actual plot of the life-cycle costs, the cumulative annualized cost plot is a smooth curve representing equal annual costs (in present values) over the entire operating period. The two curves result in the same final life-cycle cost at Year 20.

Annualized Costs

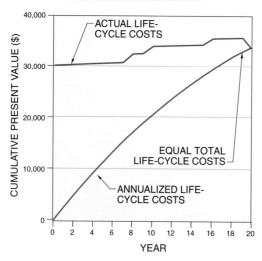

Figure 15-16. *Annualizing costs spreads costs evenly over the operating period, but results in the same total life-cycle cost for a system.*

Unit Costs. From the annualized cost, a unit cost can be determined. For example, the annualized present value cost of the PV system is $2444. If the system meets the load requirements by producing 7800 kWh per year, then the unit cost is about $0.31/kWh ($2444 ÷ 7800 kWh = $0.31/kWh). This value is much higher than the typical $0.10/kWh rate for utility power. However, if utility power is unavailable or requires a prohibitively expensive line extension, the PV system will likely compare favorably. For example, the utility connection with the line extension yields a unit cost of $0.35/kWh ($36,606 ÷ 13.59 ÷ 7800 kWh = $0.35/kWh) and the engine-generator system yields a unit cost of $0.45/kWh ($47,755 ÷ 13.59 ÷ 7800 kWh = $0.45/kWh).

Financial Payback

Assuming that a PV system is less expensive over the operating period than an alternate power source, then at some point the PV system will achieve financial payback. The *payback period* is the amount of time (typically in years) required for the avoided costs from an alternate energy source to match the cost of the chosen energy-production system. At that point, the chosen system has paid back its

cost with funds that would have otherwise been used to pay for energy from the alternate system. From that point onward, further avoided costs represent a net savings.

For example, consider a PV system that costs $5000, but avoids $500 in costs each year. That is, the alternate power-source option would have cost $500 per year if it was chosen instead, but since it is not needed, the $500 is saved. One might assume that these avoided costs accumulate each year; the total avoided costs would be $1000 in Year 2, $1500 in Year 3, $2000 in Year 4, and so on. If at Year 10, the total avoided cost matches the total system cost, then the payback period is 10 years. **See Figure 15-17.**

Payback Point

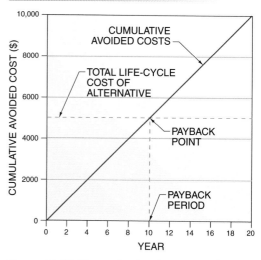

Figure 15-17. *The payback point occurs when the cumulative avoided cost of one system matches the total life-cycle cost of another system.*

However, accurate financial payback calculations are rarely so simple, because there are multiple ways of analyzing the cumulative life-cycle cost over the operating period and each must account for the change in the value of money over time.

Actual-Cost Payback. One way to determine the payback point is a comparison of the actual costs accumulating throughout the operating period for the various system options. The payback point is easily seen when the life-cycle costs are tabulated or plotted, but the preparation of these numbers is tedious. The present values of the additional costs for each year and each system must be calculated. Cumulative values are then totaled for each year for all costs up to and including that year. When plotted, the variations between the large initial cost and smaller periodic expenses are observable, resulting in stepped cumulative life-cycle cost plots. **See Figure 15-18.**

The *payback point* is the point at which the accumulated life-cycle cost of one system option matches the total life-cycle cost of the other system option. For example, the PV system has a total life-cycle cost of $33,214. The cumulative costs of the utility connection reach this amount at about Year 13. That is, the avoided costs up to that point equal the total life-cycle cost of the PV system over 20 years. Likewise, the PV system achieves a payback against the engine generator option, which occurs even earlier, at about 10 years.

Annualized-Cost Payback. An annualized cost for each system option can be used to determine the payback point in a manner similar to actual-cost payback analysis. The advantage is that the calculations are typically simpler, though, because of the assumptions inherent to annualizing, they may be less accurate. The utility connection with the line extension has an annualized cost of $2694 ($36,606 ÷ 13.59 = $2694). Each year this avoided cost is discounted to its present value. The cumulative avoided cost rises quickly, until it eventually equals the total life-cycle cost of the PV system, which occurs during Year 17. **See Figure 15-19.** For the engine-generator system, the payback point occurs even earlier, during Year 12, since its annualized avoided cost is higher at $3514.

There will often be small differences between the payback periods determined from the actual-cost and annualized-cost methods. The actual-cost method may be considered more accurate, but the payback point can sometimes be difficult to determine because of the stepped life-cycle plots. Compared to the many other assumptions and estimates used in cost analyses, the variance is usually considered to be within acceptable limits.

Overall, both methods of payback analysis are considered legitimate, and best practices often include citing a payback range. For example, in comparison to the engine-generator system, the PV system is expected to achieve payback in 10 to 13 years.

Actual-Cost Payback Analysis

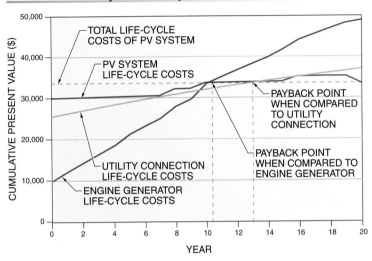

Figure 15-18. *The payback point can be determined by comparing the actual life-cycle costs of the various system options.*

Annualized-Cost Payback Analysis

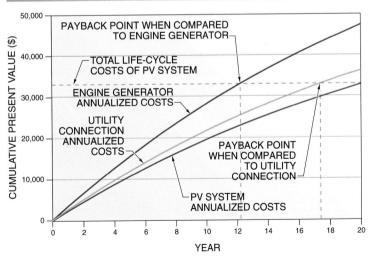

Figure 15-19. *The annualized-cost method of determining the payback period is relatively simple, though it may be less accurate than the actual-cost method.*

Operating a PV system requires no input of fuel, which also means that an operating PV system has no impact on pollution and carbon dioxide emissions and requires no fossil-fuel resources. However, manufacturing the PV system does require energy, which likely comes from nonrenewable resources. Therefore, it is important to analyze whether a PV system can reimburse this energy within its lifetime and, if so, how long that will take.

Energy payback is the amount of time a PV system must operate to produce enough energy to offset the energy required to manufacture the PV system. Energy payback is the ratio of the energy required to manufacture a unit of a PV array to the energy return the unit will produce in an average year. The ratio equals the number of years required for energy payback. For example, if a module requires 300 kWh of energy to manufacture, and once in operation it produces 100 kWh per year, then the energy payback is 3 years (300 kWh ÷ 100 kWh/yr = 3 yr).

Energy Investment

The total energy investment of manufacturing PV modules includes refining raw silicon, crystallizing silicon ingots, processing silicon wafers into PV cells, and assembling the modules. The energy required to manufacture components such as conductors, glass, and frames should also be considered. Each step in the manufacturing process contributes to the total energy cost of the module.

With so many steps and variations between cell technologies and manufacturers, it is extremely difficult to estimate these energy costs. Also, there is the question of whether to include the raw silicon refining in the total, since most module manufacturers recrystallize waste silicon left over from the electronics industry, which is virtually free. However, researchers have attempted to quantify these figures.

Monocrystalline modules are estimated to require between 600 kWh/m² and 1000 kWh/m². Polycrystalline cells require about 420 kWh/m² to 650 kWh/m². Thin-film modules, which use a fraction of the amount of silicon used by crystalline cells, require about 240 kWh/m². (These figures do not include the raw silicon refinement step.)

Energy Return

The amount of energy a module will produce is determined by the module efficiency and the receivable insolation, which can be found in the solar radiation data set for the nearest location. The daily insolation is multiplied by 365 to determine the annual insolation. For example, annual insolation for Portland, OR, for a latitude-tilt array is 1423 kWh/m²/yr (3.9 kWh/m²/day × 365 days = 1423 kWh/m²/yr). The average solar resource for the United States is about 1800 kWh/m²/yr. The energy payback period can then be calculated with the following formula:

$$t_{EPB} = \frac{E_{man}}{H_{ann} \times \eta}$$

where

t_{EPB} = energy payback period (in yr)

E_{man} = manufacturing energy (in kWh/m²)

H_{ann} = average annual receivable irradiation (in kWh/m²/yr)

η_m = module efficiency

For example, what is the energy payback period for an array of polycrystalline modules that are 12% efficient and are estimated to require 600 kWh/m² to manufacture?

$$t_{EPB} = \frac{E_{man}}{H_{ann} \times \eta}$$

$$t_{EPB} = \frac{600}{1423 \times 0.12}$$

$$t_{EPB} = \frac{600}{171}$$

$$t_{EPB} = \textbf{3.5 yr}$$

Given the wide range of estimates for manufacturing energy investment and the levels of insolation available to most of the United States, most calculations result in energy payback periods of between 2 yr and 5 yr. However, the length of payback periods is expected to fall in coming years as manufacturing processes become more efficient and module efficiencies improve, even as PV demand grows to require dedicated silicon supply chains, requiring silicon refining to be included in calculations.

Incentive Adjustments

These life-cycle cost analyses have neglected incentives, which may have a significant effect on final life-cycle cost. With incentives, the unit cost of PV energy may fall to the point where it is comparable to utility rates. The different types of incentives must be treated differently, however, depending on how they are applied. **See Figure 15-20.**

Rebates and grants are one-time payments that reduce initial costs and are therefore the easiest incentives to account for. For example, if the initial cost of a 4.5 kW system is partially offset by a rebate of $2 per watt, then $9000 is deducted from the initial costs (with no need to adjust it to present value). However, if there were a significant lag between the time of installation and the funding of the rebate, such as with a performance verification period, it would be appropriate to convert the rebate amount to the present value for Year 0. For example, a $4000 rebate that is applied in Year 3 with a 4% discount rate has a present value of $3544 ($4000 × 0.889 = $3544).

Production-based incentives and renewable energy certificates are recurring payments. If energy export to the utility is relatively steady every year with no downtimes, the incentive income can be considered approximately equal for each year of operation. For example, if a system owner receives approximately $200 per year for 10 years at a 4% discount rate, this amounts to $1622 in present value ($200 × 8.11 = $1622) that can be deducted from the life-cycle cost.

Life-Cycle Costs with Incentives

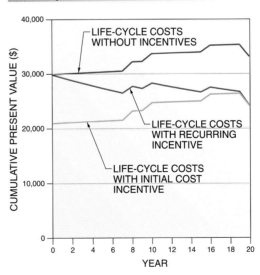

Figure 15-20. *Incentives affect the total life-cycle cost differently, depending on how and when they are applied.*

Loans, on the other hand, rather than lowering the life-cycle cost, add additional cost. Any interest paid is a cost that should be included in the life-cycle cost. (The exception would be no-interest loans, which have no effect on the life-cycle cost.) However, calculations involving loans can be complicated. The annual loan payments are usually the same, but the portion of payment that goes toward paying interest is dynamic. The first few payments are applied almost entirely to interest. The ratio of principal to interest owed changes gradually, until the last payments include very little interest. Because of the changing amounts of interest and principal, the recurring present value factor cannot be used. Financial analysis software is the best tool for accurately determining the costs of loans.

The accounting for tax incentives varies based on the type of incentive. Any sales tax (or lack thereof, for exempt equipment) is included in the purchase cost of the equipment and therefore does not require separate attention. Income tax incentives are typically a one-time savings, since a certain amount of income tax would normally have been paid on the income that was used to purchase the PV system. This amount acts like a credit that can be applied against the initial costs. Property taxes are more complicated. If the value added by the PV system to the property is exempt from property tax, then there are no additional costs. However, without a full property tax exemption, any additional taxes on the property are a cost that must be included in the life-cycle cost analysis. This value and the tax rate can change significantly over 20 years, so the present value of the cost is difficult to quantify. With research and software tools, an owner may be able to estimate this cost.

As production quantity and efficiency of PV modules and other major components continues to rise, PV systems are becoming more affordable and financially competitive with alternate energy sources. Currently, incentives are available to make up much, if not all, of the gap. Eventually, costs will fall to the point where incentives are no longer needed, with PV systems comparable to, or even less expensive than, traditional options.

Refer to Quick Quiz® on CD-ROM

Summary

- Incentives stimulate the renewable-energy technology industries because more-affordable systems increase demand.

- At some point, the costs of renewable energy will be competitive with traditional fossil-fuel sources, so there will no longer be a need for subsidies.

- Rebates are the most common type of incentive and are typically based on system capacity or a percentage of the purchase price.

- Grant programs may require documentation that a project is fulfilling its objectives.

- Production incentives are based on the idea that payments based on performance rather than capital investments can be a more effective mechanism for ensuring quality projects.

- Renewable energy certificates (RECs) are tradable certificates that give the buyer the right to claim the use of renewable energy.

- RECs and utility interconnection credits for exported electricity are not mutually exclusive.

- A standardized life-cycle cost analysis is used to analyze PV systems against alternate energy sources.

- Financial costs included in the total life-cycle cost include initial costs, maintenance costs, energy costs, and repair and replacement costs. The salvage value of the system at the end of its life is a positive value.

- Life-cycle costs include monetary values at the beginning, middle, and end of a system's lifetime.

- In order to compare a future monetary value to the present value, the future value is discounted by a certain rate.

- If the discount rate is too low, the value of future costs will be exaggerated, but if the discount rate is too high, the value of future costs will be underestimated.

- Multiple life-cycle cost calculations based on different cost estimates and/or discount rates give the final values a reasonable margin of error.

- The life-cycle cost for an energy system is the sum of the present values of the various types of costs over its expected lifetime, minus the present value of the system at the end of its lifetime.

- The system with the lower life-cycle cost is the better long-term value, assuming all other factors, such as reliability or environmental impact, are equal.

- Incentives can have a significant affect on final life-cycle cost.

- The easiest way to compare the cost effectiveness of energy-supply options is to compare the unit costs.

- At the point of the financial payback, the renewable-energy system has paid back its cost with the funds that would have otherwise been used to pay for energy from an alternate source.

- With incentives, unit costs of PV energy can fall to the point where they are comparable to utility rates.

Definitions

- An *incentive* is a monetary inducement to invest in a certain type of capital improvement, such as an energy-generating system or energy-conservation measure.

- A *rebate* is a one-time refund for a portion of an original purchase price.

- A *grant* is a one-time monetary payment for certain types of projects.

- A *tax incentive* is a measure to reduce owed taxes as an inducement for investment.

- A *production incentive* is a recurring payment program that provides project owners with cash payments based on electricity production.

- A *renewable energy certificate (REC)* is a tradable commodity that represents a certain amount of electricity generated from renewable resources.

- A *cost analysis* is a comparison of two or more options that considers both the cost and value of each.

- The *life-cycle cost* is the total cost of all the expenses incurred over the life of an electricity-generating system.

- An *initial cost* is an expense related to the design, engineering, equipment, and installation of a system.

- A *maintenance cost* is an expense related to the operation and maintenance of a system.

- An *energy cost* is an expense related to the energy input of an electricity-generating system.

- A *replacement cost* is an expense related to the periodic replacement of components with expected lifetimes shorter than the expected lifetime of the entire system.

- A *repair cost* is an expense related to the restoration of a component or system to its nominal operating state following damage or wear.

- The *salvage value* is the monetary value of a system at the end of its expected useful lifetime.

- The *discount rate* is the rate at which the value of future money is reduced to its present value.

- The *single present value factor* is a multiplication factor for determining the present value of a single future cost for a given discount rate and period.

- The *recurring present value factor* is a multiplication factor for determining the present value of an annually recurring cost for a given discount rate.

- The *annualized cost* is a portion of the total life-cycle cost attributed equally to each year of operation.

- The *payback period* is the amount of time (typically in years) required for the avoided costs from an alternate energy source to match the cost of the chosen energy-production system.

- The *payback point* is the point at which the accumulated life-cycle cost of one system option matches the total life-cycle cost of the other system option.

- *Energy payback* is the amount of time a PV system must operate to produce enough energy to offset the energy required to manufacture the PV system.

Review Questions

1. How do incentives improve the cost effectiveness of future renewable-energy systems?

2. Compare the common types of incentives.

3. Explain the difference between the two primary types of tax incentives.

4. How do production incentives encourage quality and efficient projects?

5. How do renewable energy certificates (RECs) help utilities fulfill renewable energy quotas?

6. How are life-cycle cost analyses used to help choose the energy-producing system with the best long-term value?

7. What are the requirements of a valid life-cycle cost analysis?

8. Why must future life-cycle costs be discounted to determine their present values?

9. What are two reasons why a simple annualization may not accurately profile a system's life-cycle costs?

10. How is financial payback determined?

11. How must the various types of incentives be treated differently when including them in life-cycle cost?

Chapter 1. Introduction to Photovoltaic Systems

1. If an alternate power source is very expensive, such as a utility connection to a remote location, then a PV system can save money. A PV system also produces environmentally friendly energy, is very flexible and adaptable, and can offer energy independence. However, the system's initial cost can be high and it may require a large surface area to produce a significant amount of power. The available solar resource may affect the feasibility of the system. There is also a lack of knowledge among some potential system owners, installers, and local officials that could impede a successful system installation.

3. Utility-scale power plants using PV technology are relatively simple in design and may have no moving parts. The plants can be located closer to populated areas because they do not involve hazardous materials or cause any air, water, or noise pollution. Also, unlike conventional power plants, PV power plants can be easily expanded incrementally as demand increases, as long as space allows.

5. Installers are the most visible level of the PV industry to the consumers, so it is vital that installers be professional and qualified individuals. Also, safe and quality PV system installations are essential for the success and acceptance of PV technology. Installers, both the contractors and the individual electricians, have important roles in ensuring quality PV installations.

7. Concentrating collectors gather solar radiation from a large area and focus it down to a small area for energy conversion, such as a small PV module or heat sink. Since there is more power created per area of conversion material, the system is more efficient. Flat-plate collectors do not concentrate solar radiation in any way.

Chapter 2. Solar Radiation

1. Irradiance is a measure of power and is an instantaneous value. That is, it is not measured over any length of time and can change from moment to moment. Irradiation is a measure of energy and accumulates over time. Irradiation is irradiance over a period of time, such as an hour or day, resulting in a total amount of energy. Greater irradiance results in greater irradiation.

3. Direct radiation is the component of the total global radiation inside Earth's atmosphere that comes directly from the sun. It is not scattered or otherwise changed in direction before reaching the surface of Earth. Diffuse radiation is the portion of the total global radiation that becomes scattered by molecules in the atmosphere. These scattered rays may still reach the ground, but can come from any portion of the sky.

5. The amount of solar radiation that is absorbed or scattered in the atmosphere depends on how much atmosphere it passes through before reaching Earth's surface. As the zenith angle increases (the sun approaches the horizon), the sun's rays must pass through a greater amount of atmosphere to reach Earth's surface. This reduces

the quantity of solar radiation, and also changes its wavelength composition. Air mass (AM) is a representation of the relative thickness of atmosphere that solar radiation must pass through to reach a point on Earth's surface. Air mass depends upon the time of day, the time of year, and the altitude and latitude of the specified location.

7. Pyranometers measure solar irradiance from the entire sky (180° field of view), so their measurements include both the direct and diffuse components. By shading the very narrow field of view of the direct component, the pyranometer measures only the diffuse component. The diffuse measurement can be subtracted from the total component to yield the direct component.

9. The longitude of a location determines the time difference from the standard meridian. This difference is equal to 4 min for each degree of longitude. Solar time is early for locations east of the standard meridian and late for locations west of the standard meridian. The Equation of Time value is another factor that contributes to the difference between solar time and standard time. The Equation of Time value is a correction factor that takes into account small variations in Earth's orbit. This value changes throughout the year.

11. Arrays receive the maximum amount of available solar radiation when mounted perpendicular to the direct radiation rays. In the summer, the sun is higher in the sky than in the winter. To maintain a perpendicular relationship, for at least part of the day, an array must have a smaller tilt angle in the summer and a larger tilt angle in the winter.

Chapter 3. Site Surveys and Preplanning

1. The solar radiation resource should be researched before visiting the site. Solar radiation resource information is available as data sets for specific locations, and provides average daily solar radiation for each month on various surfaces. Where no data exists for a certain area, the local resource can be estimated by comparing the data for the closest sites. Solar radiation measurement equipment can be used at the site, though meaningful resource data requires more than a year of collection.

3. OSHA requires fall protection in many of the areas where PV modules are installed. Fall protection can be in the form of equipment worn by a worker to reduce the potential for injury from a fall, or precautions taken in the work area to prevent a fall. A full-body harness is a fall protection device that evenly distributes fall-arresting forces throughout the body to prevent injury. Guardrail systems consist of three railings that must be secured to the personnel side of upright structural members. On large, flat roofs, a warning line system is a high-visibility rope marking a controlled-access zone near the edge of the roof.

5. It may be desirable to face an array other than due south when more solar energy is available in the morning or afternoon due to the local climate, or to match demand at certain times of the day. Utility-owned PV systems are sometimes installed facing southwest, shifting peak power output to later in the afternoon to coincide with peak demand.

7. PV arrays are much more sensitive to shading than other solar energy systems. Depending on the magnitude and location of the shading, the reduction in output can be disproportionately higher than the percentage of array area shaded. In the worst cases, even a 10% shading can cause the loss of most of the output. For this reason, installers must carefully assess the shading potential at an installation site, and be prepared to adjust array location and/or orientation to minimize shading.

9. Problems with roof coverings are determined through visual inspections. These issues include brittleness, cracking, warping, misalignment, flaking, and rusting. Problems with the underlying structural supports involve inspecting the rafters or trusses from underneath for warping and signs of leaks. From the top of the roof, flatness can be checked by stretching a string line across the roof in various directions to reveal dips. Movement from walking across the roof indicates weakened roof supports that must be further inspected.

11. Some PV systems are specifically designed to produce a certain amount of energy or power, such as stand-alone systems or interactive systems with battery backup. These systems are sized from a minimum energy requirement for the required load scaled up to account for power losses in the system. Interactive systems without a battery backup, however, have no minimum power output requirement. These systems are often sized based on space and/or cost limitations with a performance estimate that is the peak rating of the array scaled down to account for power losses in the system.

Chapter 4. System Components and Configurations

1. Electrical power demand usually fluctuates considerably but PV power is produced in a relatively steady process. Therefore, the electricity supply does not always coincide with when it is needed. Excess energy must be stored for later use, and then recovered when it is needed at high demand. Energy-storage systems level out energy production and demand.

3. Inverters convert DC power from battery systems or arrays to utility-grade AC power for AC loads or export to the utility grid. Inverters may be individual devices or may be combined with other devices into a single power conditioning unit (PCU). The other devices included in power conditioning units may include charge controllers, rectifiers, or maximum power point trackers. However, since an inverter is commonly the major component in a multifunction PCU, the unit is still often called an "inverter."

5. The voltage and current output from an array can vary with temperature, irradiance, and load, and produce power outputs anywhere up to the rated (maximum) power level. For any combination of temperature and irradiance, there is a maximum possible power output that corresponds to a certain voltage and current. A maximum power point tracker (MPPT) uses electronics to continually adjust the load on a PV device under changing temperature and irradiance conditions to keep it operating at its maximum power point, maximizing the array output.

7. Stand-alone PV systems are sized and designed to power a specific electrical load using the solar radiation resource at a given location. Stand-alone systems, since they have no other power source to rely on, are often sized for the worst-case scenario: highest load demand with lowest average insolation. If the system includes a battery bank, the array must be sized to both charge the battery bank and operate the load.

9. Utility-interactive systems make a bi-directional interface with the utility at the distribution panel or electrical service entrance. When the PV system does not produce enough power to meet system loads, additional power is imported from the utility. If there is an excess of PV power, the excess power is fed back (exported) to the grid. For exporting, the inverter must monitor the utility's power to match the voltage, phase, and other parameters. If the inverter is bimodal, then it also senses utility outages and then switches to back-up mode.

11. Hybrid systems offer greater system reliability and flexibility in meeting variable loads because they use multiple energy sources. Common energy sources used in hybrid systems include engine generators, wind turbines, and micro-hydroelectric generators. Since the other energy sources may be active at night or on demand, unlike PV arrays, hybrid systems are better able to meet varying electricity demands.

Chapter 5. Cells, Modules, and Arrays

1. Crystalline silicon cell wafers are produced in three basic types: monocrystalline, polycrystalline, and ribbon silicon. The wafer is etched and placed into a diffusion furnace, where phosphorous gas penetrates the outer surfaces of the cell, creating a thin n-type semiconductor layer surrounding the original p-type semiconductor material. The edge of the wafer is then abraded and grid patterns are screen printed on the top surface of the cell to provide electrical connections. Finally, the entire back surface of the cell is coated with a thin layer of aluminum that alloys with the silicon and neutralizes the n-type semiconductor layer on the back surface.

3. The photovoltaic effect describes how photons free electrons in PV cells and induce them to flow. A short-circuit condition typically allows uncontrolled current from an infinite supply of electrons. However, the electrons in a PV device cannot be released without the initial energy from the photons, so there is a finite supply. Therefore, the short-circuit current is limited to the quantity of electrons made available by the photons.

5. PV devices with higher efficiencies require less surface area to produce each watt of power, which saves some costs for raw materials, mounting structures, and other equipment. However, higher efficiency modules are generally no less expensive than less efficient ones, because the price for modules is generally based on the maximum power rating and not on the size.

7. Changes in solar irradiance have a small effect on voltage, but a significant effect on the current output of PV devices. The current of a PV device increases proportionally with increasing incident solar irradiance. Consequently, since the voltage remains nearly the same, the power also increases proportionally.

9. Cell temperature must first be measured or estimated using the temperature-rise coefficient. Temperature coefficients for voltage, current, and power must also be determined, either from manufacturer's documentation or by converting common percentage coefficients. Values for voltage, current, and power parameters, rated at a certain solar irradiance and cell temperature, are used as reference points. The coefficients are used with equations to determine the amount of change from the reference values due to temperatures other than the rated cell temperature.

11. When PV devices with dissimilar current outputs are connected in series, current output is limited to the current of the lowest-current output device in the entire string. The voltage output of the circuit equals the sum of the voltages of the individual devices. PV devices with dissimilar current output may be connected in parallel. The resulting current is the sum of the individual currents and the overall voltage is the same as the average voltage of all the devices.

13. Solar irradiance and module cell temperature have great effects on module performance, and many factors can affect the cell temperature, such as ambient temperature, wind speed, location, materials, and mounting configuration. Ideal test conditions can produce very different results than more-realistic field conditions. Therefore, it is important to understand the test conditions used to evaluate performance so that realistic predictions can be made about actual system performance.

Chapter 6. Batteries

1. A battery is a collection of electrochemical cells that are contained in the same case and connected together electrically to produce a desired voltage. A battery cell is the basic unit in a battery that stores electrical energy in chemical bonds and delivers this energy through chemical reactions. A battery cell consists of one or more sets of positive and negative plates immersed in an electrolyte solution.

3. Capacity is the measure of the electrical energy storage potential of a cell or battery. Several physical factors affect the capacity, including the quantity of active material; the number, design, and dimensions of the plates; and the electrolyte concentration. Operational factors affecting capacity include discharge rate, charging method, temperature, age, and condition of the cell or battery.

5. A traction battery is a class of battery designed for repeated deep-discharge cycle service. Traction batteries are very popular in PV systems for their deep-cycle capability, long life, and durability.

7. Autonomy is the number of days a battery system can operate the system loads without further charging. A low average daily DOD results in a longer autonomy because a smaller portion of battery capacity is drained each day. Therefore, a given capacity can support a greater number of days at the average daily DOD. A high average daily DOD reduces autonomy because a larger amount of battery capacity is drained each day, so the capacity can support fewer days at this rate.

9. A battery bank is a group of batteries connected together with series and parallel connections to provide a specific voltage and capacity. Connecting batteries in series increases system voltage. The resulting voltage is the sum of the voltages of the individual batteries. Batteries do not need to be the same voltage, but total capacity will be limited to the battery with the lowest individual capacity. Connecting batteries in parallel increases system capacity. The resulting capacity is the sum of the capacities of the individual batteries.

Chapter 7. Charge Controllers

1. Charge acceptance is the ratio of the increase in battery charge to the amount of charge supplied to the battery. Since a given amount of capacity always requires more charge to overcome losses and inefficiencies, charge acceptance is always less than 100%. Overcharge is the ratio of applied charge to the resulting increase in battery charge. Overcharge is the inverse of charge acceptance.

3. Interrupting-type charge controllers regulate charging current by switching the charging current ON and OFF. Interrupting-type charge controllers are simple designs, but not widely used in PV systems. Since the array current is switched abruptly, they have difficulty avoiding excessive overcharge and are best used with the more tolerant flooded, open-vent batteries. A linear-type charge controller limits the charging current in a linear or gradual manner with high-speed switching or linear control. Linear-type charge controllers provide a more consistent charging process, so they can generally charge batteries more efficiently and faster. Linear-type controllers are compatible with more types of batteries.

5. Pulse-width modulation (PWM) simulates any level of current by switching a full current ON and OFF in pulses. The switching is done at very high speed (several hundred hertz) so that the effect appears smooth. The length of time that the current is ON is the width of the pulse and determines the simulated current level. When the battery is partially charged, the current pulse is essentially ON all the time. To simulate a lower charging current as the voltage rises, the pulse width is decreased. For example, if the pulses switch the full charging current so that it is ON half the time and OFF half the time, the result effectively simulates a current at 50% of the full current.

7. A charge controller setpoint is a battery voltage at which a charge controller performs charge regulation or load control functions. The voltage regulation (VR) setpoint is the voltage that triggers the onset of battery charge regulation. The array reconnect voltage (ARV) setpoint is the voltage at which a charge controller reconnects the array to the battery and resumes charging. The low-voltage disconnect (LVD) setpoint is the voltage that triggers the disconnection of system loads to prevent battery overdischarge. The load reconnect voltage (LRV) setpoint is the voltage at which a charge controller reconnects loads to the battery system.

9. Temperature affects battery charging characteristics, so some charge controllers automatically adjust setpoints to compensate for this factor. When battery temperature is low, temperature compensation increases the VR setpoint to allow the battery to reach a moderate gassing level and fully charge. When battery temperature is high, the VR setpoint is lowered to minimize excessive overcharge and electrolyte loss. Temperature compensation helps ensure that a battery is fully charged during cold weather and not overcharged during hot weather.

11. Low-voltage modules are composed of fewer cells than most modules, affecting the scale of the I-V curve. As the batteries near full charge, the voltage of the charging circuit rises. With low-voltage modules, the increase in voltage within the normal range of battery voltages produces a corresponding decrease in current, which effectively limits the battery charging to a float-charging level.

Chapter 8. Inverters

1. The rotating generators that provide most of the electrical power on the utility grid naturally produce sine waves, so most loads are designed to operate from sinusoidal AC power. Therefore, interactive inverters must produce sine waves for utility synchronization. Other waveforms may damage loads.

3. Power-quality problems can be caused by the power source, but they can also be caused by loads on the electrical system. Both of these causes can be true for inverter output circuits. Fortunately, most inverters self-monitor for power quality automatically, so problems are usually easy to identify. Harmonics are present in square waves and modified square waves, so harmonics issues are especially important for inverters producing these types of waveforms.

5. Thyristors and transistors are solid-state electronic components used in inverters. Both types of components perform switching functions when activated by a control signal. However, thyristors are activated by a small current, while transistors are activated by a small voltage. Also, thyristors can be only completely ON or completely OFF, while transistors can be activated by any degree in between, like a dimmer switch.

7. Pulse-width modulation (PWM) simulates waveforms by switching a series device ON and OFF at high frequency and for variable lengths of time. When the pulses are narrow, the current is OFF most of the time, which simulates a low voltage. When pulses are wide, the current is ON most of the time, which simulates a high voltage. By using very high frequencies and gradual changes in pulse width, the PWM output appears to be a smooth, true sine wave.

9. Solid-state switching devices are capable of handling only so much current before they become overheated and fail. Therefore, thermal management in electronic inverters is a major concern, and temperature is the primary limiting factor for inverter power ratings. Many inverters use heat sinks and/or ventilation fans to regulate temperature. Interactive inverters control high temperatures by forcing a higher DC input voltage, which reduces the input current.

11. Most modern inverters provide features for data monitoring and communications. Interfaces may include displays and controls on the inverter itself, while others interface with remote units or computers. Inverter interfaces typically provide basic system information, including interconnection status, AC output voltage and power, DC input voltage, MPPT status, error codes, fault conditions, and other parameters. Many inverters record energy production on daily and cumulative bases.

Chapter 9. System Sizing

1. Sizing interactive systems begins with the specifications of a PV module chosen for the system. Module ratings at Standard Test

Conditions (STC) are used to calculate the total expected array DC power output per peak sun hour. This is then derated for various losses and inefficiencies in the system, which includes guaranteed module output that is less than 100%, array operating temperature, array wiring and mismatch losses, inverter power conversion efficiency, and inverter MPPT efficiency. The result is a final AC power output that is substantially lower but realistically accounts for expected real-world conditions.

3. Bimodal systems normally operate as interactive systems, but can operate as stand-alone systems during utility outages. Therefore, the bimodal systems are typically sized according to the stand-alone methodology. However, a significant difference between bimodal systems and true stand-alone systems is that bimodal systems typically supply only a few select critical loads while in stand-alone mode.

5. Sizing PV systems for stand-alone operation involves four sets of calculations. First, a load analysis determines the electrical load requirements. Then, monthly load requirements are compared to the local insolation data to determine the critical design parameters. Next, the battery bank is sized to be able to independently supply the loads for a length of time if cloudy weather reduces array output. Finally, the PV array is sized to fully charge the battery bank according to the critical design parameters.

7. Several factors must be considered when selecting an inverter. The inverter must have a maximum continuous power output rating at least as great as the largest single AC load and be able to supply surge currents. For voltage output, most stand-alone inverters produce either 120 V single-phase output or 120/240 V split-phase output. Some higher-power inverters output three-phase power. The inverter DC-input voltage must also correspond with either the array voltage (for interactive systems) or the battery-bank voltage (for stand-alone systems).

9. Most stand-alone systems are sized for a system availability up to about 95% for noncritical applications and to 99% or greater for critical applications. Each percentage-point increase in system availability costs is increasingly more expensive for larger battery banks and arrays, which is impractical from an economic standpoint for all but the most critical applications. Sizing of stand-alone systems must achieve an acceptable balance between system availability and cost goals for a given application.

11. Just as with battery banks, certain factors reduce the array output from the factory ratings to actual output values. Array current output is reduced when dust and debris collect on the array surface, blocking solar radiation. Another two factors determine the required array voltage rating: high temperature and a necessary increase in voltage in order to charge the batteries.

Chapter 10. Mechanical Integration

1. Cooler arrays generally are more reliable, last longer, operate with greater efficiency, and produce more power, so array temperatures should be minimized wherever possible. Active cooling means, such as fans and water-circulating pumps, may be used with some concentrating arrays, but are not practical for flat-plate modules. Only passive cooling means are employed for flat-plate modules, such as mounting the array in a way that allows air circulation around the modules. Keeping modules and arrays clear of obstructions also promotes natural cooling.

3. While the outward appearance of arrays and overall installations has little to do with system functionality or performance, it has a notable influence on consumer acceptance of PV technology. For arrays mounted on sloped roofs, the lines and location of the array should be consistent with building features. Color may be a consideration in choosing and integrating modules to complement other building colors. Finally, quality workmanship on the overall system installation improves the aesthetic appearance.

5. A BIPV array replaces some conventional building materials with modules that also perform a structural or cladding function. BIPV systems replace conventional building materials such as roof shingles, windows, skylights, awnings, and facades. The cost of the array is partially offset by avoiding the cost of some of the conventional building materials, though high engineering and architectural costs often outweigh the savings.

7. Sun-tracking arrays increase the array's power output, but are complex systems and require a greater investment in time and expense. In order for active tracking mounts to be effective, the solar energy gained by tracking must more than compensate for the electrical energy used by active tracking motors. The energy gain must also offset the increased maintenance and troubleshooting likely with these systems.

9. The principal types of structural loads are dead loads, live loads, wind loads, snow loads, and vibration loads. These loads are either static or dynamic. Static loads are constant loads, while dynamic loads change in magnitude and sometimes in direction. Dead loads and snow loads are static loads. Live loads, wind loads, and vibration loads are dynamic loads.

11. Lag screws must be secured into thick, solid pieces of wood, such as rafters. When attachment points must be made where there are no rafters underneath, doubled-up blocking can be used. Blocking is the addition of lumber under a roof surface and between trusses or rafters as supplemental structural support.

Chapter 11. Electrical Integration

1. Article 690 addresses requirements for all PV installations covered under the scope of the NEC®. As an electrical system, though, many other articles may apply. Some are directly referenced from Article 690 as sources for further general information on system equipment and installation regulations. Other articles may not be directly referenced, but may apply in general to all electrical systems, or may apply for certain configurations. Whenever the requirements of Article 690 and other articles differ, the requirements of Article 690 apply.

3. A conductor's material, insulation type, size, and application (such as direct burial, conduit, or free air) determine its nominal ampacity. Since temperature affects a conductor's ampacity, this nominal ampacity is derated (reduced) for ambient application temperatures higher than the nominal 30°C. When more than three current-carrying conductors are installed together in a conduit or raceway longer than 24′, conductor ampacities must be further derated.

5. PV arrays and their corresponding electrical connections are usually installed with full exposure to the elements, including temperature extremes, sunlight (UV) exposure, and precipitation. Consequently, any conductors used for these circuits may need to be rated for outdoor applications with high temperature, sunlight, and moisture resistance. Conductors may also need to be rated for at least 90°C (194°F). Conductors in array tracking systems must be especially flexible to withstand repeated movement, which can weaken or crack regular insulation and break conductor strands. These conductors must be rated as "hard-service," flexible cables, in addition to being listed for outdoor use.

7. Junction boxes are protective enclosures used to terminate, combine, and connect various circuits or components together. Junction boxes on the back surface of a module contain and protect the module's terminals and bypass diodes. Combiner boxes are used to combine multiple parallel array source circuits into the PV output circuit.

9. Conductors and overcurrent protection devices are sized to handle 125% of maximum possible current. The maximum current is based on the short-circuit current of the PV circuit (either a source circuit or the output circuit), which is increased by 125% to account for irradiance higher than the rating conditions. Therefore, conductor ampacity and overcurrent protection device ratings are equal to 125% of 125%, or 156% (125% × 125% = 156%), of the PV circuit's short-circuit current.

11. Supplementary overcurrent protection devices are intended to protect an individual component and to be used in addition to higher-rated branch-circuit overcurrent protection devices. Both PV source circuits and the PV output circuit must include overcurrent protection, but since the source-circuit protection is in addition to the PV output circuit protection, the source-circuit devices may be rated as supplementary. In this case, the PV source circuit, composed of a series string of modules, is treated as a single component.

13. Most PV systems involve both AC and DC, which involves the incorporation of the grounding requirements for these portions of the PV system. These two grounding systems must be bonded together. There are two acceptable methods of meeting this requirement. In the first method, separate AC and DC grounding electrodes are connected with the bonding conductor. In the second method, the DC grounding system is bonded to the AC grounding system, which is then connected to a single grounding electrode through an AC grounding electrode conductor. This method is more common for grounding interactive PV systems.

15. PV systems with batteries must also include some method of controlling the charge applied to the batteries. In the case of self-regulated systems, charge control is accomplished through careful sizing and matching of the array to the battery bank. Most systems, however, which have a potential charging current greater than C/33 (3% of the battery capacity per hour), must include an active means of charge control.

Chapter 12. Utility Interconnection

1. Various parameters, such as phase sequence, voltage, frequency, and phase balance, must be synchronized between two AC electrical systems before they are interconnected. For rotating generators, this synchronizing involves modifying the speed and power of the mechanical rotation and requires external controls. Solid-state inverters include microprocessors to monitor and automatically adjust the output, requiring no external devices. Also, inverters have no moving parts.

3. IEEE 1547 is a family of standards, guides, and recommended practices. ANSI/IEEE 1547-2003, *Standard for Interconnecting Distributed Resources with Electric Power Systems,* is a broad interconnection standard addressing requirements for all types of distributed power sources, including PV systems, fuel cells, wind turbines, engine generators, and large combustion turbines. It establishes requirements for testing, performance, maintenance, and safety of interconnections, as well as responses to abnormal events, islanding, and power quality. Additional, specific technical issues are addressed in the IEEE 1547. X series.

5. The point of connection is the location at which an interactive distributed-generation system makes its interconnection with the electric utility system. The NEC® permits the output of interactive PV inverters to be connected to either the load side (customer side) or supply side (utility side) of the service disconnect. For many smaller systems, the point of connection is usually made on the load side, usually at a circuit breaker in the distribution panel. When the requirements for a load-side connection are not possible (such as due to the size of the PV system), interactive systems may be connected to the supply side.

7. The interactive inverter must display a label that it is listed to applicable standards for interactive operation. A directory of all the power sources for a building must also be placed in a central location. Additionally, if the interconnection is on the load side of the main service disconnect, then the panelboard must also be labeled showing all sources of power.

9. When interconnecting a distributed-generation power source to the electric utility, the utility company assumes some of the risks and responsibilities of the customer's system. The utility must ensure that the system is safe for the customer, their neighbors, and the utility lineworkers who may work on or near the system. They must also be sure that the system will not adversely affect the operation of loads anywhere else on the system. Insurance requirements cover the customer and the utility in the event of a problem or accident with the system. Inspection rights allow the utility to check for listed equipment and proper installation before interconnection. Interconnection fees offset some of the additional costs utilities incur from this process.

Chapter 13. Permitting and Inspection

1. Local jurisdictions may choose to develop their own building code or they can adopt a model building code. They can also make small modifications to a model building code to suit their locale. Therefore, while some aspects of building codes are common in most jurisdictions, there can still be differences between jurisdictions.

3. Classified products can be tested to the same standards as listed products and can be of equal quality. The difference is that listed products are evaluated for all reasonably foreseeable situations applicable to the standard, while classified products are evaluated for only a certain set of circumstances.

5. Permit applications require certain information about the scope and specifications of the work, typically including the construction tasks, location, permit applicant, expiration date, and inspection requirements. Permit applications for PV systems should contain site drawings, electrical diagrams, specifications for major components and equipment, and array mounting information. This documentation is the basis for the plans review and subsequent on-site inspections.

7. An inspection checklist provides an extensive list of code references for PV systems, based on the NEC® and industry standards. NEC® section citations are included for reference. An inspection checklist is not intended to include every applicable requirement, but can be a useful guide to common PV-system issues for inspectors, as well as PV-system designers, installers, and operators.

Chapter 14. Commissioning, Maintenance, and Troubleshooting

1. One at a time, each disconnect is closed and each component is switched ON, and voltage and current are compared with expected values. (Array and battery bank voltages should be confirmed before closing their disconnects.) Meters, indicators, and displays on charge controllers and inverters are used for measuring and verifying system parameters. If there are any installation problems, this reduces the chances of safety hazards and damaged equipment. If at any step the system parameters are outside of the acceptable range, the startup should be aborted and troubleshooting procedures should be used to find the cause of the problem.

3. A complete documentation package should include system and equipment manuals, permitting documents, and any warranty or service information. System manuals generally include safety requirements, electrical and mechanical drawings, parts lists with sources, system warranty information, and procedures for operation, maintenance, and troubleshooting. A documentation package should include the manuals for all major components, including PV modules, inverters, and charge controllers. Copies of permits, certificates of inspection, and utility interconnection agreements should also be included.

5. Batteries are usually the most maintenance-intensive components in any PV system. Open-vent batteries are the most maintenance-intensive type of batteries because they need periodic water additions and cleaning. Sealed batteries require much less maintenance. Battery maintenance tasks include cleaning, tightening terminals, watering, and checking battery health and performance. Performance checks include specific gravity measurements, load tests, and capacity tests.

7. The specific gravity of the electrolyte in open-vent batteries should be checked as part of a regular maintenance schedule, and more often if problems are suspected. Specific gravity can be used to estimate battery state of charge in lead-acid batteries (though not in nickel-cadmium batteries). Relatively small variations between individual cells indicate that an equalization charge may be required. Significantly low specific gravity of individual cells indicates cell failure or shorts and likely requires battery or cell replacement.

9. Using computer software, information from data-acquisition systems can be plotted to illustrate the changes of values over time. This facilitates spotting trends and correlations between parameters. When troubleshooting, data-acquisition systems are invaluable tools. They allow the technician to review past records and identify slow trends over time or short-term events that may have impaired system performance.

11. Observation involves determining what part of the system is not working properly and observing the conditions that may have contributed to the cause. This includes obvious problems such as broken or burned equipment, water intrusion, and blown fuses. All equipment with monitoring, such as inverters and charge controllers, should be checked for error or fault indicators. Some indicators will provide very detailed information. Observation also includes attention to more subtle clues, such as any recent changes in the load requirements, maintenance plan, or weather patterns.

Chapter 15. Economic Analysis

1. Incentives stimulate the renewable-energy technology industries because more-affordable systems increase demand. Higher demand increases production and promotes research into more-efficient systems and manufacturing methods, both of which lower system costs. At some point, the costs of renewable energy will be competitive with traditional fossil-fuel sources, so there will no longer be a need for subsidies; incentive programs are intended as a relatively short-lived effort.

3. Tax incentives are typically categorized as deductions or credits. Exemptions are fundamentally similar to deductions. When an incentive is offered as a tax deduction, an amount is removed from a taxable value. Since the tax owed is calculated as a percentage, the smaller taxable value results in less tax owed. In contrast, tax credits do not reduce a taxable value. Instead, they cancel a portion of actual taxes owed. Credits may be based on a set dollar amount, a percentage of the cost of the system, or a certain amount per unit of energy production, such as cents per kilowatt-hour. Neither deductions (exemptions) nor credits are automatically better than the other. Their ultimate value to the consumer depends on several factors.

5. In states with renewables portfolio standards (RPSs), which require that utilities produce a minimum amount of their electricity with renewable resources, those utilities that fail to meet their quota can purchase RECs. Purchasing RECs gives the buyer the right to claim excess "green" energy produced elsewhere as their own to fulfill their quota. In a way, the price paid for the RECs acts as a penalty mechanism. In return for an REC, the renewable-energy producer receives funds that help offset the costs of the initial system investment and/or contribute to future investment in renewable-energy technologies.

7. In order to make a valid comparison, the systems analyzed must meet the same load demands with the same reliability over the same lifetime. This may require adding additional components, such as an energy-storage system, to match system power output to power demand. Components for system redundancy may also be required to make the two systems equivalent in reliability and system availability.

9. The annualized cost is a portion of the total life-cycle cost attributed equally to each year of operation. However, simply dividing the total life-cycle cost by the number of operating years assumes that costs are spread evenly over the operating period, which is rarely the case. From a cash-flow perspective, this inaccurately profiles the life-cycle cost. Also, the simple division assumes that the costs are numerically equal for every year, which is also not the case. The annualized cost method assigns an equal present value for each year, but because of the time value of money, the numerical values for future costs will vary.

11. The different types of incentives must be treated differently, depending on how they are applied. Rebates and grants are one-time payments that reduce the initial costs. Production-based incentives and renewable energy certificates are recurring payments. If the incentive income can be considered approximately equal for each year of operation, the recurring present value factor can be used to calculate their present values. Loans add additional cost in yearly increments, but since the values of principal and interest change each year, the recurring present value factor cannot be used. Financial analysis software is the best tool for accurately determining the costs of loans. The application of tax incentives to life-cycle cost varies, though most are one-time savings credited to the initial costs.

Photovoltaic Systems

Appendix

*Additional solar radiation data sets and sun path charts are available on the included CD-ROM.

Math Review

EXPONENTS

An exponent is a number placed at the upper right-hand side of another (base) number. An exponent indicates the power to which the base number is raised; that is, it indicates how many times the base number is used as a multiplication factor. Exponents can also be applied to letters that represent variables.

For example:

$3^4 = 3 \times 3 \times 3 \times 3$

$a^3 = a \times a \times a$

There are two special cases of exponents. The first special case is an exponent of 1. Any base number with an exponent of 1 is equal to the number. The other special case is an exponent of 0 (zero). Any base number with an exponent of 0 (zero) equals 1.

For example:

$7^1 = 7$

$x^1 = x$

$13^0 = y^0 = 1099^0 = 1$

Negative Exponents

A number with a negative exponent is equivalent to 1 divided by the result of raising the base number to the indicated power.

For example:

$$3^{-1} = \frac{1}{3} = 0.33$$

$$3^{-2} = \frac{1}{3^2} = \frac{1}{9} = 0.11$$

$$3^{-3} = \frac{1}{3^3} = \frac{1}{27} = 0.04$$

Fractional Exponents

A base number with a fractional exponent is equal to the nth root of the base number, where n is the number in the denominator (bottom number) of the fraction. For example, a "2" in the denominator indicates a square root, and a "3" indicates a cube root. The numerator (top number) of the fraction is treated in the same way as a whole number exponent, except that it is combined with the root function. The order in which the powers and roots are calculated does not affect the answer.

For example:

$$16^{1/2} = \sqrt[2]{16} = 4$$

$$27^{1/3} = \sqrt[3]{27} = 3$$

$$27^{2/3} = \sqrt[3]{27^2} = \sqrt[3]{729} = 9$$

$$27^{2/3} = \left(\sqrt[3]{27}\right)^2 = 3^2 = 9$$

Math Operations

Numbers with exponents must be evaluated before adding or subtracting. Numbers with exponents may be multiplied or divided before evaluating the exponents only if the base numbers are identical. When multiplying numbers with the same base, the exponents are added together. When dividing numbers with the same base, the exponent of the denominator is subtracted from the exponent of the numerator.

For example:

$$4^2 \times 4^3 = 4 \times 4 \times 4 \times 4 \times 4 = 4^{2+3} = 4^5 = 1024$$

$$\frac{5^6}{5^4} = \frac{5 \times 5 \times 5 \times 5 \times 5 \times 5}{5 \times 5 \times 5 \times 5} = 5^{6-4} = 5^2 = 25$$

TRIGONOMETRY

Trigonometry includes the study of the relationships between angles and side lengths of triangles. An understanding of basic trigonometry is useful for working with solar declination, tilt, azimuth, and profile angles. The term "trigonometry" is often shortened to "trig" in common usage.

Angles

An angle is a measure of the rotation between two lines. Angles are commonly measured in degrees, designated with a degree symbol (°). A right angle is a 90° angle and is marked with a small square. The sum of the three angles in a triangle always equals 180°. For example, if a triangle contains a 90° angle and a 30° angle, then the third angle must be 60°. This is true because 90 + 30 + 60 = 180.

Right Triangles

The right triangle is a fundamental part of trigonometry. A right triangle is a triangle where one of the angles is exactly 90°. When performing calculations with a right triangle, an angle other than the 90° angle is chosen as the reference angle. The reference angle is often designated with a lowercase Greek letter such as alpha (α) or theta (θ). **See Figure A-1.**

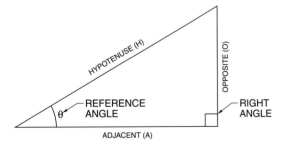

Figure A-1.

The sides of the triangle are given names based on their location relative to the reference angle. The side of the triangle next to the reference angle and connecting to the right angle is called the adjacent side (A). The side of the angle opposite the right angle is called the hypotenuse (H). The side opposite the reference angle is called the opposite side (O).

Trigonometric Functions

A function is a mathematical equation used to solve for unknown values. There are three basic trigonometric functions based on the right triangle. These functions are the sine (sin), cosine (cos), and tangent (tan) functions.

The functions are based on the ratios of the lengths of the sides of the triangle and are defined relative to the reference angle. For example, the sine function is the ratio of the length of the opposite side to the length of the hypotenuse. The sine function is conventionally written as follows:

$$\sin(\theta) = \frac{opposite}{hypotenuse} = \frac{O}{H}$$

Likewise, the cosine function is the ratio of the length of the adjacent side to the length of the hypotenuse. The cosine function is conventionally written as follows:

$$\cos(\theta) = \frac{adjacent}{hypotenuse} = \frac{A}{H}$$

The tangent function is the ratio of the length of the opposite side to the length of the adjacent side. The tangent function is conventionally written as follows:

$$\tan(\theta) = \frac{opposite}{adjacent} = \frac{O}{A}$$

The value of a trigonometric function of a certain angle in a right triangle is always the same, regardless of the size of the triangle. **See Figure A-2.** These values are usually determined with a scientific calculator that includes trigonometry functions, though they can also be looked up in trigonometric tables organized by angle.

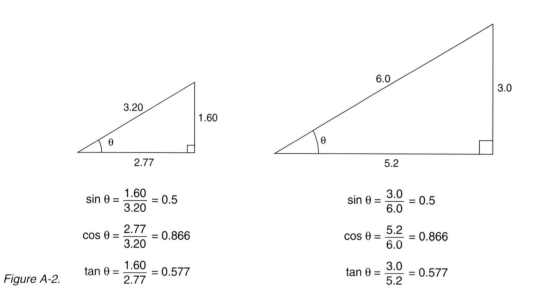

Figure A-2.

Using Trigonometric Functions

The predictable relationship between angles and side lengths means that trigonometric functions can be used to solve for unknown values. The unknown value can either be the length of a side or the degrees of an angle. There must be at least two known values in order to form a trigonometric formula with an unknown value. The appropriate trigonometric function is chosen based on the known and unknown information.

For example, a right triangle with a reference angle of 30° has one side length equal to 5. **See Figure A-3.** The other side lengths are unknown. Since the known side is adjacent to the angle and the desired side is opposite to the angle, the tangent function is chosen.

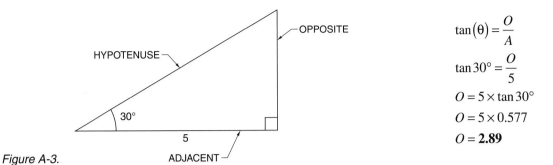

Figure A-3.

$$\tan(\theta) = \frac{O}{A}$$

$$\tan 30° = \frac{O}{5}$$

$$O = 5 \times \tan 30°$$

$$O = 5 \times 0.577$$

$$O = \mathbf{2.89}$$

If the hypotenuse side were known instead of the adjacent side, the sine function would have been appropriate. Alternatively, if the same sides were used with the third angle, the tangent function would still be appropriate but the formula would be arranged differently. Since the third angle can be easily determined in this case ($180° - 90° - 30° = 60°$), the calculation can be checked in this way. **See Figure A-4.** Using the third angle as a reference angle, the known side equal to 5 is the opposite side and the desired side is the adjacent side.

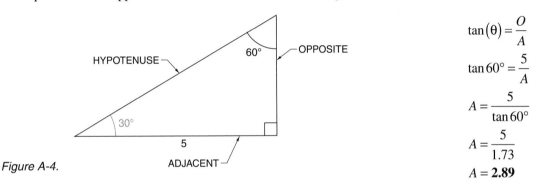

Figure A-4.

$$\tan(\theta) = \frac{O}{A}$$

$$\tan 60° = \frac{5}{A}$$

$$A = \frac{5}{\tan 60°}$$

$$A = \frac{5}{1.73}$$

$$A = \mathbf{2.89}$$

When two sides are known and an angle is the unknown value, the angle must be determined from the ratio of two appropriate sides. Therefore, the trigonometric function must be done in reverse. The inverse of a trigonometric function is typically designated with the prefix "arc." For example, if sin (θ) = y, then θ = arcsin (y). Another way to indicate the inverse of a trigonometric function is to include an exponent of –1 immediately after the function. For example, $\sin^{-1}$ (y) is equivalent to arcsin (y). (This should not be confused with $[\sin(y)]^{-1}$, which is equal to $\frac{1}{\sin(y)}$.)

For example, a right triangle has a hypotenuse equal to 6 and one side equal to 3. **See Figure A-5.** Because of the arrangement of the known side lengths and the desired angle, the sine function is chosen.

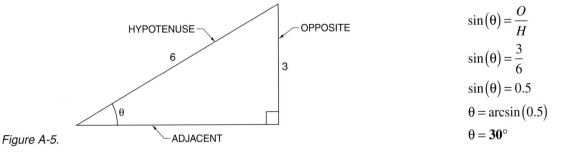

Figure A-5.

$$\sin(\theta) = \frac{O}{H}$$

$$\sin(\theta) = \frac{3}{6}$$

$$\sin(\theta) = 0.5$$

$$\theta = \arcsin(0.5)$$

$$\theta = \mathbf{30°}$$

The angle is determined by using the inverse trigonometric capability on a calculator or by looking up the angle associated with the sine value of 0.5 in trigonometric tables. In this case, the angle is 30°.

ALBUQUERQUE, NM (35.1°N, 106.6°W)   Elevation: 1619 m   Pressure: 838 mb

SOLAR RADIATION FOR FLAT-PLATE COLLECTORS FACING SOUTH AT A FIXED TILT*

Tilt		Jan	Feb	Mar	Apr	May	Jun	Jul	Aug	Sep	Oct	Nov	Dec	Year
0°	Average	3.2	4.2	5.4	6.8	7.7	8.1	7.5	6.9	5.9	4.7	3.5	2.9	**5.6**
	Minimum	2.6	3.4	4.4	6.2	6.9	7.4	6.8	5.9	5.4	4.0	2.8	2.2	**5.1**
	Maximum	3.7	4.7	6.2	7.3	8.3	8.7	7.9	7.4	6.6	5.3	3.9	3.3	**5.9**
Lat − 15°	Average	4.6	5.4	6.3	7.3	7.7	7.8	7.4	7.2	6.6	5.9	4.8	4.3	**6.3**
	Minimum	3.6	4.3	5.1	6.5	6.8	7.2	6.7	6.1	6.0	4.9	3.8	3.1	**5.7**
	Maximum	5.4	6.1	7.3	7.8	8.3	8.3	7.8	7.7	7.4	6.8	5.5	5.0	**6.6**
Lat°	Average	5.3	6.0	6.5	7.2	7.2	7.1	6.9	6.9	6.8	6.5	5.5	5.0	**6.4**
	Minimum	4.1	4.7	5.3	6.4	6.4	6.6	6.3	5.8	6.1	5.2	4.3	3.6	**5.8**
	Maximum	6.3	6.8	7.7	7.7	7.7	7.5	7.2	7.4	7.6	7.4	6.4	6.0	**6.8**
Lat + 15°	Average	5.8	6.2	6.5	6.6	6.3	6.1	6.0	6.3	6.5	6.6	5.9	5.5	**6.2**
	Minimum	4.4	4.8	5.1	5.9	5.7	5.6	5.5	5.3	5.8	5.3	4.6	3.8	**5.5**
	Maximum	6.9	7.1	7.6	7.1	6.8	6.4	6.3	6.7	7.3	7.7	6.9	6.6	**6.5**
90°	Average	5.2	5.1	4.5	3.7	2.8	2.4	2.5	3.2	4.2	5.1	5.2	5.1	**4.1**
	Minimum	3.9	3.9	3.5	3.4	2.5	2.2	2.3	2.8	3.7	4.0	3.9	3.5	**3.5**
	Maximum	6.4	5.8	5.4	4.0	3.0	2.5	2.7	3.4	4.6	6.0	6.2	6.2	**4.4**

SOLAR RADIATION FOR 1-AXIS TRACKING FLAT-PLATE COLLECTORS WITH A NORTH-SOUTH AXIS*

Axis Tilt		Jan	Feb	Mar	Apr	May	Jun	Jul	Aug	Sep	Oct	Nov	Dec	Year
0°	Average	4.9	6.2	7.6	9.6	10.6	10.9	9.9	9.2	8.3	7.0	5.3	4.5	**7.8**
	Minimum	3.7	4.7	5.7	8.4	9.0	9.5	8.5	7.4	7.0	5.5	4.1	3.2	**6.9**
	Maximum	5.9	7.1	9.3	10.6	11.8	12.1	10.8	10.3	9.6	8.4	6.3	5.4	**8.3**
Lat − 15°	Average	5.9	7.1	8.3	10.0	10.6	10.8	9.9	9.5	8.8	7.9	6.3	5.5	**8.4**
	Minimum	4.4	5.4	6.2	8.7	9.1	9.5	8.5	7.6	7.5	6.1	4.8	3.8	**7.4**
	Maximum	7.2	8.1	10.2	11.0	11.9	12.0	10.8	10.6	10.4	9.5	7.5	6.7	**8.9**
Lat°	Average	6.5	7.5	8.6	9.9	10.3	10.4	9.5	9.3	9.0	8.3	6.8	6.1	**8.5**
	Minimum	4.8	5.7	6.4	8.7	8.8	9.1	8.1	7.5	7.6	6.3	5.2	4.2	**7.5**
	Maximum	7.9	8.7	10.5	10.9	11.6	11.5	10.4	10.4	10.5	10.0	8.1	7.4	**9.1**
Lat + 15°	Average	6.9	7.7	8.5	9.5	9.7	9.7	8.9	8.9	8.8	8.4	7.1	6.5	**8.4**
	Minimum	5.1	5.8	6.3	8.3	8.3	8.4	7.6	7.1	7.4	6.4	5.3	4.4	**7.3**
	Maximum	8.4	8.9	10.4	10.6	10.9	10.7	9.8	9.9	10.3	10.1	8.5	7.9	**8.9**

SOLAR RADIATION FOR 2-AXIS TRACKING FLAT-PLATE COLLECTORS*

		Jan	Feb	Mar	Apr	May	Jun	Jul	Aug	Sep	Oct	Nov	Dec	Year
2-Axis Tracking	Average	6.9	7.7	8.6	10.0	10.8	11.1	10.0	9.5	9.0	8.4	7.2	6.6	**8.8**
	Minimum	5.1	5.8	6.4	8.8	9.2	9.7	8.6	7.7	7.6	6.4	5.3	4.4	**7.7**
	Maximum	8.5	8.9	10.5	11.1	12.1	12.3	11.0	10.6	10.5	10.1	8.6	8.1	**9.4**

DIRECT BEAM SOLAR RADIATION FOR CONCENTRATING COLLECTORS*

		Jan	Feb	Mar	Apr	May	Jun	Jul	Aug	Sep	Oct	Nov	Dec	Year
1-Axis (E-W) Tracking	Average	4.5	4.6	4.6	5.3	5.9	6.3	5.5	5.2	4.9	5.1	4.6	4.4	**5.1**
	Minimum	3.1	3.0	3.0	4.3	4.7	5.3	4.3	3.7	3.9	3.3	2.8	2.6	**4.3**
	Maximum	5.8	5.7	6.2	6.1	7.1	7.4	6.4	6.0	6.2	6.5	5.9	5.7	**5.5**
1-Axis (N-S) Tracking, Tilt = 0°	Average	3.7	4.6	5.6	7.2	8.0	8.4	7.2	6.7	6.2	5.4	4.0	3.4	**5.9**
	Minimum	2.5	3.0	3.4	5.7	6.1	6.8	5.4	4.7	4.7	3.4	2.4	2.0	**4.8**
	Maximum	4.8	5.7	7.6	8.4	9.7	10.0	8.4	7.9	7.9	7.2	5.2	4.5	**6.4**
1-Axis (N-S) Tracking, Tilt = Lat°	Average	5.1	5.7	6.3	7.4	7.7	7.8	6.8	6.8	6.7	6.5	5.4	4.8	**6.4**
	Minimum	3.4	3.7	3.9	5.8	5.9	6.4	5.1	4.7	5.1	4.1	3.2	2.8	**5.3**
	Maximum	6.6	7.2	8.6	8.6	9.3	9.4	7.9	8.0	8.6	8.6	6.9	6.3	**7.0**
2-Axis Tracking	Average	5.4	5.9	6.3	7.5	8.1	8.5	7.3	7.0	6.8	6.6	5.7	5.2	**6.7**
	Minimum	3.7	3.8	3.9	5.9	6.2	6.9	5.5	4.9	5.2	4.2	3.4	3.1	**5.6**
	Maximum	7.1	7.3	8.6	8.8	9.9	10.2	8.5	8.2	8.7	8.6	7.3	6.8	**7.3**

AVERAGE CLIMATIC CONDITIONS

	Jan	Feb	Mar	Apr	May	Jun	Jul	Aug	Sep	Oct	Nov	Dec	Year
Average Temperature (°C)	1.2	4.4	8.3	12.9	17.9	23.4	25.8	24.4	20.3	13.9	6.8	1.8	**13.4**
Average Low (°C)	-5.7	-3.1	0.1	4.2	9.2	14.6	18.0	17.0	12.9	6.1	-0.4	-4.9	**5.7**
Average High (°C)	8.2	11.9	16.3	21.6	26.5	32.2	33.6	31.7	27.7	21.7	14.1	8.6	**21.2**
Record Low (°C)	-27.2	-20.6	-13.3	-7.2	-2.2	4.4	11.1	11.1	2.8	-6.1	-21.7	-21.7	**-27.2**
Record High (°C)	20.6	24.4	29.4	31.7	36.7	40.6	40.6	38.3	37.8	32.8	25.0	22.2	**40.6**
Heating Degree Days†	531	389	312	167	49	0	0	0	10	144	345	512	**2458**
Cooling Degree Days†	0	0	0	4	36	155	233	188	70	6	0	0	**691**
Relative Humidity (%)	56	50	40	33	31	30	42	47	48	45	50	57	**44**
Wind Speed (m/s)	3.7	3.9	4.5	4.9	4.8	4.5	4.0	3.8	3.8	3.6	3.6	3.5	**4.1**

* in kWh/m²/day, ±9%
† based on 18.3°C (65°F)

CARIBOU, ME (46.9°N, 68.0°W)

Elevation: 190 m   Pressure: 991 mb

SOLAR RADIATION FOR FLAT-PLATE COLLECTORS FACING SOUTH AT A FIXED TILT*

Tilt		Jan	Feb	Mar	Apr	May	Jun	Jul	Aug	Sep	Oct	Nov	Dec	Year
0°	Average	1.6	2.6	3.8	4.6	5.2	5.7	5.6	4.8	3.6	2.3	1.4	1.2	**3.6**
	Minimum	1.3	1.9	3.0	3.5	3.6	4.7	4.6	4.0	3.1	1.8	1.1	0.9	**3.1**
	Maximum	1.8	2.9	4.3	5.4	6.2	6.4	6.5	5.7	4.2	2.7	1.7	1.5	**3.7**
Lat – 15°	Average	2.9	3.9	5.0	5.1	5.3	5.6	5.6	5.2	4.3	3.1	2.2	2.2	**4.2**
	Minimum	2.0	2.7	3.6	3.8	3.6	4.4	4.5	4.2	3.6	2.3	1.6	1.4	**3.6**
	Maximum	3.7	4.7	5.6	6.2	6.3	6.2	6.6	6.2	5.2	3.9	2.9	3.1	**4.5**
Lat°	Average	3.3	4.3	5.2	5.0	4.9	5.1	5.2	4.9	4.3	3.3	2.4	2.5	**4.2**
	Minimum	2.3	2.9	3.7	3.6	3.4	4.1	4.2	4.0	3.5	2.4	1.8	1.6	**3.6**
	Maximum	4.3	5.2	5.9	6.2	5.9	5.7	6.1	6.0	5.2	4.1	3.2	3.6	**4.5**
Lat + 15°	Average	3.5	4.5	5.2	4.7	4.4	4.5	4.6	4.5	4.1	3.3	2.5	2.7	**4.0**
	Minimum	2.4	3.0	3.6	3.3	3.0	3.5	3.6	3.6	3.3	2.3	1.8	1.6	**3.4**
	Maximum	4.7	5.5	5.9	5.9	5.2	4.9	5.3	5.4	5.0	4.2	3.4	3.9	**4.3**
90°	Average	3.4	4.2	4.6	3.7	2.9	2.8	2.9	3.0	3.2	2.8	2.3	2.7	**3.2**
	Minimum	2.3	2.8	3.1	2.4	2.0	2.3	2.3	2.5	2.5	2.0	1.6	1.5	**2.6**
	Maximum	4.7	5.2	5.3	4.8	3.4	3.0	3.2	3.7	3.8	3.6	3.2	3.9	**3.5**

SOLAR RADIATION FOR 1-AXIS TRACKING FLAT-PLATE COLLECTORS WITH A NORTH-SOUTH AXIS*

Axis Tilt		Jan	Feb	Mar	Apr	May	Jun	Jul	Aug	Sep	Oct	Nov	Dec	Year
0°	Average	2.5	3.8	5.4	6.1	6.8	7.3	7.3	6.4	5.0	3.1	1.9	1.9	**4.8**
	Minimum	1.8	2.7	3.7	4.1	4.2	5.4	5.8	5.1	3.9	2.3	1.5	1.1	**4.0**
	Maximum	3.2	4.6	6.2	7.6	8.1	8.6	9.2	7.8	5.8	4.0	2.6	2.6	**5.1**
Lat – 15°	Average	3.4	4.8	6.3	6.6	7.0	7.4	7.4	6.7	5.5	3.8	2.5	2.6	**5.3**
	Minimum	2.4	3.3	4.3	4.4	4.2	5.4	5.8	5.4	4.3	2.7	1.9	1.5	**4.4**
	Maximum	4.6	5.9	7.3	8.3	8.3	8.6	9.4	8.3	6.7	4.8	3.5	3.8	**5.7**
Lat°	Average	3.7	5.1	6.5	6.5	6.7	7.1	7.2	6.6	5.6	3.9	2.7	2.8	**5.4**
	Minimum	2.6	3.4	4.3	4.3	4.1	5.1	5.6	5.2	4.3	2.8	2.0	1.7	**4.4**
	Maximum	5.1	6.4	7.5	8.3	8.1	8.3	9.1	8.2	6.7	5.1	3.8	4.2	**5.8**
Lat + 15°	Average	3.9	5.3	6.5	6.3	6.3	6.6	6.7	6.3	5.4	3.9	2.8	3.0	**5.2**
	Minimum	2.7	3.5	4.3	4.1	3.8	4.8	5.3	5.0	4.2	2.8	2.0	1.7	**4.2**
	Maximum	5.4	6.6	7.5	8.1	7.6	7.8	8.6	7.8	6.5	5.1	3.9	4.4	**5.7**

SOLAR RADIATION FOR 2-AXIS TRACKING FLAT-PLATE COLLECTORS*

		Jan	Feb	Mar	Apr	May	Jun	Jul	Aug	Sep	Oct	Nov	Dec	Year
2-Axis Tracking	Average	4.0	5.2	6.5	6.6	7.1	7.6	7.6	6.8	5.6	4.0	2.8	3.0	**5.6**
	Minimum	2.7	3.5	4.3	4.4	4.3	5.6	6.0	5.4	4.3	2.8	2.0	1.7	**4.5**
	Maximum	5.4	6.5	7.5	8.3	8.4	8.9	9.6	8.4	6.7	5.1	3.9	4.5	**6.0**

DIRECT BEAM SOLAR RADIATION FOR CONCENTRATING COLLECTORS*

		Jan	Feb	Mar	Apr	May	Jun	Jul	Aug	Sep	Oct	Nov	Dec	Year
1-Axis (E-W) Tracking	Average	2.1	2.5	2.7	2.6	2.9	3.2	3.2	2.9	2.5	1.9	1.4	1.6	**2.5**
	Minimum	1.1	1.4	1.5	1.4	1.5	1.7	2.4	2.2	1.7	1.0	0.8	0.8	**1.9**
	Maximum	3.5	3.6	3.7	3.6	4.0	4.1	4.7	4.1	3.4	2.8	2.5	2.7	**2.8**
1-Axis (N-S) Tracking, Tilt = 0°	Average	1.3	2.1	2.9	3.3	3.9	4.2	4.3	3.8	3.0	1.8	0.9	0.9	**2.7**
	Minimum	0.7	1.1	1.5	1.7	1.8	2.3	3.2	2.9	2.0	0.9	0.5	0.4	**2.0**
	Maximum	2.2	2.9	4.0	4.5	5.4	5.5	6.4	5.3	4.0	2.7	1.8	1.6	**3.0**
1-Axis (N-S) Tracking, Tilt = Lat°	Average	2.3	3.0	3.7	3.6	3.9	4.0	4.2	4.0	3.5	2.4	1.5	1.7	**3.1**
	Minimum	1.3	1.6	1.9	1.9	1.8	2.2	3.1	3.1	2.3	1.3	0.9	0.8	**2.3**
	Maximum	3.7	4.3	4.9	5.0	5.4	5.3	6.2	5.6	4.7	3.6	2.8	2.9	**3.5**
2-Axis Tracking	Average	2.4	3.1	3.7	3.7	4.1	4.4	4.5	4.1	3.5	2.4	1.6	1.8	**3.3**
	Minimum	1.3	1.7	2.0	1.9	1.9	2.3	3.3	3.2	2.3	1.3	0.9	0.9	**2.4**
	Maximum	4.0	4.4	4.9	5.1	5.7	5.8	6.7	5.8	4.7	3.7	3.0	3.1	**3.7**

AVERAGE CLIMATIC CONDITIONS

	Jan	Feb	Mar	Apr	May	Jun	Jul	Aug	Sep	Oct	Nov	Dec	Year
Average Temperature (°C)	-12.8	-11.2	-4.1	3.3	10.5	15.8	18.6	17.1	12.1	6.2	-0.7	-9.6	**3.8**
Average Low (°C)	-18.7	-17.4	-9.5	-1.7	4.5	9.5	12.5	11.2	6.2	1.3	-4.6	-14.7	**-1.8**
Average High (°C)	-7.0	-5.0	1.3	8.2	16.5	22.2	24.7	23.1	17.8	11.1	3.1	-4.4	**9.3**
Record Low (°C)	-35.6	-40.6	-28.9	-18.9	-7.8	-1.1	2.2	1.1	-5.0	-10.0	-20.6	-35.0	**-40.6**
Record High (°C)	11.1	11.1	22.8	30.0	35.6	35.6	35.0	35.0	32.8	26.1	20.0	14.4	**35.6**
Heating Degree Days†	966	826	696	452	243	79	34	64	191	376	572	864	**5362**
Cooling Degree Days†	0	0	0	0	0	4	42	26	0	0	0	0	**73**
Relative Humidity (%)	72	70	69	69	66	70	75	78	78	77	80	77	**73**
Wind Speed (m/s)	5.1	5.0	5.3	5.1	4.8	4.4	4.1	3.8	4.2	4.5	4.7	4.9	**4.6**

* in kWh/m²/day, ±9%
† based on 18.3°C (65°F)

CHICAGO, IL (41.8°N, 87.8°W)

Elevation: 190 m   Pressure: 992 mb

SOLAR RADIATION FOR FLAT-PLATE COLLECTORS FACING SOUTH AT A FIXED TILT*

Tilt		Jan	Feb	Mar	Apr	May	Jun	Jul	Aug	Sep	Oct	Nov	Dec	Year
0°	Average	1.8	2.6	3.5	4.6	5.7	6.3	6.1	5.4	4.2	3.0	1.8	1.5	**3.9**
	Minimum	1.5	2.2	3.1	4.0	4.8	5.7	5.4	4.8	3.6	2.5	1.4	1.2	**3.7**
	Maximum	2.1	3.0	3.9	5.3	6.8	7.2	6.6	6.1	4.9	3.5	2.1	1.7	**4.2**
Lat − 15°	Average	2.7	3.5	4.1	5.0	5.8	6.1	6.1	5.7	4.9	3.9	2.5	2.2	**4.4**
	Minimum	2.0	2.8	3.6	4.2	4.8	5.5	5.4	5.0	4.0	3.0	1.7	1.6	**4.2**
	Maximum	3.1	4.2	4.8	5.9	6.9	7.1	6.6	6.6	5.9	4.7	3.3	2.9	**4.8**
Lat°	Average	3.1	3.8	4.2	4.9	5.4	5.7	5.6	5.5	4.9	4.2	2.8	2.4	**4.4**
	Minimum	2.1	3.0	3.6	4.1	4.4	5.1	5.0	4.8	4.0	3.1	1.8	1.8	**4.1**
	Maximum	3.5	4.6	4.9	5.8	6.5	6.6	6.2	6.4	6.1	5.1	3.7	3.3	**4.8**
Lat + 15°	Average	3.3	3.9	4.1	4.5	4.8	4.9	4.9	5.0	4.7	4.2	2.9	2.6	**4.1**
	Minimum	2.2	3.0	3.5	3.8	4.0	4.4	4.3	4.3	3.8	3.1	1.8	1.8	**3.8**
	Maximum	3.8	4.7	4.9	5.4	5.7	5.7	5.4	5.8	5.8	5.2	3.8	3.6	**4.5**
90°	Average	3.1	3.5	3.2	3.0	2.8	2.6	2.7	3.1	3.4	3.4	2.5	2.4	**3.0**
	Minimum	2.0	2.5	2.7	2.5	2.4	2.5	2.5	2.8	2.7	2.5	1.5	1.7	**2.7**
	Maximum	3.7	4.3	3.8	3.6	3.2	3.0	2.9	3.5	4.2	4.3	3.5	3.5	**3.3**

SOLAR RADIATION FOR 1-AXIS TRACKING FLAT-PLATE COLLECTORS WITH A NORTH-SOUTH AXIS*

Axis Tilt		Jan	Feb	Mar	Apr	May	Jun	Jul	Aug	Sep	Oct	Nov	Dec	Year
0°	Average	2.5	3.5	4.5	5.9	7.3	8.0	7.9	7.0	5.6	4.1	2.4	1.9	**5.1**
	Minimum	1.7	2.7	3.9	4.7	5.8	7.1	6.9	5.9	4.4	3.0	1.6	1.5	**4.7**
	Maximum	2.9	4.2	5.4	7.3	9.1	9.6	8.9	8.5	6.9	5.0	3.0	2.6	**5.7**
Lat − 15°	Average	3.2	4.2	5.0	6.2	7.4	8.0	7.9	7.3	6.2	4.8	2.9	2.5	**5.5**
	Minimum	2.2	3.2	4.3	5.0	5.9	7.1	6.9	6.1	4.8	3.5	1.8	1.8	**5.1**
	Maximum	3.7	5.2	6.1	7.8	9.4	9.6	9.0	9.0	7.7	6.0	3.9	3.4	**6.2**
Lat°	Average	3.5	4.5	5.1	6.1	7.2	7.7	7.7	7.2	6.2	5.0	3.1	2.7	**5.5**
	Minimum	2.3	3.3	4.3	4.9	5.7	6.8	6.7	6.0	4.8	3.6	1.9	1.9	**5.1**
	Maximum	4.1	5.5	6.2	7.8	9.1	9.2	8.7	8.8	7.8	6.3	4.2	3.8	**6.2**
Lat + 15°	Average	3.7	4.5	5.0	5.9	6.8	7.2	7.2	6.8	6.0	5.0	3.2	2.9	**5.3**
	Minimum	2.4	3.3	4.2	4.6	5.4	6.3	6.2	5.7	4.6	3.5	1.9	2.0	**4.9**
	Maximum	4.3	5.6	6.1	7.4	8.6	8.6	8.1	8.4	7.6	6.3	4.4	4.1	**6.0**

SOLAR RADIATION FOR 2-AXIS TRACKING FLAT-PLATE COLLECTORS*

		Jan	Feb	Mar	Apr	May	Jun	Jul	Aug	Sep	Oct	Nov	Dec	Year
2-Axis Tracking	Average	3.7	4.6	5.1	6.2	7.5	8.2	8.1	7.4	6.2	5.0	3.2	2.9	**5.7**
	Minimum	2.4	3.4	4.3	5.0	6.0	7.2	7.1	6.2	4.8	3.6	1.9	2.0	**5.2**
	Maximum	4.3	5.6	6.2	7.8	9.5	9.8	9.2	9.0	7.8	6.3	4.4	4.1	**6.4**

DIRECT BEAM SOLAR RADIATION FOR CONCENTRATING COLLECTORS*

		Jan	Feb	Mar	Apr	May	Jun	Jul	Aug	Sep	Oct	Nov	Dec	Year
1-Axis (E-W) Tracking	Average	2.0	2.2	2.2	2.6	3.3	3.7	3.6	3.3	2.9	2.6	1.7	1.5	**2.6**
	Minimum	1.1	1.3	1.6	1.8	2.1	3.0	2.8	2.3	1.9	1.5	0.7	0.9	**2.3**
	Maximum	2.5	3.3	3.1	3.8	4.9	4.9	4.4	4.4	4.1	3.6	2.6	2.5	**3.2**
1-Axis (N-S) Tracking, Tilt = 0°	Average	1.3	1.9	2.4	3.3	4.3	4.8	4.8	4.2	3.4	2.5	1.3	1.0	**2.9**
	Minimum	0.7	1.1	1.6	2.2	2.8	4.0	3.8	3.0	2.2	1.4	0.5	0.6	**2.6**
	Maximum	1.7	2.8	3.4	4.8	6.4	6.4	5.9	5.9	4.7	3.5	1.9	1.6	**3.6**
1-Axis (N-S) Tracking, Tilt = Lat°	Average	2.1	2.6	2.9	3.6	4.3	4.6	4.6	4.3	3.9	3.2	1.9	1.6	**3.3**
	Minimum	1.2	1.5	2.0	2.4	2.8	3.7	3.7	3.1	2.5	1.9	0.8	0.9	**2.9**
	Maximum	2.7	3.9	4.1	5.2	6.3	6.1	5.6	6.1	5.4	4.5	2.9	2.6	**4.0**
2-Axis Tracking	Average	2.3	2.7	2.9	3.6	4.5	5.0	4.9	4.5	3.9	3.3	2.0	1.7	**3.4**
	Minimum	1.3	1.6	2.0	2.4	3.0	4.1	3.9	3.2	2.5	1.9	0.8	1.0	**3.0**
	Maximum	2.9	4.0	4.1	5.2	6.7	6.6	6.1	6.2	5.4	4.5	3.0	2.8	**4.2**

AVERAGE CLIMATIC CONDITIONS

	Jan	Feb	Mar	Apr	May	Jun	Jul	Aug	Sep	Oct	Nov	Dec	Year
Average Temperature (°C)	-6.1	-3.7	2.9	9.2	14.9	20.3	22.9	22.1	18.0	11.6	4.4	-3.0	**9.4**
Average Low (°C)	-10.6	-8.2	-1.9	3.7	8.7	14.2	17.0	16.4	12.2	5.7	-0.2	-7.2	**4.2**
Average High (°C)	-1.7	0.8	7.7	14.8	21.2	26.4	28.7	27.7	23.8	17.4	9.1	1.1	**14.8**
Record Low (°C)	-32.8	-27.2	-22.2	-13.9	-4.4	2.2	4.4	5.0	-2.2	-8.3	-17.2	-31.7	**-32.8**
Record High (°C)	18.3	21.7	31.1	32.8	33.9	40.0	38.9	38.3	37.2	32.8	25.6	21.7	**40.0**
Heating Degree Days[†]	758	616	479	273	131	19	3	11	47	217	417	661	**3631**
Cooling Degree Days[†]	0	0	0	0	26	79	144	126	37	7	0	0	**418**
Relative Humidity (%)	72	72	70	65	64	66	68	71	71	69	73	76	**70**
Wind Speed (m/s)	5.2	5.1	5.4	5.3	4.6	4.2	3.7	3.6	4.0	4.5	4.8	4.9	**4.6**

* in kWh/m²/day, ±9%
[†] based on 18.3°C (65°F)

PORTLAND, OR (45.6°N, 122.6°W)

Elevation: 12 m   Pressure: 1017 mb

SOLAR RADIATION FOR FLAT-PLATE COLLECTORS FACING SOUTH AT A FIXED TILT*

Tilt		Jan	Feb	Mar	Apr	May	Jun	Jul	Aug	Sep	Oct	Nov	Dec	Year
0°	Average	1.2	1.9	3.0	4.2	5.3	5.9	6.3	5.4	4.1	2.5	1.4	1.0	3.5
	Minimum	0.9	1.4	2.4	3.4	4.3	4.8	5.0	4.6	3.2	1.8	0.9	0.7	3.1
	Maximum	1.5	2.4	3.8	4.8	6.2	6.9	7.2	6.3	4.9	3.2	1.7	1.3	3.8
Lat − 15°	Average	1.7	2.5	3.6	4.6	5.4	5.8	6.3	5.9	5.1	3.4	1.9	1.4	4.0
	Minimum	1.1	1.4	2.7	3.5	4.3	4.7	4.9	5.0	3.7	2.3	1.1	1.0	3.5
	Maximum	2.7	3.6	5.0	5.4	6.3	6.8	7.2	7.0	6.2	4.6	2.8	2.5	4.3
Lat°	Average	1.9	2.6	3.7	4.5	5.0	5.3	5.8	5.6	5.1	3.6	2.1	1.6	3.9
	Minimum	1.2	1.4	2.6	3.4	4.0	4.3	4.6	4.8	3.6	2.3	1.2	1.0	3.4
	Maximum	3.0	3.9	5.2	5.3	5.9	6.2	6.7	6.7	6.4	5.0	3.1	2.9	4.2
Lat + 15°	Average	1.9	2.7	3.5	4.1	4.4	4.6	5.1	5.1	4.9	3.6	2.1	1.6	3.6
	Minimum	1.2	1.3	2.5	3.1	3.5	3.8	4.0	4.3	3.4	2.2	1.2	1.0	3.2
	Maximum	3.2	4.0	5.1	4.9	5.2	5.3	5.8	6.1	6.1	5.0	3.2	3.1	3.9
90°	Average	1.8	2.3	2.8	2.9	2.8	2.7	3.1	3.4	3.7	3.0	1.9	1.5	2.6
	Minimum	1.0	1.0	1.9	2.2	2.3	2.3	2.5	2.9	2.6	1.8	1.0	0.9	2.3
	Maximum	3.1	3.5	4.1	3.5	3.3	3.1	3.4	4.0	4.7	4.3	2.9	3.0	2.9

SOLAR RADIATION FOR 1-AXIS TRACKING FLAT-PLATE COLLECTORS WITH A NORTH-SOUTH AXIS*

Axis Tilt		Jan	Feb	Mar	Apr	May	Jun	Jul	Aug	Sep	Oct	Nov	Dec	Year
0°	Average	1.5	2.4	3.8	5.2	6.6	7.4	8.3	7.3	5.7	3.4	1.7	1.2	4.5
	Minimum	0.9	1.3	2.7	3.7	4.8	5.8	6.1	5.7	3.8	2.2	1.0	0.8	3.9
	Maximum	2.3	3.4	5.4	6.4	8.0	9.2	9.9	9.0	7.2	4.7	2.4	2.0	5.0
Lat − 15°	Average	1.9	2.9	4.3	5.6	6.8	7.5	8.4	7.7	6.4	4.1	2.1	1.6	4.9
	Minimum	1.1	1.4	2.9	3.9	4.9	5.8	6.1	6.1	4.2	2.5	1.2	1.0	4.2
	Maximum	3.2	4.3	6.3	6.9	8.3	9.3	10.1	9.6	8.2	5.8	3.2	2.9	5.4
Lat°	Average	2.0	3.0	4.3	5.5	6.5	7.2	8.1	7.5	6.5	4.3	2.2	1.7	4.9
	Minimum	1.2	1.4	2.9	3.8	4.7	5.6	5.8	5.9	4.2	2.5	1.2	1.0	4.2
	Maximum	3.5	4.6	6.4	6.8	8.0	8.9	9.8	9.5	8.4	6.1	3.5	3.3	5.4
Lat + 15°	Average	2.1	3.0	4.2	5.3	6.1	6.7	7.6	7.2	6.3	4.2	2.3	1.7	4.7
	Minimum	1.2	1.3	2.8	3.6	4.4	5.2	5.5	5.6	4.0	2.5	1.2	1.0	4.0
	Maximum	3.6	4.6	6.3	6.5	7.5	8.3	9.2	9.0	8.2	6.1	3.6	3.4	5.2

SOLAR RADIATION FOR 2-AXIS TRACKING FLAT-PLATE COLLECTORS*

		Jan	Feb	Mar	Apr	May	Jun	Jul	Aug	Sep	Oct	Nov	Dec	Year
2-Axis Tracking	Average	2.1	3.0	4.4	5.6	6.9	7.7	8.6	7.7	6.5	4.3	2.3	1.8	5.1
	Minimum	1.2	1.4	2.9	3.9	5.0	6.0	6.2	6.1	4.2	2.5	1.3	1.0	4.3
	Maximum	3.7	4.7	6.4	6.9	8.4	9.5	10.3	9.7	8.4	6.1	3.6	3.5	5.6

DIRECT BEAM SOLAR RADIATION FOR CONCENTRATING COLLECTORS*

		Jan	Feb	Mar	Apr	May	Jun	Jul	Aug	Sep	Oct	Nov	Dec	Year
1-Axis (E-W) Tracking	Average	1.0	1.4	1.9	2.2	2.8	3.4	4.2	3.8	3.3	2.2	1.1	0.8	2.3
	Minimum	0.2	0.2	0.9	1.1	1.4	2.2	2.2	2.5	1.5	0.9	0.4	0.2	1.9
	Maximum	2.4	2.7	3.4	3.1	3.9	4.8	6.0	5.2	4.7	3.6	2.2	2.2	2.9
1-Axis (N-S) Tracking, Tilt = 0°	Average	0.6	1.1	1.9	2.7	3.6	4.4	5.4	4.8	3.7	2.0	0.7	0.5	2.6
	Minimum	0.2	0.2	0.9	1.3	1.8	2.9	2.9	3.1	1.6	0.8	0.3	0.1	2.1
	Maximum	1.5	2.2	3.7	3.8	5.1	6.3	8.1	6.7	5.3	3.3	1.4	1.3	3.2
1-Axis (N-S) Tracking, Tilt = Lat°	Average	1.1	1.6	2.4	3.0	3.6	4.2	5.3	5.0	4.4	2.7	1.2	0.9	3.0
	Minimum	0.3	0.2	1.1	1.4	1.8	2.7	2.8	3.3	1.9	1.1	0.5	0.2	2.3
	Maximum	2.6	3.2	4.5	4.2	5.1	6.1	7.8	7.0	6.3	4.4	2.3	2.3	3.6
2-Axis Tracking	Average	1.2	1.7	2.4	3.0	3.8	4.5	5.7	5.2	4.4	2.7	1.3	0.9	3.1
	Minimum	0.3	0.2	1.1	1.4	1.9	3.0	3.0	3.4	2.0	1.1	0.5	0.2	2.4
	Maximum	2.7	3.2	4.5	4.3	5.4	6.6	8.4	7.3	6.3	4.5	2.5	2.5	3.8

AVERAGE CLIMATIC CONDITIONS

	Jan	Feb	Mar	Apr	May	Jun	Jul	Aug	Sep	Oct	Nov	Dec	Year
Average Temperature (°C)	4.2	6.4	8.5	10.6	13.9	17.5	20.1	20.3	17.4	12.5	7.8	4.6	12.0
Average Low (°C)	0.9	2.3	3.7	5.2	8.3	11.6	13.6	13.8	11.1	7.2	4.2	1.6	6.9
Average High (°C)	7.4	10.6	13.3	15.9	19.5	23.3	26.6	26.8	23.7	17.8	11.4	7.6	17.0
Record Low (°C)	-18.9	-19.4	-7.2	-1.7	-1.7	3.9	6.1	6.7	1.1	-3.3	-10.6	-14.4	-19.4
Record High (°C)	17.2	21.7	26.7	30.6	37.8	37.8	41.7	41.7	40.6	33.3	22.8	17.8	41.7
Heating Degree Days[†]	437	333	305	233	138	51	16	19	57	181	315	427	2512
Cooling Degree Days[†]	0	0	0	0	0	26	71	82	28	0	0	0	206
Relative Humidity (%)	81	78	75	72	69	66	63	65	69	78	82	83	73
Wind Speed (m/s)	4.4	4.2	3.8	3.4	3.4	3.4	3.5	3.3	3.1	3.0	4.0	4.2	3.6

* in kWh/m²/day, ±9%
† based on 18.3°C (65°F)

SAN FRANCISCO, CA (37.6°N, 122.4°W)
Elevation: 5 m   Pressure: 1017 mb

SOLAR RADIATION FOR FLAT-PLATE COLLECTORS FACING SOUTH AT A FIXED TILT*

Tilt		Jan	Feb	Mar	Apr	May	Jun	Jul	Aug	Sep	Oct	Nov	Dec	Year
	Average	2.2	3.0	4.2	5.7	6.7	7.2	7.3	6.5	5.4	3.9	2.5	2.0	**4.7**
0°	Minimum	1.8	2.3	3.3	4.4	5.7	6.1	6.9	5.7	4.7	3.2	2.1	1.4	**4.4**
	Maximum	2.5	3.9	5.3	6.4	7.4	7.9	7.9	7.3	6.0	4.4	2.9	2.4	**4.9**
	Average	3.1	3.9	5.0	6.2	6.8	7.0	7.3	6.9	6.2	5.0	3.5	2.9	**5.3**
Lat – 15°	Minimum	2.4	2.9	3.8	4.7	5.7	5.9	6.9	6.0	5.3	4.0	2.8	1.9	**4.9**
	Maximum	3.8	5.5	6.5	7.0	7.5	7.7	7.8	7.7	7.1	5.8	4.3	3.9	**5.6**
	Average	3.5	4.2	5.2	6.1	6.4	6.5	6.8	6.7	6.4	5.4	3.9	3.4	**5.4**
Lat°	Minimum	2.7	3.1	3.9	4.6	5.4	5.4	6.4	5.8	5.4	4.3	3.1	2.0	**4.9**
	Maximum	4.3	6.1	6.8	6.9	7.1	7.1	7.2	7.4	7.3	6.4	4.9	4.6	**5.7**
	Average	3.7	4.4	5.1	5.6	5.7	5.6	5.9	6.1	6.1	5.5	4.1	3.6	**5.1**
Lat + 15°	Minimum	2.8	3.1	3.8	4.2	4.8	4.7	5.6	5.3	5.2	4.3	3.2	2.1	**4.6**
	Maximum	4.7	6.4	6.8	6.4	6.3	6.1	6.3	6.7	7.0	6.5	5.2	5.0	**5.5**
	Average	3.3	3.6	3.7	3.4	2.8	2.5	2.7	3.3	4.1	4.3	3.6	3.3	**3.4**
90°	Minimum	2.5	2.4	2.7	2.6	2.5	2.3	2.6	3.0	3.5	3.4	2.8	1.9	**2.9**
	Maximum	4.2	5.4	4.9	3.8	3.0	2.6	2.8	3.6	4.7	5.2	4.6	4.7	**3.7**

SOLAR RADIATION FOR 1-AXIS TRACKING FLAT-PLATE COLLECTORS WITH A NORTH-SOUTH AXIS*

Axis Tilt		Jan	Feb	Mar	Apr	May	Jun	Jul	Aug	Sep	Oct	Nov	Dec	Year
	Average	3.0	4.1	5.7	7.6	8.8	9.3	9.7	8.7	7.4	5.5	3.5	2.8	**6.3**
0°	Minimum	2.3	2.8	4.1	5.6	7.0	7.5	9.0	7.4	6.2	4.3	2.8	1.7	**5.7**
	Maximum	3.7	6.0	7.7	8.9	10.1	10.8	10.7	10.2	8.8	6.5	4.4	3.8	**6.7**
	Average	3.7	4.7	6.3	8.0	8.9	9.2	9.7	9.0	8.1	6.3	4.3	3.5	**6.8**
Lat – 15°	Minimum	2.8	3.3	4.5	5.8	7.1	7.4	9.0	7.7	6.7	4.8	3.3	2.0	**6.1**
	Maximum	4.6	7.1	8.6	9.4	10.3	10.7	10.8	10.6	9.6	7.6	5.4	4.8	**7.2**
	Average	4.0	5.0	6.5	8.0	8.7	8.9	9.4	8.8	8.2	6.6	4.6	3.9	**6.9**
Lat°	Minimum	3.0	3.4	4.6	5.7	6.8	7.1	8.7	7.5	6.8	5.1	3.5	2.2	**6.1**
	Maximum	5.1	7.6	8.9	9.4	10.0	10.3	10.4	10.4	9.7	8.0	5.9	5.4	**7.3**
	Average	4.2	5.1	6.4	7.7	8.1	8.2	8.7	8.4	8.0	6.7	4.8	4.1	**6.7**
Lat + 15°	Minimum	3.1	3.4	4.5	5.5	6.4	6.6	8.1	7.1	6.6	5.1	3.6	2.3	**5.9**
	Maximum	5.4	7.8	8.8	9.0	9.4	9.6	9.7	9.9	9.5	8.1	6.1	5.7	**7.1**

SOLAR RADIATION FOR 2-AXIS TRACKING FLAT-PLATE COLLECTORS*

		Jan	Feb	Mar	Apr	May	Jun	Jul	Aug	Sep	Oct	Nov	Dec	Year
	Average	4.2	5.1	6.5	8.1	9.0	9.4	9.9	9.0	8.2	6.7	4.8	4.1	**7.1**
2-Axis Tracking	Minimum	3.1	3.4	4.6	5.8	7.2	7.6	9.2	7.7	6.8	5.1	3.7	2.3	**6.3**
	Maximum	5.4	7.8	8.9	9.5	10.4	11.0	11.0	10.6	9.7	8.1	6.2	5.8	**7.5**

DIRECT BEAM SOLAR RADIATION FOR CONCENTRATING COLLECTORS*

		Jan	Feb	Mar	Apr	May	Jun	Jul	Aug	Sep	Oct	Nov	Dec	Year
1-Axis (E-W) Tracking	Average	2.5	2.8	3.3	4.0	4.7	5.1	5.5	4.9	4.6	3.9	2.8	2.5	**3.9**
	Minimum	1.6	1.5	1.9	2.4	3.3	3.6	4.9	3.7	3.3	2.6	1.8	1.1	**3.2**
	Maximum	3.5	4.9	5.2	5.2	5.7	6.3	6.4	6.1	5.6	5.0	3.9	3.9	**4.2**
1-Axis (N-S) Tracking, Tilt = 0°	Average	1.9	2.5	3.8	5.2	6.1	6.5	7.1	6.3	5.4	3.9	2.3	1.8	**4.4**
	Minimum	1.2	1.3	2.1	3.1	4.2	4.6	6.2	4.7	4.0	2.5	1.5	0.8	**3.7**
	Maximum	2.6	4.6	6.0	6.7	7.5	8.3	8.3	8.0	6.8	5.0	3.2	2.8	**4.7**
1-Axis (N-S) Tracking, Tilt = Lat°	Average	2.7	3.3	4.4	5.4	5.9	6.1	6.8	6.4	6.0	4.8	3.2	2.7	**4.8**
	Minimum	1.8	1.7	2.5	3.2	4.1	4.3	5.9	4.8	4.4	3.2	2.0	1.2	**4.0**
	Maximum	3.8	6.0	6.9	7.0	7.3	7.7	7.9	8.1	7.6	6.3	4.4	4.2	**5.3**
2-Axis Tracking	Average	2.9	3.4	4.4	5.5	6.3	6.7	7.3	6.6	6.1	4.9	3.3	2.9	**5.0**
	Minimum	1.9	1.7	2.5	3.3	4.3	4.7	6.4	5.0	4.4	3.2	2.2	1.3	**4.2**
	Maximum	4.0	6.1	6.9	7.1	7.7	8.4	8.5	8.3	7.6	6.3	4.7	4.5	**5.5**

AVERAGE CLIMATIC CONDITIONS

	Jan	Feb	Mar	Apr	May	Jun	Jul	Aug	Sep	Oct	Nov	Dec	Year
Average Temperature (°C)	9.3	11.2	11.8	13.1	14.5	16.4	17.1	17.6	18.1	16.1	12.7	9.7	**13.9**
Average Low (°C)	5.4	7.2	7.7	8.4	9.8	11.4	12.2	12.8	12.9	11.0	8.4	5.9	**9.4**
Average High (°C)	13.1	15.2	16.0	17.7	19.2	21.3	22.0	22.4	23.1	21.2	16.9	13.4	**18.4**
Record Low (°C)	-4.4	-3.9	-1.1	-0.6	2.2	5.0	6.1	5.6	3.3	1.1	-3.9	-6.7	**-6.7**
Record High (°C)	22.2	25.6	29.4	33.3	36.1	41.1	40.6	36.7	39.4	37.2	29.4	23.9	**41.1**
Heating Degree Days†	281	199	202	159	121	67	51	38	44	75	170	269	**1676**
Cooling Degree Days†	0	0	0	3	0	9	12	16	36	6	0	0	**81**
Relative Humidity (%)	78	76	73	71	71	72	73	74	72	72	75	77	**74**
Wind Speed (m/s)	3.3	4.0	4.8	5.5	6.2	6.2	6.2	5.7	5.0	4.3	3.7	3.5	**4.9**

* in kWh/m²/day, ±9%
† based on 18.3°C (65°F)

TAMPA, FL (28.0°N, 82.5°W)

Elevation: 3 m   Pressure: 1018 mb

SOLAR RADIATION FOR FLAT-PLATE COLLECTORS FACING SOUTH AT A FIXED TILT*

Tilt		Jan	Feb	Mar	Apr	May	Jun	Jul	Aug	Sep	Oct	Nov	Dec	Year
	Average	3.2	4.0	5.1	6.2	6.4	6.1	5.8	5.5	4.9	4.4	3.6	3.1	**4.9**
0°	Minimum	2.6	3.4	4.3	5.8	5.7	5.1	5.1	4.6	4.3	3.6	3.0	2.7	**4.5**
	Maximum	3.7	4.5	5.6	6.8	7.1	6.9	6.4	6.3	5.4	4.9	4.0	3.5	**5.2**
	Average	3.9	4.6	5.5	6.4	6.4	5.9	5.7	5.5	5.2	5.0	4.2	3.8	**5.2**
Lat − 15°	Minimum	3.0	3.8	4.7	5.9	5.7	5.0	5.0	4.7	4.5	4.0	3.4	3.1	**4.7**
	Maximum	4.6	5.3	6.1	7.1	7.0	6.7	6.3	6.3	5.8	5.6	4.8	4.4	**5.5**
	Average	4.5	5.1	5.8	6.3	6.0	5.5	5.3	5.4	5.2	5.4	4.8	4.4	**5.3**
Lat°	Minimum	3.3	4.2	4.8	5.9	5.4	4.7	4.7	4.5	4.5	4.1	3.8	3.5	**4.8**
	Maximum	5.3	5.9	6.4	7.0	6.6	6.2	5.9	6.2	5.9	6.1	5.6	5.2	**5.7**
	Average	4.8	5.3	5.7	5.9	5.3	4.8	4.7	4.9	5.0	5.5	5.1	4.7	**5.1**
Lat + 15°	Minimum	3.5	4.3	4.7	5.4	4.8	4.1	4.2	4.2	4.3	4.1	4.0	3.7	**4.6**
	Maximum	5.8	6.2	6.4	6.5	5.8	5.3	5.2	5.6	5.7	6.2	6.0	5.7	**5.5**
	Average	4.0	4.0	3.5	2.8	2.0	1.7	1.8	2.2	2.9	3.9	4.2	4.1	**3.1**
90°	Minimum	2.8	3.2	3.0	2.6	1.9	1.6	1.7	2.0	2.4	2.8	3.1	3.1	**2.7**
	Maximum	5.0	4.8	4.0	3.0	2.1	1.8	1.9	2.4	3.2	4.4	5.0	5.0	**3.3**

SOLAR RADIATION FOR 1-AXIS TRACKING FLAT-PLATE COLLECTORS WITH A NORTH-SOUTH AXIS*

Axis Tilt		Jan	Feb	Mar	Apr	May	Jun	Jul	Aug	Sep	Oct	Nov	Dec	Year
	Average	4.5	5.5	6.8	8.3	8.3	7.5	7.1	6.8	6.2	6.0	4.9	4.2	**6.3**
0°	Minimum	3.3	4.3	5.5	7.5	7.1	6.0	6.0	5.6	5.2	4.4	3.9	3.3	**5.6**
	Maximum	5.4	6.6	7.9	9.4	9.4	8.8	8.1	8.1	7.2	6.8	5.8	5.2	**6.9**
	Average	5.0	5.9	7.2	8.4	8.2	7.4	7.1	6.9	6.5	6.4	5.4	4.8	**6.6**
Lat − 15°	Minimum	3.5	4.6	5.8	7.7	7.1	5.9	6.0	5.7	5.3	4.7	4.2	3.7	**5.8**
	Maximum	6.0	7.1	8.3	9.6	9.3	8.7	8.0	8.2	7.5	7.3	6.4	5.8	**7.2**
	Average	5.4	6.3	7.3	8.4	8.0	7.1	6.8	6.8	6.5	6.7	5.9	5.2	**6.7**
Lat°	Minimum	3.8	4.9	5.9	7.6	6.9	5.7	5.7	5.6	5.3	4.8	4.5	4.0	**5.9**
	Maximum	6.6	7.6	8.5	9.6	9.1	8.4	7.7	8.0	7.6	7.7	6.9	6.4	**7.3**
	Average	5.6	6.4	7.3	8.1	7.5	6.6	6.3	6.4	6.4	6.8	6.1	5.5	**6.6**
Lat + 15°	Minimum	3.9	4.9	5.8	7.3	6.4	5.3	5.3	5.3	5.2	4.8	4.6	4.2	**5.8**
	Maximum	7.0	7.8	8.5	9.2	8.5	7.8	7.2	7.6	7.4	7.8	7.3	6.8	**7.2**

SOLAR RADIATION FOR 2-AXIS TRACKING FLAT-PLATE COLLECTORS*

		Jan	Feb	Mar	Apr	May	Jun	Jul	Aug	Sep	Oct	Nov	Dec	Year
	Average	5.7	6.4	7.4	8.5	8.3	7.6	7.2	6.9	6.5	6.8	6.1	5.6	**6.9**
2-Axis Tracking	Minimum	3.9	5.0	5.9	7.7	7.2	6.0	6.0	5.7	5.4	4.9	4.6	4.3	**6.1**
	Maximum	7.1	7.8	8.5	9.7	9.4	8.9	8.1	8.2	7.6	7.8	7.3	7.0	**7.6**

DIRECT BEAM SOLAR RADIATION FOR CONCENTRATING COLLECTORS*

		Jan	Feb	Mar	Apr	May	Jun	Jul	Aug	Sep	Oct	Nov	Dec	Year
1-Axis (E-W) Tracking	Average	3.3	3.6	3.8	4.3	4.1	3.4	3.1	3.0	2.9	3.5	3.5	3.4	**3.5**
	Minimum	1.9	2.4	2.6	3.5	3.1	2.2	2.3	2.0	2.0	2.0	2.3	2.2	**2.8**
	Maximum	4.6	4.8	4.8	5.3	5.0	4.6	3.9	4.0	3.7	4.4	4.6	4.5	**4.1**
1-Axis (N-S) Tracking, Tilt = 0°	Average	3.0	3.7	4.7	5.7	5.4	4.4	4.0	3.9	3.6	3.9	3.3	2.9	**4.0**
	Minimum	1.6	2.4	3.2	4.7	4.1	2.8	3.0	2.7	2.5	2.3	2.2	1.8	**3.2**
	Maximum	4.1	5.1	5.9	7.1	6.7	5.9	5.0	5.3	4.8	4.9	4.3	3.9	**4.7**
1-Axis (N-S) Tracking, Tilt = Lat°	Average	3.8	4.4	5.1	5.8	5.1	4.1	3.8	3.8	3.8	4.5	4.1	3.7	**4.3**
	Minimum	2.1	2.8	3.5	4.8	3.9	2.6	2.8	2.7	2.7	2.6	2.8	2.3	**3.4**
	Maximum	5.2	6.0	6.5	7.2	6.3	5.5	4.7	5.2	5.0	5.6	5.3	5.0	**5.1**
2-Axis Tracking	Average	4.0	4.5	5.1	5.9	5.4	4.5	4.0	3.9	3.8	4.5	4.3	4.0	**4.5**
	Minimum	2.2	2.9	3.5	4.8	4.1	2.8	3.0	2.7	2.7	2.6	2.9	2.5	**3.6**
	Maximum	5.5	6.2	6.5	7.3	6.7	6.0	5.1	5.4	5.0	5.7	5.6	5.4	**5.3**

AVERAGE CLIMATIC CONDITIONS

	Jan	Feb	Mar	Apr	May	Jun	Jul	Aug	Sep	Oct	Nov	Dec	Year
Average Temperature (°C)	15.5	16.4	19.1	21.8	25.1	27.2	27.8	27.8	27.2	23.8	19.8	16.8	**22.4**
Average Low (°C)	10.0	10.9	13.6	16.0	19.7	22.7	23.6	23.6	22.7	18.4	14.0	11.3	**17.2**
Average High (°C)	21.0	21.9	24.8	27.6	30.7	31.9	32.3	32.3	31.7	29.1	25.4	22.3	**27.6**
Record Low (°C)	-6.1	-4.4	-1.7	4.4	9.4	11.7	17.2	19.4	13.9	4.4	-5.0	-7.8	**-7.8**
Record High (°C)	30.0	31.1	32.8	33.9	36.7	37.2	36.1	36.7	35.6	34.4	32.2	30.0	**37.2**
Heating Degree Days†	130	89	47	4	0	0	0	0	0	0	39	95	**403**
Cooling Degree Days†	42	34	71	107	210	267	294	294	267	171	82	47	**1887**
Relative Humidity (%)	75	73	72	69	70	74	77	78	77	74	75	75	**74**
Wind Speed (m/s)	3.9	4.1	4.2	4.1	3.9	3.6	3.3	3.1	3.4	3.8	3.8	3.8	**3.8**

* in kWh/m²/day, ±9%
† based on 18.3°C (65°F)

SUN PATH CHART FOR 20°N

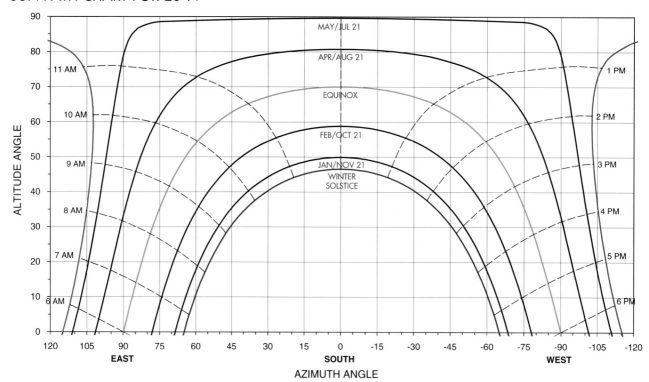

SUN PATH CHART FOR 30°N

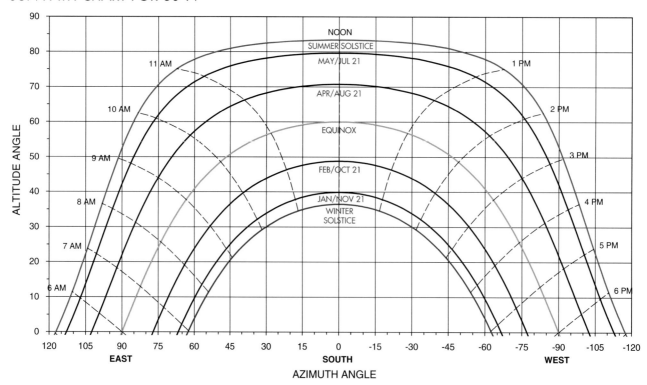

SUN PATH CHART FOR 40°N

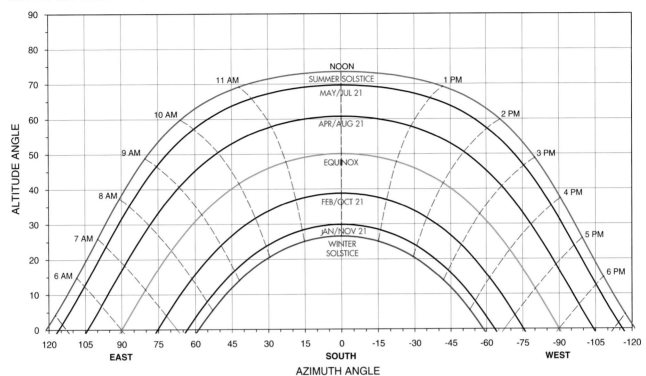

SUN PATH CHART FOR 50°N

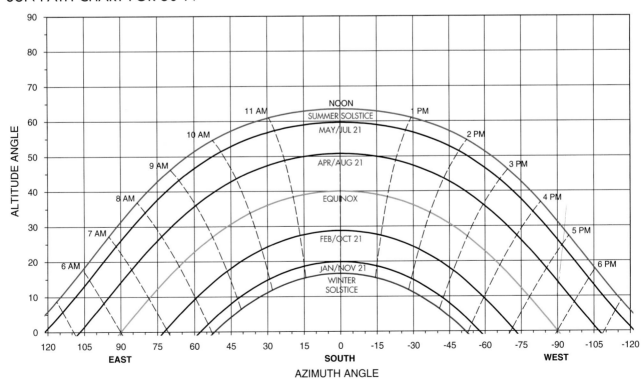

SUN PATH CHART FOR 60°N

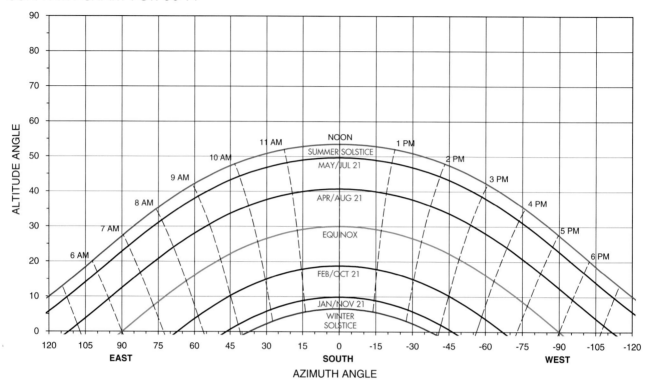

SUN PATH CHART FOR 70°N

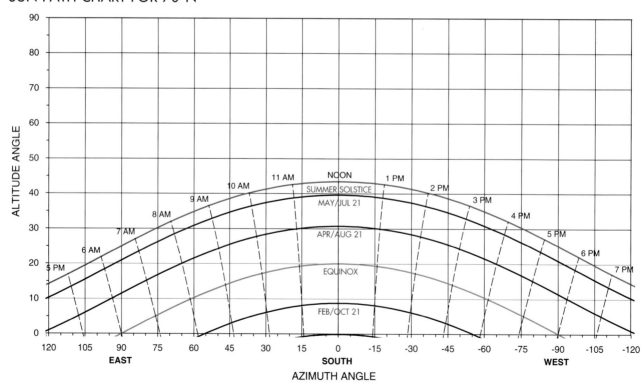

INTERACTIVE SYSTEM SIZING

❶ P_{mp} Enter the maximum DC power rating (at STC) for the chosen module.

❷ C_g Enter a derating factor for manufacturer power guarantee. Due to small differences between cells, not every module designed for a certain rating will exactly equal that rating. It may be slightly less than the rating, but the manufacturer may guarantee that it will be at least a certain percentage, typically 90% (0.90), of the rating.

❸ n_m Enter the number of modules in the array.

❹ P_{arr-g} Calculate the minimum guaranteed power output of the array.

$$P_{arr-g} = P_{mp} \times C_g \times n_m$$
❹ = **❶** × **❷** × **❸**

❺ T_{amb} Enter the average expected module operating temperature.

❻ $C_{\%P}$ Enter the relative temperature coefficient for power. This value may be given by the module manufacturer. Otherwise, typical values range from $-0.4\%/°C$ to $-0.5\%/°C$ ($-0.004/°C$ to $-0.005/°C$).

❼ P_{arr-T} Calculate the temperature-corrected array power output.

$$P_{arr-T} = P_{arr-g} + (P_{arr-g} \times C_{\%P} \times [T_{amb} - 25])$$
❼ = **❹** + (**❹** × **❻** × [**❺** − 25])

❽ C_{loss} Enter a factor to account for power losses due to wiring and module mismatch. A typical value is 3% (0.03).

❾ $P_{arr-net}$ Calculate the net array power output.

$$P_{arr-net} = P_{arr-T} - (P_{arr-T} \times C_{loss})$$
❾ = **❼** − (**❼** × **❽**)

❿ P_{inv-DC} Enter the inverter maximum DC power input rating. This value should be given in the inverter specifications.

⓫ η_{inv} Enter the inverter power conversion efficiency. This value should be given in the inverter specifications. Typical values range from 90% to 95% (0.90 to 0.95).

⓬ η_{MPPT} Enter the inverter MPPT efficiency. This value can be as high as 100% (1.00)

⓭ P_{inv-AC} Calculate the inverter maximum AC power output. This calculation depends on its input rating.

If the inverter maximum DC power input rating is greater than or equal to the maximum array output, the array output is used to calculate the inverter output.

If $P_{inv-DC} \geq P_{arr-net}$, then $P_{inv-AC} = P_{arr-net} \times \eta_{inv} \times \eta_{MPPT}$
If **❿** ≥ **❾**, then **⓭** = **❾** × **⓫** × **⓬**

Interactive System Sizing Worksheet

PV-Module Rated DC Power Output	**❶**	W
Manufacturer Power Guarantee	**❷**	
Number of Modules in Array	**❸**	
Array Guaranteed Power Output	**❹**	W
Array Avg Operating Temperature	**❺**	°C
Temperature Coefficient for Power	**❻**	/°C
Temperature-Corrected Array Power Output	**❼**	W
Array Wiring and Mismatch Losses	**❽**	
Net Array Power Output	**❾**	W
Inverter Maximum DC Power Rating	**❿**	W
Inverter Power Conversion Efficiency	**⓫**	
Inverter MPPT Efficiency	**⓬**	
Inverter Maximum AC Power Output	**⓭**	W
Average Daily Insolation	**⓮**	PSH/day
Average Daily Energy Production	**⓯**	Wh/day
	⓯	kWh/day

⬤ (gray) Input values that are entered into calculations
⬤ (black) Output values that must be calculated or determined

---

If the inverter maximum DC power input rating is less than the maximum array output, the inverter DC input rating is used to calculate the inverter output.

If $P_{inv-DC} < P_{arr-net}$, then $P_{inv-AC} = P_{inv-DC} \times \eta_{inv} \times \eta_{MPPT}$
If **❿** < **❾**, then **⓭** = **❿** × **⓫** × **⓬**

⓮ PSH Enter the average daily insolation for the location and array orientation. This may be an average for either a certain month or the entire year.

⓯ E Calculate average daily energy production during the chosen period (month or year). To convert from watt-hours to kilowatt-hours, divide by 1000.

$$E = P_{inv-AC} \times PSH$$
⓯ = **⓭** × **⓮**

If the final system output is not within the desired range, such as above a minimum size requirement for an incentive program, different module and/or inverter choices can be made that change the calculation results.

STAND-ALONE SYSTEM SIZING

Load Analysis

1 n_{load} Enter number of loads of a given description. All loads within a description must use the same power type (AC or DC) and have the same power rating and operation time. AC loads and DC loads are listed separately, but require the same types of information.

2 P_{load} Enter the power rating of the load type.

3 t_{load} Enter the operating time of the load type.

4 E_{load} Calculate the daily energy consumption of each load type individually.

$$E_{load} = n_{load} \times P_{load} \times t_{load}$$

4 = **1** × **2** × **3**

5 P_{AC} Calculate the total AC power requirement, if every load operated simultaneously.

P_{AC} = sum of all P_{load} for AC loads

5 = sum of all **2** for AC loads

6 P_{DC} Calculate the total DC power requirement, if every load operated simultaneously.

P_{DC} = sum of all P_{load} for DC loads

6 = sum of all **2** for DC loads

7 E_{AC} Calculate the total energy consumption by all AC loads.

E_{AC} = sum of all E_{load} for AC loads

7 = sum of all **4** for AC loads

8 E_{DC} Calculate the total energy consumption by all DC loads.

E_{DC} = sum of all E_{load} for DC loads

8 = sum of all **4** for DC loads

9 η_{inv} Enter the inverter power conversion efficiency. This value should be in the inverter specifications. Typical values are 90% to 95% (0.90 to 0.95).

10 t_{op} Calculate the weighted average operating time for all of the loads. Each load's energy consumption and operating time is included. If the system includes AC loads, their AC energy consumption is first converted to equivalent DC energy consumption by dividing by the inverter efficiency.

$$t_{op} = \frac{(E_1 \times t_1) + (E_2 \times t_2) + \ldots + (E_n \times t_n)}{E_1 + E_2 + \ldots + E_n}$$

10 = $\dfrac{(\textbf{4} \times \textbf{3}) + (\textbf{4} \times \textbf{3}) + \ldots + (\textbf{4} \times \textbf{3})}{\textbf{4} + \textbf{4} + \ldots + \textbf{4}}$

11 E_{SDC} Calculate the total daily DC energy required to supply all of the loads during this month.

$$E_{SDC} = \frac{E_{AC}}{\eta_{inv}} + E_{DC}$$

11 = $\dfrac{\textbf{7}}{\textbf{9}}$ + **8**

A load analysis must be done for each month individually, unless the load operations are constant throughout the year.

Load Analysis Worksheet

AC LOADS   Month:

Load Description	Qty	Power Rating (W)	Operating Time (hr/day)	Energy Consumption (Wh/day)
	1	**2**	**3**	**4**

DC LOADS

	Qty	Power Rating	Operating Time	Energy Consumption
	1	**2**	**3**	**4**

Total AC Power	**5**	W
Total DC Power	**6**	W
Total Daily AC Energy Consumption	**7**	Wh/day
Total Daily DC Energy Consumption	**8**	Wh/day
Inverter Efficiency	**9**	
Weighted Operating Time	**10**	hr/day
Average Daily DC Energy Consumption	**11**	Wh/day

STAND-ALONE SYSTEM SIZING

Critical Design Analysis

11 E_{SDC} Enter the total required daily system DC energy calculated from the load analyses for each month.

12 *PSH*   Enter the average daily insolation for the location and possible array orientations for each month. This information may be found in the solar radiation data set for the nearest location.

13 *design ratio*

Calculate the design ratio for each month and each array orientation.

$$design\, ratio = \frac{E_{SDC}}{PSH}$$

$$❸ = \frac{⓫}{⓬}$$

14 *critical design month*

Determine the critical design month. For each orientation, identify the month with the highest design ratio. This is the critical design month for this orientation. If there are multiple orientations in the analysis, identify the orientation corresponding to the lowest critical design month design ratio. This is the selected array orientation. This orientation's critical design month is the selected critical design month.

15 E_{crit} Enter the total required daily system DC energy for the critical design month.

16 t_{PSH} Enter the average daily insolation for the selected array orientation during the critical design month.

Critical Design Analysis Worksheet

Month	Average Daily DC Energy Consumption (Wh/day)	Array Orientation 1		Array Orientation 2		Array Orientation 3	
		Insolation (PSH/day)	Design Ratio	Insolation (PSH/day)	Design Ratio	Insolation (PSH/day)	Design Ratio
January	⓫	⓬	❸	⓬	❸	⓬	❸
February							
March							
April							
May							
June							
July							
August							
September							
October							
November							
December							

Critical Design Month   **14**

Optimal Orientation   **14**

Average Daily DC Energy Consumption   **15**   Wh/day

Insolation   **16**   PSH/day

STAND-ALONE SYSTEM SIZING

Battery-Bank Sizing

⑮ E_{crit} Enter the total required daily system DC energy for the critical design month.

⑰ V_{SDC} Enter the nominal DC-system voltage.

⑱ t_a Enter the desired system autonomy.

⑲ B_{out} Calculate the required battery-bank output.

$$B_{out} = \frac{E_{crit} \times t_a}{V_{SDC}}$$

$$⑲ = \frac{⑮ \times ⑱}{⑰}$$

⑳ DOD_a Enter an allowable depth of discharge. For deep-cycle lead-acid batteries, a typical value if 80% (0.80).

⑩ t_{op} Enter the weighted average operating time for all of the loads. This value is calculated in the load analysis for the critical design month.

㉑ r_d Calculate the average battery-bank discharge rate.

$$r_d = \frac{t_{op} \times t_a}{DOD_a}$$

$$㉑ = \frac{⑩ \times ⑱}{⑳}$$

㉒ T_{min} Enter the minimum expected operating temperature for the battery bank.

㉓ $C_{T,rd}$ Determine the derating factor associated with the batteries' minimum operating temperature and discharge rate. This information is typically found on charts or graphs available from the battery manufacturer.

㉔ B_{rated} Calculate the rated capacity required for the battery bank.

$$B_{rated} = \frac{B_{out}}{DOD_a \times C_{T,rd}}$$

$$㉔ = \frac{⑲}{⑳ \times ㉓}$$

㉕ V_{batt} Enter the nominal voltage of the selected battery.

㉖ B_{batt} Enter the rated capacity of the selected battery.

㉗ n_{series} Calculate the number of batteries required to be connected in series.

$$n_{series} = \frac{V_{SDC}}{V_{batt}}$$

$$㉗ = \frac{⑰}{㉕}$$

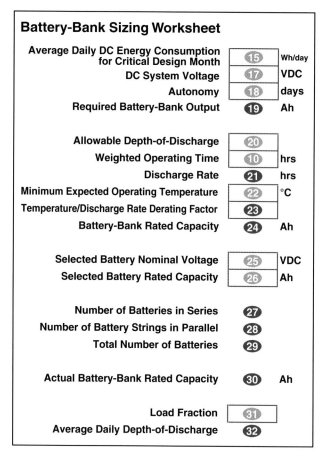

Battery-Bank Sizing Worksheet

Average Daily DC Energy Consumption for Critical Design Month	⑮	Wh/day
DC System Voltage	⑰	VDC
Autonomy	⑱	days
Required Battery-Bank Output	⑲	Ah
Allowable Depth-of-Discharge	⑳	
Weighted Operating Time	⑩	hrs
Discharge Rate	㉑	hrs
Minimum Expected Operating Temperature	㉒	°C
Temperature/Discharge Rate Derating Factor	㉓	
Battery-Bank Rated Capacity	㉔	Ah
Selected Battery Nominal Voltage	㉕	VDC
Selected Battery Rated Capacity	㉖	Ah
Number of Batteries in Series	㉗	
Number of Battery Strings in Parallel	㉘	
Total Number of Batteries	㉙	
Actual Battery-Bank Rated Capacity	㉚	Ah
Load Fraction	㉛	
Average Daily Depth-of-Discharge	㉜	

㉘ $n_{parallel}$ Calculate the number of battery strings required to be connected in parallel. Round the result up to the nearest whole number.

$$n_{parallel} = \frac{B_{rated}}{B_{batt}}$$

$$㉘ = \frac{㉔}{㉖}$$

㉙ n_{batt} Calculate the total number of batteries required for the battery bank.

$$n_{batt} = n_{series} \times n_{parallel}$$

$$㉙ = ㉗ \times ㉘$$

㉚ B_{actual} Calculate the actual battery-bank rated capacity.

$$B_{actual} = B_{batt} \times n_{parallel}$$

$$㉚ = ㉖ \times ㉘$$

㉛ LF Determine the load fraction for the battery-bank operation. A typical value is 0.75

㉜ DOD_{avg} Calculate the average daily depth of discharge.

$$DOD_{avg} = \frac{LF \times E_{day}}{B_{rated} \times V_{SDC}}$$

$$㉜ = \frac{㉛ \times ⑮}{㉔ \times ⑰}$$

STAND-ALONE SYSTEM SIZING

Array Sizing

15 E_{crit} Enter the total required daily system DC energy for the critical design month. This information is from the critical design analysis.

16 t_{PSH} Enter the average daily insolation, in peak sun hours, for the selected array orientation during the critical design month. This information is from the critical design analysis.

17 V_{SDC} Enter the nominal DC-system voltage.

33 η_{batt} Enter battery charging efficiency. Typical values for most battery systems range from 85% to 95% (0.85 to 0.95).

34 I_{array} Calculate the required array maximum-power current.

$$I_{array} = \frac{E_{crit}}{\eta_{batt} \times V_{SDC} \times t_{PSH}}$$

$$\boxed{34} = \frac{\boxed{15}}{\boxed{33} \times \boxed{17} \times \boxed{16}}$$

35 C_S Enter the derating factor for array soiling. Typical values for most systems range from 90% to 95% (0.90 to 0.95).

36 I_{rated} Calculate the required array maximum-power current rating.

$$I_{rated} = \frac{I_{array}}{C_S}$$

$$\boxed{36} = \frac{\boxed{34}}{\boxed{35}}$$

37 $C_{\%V}$ Enter the relative temperature coefficient for voltage. This value may be given by the module manufacturer. Otherwise, a typical value is $-0.4\%/°C$ ($-0.004/°C$).

38 T_{max} Enter the maximum expected ambient temperature for the array location. This information can be found in the solar radiation data set for the nearest location.

39 T_{ref} Enter the reference temperature associated with the maximum-power voltage rating. This value is typically 25°C.

40 V_{rated} Calculate the required array maximum-power voltage rating.

$$V_{rated} = 1.2 \times \{V_{SDC} + [V_{SDC} \times C_{\%V} \times (T_{max} - T_{ref})]\}$$

$$\boxed{40} = 1.2 \times \{\boxed{17} + [\boxed{17} \times \boxed{37} \times (\boxed{38} - \boxed{39})]\}$$

41 I_{mp} Enter the maximum-power current rating for the selected module.

42 V_{mp} Enter the maximum-power voltage rating for the selected module.

43 P_{mp} Enter the maximum power rating for the selected module.

Array Sizing Worksheet

Average Daily DC Energy Consumption for Critical Design Month	**15**	Wh/day
Critical Design Month Insolation	**16**	PSH/day
DC System Voltage	**17**	VDC
Battery Charging Efficiency	**33**	
Required Array Maximum-Power Current	**34**	A
Soiling Factor	**35**	
Rated Array Maximum-Power Current	**36**	A
Temperature Coefficient for Voltage	**37**	/°C
Maximum Expected Module Temperature	**38**	°C
Rating Reference Temperature	**39**	°C
Rated Array Maximum-Power Voltage	**40**	VDC
Module Rated Maximum-Power Current	**41**	A
Module Rated Maximum-Power Voltage	**42**	VDC
Module Rated Maximum Power	**43**	W
Number of Modules in Series	**44**	
Number of Module Strings in Parallel	**45**	
Total Number of Modules	**46**	
Actual Array Rated Power	**47**	W

44 n_{series} Calculate the number of modules required to be connected in series. Round the result up to the nearest whole number.

$$n_{series} = \frac{V_{rated}}{V_{mp}}$$

$$\boxed{44} = \frac{\boxed{40}}{\boxed{42}}$$

45 $n_{parallel}$ Calculate the number of module strings required to be connected in parallel. Round the result up to the nearest whole number.

$$n_{parallel} = \frac{I_{rated}}{I_{mp}}$$

$$\boxed{45} = \frac{\boxed{36}}{\boxed{41}}$$

46 n_{mod} Calculate the total number of modules in the array.

$$n_{mod} = n_{series} \times n_{parallel}$$

$$\boxed{46} = \boxed{44} \times \boxed{45}$$

47 P_{actual} Calculate the actual array rated power.

$$P_{actual} = P_{mp} \times n_{mod}$$

$$\boxed{47} = \boxed{43} \times \boxed{46}$$

ELECTRICAL ABBREVIATIONS

A	Ampere		k	Kilo (1000)
A/C	Air Conditioner		kcmil	1000 Circular Mils
AC	Alternating Current		kFT	1000 Feet
AEGCP	Assured Equipment Grounding Program		kVA	Kilovolt-Ampere
A/H	Air Handler		kW	Kilowatt
AHCC	Ambulatory Health Care Center		kWh	Kilowatt-Hour
AHJ	Authority Having Jurisdiction		L	Line
AIR	Ampere Interrupting Rating		LFLI	Less-Flammable, Liquid-Insulated
Al	Aluminum		LP	Lighting Panel
ANSI	American National Standards Institute		LRA	Locked-Rotor Ampacity
ATCB	Adjustable-Trip Circuit Breaker		LRC	Locked-Rotor Current
AWG	American Wire Gauge		MBCSCGF	Motor Branch-Circuit, Short-Circuit, Ground-Fault
BJ	Bonding Jumper		MBJ	Main Bonding Jumper
C	Celsius		MCC	Main Control Center
CATV	Cable Antenna Television		N	Neutral
CB	Circuit Breaker		NATCB	Nonadjustable-Trip Circuit Breaker
CH	Chapter		NEC®	National Electrical Code®
CM	Circular Mils		NEMA	National Electrical Manufacturers Association
CMP	Code-making Panel		NESC	National Electrical Safety Code
CU	Copper		NFPA	National Fire Protection Association
DC	Direct Current		NO.	Number
DIA	Diameter		NTDF	Non-Time Delay Fuse
DP	Double Pole		OCPD	Overcurrent Protection Device
E	Voltage		OL	Overload(s)
EBJ	Equipment Bonding Jumper		OSHA	Occupational Safety and Health Administration
E_{ff}	Efficiency		P	Power
EGC	Equipment Grounding Conductor		PC	Personal Computer
EX.	Exception		PF	Power Factor
F	Fahrenheit		R	Resistance; Resistor
FLA	Full-Load Amperes		RMS	Root Mean Square
FLC	Full-Load Current		ROC	Receipt of Comments
FPN	Fine Print Note		ROP	Receipt of Proposals
FR	Frame		SF	Service Factor
G	Ground		SP	Single-Pole
GEC	Grounding Electrode Conductor		SPCB	Single-Pole Circuit Breaker
GES	Grounding Electrode System		SQ FT	Square Foot (Feet)
GFCI	Ground Fault Circuit Interrupter		SWD	Switched Disconnect
GFPE	Ground Fault Protection of Equipment		T	Time
GR	Green		TDF	Time-Delay Fuse
HACR	Heating, Air-Conditioning, Refrigeration		TP	Thermally Protected
HP	Horsepower		UF	Underground Feeder
HRS	Hours		UL	Underwriter's Laboratory
I	Current		V	Volts
IDCI	Immersion Detection Circuit Interrupter		VA	Volt-Ampere
IG	Isolated Ground		VAC	Volts Alternating Current
IN.	Inch		VD	Voltage Drop
ITB	Instantaneous-Trip Circuit Breaker		W	Watts
ITCB	Inverse-Time Circuit Breaker		W	White
K	Conductor Resistivity		WP	Weatherproof

METRIC TO ENGLISH EQUIVALENTS

	Unit	British Equivalent		
LENGTH	kilometer	0.62 mi		
	hectometer	109.36 yd		
	dekameter	32.81″		
	meter	39.37″		
	decimeter	3.94″		
	centimeter	0.39″		
	millimeter	0.039″		
AREA	square kilometer	0.3861 sq mi		
	hectacre	2.47 A		
	acre	119.60 sq yd		
	square centimeter	0.155 sq in.		
VOLUME	cubic centimeter	0.061 cu in.		
	cubic decimeter	61.023 cu in.		
	cubic meter	1.307 cu yd		
		cubic	*dry*	*liquid*
CAPACITY	kiloliter	1.31 cu yd		
	hectoliter	3.53 cu ft	2.84 bu	
	dekaliter	0.35 cu ft	1.14 pk	2.64 gal.
	liter	61.02 cu in.	0.908 qt	1.057 qt
	cubic decimeter	61.02 cu in.	0.908 qt	1.057 qt
	deciliter	6.1 cu in.	0.18 pt	0.21 pt
	centiliter	0.61 cu in.		338 fl oz
	milliliter	0.061 cu in.		0.27 fl dr
MASS AND WEIGHT	metric ton	1.102 t		
	kilogram	2.2046 lb		
	hectogram	3.527 oz		
	dekagram	0.353 oz		
	gram	0.353 oz		
	decigram	1.543 gr		
	centigram	0.154 gr		
	milligram	0.015 gr		

ENGLISH TO METRIC EQUIVALENTS

		Unit	Metric Equivalent
LENGTH		mile	1.609 km
		rod	5.029 m
		yard	0.9144 m
		foot	30.48 cm
		inch	2.54 cm
AREA		square mile	2.590 km^2
		acre	.405 hectacre, 4047 m^2
		square rod	25.293 m^2
		square yard	0.836 m^2
		square foot	0.093 m^2
		square inch	6.452 cm^2
VOLUME		cubic yard	0.765 m^3
		cubic foot	0.028 m^3
		cubic inch	16.387 cm^3
CAPACITY	*U.S. liquid measure*	gallon	3.785 l
		quart	0.946 l
		pint	0.473 l
		gill	118.294 ml
		fluidounce	29.573 ml
		fluidram	3.697 ml
		minim	0.61610 ml
	U.S. dry measure	bushel	35.239 l
		peck	8.810 l
		quart	1.101 l
		pint	0.551 l
	British imperial liquid and dry measure	bushel	0.036 m^3
		peck	0.0091 m^3
		gallon	4.546 l
		quart	1.136 l
		pint	568.26 cm^3
		gill	142.066 cm^3
		fluidounce	28.412 cm^3
		fluidram	3.5516 cm^3
		minim	0.59194 cm^3
MASS AND WEIGHT	*avoirdupois*	short ton	0.907 t
		long ton	1.016 t
		pound	0.454 kg
		ounce	28.350 g
		dram	1.772 g
		grain	0.0648 g
	troy	pound	0.373 kg
		ounce	31.103 g
		pennyweight	1.555 g
		grain	0.648 g
	apothecaries'	pound	0.373 kg
		ounce	31.103 g
		dram	3.888 g
		scruple	1.296 g
		grain	0.648 g

METRIC PREFIXES

Multiple	Prefixes	Symbols	Meaning
$1,000,000,000,000 = 10^{12}$	tera	T	trillion
$1,000,000,000 = 10^{9}$	giga	G	billion
$1,000,000 = 10^{6}$	mega	M	million
$1000 = 10^{3}$	kilo	k	thousand
$100 = 10^{2}$	hecto	h	hundred
$10 = 10^{1}$	deka	da	ten
$1 = 10^{0}$	——	——	unit
$0.1 = 10^{-1}$	deci	d	tenth
$0.01 = 10^{-2}$	centi	c	hundredth
$0.001 = 10^{-3}$	milli	m	thousandth
$0.000001 = 10^{-6}$	micro	μ	millionth
$0.000000001 = 10^{-9}$	nano	n	billionth
$0.000000000001 = 10^{-12}$	pico	p	trillionth

METRIC CONVERSIONS

Initial Units	Final Units											
	giga	mega	kilo	hecto	deka	base unit	deci	centi	milli	micro	nano	pico
giga		3R	6R	7R	8R	9R	10R	11R	12R	15R	18R	21R
mega	3L		3R	4R	5R	6R	7R	8R	9R	12R	15R	18R
kilo	6L	3L		1R	2R	3R	4R	5R	6R	9R	12R	15R
hecto	7L	4L	1L		1R	2R	3R	4R	5R	8R	11R	14R
deka	8L	5L	2L	1L		1R	2R	3R	4R	7R	10R	13R
base unit	9L	6L	3L	2L	1L		1R	2R	3R	6R	9R	12R
deci	10L	7L	4L	3L	2L	1L		1R	2R	5R	8R	11R
centi	11L	8L	5L	4L	3L	2L	1L		1R	4R	7R	10R
milli	12L	9L	6L	5L	4L	3L	2L	1L		3R	6R	9R
micro	15L	12L	9L	8L	7L	6L	5L	4L	3L		3R	6R
nano	18L	15L	12L	11L	10L	9L	8L	7L	6L	3L		3H
pico	21L	18L	15L	14L	13L	12L	11L	10L	9L	6L	3L	

THREE-PHASE VOLTAGE VALUES
For 208 V × 1.732, use 360
For 230 V × 1.732, use 398
For 240 V × 1.732, use 416
For 440 V × 1.732, use 762
For 460 V × 1.732, use 797
For 480 V × 1.732, use 831
For 2400 V × 1.732, use 4157
For 4160 V × 1.732, use 7205

Ohm's Law

Ohm's law is the relationship between the voltage, current, and resistance in an electrical circuit. Ohm's law states that current in a circuit is proportional to the voltage and inversely proportional to the resistance.

Power Formula

The power formula is the relationship between the voltage, current, and power in an electrical circuit. The power formula states that the power in a circuit is proportional to the voltage and the current.

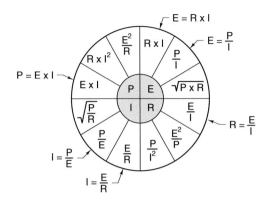

VALUES IN INNER CIRCLE
ARE EQUAL TO VALUES
IN CORRESPONDING
OUTER CIRCLE

OHM'S LAW AND POWER FORMULA

POWER FORMULAS — 1ϕ, 3ϕ					
Phase	To Find	Use Formula	Example		
			Given	Find	Solution
1ϕ	I	$I = \dfrac{VA}{V}$	32,000 VA, 240 V	I	$I = \dfrac{VA}{V}$ $I = \dfrac{32,000 \text{ VA}}{240 \text{ V}}$ $I = $ **133 A**
1ϕ	VA	$VA = I \times V$	100 A, 240 V	VA	$VA = I \times V$ $VA = 100 \text{ A} \times 240 \text{ V}$ $VA = $ **24,000 VA**
1ϕ	V	$V = \dfrac{VA}{I}$	42,000 VA, 350 A	V	$V = \dfrac{VA}{I}$ $V = \dfrac{42,000 \text{ VA}}{350 \text{ A}}$ $V = $ **120 V**
3ϕ	I	$I = \dfrac{VA}{V \times \sqrt{3}}$	72,000 VA, 208 V	I	$I = \dfrac{VA}{V \times \sqrt{3}}$ $I = \dfrac{72,000 \text{ VA}}{360 \text{ V}}$ $I = $ **200 A**
3ϕ	VA	$VA = I \times V \times \sqrt{3}$	2 A, 240 V	VA	$VA = I \times V \times \sqrt{3}$ $VA = 2 \times 416$ $VA = $ **832 VA**

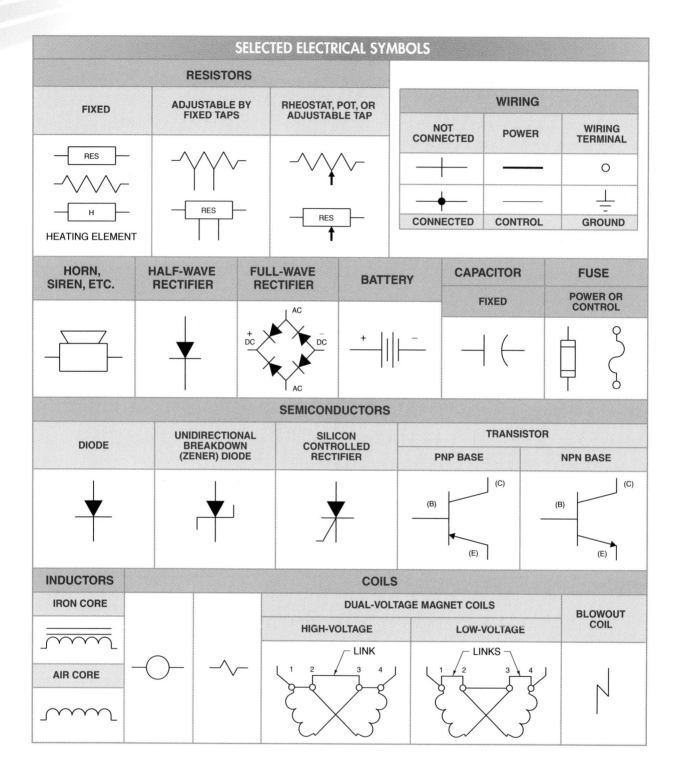

SELECTED ELECTRICAL SYMBOLS

ENCLOSURE TYPES

Type	Use	Service Conditions	Tests	Comments
1	Indoor	No unusual	Rod entry, rust resistance	
3	Outdoor	Windblown dust, rain, sleet, and ice on enclosure	Rain, external icing, dust, and rust resistance	Do not provide protection against internal condensation or internal icing
3R	Outdoor	Falling rain and ice on enclosure	Rod entry, rain, external icing, and rust resistance	Do not provide protection against dust, internal condensation, or internal icing
4	Indoor/outdoor	Windblown dust and rain, splashing water, hose-directed water, and ice on enclosure	Hosedown, external icing, and rust resistance	Do not provide protection against internal condensation or internal icing
4X	Indoor/outdoor	Corrosion, windblown dust and rain, splashing water, hose-directed water, and ice on enclosure	Hosedown, external icing, and corrosion resistance	Do not provide protection against internal condensation or internal icing
6	Indoor/outdoor	Occasional temporary submersion at a limited depth		
6P	Indoor/outdoor	Prolonged submersion at a limited depth		
7	Indoor locations classified as Class I, Groups A, B, C, or D, as defined in the NEC®	Withstand and contain an internal explosion of specified gases, contain an explosion sufficiently so an explosive gas-air mixture in the atmosphere is not ignited	Explosion, hydrostatic, and temperature	Enclosed heat-generating devices shall not cause external surfaces to reach temperatures capable of igniting explosive gas-air mixtures in the atmosphere.
9	Indoor locations classified as Class II, Groups E or G, as defined in the NEC®	Dust	Dust penetration, temperature, and gasket aging	Enclosed heat-generating devices shall not cause external surfaces to reach temperatures capable of igniting explosive gas-air mixtures in the atmosphere
12	Indoor	Dust, falling dirt, and dripping noncorrosive liquids	Drip, dust, and rust resistance	Do not provide protection against internal condensation
13	Indoor	Dust, spraying water, oil, and noncorrosive coolant	Oil explosion and rust resistance	Do not provide protection against internal condensation

AC/DC FORMULAS

To Find	DC	AC		
		1φ, 115 or 220V	1φ, 208, 230, or 240 V	3φ—All Voltages
I, HP known	$\dfrac{HP \times 746}{E \times E_{ff}}$	$\dfrac{HP \times 746}{E \times E_{ff} \times PF}$	$\dfrac{HP \times 746}{E \times E_{ff} \times PF}$	$\dfrac{HP \times 746}{1.73 \times E \times E_{ff} \times PF}$
I, kW known	$\dfrac{kW \times 1000}{E}$	$\dfrac{kW \times 1000}{E \times PF}$	$\dfrac{kW \times 1000}{E \times PF}$	$\dfrac{kW \times 1000}{1.73 \times E \times PF}$
I, kVA known		$\dfrac{kVA \times 1000}{E}$	$\dfrac{kVA \times 1000}{E}$	$\dfrac{kVA \times 1000}{1.763 \times E}$
kW	$\dfrac{I \times E}{1000}$	$\dfrac{I \times E \times PF}{1000}$	$\dfrac{I \times E \times PF}{1000}$	$\dfrac{I \times E \times 1.73 \times PF}{1000}$
kVA		$\dfrac{I \times E}{1000}$	$\dfrac{I \times E}{1000}$	$\dfrac{I \times E \times 1.73}{1000}$
HP (output)	$\dfrac{I \times E \times E_{ff}}{746}$	$\dfrac{I \times E \times E_{ff} \times PF}{746}$	$\dfrac{I \times E \times E_{ff} \times PF}{746}$	$\dfrac{I \times E \times 1.73 \times E_{ff} \times PF}{746}$

E_{ff} = efficiency

VOLTAGE UNBALANCE

$$V_u = \frac{V_d}{V_a} \times 100$$

where
V_u = voltage unbalance (%)
V_d = voltage deviation (V)
V_a = voltage average (V)
100 = constant

STANDARD SIZES OF FUSES AND CBs

NEC® 240-6(a) lists standard ampere ratings of fuses and fixed-trip CBs as follows:
15, 20, 25, 30, 35, 40, 45,
50, 60, 70, 80, 90, 100, 110,
125, 150, 175, 200, 225,
250, 300, 350, 400, 450,
500, 600, 700, 800,
1000, 1200, 1600,
2000, 2500, 3000, 4000, 5000, 6000

RACEWAYS

EMT	Electrical Metallic Tubing
ENT	Electrical Nonmetallic Tubing
FMC	Flexible Metal Conduit
FMT	Flexible Metallic Tubing
IMC	Intermediate Metal Conduit
LFMC	Liquidtight Flexible Metal Conduit
LFNC	Liquidtight Flexible Nonmetallic Conduit
RMC	Rigid Metal Conduit
RNC	Rigid Nonmetallic Conduit

Glossary

A

absorption charging: Battery charging following bulk charging that reduces the charge current to maintain the battery voltage at a regulation voltage for a certain period.

AC bus hybrid system: A hybrid system that supplies loads with AC power from multiple energy sources.

AC module: A PV module that outputs AC power through an interactive inverter attached in place of the normal DC junction box.

active material: The chemically reactive compound on a battery cell plate.

active tracking mount: An array mounting system that uses electric motors and gear drives to automatically direct the array toward the sun.

adjustable mounting system: A variation of a fixed-tilt array mounting system that permits manual adjustment of the tilt and/or azimuth angles to increase the array output.

air mass (AM): A representation of the relative thickness of atmosphere that solar radiation must pass through to reach a point on Earth's surface.

allowable depth of discharge: The maximum percentage of total capacity that is permitted to be withdrawn from a battery.

allowable withdrawal load: The force required to remove a screw from a material by tensile (pulling) force only.

alternating current (AC): Electrical current that changes between positive and negative directions.

ampacity: The current that a conductor can carry continuously under the conditions of use without exceeding its temperature rating.

ampere-hour integrating charge controller: A charge controller that counts the total amount of charge (in ampere-hours) into and out of a battery and regulates charging current based on a preset amount of overcharge.

analemma: A diagram of solar declination against the Equation of Time.

annualized cost: A portion of the total life-cycle cost attributed equally to each year of operation.

apparent power: A combination of true and reactive power and is given in units of volt-amperes (VA).

array: A complete PV power-generating unit consisting of a number of individual electrically and mechanically integrated modules with structural supports, trackers, or other components.

array azimuth angle: The horizontal angle between a reference direction and the direction an array surface faces.

array reconnect voltage (ARV) setpoint: The voltage at which an interrupting-type charge controller reconnects the array to the battery and resumes charging.

array tilt angle: The vertical angle between horizontal and the array surface.

authority having jurisdiction (AHJ): An organization, office, or individual designated by local government with legal powers to administer, interpret, and enforce building codes.

autonomy: The amount of time a fully charged battery system can supply power to system loads without further charging.

average daily depth of discharge: The average percentage of the total capacity that is withdrawn from a battery each day.

avoided cost: The cost that a utility would normally incur to generate a given amount of power, often synonymous with the wholesale market value of electricity.

B

back-fed circuit breaker: A circuit breaker that allows current flow in either direction.

balance-of-system (BOS) component: An electrical or structural component, aside from a major component, that is required to complete a PV system.

basic wind speed: The maximum value of a 3 sec gust at 33′ (10 m) elevation, which is used in wind load calculations.

battery: A collection of electrochemical cells that are contained in the same case and connected together electrically to produce a desired voltage.

battery bank: A group of batteries connected together with series and parallel connections to provide a specific voltage and capacity.

battery cell: The basic unit in a battery that stores electrical energy in chemical bonds and delivers this energy through chemical reactions.

battery load tester: A test instrument that indicates battery health by drawing a high discharge current from a battery for a short period.

bimodal system: A PV system that can operate in either utility-interactive or stand-alone mode, and uses battery storage.

blocking: The addition of lumber under a roof surface and between trusses or rafters as supplemental structural support.

blocking diode: A diode used in PV source circuits to prevent reverse current flow.

breakdown voltage: The minimum reverse-bias voltage that results in a rapid increase in current through an electronic device.

building-integrated photovoltaic (BIPV) array: A fixed array that replaces conventional building materials with specially designed modules that perform an architectural function in addition to producing power.

building code: A set of regulations that prescribes the materials, standards, and methods to be used in the construction, maintenance, and repair of buildings and other structures.

bulk charging: The battery charging at a relatively high charge rate that charges the battery up to a regulation voltage, resulting in a state of charge of about 80% to 90%.

bypass diode: A diode used to pass current around, rather than through, a group of PV cells.

C

capacity: The measure of the electrical energy storage potential of a cell or battery.

captive electrolyte: Electrolyte that is immobilized.

catalyst: A substance that causes other substances to chemically react but does not participate in the reaction.

catalytic recombination cap (CRC): A vent cap that reduces electrolyte loss from an open-vent flooded battery by recombining vented gases into water.

charge acceptance: The ratio of the increase in battery charge to the amount of charge applied to the battery.

charge control algorithm: A programmed series of functions that a charge controller uses to control current and/or voltage in order to maintain battery state of charge.

charge controller: A device that regulates battery charge by controlling the charging voltage and/or current from a DC power source, such as a PV array.

charge controller setpoint: A battery condition, commonly the voltage, at which a charge controller performs regulation or switching functions.

charger: A device that combines a rectifier with filters, transformers, and other components to condition DC power for the purpose of battery charging.

charging: The process of a cell or battery receiving current and converting the electrical energy into chemical energy.

circuit breaker: An electrical switch that automatically opens as a means of overcurrent protection, and that can be manually opened as a disconnecting means.

clamping voltage: The voltage at which a surge arrestor initiates its transient protection.

combiner box: A junction box used as the parallel connection point for two or more circuits.

commissioning: The starting and operation of a PV system for the first time.

concentrating collector: A solar energy collector that enhances solar energy by focusing it on a smaller area through reflective surfaces or lenses.

concentrating solar power (CSP): A technology that uses mirrors and lenses to reflect and concentrate solar radiation from a large area onto a small area.

construction bond: A contract in which a surety company assures that a contractor will complete their work in accordance with contracting laws.

cost analysis: A comparison of two or more options that considers both the cost and value of each.

critical design ratio: The ratio of electrical energy demand to average insolation during a period.

critical design month: The month with the highest critical design ratio.

current unbalance: The unbalance that occurs when current is not equal on the three power lines of a three-phase system.

current-voltage (I-V) characteristic: The basic electrical output profile of a PV device.

cutoff voltage: The minimum battery voltage specified by the manufacturer that establishes the battery capacity at a specific discharge rate.

cycle: 1. A battery discharge followed by a charge. **2.** The interval of time between the beginnings of each waveform pattern.

D

data acquisition: The recording and processing of data from a monitoring system.

DC bus hybrid system: A hybrid system that combines DC power output from all energy sources, including the PV array, for charging the battery bank.

DC-DC converter: A device that converts DC power from one voltage to another.

dead load: A static structural load due to the weight of permanent building members, supported structure, and attachments.

dedicated space: The clear space reserved around electrical equipment for the existing equipment and potential future additions.

deficit charge: A charge cycle in which less charge is returned to a battery bank than what was withdrawn on discharge.

delamination: The separation of the bonded layers of glass and/or plastic encasing the PV cells of a module.

depth of discharge (DOD): The percentage of withdrawn energy in a battery compared to the fully charged capacity.

design load: A calculated structural load used to evaluate the strength of a structure to failure.

diffuse radiation: Solar radiation that is scattered by the atmosphere and clouds.

direct-coupled PV system: A type of stand-alone system where the output of a PV module or array is directly connected to a DC load.

direct current (DC): Electrical current that flows in one direction, either positive or negative.

direct mount: A type of fixed-tilt array mounting system where modules are affixed directly to an existing finished rooftop or other building surface, with little or no space between a module and the surface.

direct radiation: Solar radiation directly from the sun that reaches Earth's surface without scattering.

discharging: The process of a cell or battery converting chemical energy to electrical energy and delivering current.

disconnect: A device used to isolate equipment and conductors from sources of electricity for the purpose of installation, maintenance, or service.

discount rate: The rate at which the value of future money is reduced to its present value.

distributed generation: A system in which many smaller power-generating systems create electrical power near the point of consumption.

diversionary charge controller: A charge controller that regulates charging current to a battery system by diverting excess power to an auxiliary load.

diversion load: An auxiliary load that is not a critical system load, but is always available to utilize the full array power in a useful way to protect a battery from overcharge.

doping: The process of adding small amounts of impurity elements to semiconductors to alter their electrical properties.

dual-axis tracking: A sun-tracking system that rotates two axes independently to exactly follow the position of the sun.

dual metering: The arrangement that measures energy exported to and imported from the utility grid separately.

duty cycle: The percentage of time a load is operating.

E

ecliptic plane: The plane of Earth's orbit around the sun.

efficiency: The ratio of power output to power input.

electrical inspector: A local official qualified to evaluate electrical installations in the field for compliance with applicable codes and standards.

electrolyzer: An electrochemical device that uses electricity to split water into hydrogen and oxygen.

electrolyte: The conducting medium that allows the transfer of ions between battery cell plates.

electromagnetic radiation: Radiation in the form of waves with electric and magnetic properties.

electromagnetic spectrum: The range of all types of electromagnetic radiation, based on wavelength.

energy audit: A collection of information about a facility's or customer's energy use.

energy cost: An expense related to the energy input of an electricity-generating system.

energy payback: The amount of time a PV system must operate to produce enough energy to offset the energy required to manufacture the PV system.

engine generator: A combination of an internal combustion engine and a generator mounted together to produce electricity.

equalizing charging: Current-limited battery charging to a voltage higher than the bulk charging voltage, which brings each cell to a full state of charge.

Equation of Time: The difference between solar time and standard time at a standard meridian.

equatorial plane: The plane containing the earth's equator and extending outward into space.

equinox: Earth's orbital position when solar declination is zero.

equipment grounding conductor (EGC): A conductor connecting exposed metallic equipment, which might inadvertently become energized, to the grounding electrode conductor.

exposure: A wind load factor that accounts for the array height and the characteristics of the surrounding terrain.

extraterrestrial solar radiation: Solar radiation just outside Earth's atmosphere.

F

family of I-V curves: A group of I-V curves at various irradiance levels.

fill factor (FF): The ratio of maximum power to the product of the open-circuit voltage and short-circuit current.

fixed-tilt mounting system: An array mounting system that permanently secures modules in a nonmovable position at a specific tilt angle.

flat-plate collector: A solar energy collector that absorbs solar energy on a flat surface without concentrating it, and can utilize solar radiation directly from the sun as well as diffuse radiation that is reflected or scattered by clouds and other surfaces.

float charging: Battery charging at a low charge rate that maintains full battery charge by counteracting self-discharge.

flooded electrolyte: Electrolyte in the form of a liquid.

frequency: The number of waveform cycles in one second.

fuel cell: An electrochemical device that uses hydrogen and oxygen to produce DC electricity, with water and heat as byproducts.

fuse: A metallic link that melts when heated by current greater than its rating, opening the connection and providing overcurrent protection.

G

galvanic corrosion: An electrochemical process that causes electrical current to flow between two dissimilar metals, which eventually corrodes one of the materials (the anode).

gassing: The decomposition of water into hydrogen and oxygen gases as the battery charges.

gassing voltage: The voltage level at which battery gassing begins.

gas turbine: A device that compresses and burns a fuel-air mixture, which expands and spins a turbine.

generator: A device that converts mechanical energy into electricity by means of electromagnetic induction.

grant: A one-time monetary payment for certain types of projects.

grid: 1. The utility's network of conductors, substations, and equipment that distributes electricity from its central generation point to the consumer. **2.** A metal framework that supports the active material of a battery cell and conducts electricity.

grounded: The condition of something that is connected to the earth or to a conductive material that is connected to the earth.

grounded conductor: A current-carrying conductor that is intentionally grounded.

ground fault: The undesirable condition of current flowing through the grounding conductor.

ground-fault circuit interrupter (GFCI): A device that opens the ungrounded and grounded conductors when a ground fault exceeds a certain amount, typically 4 mA to 6 mA.

ground-fault protection: The automatic opening of conductors involved in a ground fault.

grounding electrode: A conductive rod, plate, or wire buried in the ground to provide a low-resistance connection to the earth.

grounding electrode conductor (GEC): The conductor connecting the grounding electrode to the rest of the electrical grounding system.

H

harmonic: A waveform component at an integer multiple of the fundamental waveform frequency.

H-bridge inverter circuit: A circuit that switches DC input into square wave AC output by using two pairs of switching devices.

hybrid battery: A battery that uses a combination of plate designs to maximize the desirable characteristics of each.

hybrid system: A stand-alone system that includes two or more distributed energy sources.

hybrid system controller: A controller with advanced features for managing multiple energy sources.

hydrometer: An instrument used to measure the specific gravity of a liquid.

I

incentive: A monetary inducement to invest in a certain type of capital improvement, such as an energy-generating system or energy-conservation measure.

incidence angle: The angle between the sun's rays and a line perpendicular to the array surface.

initial cost: An expense related to the design, engineering, equipment, and installation of a system.

insolation: The solar irradiation received over a period of time, typically one day.

installed nominal operating cell temperature (INOCT): The estimated temperature of a module operating in a specific mounting system design.

integrator: A business that designs, builds, and installs complete PV systems for particular applications by matching components from various manufacturers.

interconnection agreement: A contract between a distributed-power producer and an electric utility that establishes the terms and conditions for the interconnection.

interrupting rating: The maximum current that an overcurrent protection device is able to stop without being destroyed or causing an electric arc.

interrupting-type charge controller: A charge controller that switches the charging current ON and OFF for charge regulation.

inverse square law: A physical law that states that the amount of radiation is proportional to the inverse of the square of the distance from the source.

inverter: A device that converts DC power to AC power.

inverter efficiency: The effectiveness of an inverter at converting DC power to AC power.

islanding: The undesirable condition where a distributed-generation power source, such as a PV system, continues to transfer power to the utility grid during a utility outage.

I-V curve: The graphic representation of all possible voltage and current operating points for a PV device at a specific operating condition.

J

J-bolt: A fastener that hooks around a secure support structure and has a threaded end that is used with a nut to secure items.

junction box: A protective enclosure used to terminate, combine, and connect various circuits or components together.

L

life-cycle cost: The total cost of all the expenses incurred over the life of an electricity-generating system.

linear-type charge controller: A charge controller that limits the charging current in a linear or gradual manner with high-speed switching or linear control.

line-commutated inverter: An inverter whose switching devices are triggered by an external source.

listing: The process used by an NTRL for certifying that equipment or materials meet applicable standards.

live load: A dynamic structural load due to the weight of temporary items and people using or occupying the structure.

load: A piece of equipment that consumes electricity.

load fraction: The portion of load operating power that comes from the battery bank over the course of a day.

load reconnect voltage (LRV) setpoint: The voltage at which a charge controller reconnects loads to the battery system.

low-voltage disconnect hysteresis (LVDH): The voltage difference between the low-voltage disconnect (LVD) and load reconnect voltage (LRV) setpoints.

low-voltage disconnect (LVD) setpoint: The voltage that triggers the disconnection of system loads to prevent battery overdischarge because it is the minimum voltage a battery is allowed to reach under normal operating conditions.

M

magnetic declination: The angle between the direction a compass needle points (toward magnetic north) and true geographic north.

maintenance cost: An expense related to the operation and maintenance of a system.

maintenance log: A collection of past maintenance records.

maintenance plan: A checklist of all required regular maintenance tasks and their recommended intervals.

maximum power current (I_{mp}): The operating current on an I-V curve where the power output is at maximum.

maximum power point (P_{mp}): The operating point on an I-V curve where the product of current and voltage is at maximum.

maximum power point tracker (MPPT): A device or circuit that uses electronics to continually adjust the load on a PV device under changing temperature and irradiance conditions to keep it operating at its maximum power point.

maximum power point tracking (MPPT) charge controller: A charge controller that operates the array at its maximum power point under a range of operating conditions, as well as regulates battery charging.

maximum power voltage (V_{mp}): The operating voltage on an I-V curve where the power output is at maximum.

meridian: A plane formed by a due north-south longitude line through a location on Earth and projected out into space.

micro-hydroelectric turbine: A device that produces electricity from the flow and pressure of water.

model building code: A building code that is developed and revised by a standards organization independently of the adopting jurisdictions.

modified square wave: A synthesized, stepped waveform that approximates a true sine wave.

module: A PV device consisting of a number of individual cells connected electrically, laminated, encapsulated, and packaged into a frame.

monitoring: The repeated measurement of electrical, environmental, or battery parameters at certain intervals.

monocrystalline wafer: A silicon wafer made from a single silicon crystal grown in the form of a cylindrical ingot.

multijunction cell: A cell that maximizes efficiency by using layers of individual cells that each respond to different wavelengths of solar energy.

N

nationally recognized testing laboratory (NRTL): An OSHA-recognized, accredited safety testing organization that certifies equipment or materials to meet applicable standards.

net metering: A metering arrangement where any excess energy exported to the utility is subtracted from the amount of energy imported from it.

nominal operating cell temperature (NOCT): A reference temperature of an open-circuited module based on an irradiance level of 800 W/m², ambient temperature of 20°C (68°F), and wind speed of 1 m/s.

nominal operating conditions (NOC): A set of reference conditions that rates module performance at a solar irradiance of 800 W/m², spectral conditions of AM1.5, and at nominal operating cell temperature.

n-type semiconductor: A semiconductor that has free electrons.

O

open-circuit voltage (V_{oc}): 1. The maximum voltage on an I-V curve and is the operating point for a PV device under infinite load or open-circuit condition, and no current output. **2.** The voltage of a battery or cell when it is at steady-state.

overcharge: 1. The ratio of applied charge to the resulting increase in battery charge. **2.** The condition of a fully charged battery continuing to receive a significant charging current.

overcurrent protection device: A device that prevents conductors or devices from reaching excessively high temperatures under high-current loads by opening the circuit.

overdischarge: The condition of a battery state of charge declining to the point where it can no longer supply discharge current at a sufficient voltage without damaging the battery.

P

passive tracking mount: An array mounting system that uses nonelectrical means to automatically direct an array toward the sun.

payback period: The amount of time (typically in years) required for the avoided costs from an alternate energy source to match the cost of the chosen energy-production system.

payback point: The point at which the accumulated life-cycle cost of one system option matches the total life-cycle cost of the other system option.

peak: The maximum absolute value of a waveform.

peak sun: An estimate of maximum terrestrial solar irradiance around solar noon at sea level and has a generally accepted value of 1000 W/m².

peak sun hours: The number of hours required for a day's total solar irradiation to accumulate at peak sun condition.

peak-to-peak: A measure of the difference between positive and negative maximum values of a waveform.

period: The time it takes a periodic waveform to complete one full cycle before it repeats.

periodic waveform: A waveform that repeats the same pattern at regular intervals.

permit: Permission from the AHJ that authorizes construction work to begin and establishes the inspection requirements, but does not represent an approval of compliance with codes and standards.

phase unbalance: The unbalance that occurs when three-phase power lines are more or less than 120° out of phase.

photoelectrochemical cell: A cell that relies on chemical processes to produce electricity from light, rather than using semiconductors.

photon: A unit of electromagnetic radiation.

photovoltaic cell: A semiconductor device that converts solar radiation into direct current electricity.

photovoltaic effect: The movement of electrons within a material when it absorbs photons with energy above a certain level.

photovoltaics: A solar energy technology that uses unique properties of semiconductors to directly convert solar radiation into electricity.

photovoltaic (PV) system: An electrical system consisting of a PV module array and other electrical components needed to convert solar energy into electricity usable by loads.

plans examiner: A local official qualified to review construction plans and documentation for compliance with applicable codes and standards.

plans review: An evaluation of system-design documentation as part of the permitting process.

plate: An electrode consisting of active material supported by a grid framework.

p-n junction: The boundary of adjacent layers of p-type and n-type semiconductor materials in contact with one another.

point of connection: The location at which an interactive distributed-generation system makes its interconnection with the electric utility system.

pole mount: A type of array mounting system where modules are installed at an elevation on a pedestal.

polycrystalline wafer: A silicon wafer made from a cast silicon ingot that is composed of many silicon crystals.

power conditioning unit (PCU): A device that includes more than one power conditioning function.

power factor: The ratio of true power to apparent power and describes the displacement of voltage and current waveforms in AC circuits.

power quality: The measure of how closely the power in an electrical circuit matches the nominal values for parameters such as voltage, current, harmonics, and power factor.

primary battery: A battery that can store and deliver electrical energy but cannot be recharged.

production incentive: A recurring payment program that provides project owners with cash payments based on electricity production.

profile angle: The projection of the solar altitude angle onto an imaginary plane perpendicular to the surface of an obstruction.

p-type semiconductor: A semiconductor that has electron voids.

pulse-width modulation (PWM): A method of simulating waveforms by switching a series device ON and OFF at high frequency and for variable lengths of time.

push-pull inverter circuit: A circuit that switches DC input into AC output by using one pair of switching devices and a center-tapped transformer.

PV output circuit: The circuit connecting the PV power source to the rest of the system.

PV power source: An array or collection of arrays that generates DC power.

PV source circuit: The circuit connecting a group of modules together and to the common connection point of the DC system.

PVUSA test conditions (PTC): A set of reference conditions that rates module performance at a solar irradiance of 1000 W/m^2, ambient temperature of 20°C (68°F), and wind speed of 1 m/s.

pyranometer: A sensor that measures the total global solar irradiance in a hemispherical field of view.

pyrheliometer: A sensor that measures only direct solar radiation in the field of view of the solar disk (5.7°).

Q

qualified person: A person with skills and knowledge of the construction and operation of electrical equipment and installations and is trained in the safety hazards involved.

qualifying facility (QF): A non-utility large-scale power producer that meets the technical and procedural requirements for interconnection to the utility system.

qualifying-facility agreement: A contract between a utility and a qualifying facility that establishes the terms and conditions for interconnection and the rates or tariffs that apply.

R

rack mount: A type of fixed- or adjustable-tilt array mounting system with a triangular-shaped structure to increase the tilt angle of the array.

radiation: Energy that emanates from a source in the form of waves or particles.

reactive load: An AC load with inductive and/or capacitive elements that cause the current and voltage waveforms to become out of phase.

reactive power: The product of out-of-phase voltage and current waveforms and results in no net power flow.

rebate: A one-time refund for a portion of an original purchase price.

rectifier: A device that converts AC power to DC power.

recurring present value factor: A multiplication factor for determining the present value of an annually recurring cost for a given discount rate.

reference cell: An encapsulated PV cell that outputs a known amount of electrical current per unit of solar irradiance.

renewable energy certificate (REC): A tradable commodity that represents a certain amount of electricity generated from renewable resources.

repair cost: An expense related to the restoration of a component or system to its nominal operating state following damage or wear.

replacement cost: An expense related to the periodic replacement of components with expected lifetimes shorter than the expected lifetime of the entire system.

resistive load: A load that keeps voltage and current waveforms in phase.

resonance: The condition when a vibration frequency matches the fundamental frequency of the structure.

reverse bias: The condition of a PV device operating at negative (reverse) voltage.

ribbon wafer: A silicon wafer made by drawing a thin strip from a molten silicon mixture.

root-mean-square (RMS) value: A statistical parameter representing the effective value of a waveform.

S

sacrificial anode: A metal part, usually zinc or magnesium, that is more susceptible to galvanic corrosion than the metal structure it is attached to, so that it corrodes, rather than the structure.

salvage value: The monetary value of a system at the end of its expected useful lifetime.

secondary battery: A battery that can store and deliver electrical energy and can be charged by passing a current through it in an opposite direction to the discharge current.

self-ballasting: An attachment method that relies on the weight of the array, support structure, and ballasting material to hold the array in position.

self-commutated inverter: An inverter that can internally control the activation and duration of its switching.

self-discharge: The gradual reduction in the state of charge of a battery while at steady-state condition.

self-regulating PV system: A type of stand-alone PV system that uses no active control systems to protect the battery, except through careful design and component sizing.

semiconductor: A material that can exhibit properties of both an insulator and a conductor.

series charge controller: A charge controller that limits charging current to a battery system by open-circuiting the array.

series-interrupting charge controller: A charge controller that completely open-circuits the array, suspending current flow into the battery.

series-interrupting, pulse-width-modulated (PWM) charge controller: A charge controller that simulates a variable charging current by switching a series element ON and OFF at high frequency and for variable lengths of time.

series-linear charge controller: A charge controller that limits charging current to a battery system by gradually increasing the resistance of a series element.

short-circuit current (I_{sc}): The maximum current on an I-V curve and is the operating point for a PV device under no load or short-circuit condition, and no voltage output.

shunt charge controller: A charge controller that limits charging current to a battery system by short-circuiting the array.

shunt-interrupting charge controller: A charge controller that suspends charging current to a battery system by completely short-circuiting the array.

shunt-linear charge controller: A charge controller that limits charging current to a battery system by gradually lowering the resistance of a shunt element through which excess current flows.

sine wave: A periodic waveform the value of which varies over time according to the trigonometric sine function.

single-axis tracking: A sun-tracking system that rotates one axis to approximately follow the position of the sun.

single phasing: The complete loss of one phase on a three-phase power supply.

single present value factor: A multiplication factor for determining the present value of a single future cost for a given discount rate and period.

sinusoidal waveform: A waveform that is or closely approximates a sine wave.

site survey: A visit to the installation site to assess the site conditions and establish the needs and requirements for a potential PV system.

snow load: A static structural load due to the weight of accumulated snow.

soiling: The accumulation of dust and dirt on an array surface that shades the array and reduces electrical output.

solar altitude angle: The vertical angle between the sun and the horizon.

solar azimuth angle: The horizontal angle between a reference direction (typically due south in the Northern Hemisphere) and the sun.

solar constant: The average extraterrestrial solar power (irradiance) at a distance of 1 AU from the sun, which has a value of approximately 1366 W/m^2.

solar day: The interval of time between sun crossings of the local meridian, which is approximately 24 hr.

solar declination: The angle between the equatorial plane and the rays of the sun.

solar energy collector: A device designed to absorb solar radiation and convert it to another form, usually heat or electricity.

solar irradiance: The power of solar radiation per unit area.

solar irradiation: The total amount of solar energy accumulated on an area over time.

solar noon: The moment when the sun crosses a local meridian and is at its highest position of the day.

solar time: A timescale based on the apparent motion of the sun crossing a local meridian.

solar window: The area of sky between sun paths at summer solstice and winter solstice for a particular location.

solstice: Earth's orbital position when solar declination is at its minimum or maximum.

spanning: The addition of lumber under a roof surface and across trusses or rafters as supplemental structural support.

specific gravity: The ratio of the density of a substance to the density of water.

square wave: An alternating current waveform that switches between maximum positive and negative values every half period.

stand-alone PV system: A type of PV system that operates autonomously and supplies power to electrical loads independently of the electric utility.

standard meridian: A meridian located at a multiple of 15° east or west of zero longitude.

standard operating conditions (SOC): A set of reference conditions that rates module performance at a solar irradiance of 1000 W/m^2, spectral conditions of AM1.5, and at nominal operating cell temperature.

standard test conditions (STC): The most common and internationally accepted set of reference conditions, and rates module performance at a solar irradiance of 1000 W/m^2, spectral conditions of AM1.5, and a cell temperature of 25°C (77°F).

standard time: A timescale based on the apparent motion of the sun crossing standard meridians.

stand-by losses: The power required to operate inverter electronics and keep the inverter in a powered state.

standoff mount: A type of fixed-tilt array mounting system where modules are supported by a structure parallel to and slightly above the roof surface.

starting, lighting, and ignition (SLI) battery: A class of battery designed primarily for shallow-discharge cycle service.

state of charge (SOC): The percentage of energy remaining in a battery compared to the fully charged capacity.

stationary battery: A class of battery designed for occasional deep-discharge, limited-cycle service.

steady-state: An open-circuit condition where essentially no electrical or chemical changes are occurring.

stratification: A condition of flooded lead-acid cells in which the specific gravity of the electrolyte is greater at the bottom than at the top.

sulfation: The growth of lead sulfate crystals on the positive plate of a lead-acid cell.

sun path calculator: A device that superimposes an image of obstructions on a solar window diagram for a given location.

sun-tracking mount: An array mounting system that automatically orients the array to the position of the sun.

supplementary overcurrent protection device: An overcurrent protection device intended to protect an individual component and is used in addition to a current-limiting branch circuit overcurrent protection device.

surge arrestor: A device that protects electrical devices from transients (voltage spikes).

synchronizing: The process of connecting a generator to an energized electrical system.

system availability: The percentage of time over an average year that a stand-alone PV system meets the system load requirements.

T

tax incentive: A measure to reduce owed taxes as an inducement for investment.

temperature coefficient: The rate of change in voltage, current, or power output from a PV device due to changing cell temperature.

temperature-rise coefficient: A coefficient for estimating the rise in cell temperature above ambient temperature due to solar irradiance.

terrestrial solar radiation: Solar radiation reaching the surface of Earth.

thin-film module: A module-like PV device with its entire substrate coated in thin layers of semiconductor material using chemical vapor deposition techniques, and then laser-scribed to delineate individual cells and make electrical connections between cells.

total global radiation: All of the solar radiation reaching Earth's surface and is the sum of direct and diffuse radiation.

total harmonic distortion (THD): The ratio of the sum of all harmonic components in a waveform to the fundamental frequency component.

traction battery: A class of battery designed for repeated deep-discharge cycle service.

transformer: A device that transfers energy from one circuit to another through magnetic coupling.

transient voltage surge suppressor (TVSS): A surge-protective device that limits transient voltages by diverting or limiting surge current.

troubleshooting: A systematic method of investigating the cause of system problems and determining the best solution.

troubleshooting level: The depth of examination into the equipment or processes that compose a system.

true power: The product of in-phase voltage and current waveforms and produces useful work.

turbine: A bladed shaft that converts fluid flow into rotating mechanical energy.

U

ungrounded conductor: A current-carrying conductor that has no connection to ground.

uninterruptible power supply (UPS): A battery-based system that includes all the additional power conditioning equipment, such as inverters and charge controllers, to make a complete, self-contained power source.

utility: A company that produces and/or distributes electricity to consumers in a certain region or state.

utility-interactive system: A PV system that operates in parallel with and is connected to the electric utility grid.

V

varistor: A solid-state device that has a high resistance at low voltages and a low resistance at high voltages.

vibration load: A dynamic structural load due to periodic motion.

voltage regulation (VR) setpoint: The voltage that triggers the onset of battery charge regulation because it is the maximum voltage that a battery is allowed to reach under normal operating conditions.

voltage regulation hysteresis (VRH): The voltage difference between the voltage regulation (VR) setpoint and the array reconnect voltage (ARV) setpoint.

voltage unbalance: The unbalance that occurs when the voltages of a three-phase power supply or the terminals of a three-phase load are not equal.

W

wafer: A thin, flat disk or rectangle of base semiconductor material.

waveform: The shape of an electrical signal that varies over time.

wind load: A dynamic structural load due to wind, resulting in downward, lateral, or lifting forces.

wind turbine: A device that harnesses wind power to produce electricity.

working space: The clear space reserved around electrical equipment so that workers can inspect, operate, and maintain the equipment safely and efficiently.

Z

zenith: The point in the sky directly overhead a particular location.

zenith angle: The angle between the sun and the zenith.

Index

Page numbers in italic refer to figures.

USING THE *Photovoltaic Systems* INTERACTIVE CD-ROM

Before removing the Interactive CD-ROM from the protective sleeve, please note that the book cannot be returned for refund or credit if the CD-ROM sleeve seal is broken.

System Requirements

To use this Windows®-compatible CD-ROM, your computer must meet the following minimum system requirements:
* Microsoft® Windows Vista™, Windows 2000®, or Windows NT® operating system
* Intel® 1.3 GHz processor (or equivalent)
* 128MB of available RAM (256MB recommended)
* 335MB of available hard disc space
* 1024 x 768 monitor resolution
* CD-ROM drive (or equivalent optical drive)
* Sound output capability and speakers
* Microsoft® Internet Explorer® 6.0 or Firefox® 2.0 web browser
* Active Internet connection required for Internet links

Opening Files

Insert the Interactive CD-ROM into the computer CD-ROM drive. Within a few seconds, the home screen will be displayed allowing access to all features of the CD-ROM. Information about the usage of the CD-ROM can be accessed by clicking on Using This Interactive CD-ROM. The Quick Quizzes®, Illustrated Glossary, Solar Radiation Data Sets, Sun Path Charts, Forms and Worksheets, Solar Time Calculator, Flash Cards, Media Clips, and ATPeResources.com can be accessed by clicking on the appropriate button on the home screen. Clicking on the American Tech web site button (www.go2atp.com) accesses information on related educational products. Unauthorized reproduction of the material on this CD-ROM is strictly prohibited.

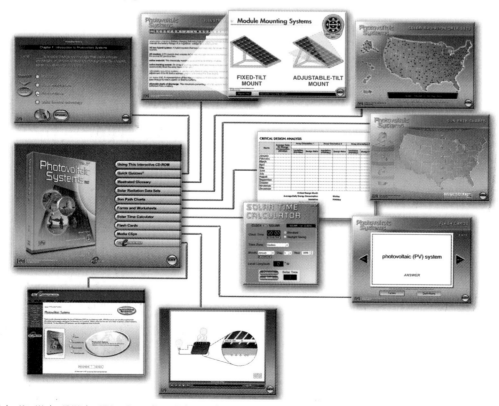